Lecture Notes in Computer Science 4616

Commenced Publication in 1973
Founding and Former Series Editors:
Gerhard Goos, Juris Hartmanis, and Jan van Leeuwen

Andreas Dress   Yinfeng Xu
Binhai Zhu (Eds.)

Combinatorial Optimization and Applications

First International Conference, COCOA 2007
Xi'an, China, August 14-16, 2007
Proceedings

 Springer

Volume Editors

Andreas Dress
CAS-MPG Partner Institute for Computational Biology
Shanghai 200031, China
and
Max Planck Institute for Mathematics in the Sciences
D-04104 Leipzig, Germany
E-mail: andreas@picb.ac.cn

Yinfeng Xu
Xi'an Jiaotong University
Xi'an, Shaanxi 710049, China
E-mail: yfxu@mail.xjtu.edu.cn

Binhai Zhu
Montana State University
Bozeman, MT 59717, USA
E-mail: bhz@cs.montana.edu

Library of Congress Control Number: 2007931336

CR Subject Classification (1998): F.2, C.2, G.2-3, I.3.5, G.1.6, E.5

LNCS Sublibrary: SL 1 – Theoretical Computer Science and General Issues

ISSN   0302-9743
ISBN-10   3-540-73555-0 Springer Berlin Heidelberg New York
ISBN-13   978-3-540-73555-7 Springer Berlin Heidelberg New York

Springer is a part of Springer Science+Business Media

springer.com

© Springer-Verlag Berlin Heidelberg 2007
Printed in Germany

Typesetting: Camera-ready by author, data conversion by Scientific Publishing Services, Chennai, India
Printed on acid-free paper   SPIN: 12088744   06/3180   5 4 3 2 1 0

Preface

The papers in this volume were presented at the 1st International Conference on Combinatorial Optimization and Applications (COCOA 2007), held August 12-15, 2007, in Xi'an, China. The topics cover most areas in combinatorial optimization and applications.

Submissions to the conference this year were conducted electronically. A total of 114 papers were submitted, of which 29 were accepted. The papers were evaluated by an International Program Committee consisting of Tetsuo Asano, Kyung-Yong Chwa, Bill Chen, Bo Chen, Andreas Dress, Pater Eades, Omer Egecioglu, Rudolf Fleischer, Bin Fu, Mordecai Golin, Ron Graham, Pavol Hell, Xiao-Dong Hu, Marek Karpinski, Minghui Jiang, Michael Langston, Hanno Lefmann, Ko-Wei Lih, Andy Mirzaian, Brendan Mumey, Mauricio G.C. Resende, Takao Nishizeki, Mike Steel, Zheng Sun, My T. Thai, Kanliang Wang, Michael Waterman, Gerhard Woeginger, Yinfeng Xu, Boting Yang, Wenan Zang, Alex Zelikovsky and Binhai Zhu. It is expected that most of the accepted papers will appear in a more complete form in scientific journals.

The submitted papers are from Australia, Canada, China, France, Germany, Greece, Hong Kong, Japan, Korea, Mexico, Poland, Romania, Russia, Switzerland, Tunisia, Turkey and USA. Each paper was evaluated by at least two Program Committee members (and in some cases by as many as seven Program Committee members), assisted in some cases by subreferees. In addition to selected papers, the conference also included two invited presentations, by Bailin Hao and Kurt Mehlhorn, and eight invited papers.

We thank all the people who made this meeting possible: the authors for submitting papers, the Program Committee members and external referees (listed in the proceedings) for their excellent work, and the two invited speakers. Finally, we thank Xi'an Jiaotong University and NSF of China for the support and local organizers and colleagues for their assistance.

August 2007
Andreas Dress
Yinfeng Xu
Binhai Zhu

Organization

Program Committee Chairs

Andreas Dress, Bielefeld University, Germany
Yinfeng Xu, Xi'an Jiaotong University, China
Binhai Zhu, Montana State University, USA

Program Committee Members

Tetsuo Asano (JAIST, Japan)
Kyung-Yong Chwa (KAIST, Korea)
Bill Chen (Nankai University, China)
Bo Chen (University of Warwick, UK)
Pater Eades (University of Sydney, Australia)
Omer Egecioglu (UC-Santa Barbara, USA)
Rudolf Fleischer (Fudan University, China)
Bin Fu (University of Texas-Pan American, USA)
Mordecai Golin (HKUST, Hong Kong)
Ron Graham (UC-San Diego, USA)
Pavol Hell (Simon Fraser University, Canada)
Xiao-Dong Hu (Chinese Academy of Sciences, China)
Marek Karpinski (University of Bonn, Germany)
Minghui Jiang (Utah State University, USA)
Michael Langston (University of Tennessee, USA)
Hanno Lefmann (TU Chemnitz, Germany)
Ko-Wei Lih (Academia Sinica, Taiwan)
Andy Mirzaian (York University, Canada)
Brendan Mumey (Montana State University, USA)
Mauricio G.C. Resende (AT&T Labs Research, USA)
Takao Nishizeki (Tohoku University, Japan)
Mike Steel (University of Canterbury, New Zealand)
Zheng Sun (Google, USA)
My T. Thai (University of Florida, USA)
Kanliang Wang (Xi'an Jiaotong University, China)
Michael Waterman (University of Southern California, USA)
Gerhard Woeginger (Eindhoven University of Technology, Netherlands)
Boting Yang (University of Regina, Canada)
Wenan Zang (HKU, Hong Kong)
Alex Zelikovsky (Georgia State University, USA)

Organizing Committee

Kanliang Wang, Xi'an Jiaotong University, China
Yinfeng Xu, Xi'an Jiaotong University, China
Binhai Zhu, Montana State University, USA

Referees

Francesc Aguilo
Irina Astrovskaya
Dumitru Brinza
Kun-Mao Chao
Chiuyuan Chen
Qiong Cheng
John Eblen
Zuzana Gedeon
Stefan Gremalschi
Christoph Helmberg
Han Hoogeveen
Takehiro Ito
Li Jiam

Anja Kohl
Yue-Kuen Kowk
James Kwok
Tak-Wah Lam
Zuopan Li
Sheng-Chyang Liaw
Zaiping Lu
Andy Perkins
Gary Perkins
Charles Phillips
Kai Plociennik
Thiruvarangan Ramaraj
Gary Rogers

Yasuhiko Takenaga
Ravi Tiwari
Vitali Voloshin
Shen Wan
Keily Westbrooks
Hong-Gwa Yeh
Xuerong Yong
Ning Zhang
Yun Zhang
Xiao Zhou
An Zhu
Xuding Zhu

Table of Contents

Invited Lecture

Matchings in Graphs Variations of the Problem 1
  Kurt Mehlhorn

Combinatorics from Bacterial Genomes 3
  Bailin Hao

Contributed Papers

An Algorithm for Computing Virtual Cut Points in Finite Metric
Spaces 4
  *Andreas W.M. Dress, Katharina T. Huber, Jacobus Koolen, and
  Vincent Moulton*

Finding the Anti-block Vital Edge of a Shortest Path Between Two
Nodes 11
  Bing Su, Qingchuan Xu, and Peng Xiao

K-Connected Target Coverage Problem in Wireless Sensor Networks . . . 20
  Deying Li, Jiannong Cao, Ming Liu, and Yuan Zheng

Searching Cycle-Disjoint Graphs 32
  Boting Yang, Runtao Zhang, and Yi Cao

An Asymptotic PTAS for Batch Scheduling with Nonidentical Job
Sizes to Minimize Makespan 44
  Yuzhong Zhang and Zhigang Cao

A New Dynamic Programming Algorithm for Multiple Sequence
Alignment 52
  Jean-Michel Richer, Vincent Derrien, and Jin-Kao Hao

Energy Minimizing Vehicle Routing Problem 62
  İmdat Kara, Bahar Y. Kara, and M. Kadri Yetis

On the On-Line k-Taxi Problem with Limited Look Ahead 72
  Weimin Ma, Ting Gao, and Ke Wang

The Minimum Risk Spanning Tree Problem 81
  Xujin Chen, Jie Hu, and Xiaodong Hu

The Size of a Minimum Critically m-Neighbor-Scattered Graph 91
  Fengwei Li and Qingfang Ye

A New Hybrid Algorithm for Feature Selection and Its Application to
Customer Recognition 102
  Luo Yan and Yu Changrui

Steiner Forests on Stochastic Metric Graphs 112
     Vangelis Th. Paschos, Orestis A. Telelis, and Vassilis Zissimopoulos

On Threshold BDDs and the Optimal Variable Ordering Problem 124
     Markus Behle

Communication Leading to Nash Equilibrium Through Robust
Messages – **S5**-Knowledge Model Case – 136
     Takashi Matsuhisa

Fundamental Domains for Integer Programs with Symmetries 146
     Eric J. Friedman

Exact Algorithms for Generalized Combinatorial Optimization
Problems 154
     Petrica C. Pop, Corina Pop Sitar, Ioana Zelina, and Ioana Taşcu

Approximation Algorithms for k-Duplicates Combinatorial Auctions
with Subadditive Bidders 163
     Wenbin Chen and Jiangtao Meng

A Grid Resource Discovery Method Based on Adaptive k-Nearest
Neighbors Clustering 171
     Yan Zhang, Yan Jia, Xiaobin Huang, Bin Zhou, and Jian Gu

Algorithms for Minimum m-Connected k-Dominating Set Problem 182
     Weiping Shang, Frances Yao, Pengjun Wan, and Xiaodong Hu

Worst Case Analysis of a New Lower Bound for Flow Shop Weighted
Completion Time Problem 191
     Danyu Bai and Lixin Tang

Scaling, Renormalization, and Universality in Combinatorial Games:
The Geometry of Chomp 200
     Eric J. Friedman and Adam Scott Landsberg

Mechanism Design by Creditability 208
     *Raphael Eidenbenz, Yvonne Anne Oswald, Stefan Schmid, and
     Roger Wattenhofer*

Infinite Families of Optimal Double-Loop Networks 220
     Xiaoping Dai, Jianqin Zhou, and Xiaolin Wang

Point Sets in the Unit Square and Large Areas of Convex Hulls of
Subsets of Points 230
     Hanno Lefmann

An Experimental Study of Compressed Indexing and Local Alignments
of DNA... 242
     *Tak-Wah Lam, Wing-Kin Sung, Siu-Lung Tam,
     Chi-Kwong Wong, and Siu-Ming Yiu*

Secure Multiparty Computations Using the 15 Puzzle (Extended
Abstract) 255
     Takaaki Mizuki, Yoshinori Kugimoto, and Hideaki Sone

A Lagrangian Relaxation Approach for the Multiple Sequence
Alignment Problem 267
  Ernst Althaus and Stefan Canzar

Single Machine Common Due Window Scheduling with Controllable
Job Processing Times 279
  Guohua Wan

A Lower Bound on Approximation Algorithms for the Closest Substring
Problem 291
  Jianxin Wang, Min Huang, and Jianer Chen

A New Exact Algorithm for the Two-Sided Crossing Minimization
Problem 301
  Lanbo Zheng and Christoph Buchheim

Improved Approximation Algorithm for Connected Facility Location
Problems (Extended Abstract).................................... 311
  Mohammad Khairul Hasan, Hyunwoo Jung, and Kyung-Yong Chwa

The Computational Complexity of Game Trees by Eigen-Distribution... 323
  ChenGuang Liu and Kazuyuki Tanaka

The Minimum All-Ones Problem for Graphs with Small Treewidth 335
  Yunting Lu and Yueping Li

An Exact Algorithm Based on Chain Implication for the Min-CVCB
Problem 343
  Jianxin Wang, Xiaoshuang Xu, and Yunlong Liu

Arc Searching Digraphs Without Jumping 354
  Brian Alspach, Danny Dyer, Denis Hanson, and Boting Yang

On the Complexity of Some Colorful Problems Parameterized by
Treewidth.. 366
  *Michael Fellows, Fedor V. Fomin, Daniel Lokshtanov,
  Frances Rosamond, Saket Saurabh, Stefan Szeider, and
  Carsten Thomassen*

A PTAS for the Weighted 2-Interval Pattern Problem over the
Preceding-and-Crossing Model 378
  Minghui Jiang

Author Index 389

Matchings in Graphs
Variations of the Problem

Kurt Mehlhorn

Max-Planck-Institut für Informatik
Stuhlsatzenhausweg 86
66123 Saarbrücken
Germany
melhorn@mpi-inf.mpg.de
http://www.mpi-inf.mpg.de/~melhorn

Many real-life optimization problems are naturally formulated as questions about matchings in (bipartite) graphs.

- We have a bipartite graph. The edge set is partitioned into classes E_1, E_2, ..., E_r. For a matching M, let s_i be the number of edges in $M \cap E_i$. A *rank-maximal matching* maximizes the vector $(s_1, s_2, \ldots, s_r)$. We show how to compute a rank-maximal matching in time $O(r\sqrt{n}m)$ [IKM$^+$06].
- We have a bipartite graph. The vertices on one side of the graph rank the vertices on the other side; there are no ties. We call a matching M more popular than a matching N if the number of nodes preferring M over N is larger than the number of nodes preferring N over M. We call a matching *popular*, if there is no matching which is more popular. We characterize the instances with a popular matching, decide the existence of a popular matching, and compute a popular matching (if one exists) in time $O(\sqrt{n}m)$ [AIKM05].
- We have a bipartite graph. The vertices on both sides rank the edges incident to them with ties allowed. A matching M is *stable* if there is no pair $(a, b) \in E \setminus M$ such that a prefers b over her mate in M and b prefers a over his mate in M or is indifferent between a and his mate. We show how to compute stable matchings in time $O(nm)$ [KMMP04].
- In a random graph, edges are present with probability p independent of other edges. We show that for $p \geq c_0/n$ and c_0 a suitable constant, every non-maximal matching has a logarithmic length augmenting path. As a consequence the *average running time of matching algorithms on random graphs* is $O(m \log n)$ [BMSH05].

References

[AIKM05] Abraham, D., Irving, R., Kavitha, T., Mehlhorn, K.: Popular Matchings. SODA, pp. 424–432 (2005)
⟨http://www.mpi-sb.mpg.de/~mehlhorn/ftp/PopularMatchings.ps⟩
[BMSH05] Bast, H., Mehlhorn, K., Schäfer, G., Tamaki, H.: Matching Algorithms are Fast in Sparse Random Graphs. Theory of Computing Systems 31(1), 3–14 (2005) preliminary version. In: Diekert, V., Habib, M. (eds.) STACS 2004. LNCS, vol. 2996, pp. 81–92. Springer, Heidelberg (2004)

A. Dress, Y. Xu, and B. Zhu (Eds.): COCOA 2007, LNCS 4616, pp. 1–2, 2007.
© Springer-Verlag Berlin Heidelberg 2007

[IKM⁺06]   Irving, R., Kavitha, T., Mehlhorn, K., Michail, D., Paluch, K.: Rank-Maximal Matchings. ACM Transactions on Algorithms 2(4), 1–9 (2006), A preliminary version appeared in SODA 2004, pp. 68–75, (2004) (http://www.mpi-sb.mpg.de/~mehlhorn/ftp/RankMaximalMatchings)

[KMMP04] Kavitha, T., Mehlhorn, K., Michail, D., Paluch, K.: Strongly Stable Matchings in Time $O(nm)$ and Extensions to the Hospitals-Residents Problem. In: Diekert, V., Habib, M. (eds.) STACS 2004. LNCS, vol. 2996, pp. 222–233. Springer, Heidelberg (2004), full version to appear in TALG (http://www.mpi-sb.mpg.de/~mehlhorn/ftp/StableMatchings.ps)

Combinatorics from Bacterial Genomes

Bailin Hao

T-Life Research Center, Fudan University, Shanghai 200433, China
and
Santa Fe Institute, Santa Fe, NM 87501, USA

By visualizing bacterial genome data we have encountered a few neat mathematical problems. The first problem concerns the number of longer missing strings (of length $K + i$, $i \geq 1$) taken away by the absence of one or more K-strings. The exact solution of the problem may be obtained by using the Golden-Jackson cluster method in combinatorics and by making use of a special kind of formal languages, namely, the factorizable language. The second problem consists in explaining the fine structure observed in one-dimensional K-string histograms of some randomized genomes. The third problem is the uniqueness of reconstructing a protein sequence from its constituent K-peptides. The latter problem has a natural connection with the number of Eulerian loops in a graph. To tell whether a protein sequence has a unique reconstruction at a given K the factorizable language again comes to our help.

References

1. Hao, B., Xie, H., Yu, Z., Chen, G.: A combinatorial problem related to avoided strings in bacterial complete genomes. Ann. Combin. 4, 247-255 (2000)
2. Xie, H., Hao, B.: Visualization of K-tuple distribution in prokaryote complete genomes and their randomized counterparts. In: Bioinformatics. CSB2002 Proceedings, pp. 31-42. IEEE Computer Society, Los Alamitos, California (2002)
3. Shen, J., Zhang, S., Lee, H.-C., Hao, B.: SeeDNA: a visualization tool for K-string content of long DNA sequences and their randomized counterparts. Genomics, Proteomics and Bioinformatics 2, 192-196 (2004)
4. Shi, X., Xie, H., Zhang, S., Hao, B.: Decomposition and reconstruction of protein sequences: the problem of uniqueness and factorizable language. J. Korean Phys. Soc. 50, 118-123 (2007)

A. Dress, Y. Xu, and B. Zhu (Eds.): COCOA 2007, LNCS 4616, p. 3, 2007.
© Springer-Verlag Berlin Heidelberg 2007

An Algorithm for Computing Virtual Cut Points in Finite Metric Spaces

Andreas W.M. Dress[1], Katharina T. Huber[2], Jacobus Koolen[3], and Vincent Moulton[4]

[1] CAS-MPG Partner Institute for Computational Biology, 320 Yue Yang Road, 200031 Shanghai, China
andreas@picb.ac.cn
[2] School of Computing Sciences, University of East Anglia, Norwich, NR4 7TJ, UK
katharina.Huber@cmp.uea.ac.uk
[3] Department of Mathematics, POSTECH, Pohang, South Korea
koolen@postech.ac.kr
[4] School of Computing Sciences, University of East Anglia, Norwich, NR4 7TJ, UK
vincent.moulton@cmp.uea.ac.uk

Abstract. In this note, we consider algorithms for computing **virtual cut points** in finite metric spaces and explain how these points can be used to study **compatible decompositions** of metrics generalizing the well-known decomposition of a **tree metric** into a sum of **pairwise compatible split metrics**.

Mathematics Subject Classification codes: 05C05, 05C12, 92B10.

1 Terminology

A **metric** D defined on a set X is a map $D : X^2 \to \mathbb{R} : (x,y) \mapsto xy$ from the set $X^2 := \{(x,y) : x,y \in X\}$ of all (ordered) pairs of elements from X into the real number field $\mathbb{R}$ such that $xx = 0$ and $xy \leq xz + yz$ (and, therefore, also $0 \leq xy = yx$) holds for all $x,y,z \in X$. A metric D is called a **proper** metric if $xy \neq 0$ holds for any two distinct points $x,y \in X$. Further, given X and D as above, we denote

(i) by $\sim_D$ the binary relation defined on X by putting $x \sim_D y \iff xy = 0$ which, in view of the fact that $xy = 0 \iff \forall_{a \in X} xa = ya$ holds for all $x,y \in X$, is clearly an equivalence relation,

(ii) by x/D the equivalence class of x relative to this equivalence relation, and

(iii) by X/D the set $\{x/D : x \in X\}$ of all such equivalence classes.

In case D is proper, the pair $M = M_D := (X,D)$ is also called a **metric space**, X is called the **point set** of that space – and every element $x \in X$ a **point** of M.

Further, given any metric space $M = (X,D)$, we denote

A. Dress, Y. Xu, and B. Zhu (Eds.): COCOA 2007, LNCS 4616, pp. 4–10, 2007.

(D1) by $[x, y]$, for all $x, y \in X$, the **interval** between x and y, i.e., the set

$$[x, y] := \{z \in X : xy = xz + zy\},$$

(D2) by $\mathrm{Prox}(M) = (X, E(M))$ the (abstract) **proximity graph** of M (sometimes also called **the underlying graph** of M, cf. [11]), that is, the graph with vertex set X and edge set

$$E(M) := \left\{\{u, v\} \in \binom{X}{2} : [u, v] = \{u, v\}\right\}.$$

(D3) by $C_x(y)$, in case x and y are two distinct points in X, the connected component of the induced graph

$$\mathrm{Prox}(M|x) := \mathrm{Prox}(M)|_{X-\{x\}} := \left(X - \{x\}, E(M) \cap \binom{X - \{x\}}{2}\right)$$

containing y, and by $\overline{C}_x(y) := C_x(y) \cup \{x\}$ the "augmented" connected component of $\mathrm{Prox}(M|x)$ containing y, i.e., the union of $C_x(y)$ and the one-point set $\{x\}$,

(D4) and we denote by

$$\pi_x := \{C_x(y) : y \in X - \{x\}\}$$

the collection of connected components of $\mathrm{Prox}(M|x)$, and by

$$\overline{\pi}_x := \{\overline{C}_x(y) : y \in X - \{x\}\}$$

the corresponding collection of "augmented" connected components of $\mathrm{Prox}(M|x)$.

Note that

(i) $[x, z] \subseteq [x, y]$ holds for all $x, y, z \in X$ with $z \in [x, y]$,

and that, in case X is finite,

(ii) $\mathrm{Prox}(M) = (X, E(M))$ is connected,

(iii) $C_x(y) = C_x(y')$ holds for all $y, y' \in X - \{x\}$ with $x \notin [y, y']$,

(iv) and $C_x(y) \cup C_y(x) = X$ holds for any two distinct $x, y \in X$ (indeed, $z \in X - C_x(y)$ implies $zx = zx + (xy - xy) = zy - xy < zy + yx$ and, hence, $y \notin [x, z]$ which in turn (cf. (iii)) implies $C_y(z) = C_y(x)$).

2   Cut Points of Metric Spaces

Given a metric space $M = (X, D)$, let $Cut(M)$ denote the set of all **cut points** of M, i.e., the set of all points $x \in M$ for which two subsets A, B of X with $A \cup B = X$ and $A \cap B = \{x\}$ of cardinality at least 2 exist such that $x \in [a, b]$ holds for all $a \in A$ and $b \in B$. Concerning cut points, one has:

Theorem 1. *Given a metric space $M = (X, D)$ of cardinality at least 3 and any point x in X, the following assertions are equivalent:*

(i) *$x \in Cut(M)$ holds,*
(ii) *there exist a pair A, B of gated[1] subsets of X of cardinality at least 2 with $A \cup B = X$ and $\{x\} = A \cap B$,*
(iii) *there exists a decomposition $D = D_1 + D_2$ of D into a sum of two non-vanishing (yet necessarily) non proper) metrics D_1, D_2 defined on X such that $X = x/D_1 \cup x/D_2$ holds.*

More precisely, there exists a canonical one-to-one correspondence between

(a) *decompositions $D = D_1 + D_2$ of D into a sum of two non-vanishing metrics D_1, D_2 such that some — necessarily unique — point $x = x_{D_1, D_2}$ in X with $X = x/D_1 \cup x/D_2$ exists, and*
(b) *pairs A, B of subsets of X with $\#A, \#B \geq 2$ such that $A \cup B = X$ holds and some — also necessarily unique — point $x = x_{A,B} \in A \cap B$ exists with $x \in [a, b]$ for all $a \in A$ and $b \in B$.*

Further, x is a cut point of M in case X is finite if and only if $\#\pi_x \geq 2$ holds, i.e., if and only if it is a cut point of the proximity graph of M.

3   Retracts

Next, given a metric space $M = (X, D)$,

(i) we define a **retraction** of M to be an idempotent map $\rho : X \to X$ such that $ab = a\rho(a) + \rho(a)\rho(b) + \rho(b)b$ holds for all $a, b \in X$ with $\rho(a) \neq \rho(b)$, and we note that the image $\rho(X)$ of ρ is a gated subset R of M and that $\rho = gate_{\rho(X)}$ always holds,
(ii) we define a **retract** of M to be any gated subset R of M such that the associated map $gate_R : X \to X$ is a retraction — so, associating to any retract of M its gate map and to any retraction of M its image, it is easily seen that this gives a canonical one-to-one correspondence between retractions and retracts of M,
(iii) we denote by D_R, for every retract R of M, the metric that is defined on X by putting $D_R(x, y) := D(gate_R(x), gate_R(y))$ for all x, y in X, and
(iv) we denote by $\mathcal{D}_{min}(M)$ the collection of all such metrics $D' = D_R$ that are associated to those retracts R of M that are members of the collection $\mathcal{R}_{min}(M)$ of all minimal retracts of M (with respect to inclusion) of cardinality > 1.

The basic facts that motivate all further work are summarized in the two following theorems:

---

[1] Recall that a gated subset of a metric space $M = (X, D)$ is a subset A of its point set X such that there exists a (necessarily unique) map $gate_A : X \to X$, the **gate map** of A (relative to M), such that $gate_A(u) \in A \cap [u, a]$ holds for all $u \in X$ and $a \in A$.

Theorem 2. *Given a metric space $M = (X, D)$ with a finite set of cut points, the following all holds:*

(i) *One has $R \in \mathcal{R}_{min}(M)$ for some subset R of X if and only if R is a maximal subset of X such that $C_x(u) = C_x(v)$ holds for every cut point $x \in X$ and any two points $u, v \in R - \{x\}$.*

(ii) *Associating furthermore, to any $R \in \mathcal{R}_{min}(M)$, the (well-defined!) map*

$$\Theta_R : Cut(M) \to \mathcal{P}(X) : x \mapsto \overline{C_x}(R)$$

from $Cut(M)$ into the power set $\mathcal{P}(X)$ of X that associates, to every cut point $x \in Cut(M)$, the union $\overline{C_x}(R)$ of $\{x\}$ and the unique connected component $C \in \pi_x$ with $R \subset C \cup \{x\}$, one has

$$R = \bigcap_{x \in Cut(M)} \Theta_R(x).$$

(iii) *More specifically, this sets up a canonical one-to-one correspondence between retracts $R \in \mathcal{R}_{min}(M)$ and maps Θ from $Cut(M)$ into $\mathcal{P}(X)$ for which $\Theta(x) \in \overline{\pi_x}$ holds for all $x \in Cut(M)$, and $\Theta(x) = \overline{C_x}(y)$ holds for all $x, y \in Cut(M)$ with $x \notin \Theta(y)$.*

Theorem 3. *Given a finite metric space $M = (X, D)$, the metric D decomposes canonically into the sum of all metrics in $\mathcal{D}_{min}(M)$, i.e., one always has*

$$D = \sum_{D' \in \mathcal{D}_{min}(M)} D'.$$

4   Virtual Cut Points of Metric Spaces

While the results collected above look quite attractive, the problem with them is that, generally, there are not many cut points in metric spaces — not even in "tree-like" metric spaces, i.e., finite metric spaces $M = (X, D)$ for which D satisfies the so-called "four-point condition" (see for instance [1,2,3,12]).

However, while this is true in general, the resulting problem can — at least to some degree — be rectified: Indeed, **tight-span theory** as devised originally (though in other terms) by John Isbell [10] (and further developed in [4,5,7] and many other papers) allows us to construct, for any metric space $M = (X, D)$, a canonical extension $\overline{M} = (T_M, D_\infty)$ whose point set T_M consists of all maps $f \in \mathbb{R}^X$ for which

$$f(x) = \sup(xy - f(y) : y \in X)$$

holds for all $x \in X$, endowed with the metric D_∞ defined on T_M by the l_∞-norm $||f, g||_\infty := \sup(|f(x) - g(x)| : x \in X)$ which — upon identifying every $x \in X$ with the corresponding **Kuratowski map** $h_x : X \to \mathbb{R} : y \mapsto xy$ in T_M (and, hence, X with the subset $T_M(real) := \{f \in T_M : 0 \in f(X)\}$ of T_M) — is indeed an "isometric extension" of X.

Furthermore, it follows easily from tight-span theory that $D_\infty(h_x, f) = f(x)$ holds for all $x \in X$ and $f \in T_M$, and that the set

$$Cut_{virt}(M) := Cut(\overline{M})$$

of **virtual cut points** of M, i.e., the set of cut points of $\overline{M}$ (which is not necessarily finite even if X is finite) can be described and computed explicitly for any finite metric space M, leading — by restriction from $\overline{M}$ to M — to the canonical **block decomposition** of D, i.e., the unique decomposition $D = D_1 + D_2 + \cdots + D_k$ of D into a sum of non-vanishing metrics $D_1, D_2, \ldots, D_k$ defined on X such that no two of them are linear multiples of each other, and there exist points $x, y \in X$ for any pair i, j of two distinct indices in $\{1, 2, \ldots, k\}$ for which $X = x/D_i \cup y/D_j$ holds, while every summand D_i is a **block metric**, i.e., no such decomposition $D_i = D_{i1} + D_{i2} + \cdots + D_{ik'}$ of D_i exist for any $i \in \{1, 2, \ldots, k\}$ (cf. [6]).

5   Virtual Cut Points and Additive Split of Metric Spaces

To compute $Cut_{virt}(M)$ for a given finite metric space $M = (X, D)$, note first that a map $f \in T_M$ is either a cut point of $\overline{M}$ or a point in $T_M(real)$ if and only if there exists a bipartition A, B of X into two disjoint proper subsets of X such that $f(a) + f(b) = ab$ holds for all $a \in A$ and $b \in B$ (cf. [4,7]).

Note further that any such bipartition A, B must be an **additive split** of M — i.e., that $aa' + bb' \le ab + a'b' = ab' + a'b$ must hold for all $a, a' \in A$ and $b, b' \in B$ — and that, conversely, given any additive split A, B of M, there exists at least one map in the set $T_M(A, B)$ consisting of all maps $f \in T_M$ for which $f(a) + f(b) = ab$ holds for all $a \in A$ and $b \in B$. For instance, such a map $f := g_A$ can be defined by first choosing two arbitrary elements $a_0 \in A$ and $b_0 \in B$ and then putting

$$g_A(b) := a_0 b - \inf_{b_1, b_2 \in B} \frac{b_1 a_0 + b_2 a_0 - b_1 b_2}{2}$$

for all $b \in B$ and

$$g_A(a) := ab_0 - g_A(b_0)$$

for all $a \in A$. It is easily shown that this map happens to be independent of the choice of a_0 and b_0 (and that it actually coincides with the unique such map for which — in addition — some elements $b, b' \in B$ with $g_A(b) + g_A(b') = bb'$ exist).

More precisely, given an additive split A, B of M, the set $T_M(A, B)$ is easily seen to coincide with the interval $[g_A, g_B]$ spanned by g_A and g_B in $\mathbb{R}^X$. So, it is a whole interval if and only if $g_A \ne g_B$ holds, which in turn holds if and only if the split A, B of M is a **block split** of M, i.e., $aa' + bb' < ab + a'b' = ab' + a'b$ holds for all $a, a' \in A$ and $b, b' \in B$, and it consists of a single point, else.

6 An Algorithm for Computing the Virtual Cut Points of Finite Metric Spaces

Thus, we have essentially reduced the problem of computing the set $Cut_{virt}(M)$ of all virtual cut points of a finite metric space $M = (X, D)$ to computing the set $\mathcal{S}_{add}(M)$ of all additive splits of M.

Unfortunately, however, this rephrasing of the problem does not appear to yield an efficient way to compute $Cut_{virt}(M)$ as there may be too many such splits: E.g., if $xy = 2$ holds for any two distinct points x, y in X, every bipartition of X is an additive split of M while the constant map that maps any point in X onto 1 is the only cut point.

So, some more care is required: To this end, we denote by $\mathcal{S}_M(block)$ the set of all block splits of M, and by $T_M(block)$ the set consisting of all maps f in T_M for which the graph

$$M(f) := (X, \{\{x, y\} \in \binom{X}{2} : f(x) + f(y) > xy\})$$

is disconnected, but not a disjoint union of 2 cliques.

It is shown in [6] (and not too difficult to see) that a map $f \in T_M$ is contained in $Cut_{virt}(M)$ if and only if it is either contained in $T_M(block)$ or it is of the form $f = \alpha g_A + \beta g_B$ where A, B is a block split of M and α and β are two non-negative numbers with $\alpha + \beta = 1$.

Thus, it suffices to compute $\mathcal{S}_M(block)$ and $T_M(block)$ to find all cut points of $\overline{M}$. To do so algorithmically, one proceeds recursively and assumes that both tasks have been solved for a subspace M' of the form

$$M' = (X', D' : X' \times X' \to \mathbb{R} : (x', y') \mapsto x'y')$$

of M, where X' is a subset of X of the form $X' = X - \{x_0\}$ for some element $x_0 \in X$, yielding a list $\mathcal{S}_{M'}(block)$ of all splits A', B' of X' for which $aa' + bb' < ab + a'b' = ab' + a'b$ holds for all $a, a' \in A'$ and all $b, b' \in B'$, and a list $T_{M'}(block)$ consisting of all maps f' in $T_{M'}$ for which the graph $M'(f')$ is disconnected, but not a disjoint union of 2 cliques.

Next, one notes that any block split A, B of X in $\mathcal{S}_M(block)$ is either of the form $X', \{x_0\}$ or it is of the form $A' \cup \{x_0\}, B'$ or $A', B' \cup \{x_0\}$ for some split A', B' of X' in $\mathcal{S}_{M'}(block)$. So, by checking which of these splits is actually in $\mathcal{S}_M(block)$, this set is easily computed.

And finally, one notes that any map f in $T_M(block)$ is either of the form g_A for some block split A, B of M (and can thus be easily computed) or it is an extension f'_* of a map f' in $T_{M'}(block)$ to a map defined on X whose value on the element $x_0 \in X$ is defined by

$$f'_*(x_0) := \sup(x_0 y - f'(y) : y \in X').$$

In particular, it follows that all of these maps may be computed quite easily in time that is a polynomial function of $\#X$.

References

1. Bandelt, H.-J., Dress, A.W.M.: Split decomposition: A new and useful approach to phylogenetic analysis of distance data. Mol. Phyl. Evol. 1, 242–252 (1992)
2. Bandelt, H.-J., Dress, A.W.M.: A canonical split decomposition theory for metrics on a finite set. Adv. Math. 92, 47–105 (1992)
3. Buneman, P.: The recovery of trees from measures of dissimilarity. In: Hodson, F.R., Kendall, D.G., Tautu, P. (eds.) Mathematics in the Archaeological and Historical Sciences, pp. 387–395. Edinburgh University Press, Edinburgh (1971)
4. Dress, A.W.M.: Trees, tight extensions of metric spaces, and the cohomological dimension of certain groups. Adv. Math. 53, 321–402 (1984)
5. Dress, A.W.M., Huber, K.T., Moulton, V.: An explicit computation of the injective hull of certain finite metric spaces in terms of their associated Buneman complex. Adv. Math. 168, 1–28 (2002)
6. Dress, A.W.M., Huber, K.T., Koolen, J., Moulton, V.: Compatible Decompositions and Block Realizations of Finite Metrics (submitted)
7. Dress, A.W.M., Moulton, V., Terhalle, W.: T-Theory. Europ. J. of Combin. 17, 161–175 (1996)
8. Dress, A.W.M., Moulton, V., Steel, M.: Trees, taxonomy and strongly compatible multi-state characters. Adv. App. Math. 19, 1–30 (1997)
9. Dress, A.W.M., Scharlau, R.: Gated sets in metric spaces. Aeq. Math. 34, 112–120 (1987)
10. Isbell, J.: Six theorems about metric spaces. Comment. Math. Helv. 39, 65–74 (1964)
11. Koolen, J., Moulton, V., Toenges, U.: A classification of the six-point prime metrics. Europ. J. of Combin. 21, 815–829 (2000)
12. Semple, C., Steel, M.A.: Phylogenetics. Oxford University Press, Oxford (2003)

Finding the Anti-block Vital Edge of a Shortest Path Between Two Nodes[*]

Bing Su[1,2], Qingchuan Xu[1,2], and Peng Xiao[1,2]

[1] School of Management, Xi'an Jiaotong University,
Xi'an, 710049, P.R. China
[2] The State Key Lab for Manufacturing Systems Engineering,
Xi'an, 710049, P.R. China
{subing,xuqch,xiaopeng}@mail.xjtu.edu.cn

Abstract. Let $P_G(s,t)$ denote a shortest path between two nodes s and t in an undirected graph G with nonnegative edge weights. A replacement path at a node $u \in P_G(s,t) = (s, \cdots, u, v, \cdots, t)$ is defined as a shortest path $P_{G-e}(u,t)$ from u to t which does not make use of (u,v). In this paper, we focus on the problem of finding an edge $e = (u,v) \in P_G(s,t)$ whose removal produces a replacement path at node u such that the ratio of the length of $P_{G-e}(u,t)$ to the length of $P_G(u,t)$ is maximum. We define such an edge as an anti-block vital edge (AVE for short), and show that this problem can be solved in $O(mn)$ time, where n and m denote the number of nodes and edges in the graph, respectively. Some applications of the AVE for two special traffic networks are shown.

1 Introduction

Suppose that a transportation network is given in which each road is associated with the time it takes to traverse it. In practice, the network is unreliable, some roads may be unsuitable at certain times (e.g. blocked by unexpected events such as traffic accidents or snowfall). From the transportation network management point of view, it is valuable to identify the result by the failure of a component. In the past, the problem of an edge removal results in the increase of the length of the shortest path between two nodes in a graph has been studied.

Corley and Sha [1] studied the MVE problem of finding an edge whose removal from the graph $G(V, E)$ resulted in the largest increase of the length of the shortest path between two given nodes. This edge is generally denoted as the most vital edge with respect to the shortest path. This problem has been solved efficiently by Malik, Mittal and Gupta [2], who gave an $O(m + n \log n)$ time algorithm by using Fibonacci heap. Nardelli, Proietti and Widmayer [3] improved the previous time bound to $O(m \cdot \alpha(m,n))$, where α is the functional inverse of the Ackermann function, and n and m denote the number of nodes and edges in the graph, respectively.

[*] The authors would like to acknowledge the support of research grant No. 70525004, 70471035, 70121001 from the NSF and No.20060401003 from the PSF Of China.

A. Dress, Y. Xu, and B. Zhu (Eds.): COCOA 2007, LNCS 4616, pp. 11–19, 2007.

Nardelli, Proiett and Widmayer [4] focused on the LD (Longest Detour) problem of finding an edge $e^* = (u^*, v^*)$ in the shortest path between two given nodes such that when this edge is removed, the length of the detour satisfies $d_{G-e^*}(u^*, t) - d_G(u^*, t) \geq d_{G-e}(u, t) - d_G(u, t)$, where $G - e = (V, E - e)$. The edge whose removal will result in the longest detour is named the detour-critical edge and [4] showed that this problem can be solved in $O(m + n \log n)$ time in undirected graphs. The same bound for undirected graphs is also achieved by Hershberger and Suri [5], who solved the Vickrey payment problem with an algorithm that also solved the detour problem. [3] improved the result to $O(m \cdot \alpha(m, n))$ time bound. [6] showed that the detour problem required $\Omega(m\sqrt{n})$ time in the worst case whenever $m = n\sqrt{n}$ in directed graphs. We refer the reader to Nardelli , Proiett and Widmayer [7], Li and Guo [8], and Bhosle [9] for extensive references to a variety of the MVE problem.

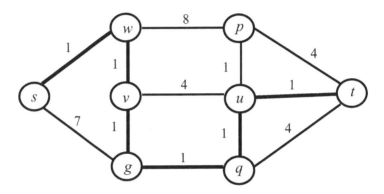

Fig. 1. MVE and the Detour-critical edge analysis. In bold, the shortest path from s to t.

Most previous studies are based on the assumption that the traveller only pays attention to the length of the replacement path minus the length of the shortest path. See Fig.1, the shortest path from s to t is $P_G(s, t) = (s, w, v, g, q, u, t)$ whose length is 6, the edge $e = (s, w)$ and the edge $e = (w, v)$ are the most vital edges and the edge $e = (w, v)$ is the detour critical edge of $P_G(s, t)$. If the edge $e = (w, v)$ is failure, the detour is $P_{G-e}(w, t) = (w, p, u, t)$ whose length is 10, the length of the detour minus the length of $P_G(w, t) = (w, v, g, q, u, t)$ is 5 and the length of the detour is 2 times of the length of $P_G(w, t)$. However, if the edge $e = (u, t)$ is failure, the length of the detour $P_{G-e}(u, t) = (u, p, t)$ minus the length of $P_G(u, t) = (u, t)$ is 4, the length of the detour is 5 times of the length of $P_G(u, t)$. From the point of view of a traveller, the edge $e = (u, t)$ is more important than the edge $e = (w, v)$.

In this paper, we define a different parameter for measuring the vitality of an edge of the shortest path $P_G(s, t)$ between the source s and destination t in $G = (V, E)$. We will face the problem of finding an edge $e = (u, v)$ in $P_G(s, t)$ whose removal produces a replacement path at node u such that the ratio of the length of the shortest path $P_{G-e}(u, t)$ in $G - e = (V, E - e)$ to the length of the shortest path $P_G(u, t)$ in G is maximum. We call such an edge an anti-block vital edge (AVE for short).

Our approach of building all the replacement paths along $P_G(s, t)$ reveals its importance in network applications. Under the assumption that a sudden blockage of an edge is possible in a transportation network, a traveller may reach a node u from which he can not continue on his path as intended, just because the outgoing edge $e = (u, v)$ to be taken is currently not operational, the traveller should route from u to t on a shortest path in $G - e = (V, E - e)$. It is important to know the ratio of the length of the replacement path $P_{G-e}(u, t)$ to the length of $P_G(u, t)$ and the anti-block vital edge in advance, and, the maximum ratio is a key parameter for measuring the competitive ratio of a strategy for the online blockage problems such as the Canadian Traverller Problem [10-12].

We show that the problem of finding the AVE can be solved in $O(mn)$ time, where n and m denote the number of nodes and edges in the graph, respectively. Some applications of the AVE for two special traffic networks are shown.

2   Problem Statement and Formulation

Let $G = (V, E)$ denote an undirected transportation network with $|V| = n$ vertices and $|E| = m$ edges, s denote the source and t the destination, $w(e)$ denote a nonnegative real weight associated to each $e \in E$. A shortest path $P_G(s, t)$ from s to t in G is defined as a path which minimizes the sum of the weights of the edges along the path from s to t and the length of $P_G(s, t)$ is denoted as $d_G(s, t)$. $P_G(u, t)$ denotes the shortest path at a node $u \in P_G(s, t) = (s, \cdots, u, v, \cdots, t)$ from u to t in G and $d_G(u, t)$ denotes the length of $P_G(u, t)$.

Definition 1. A replacement path at a node $u \in P_G(s, t) = (s, \cdots, u, v, \cdots, t)$ is a shortest path $P_{G-e}(u, t)$ from u to t which does not make use of (u, v), where $G - e = (V, E - e)$.

Definition 2. The anti-block coefficient of an edge $e = (u, v) \in P_G(s, t) = (s, \cdots, u, v, \cdots, t)$ is the ratio c_{ut} of the length of $P_{G-e}(u, t)$ to the length of $P_G(u, t)$.

Definition 3. The anti-block vital edge (AVE for short) with respect to $P_G(s, t)$ is the edge $e' = (u', v') \in P_G(s, t) = (s, \cdots, u', v', \cdots, t)$ whose removal from G results in $c_{u't} \geq c_{ut}$ for every edge $e = (u, v)$ of $P_G(s, t)$.

In order to discuss the problem, we make the following assumptions.
(1) There is only one shortest path between s and t in G.
(2) Only one edge will be blocked in the shortest path $P_G(s, t)$.
(3) G is connected even when the blocked edge is removed.

3   A Property of the Anti-block Coefficient

Let $P_G(s, t) = (v_0, v_1, \cdots, v_i, v_{i+1}, \cdots, v_{k-1}, v_k)$ denote the shortest path from s to t, where $v_0 = s$ and $v_k = t$. Let $P_G(v_i, v_j)$ denote the shortest path from

v_i to v_j and $P_{G-e}(v_i, v_j)$ denote the shortest path from v_i to v_j which does not make use of (v_i, v_{i+1}), where $v_i, v_j \in P_G(s,t)$ and $j = i+1, i+2, \cdots, k$. Let $d_G(v_i, v_j)$ denote the length of $P_G(v_i, v_j)$ and $d_{G-e}(v_i, v_j)$ denote the length of $P_{G-e}(v_i, v_j)$. Fig.2 illustrates the situation.

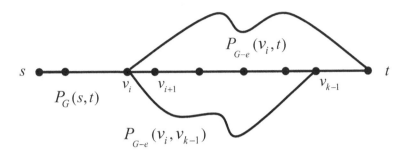

Fig. 2. Anti-block coefficient analysis in a general graph

Theorem 1. If a path $P_G(s,t) = (v_0, v_1, \cdots, v_i, v_{i+1}, \cdots, v_{k-1}, v_k)$ is the shortest path from s to t in G, where $v_0 = s$ and $v_k = t$, then $c_{v_i,v_{i+1}} \geq c_{v_i,v_{i+2}} \geq \cdots \geq c_{v_i,v_k}$ for removal of $e = (v_i, v_{i+1})$, $i = 0, 1, \cdots, k-1$.

Proof. From the definition of the anti-block coefficient of an edge, the following equality holds:

$c_{v_i,v_j} = \frac{d_{G-e}(v_i,v_j)}{d_G(v_i,v_j)}$, where $j = i+1, i+2, \cdots k$.

Since $d_{G-e}(v_i, v_j) \geq d_G(v_i, v_j)$, then

$c_{v_i,v_j} = \frac{d_{G-e}(v_i,v_j)}{d_G(v_i,v_j)} \geq \frac{d_{G-e}(v_i,v_j)+d_G(v_j,v_{j+1})}{d_G(v_i,v_j)+d_G(v_j,v_{j+1})} = \frac{d_{G-e}(v_i,v_j)+d_G(v_j,v_{j+1})}{d_G(v_i,v_{j+1})}$.

In fact, $d_{G-e}(v_i, v_j) + d_G(v_j, v_{j+1}) \geq d_{G-e}(v_j, v_{j+1})$.

Hence, $c_{v_i,v_j} \geq \frac{d_{G-e}(v_i,v_{j+1})}{d_G(v_i,v_{j+1})} = c_{v_i,v_{j+1}}$.

From the above analysis, it is known that $c_{v_i,v_{i+1}} \geq c_{v_i,v_{i+2}} \geq \cdots \geq c_{v_i,v_k}$ holds. This ends the proof.

4   Compute the Anti-block Vital Edge

We will discuss the algorithm of finding the anti-block vital edge in a general network.

Let $P_G(s,t)$ be the shortest path from s to t in G. It is quite expensive to solve the AVE problem in the way that is by sequentially removing all the edges $e = (u, v)$ along $P_G(s,t)$ and computing at each step $P_{G-e}(u,t)$. In fact, this leads to a total amount of time of $O(mn^2)$ for the m edges in $P_G(s,t)$ by using the algorithm of Dijkstra [13].

We now discuss an approach to improve the algorithm and its time complexity. We start to compute the shortest path tree rooted at t, denoted as $S_G(t)$. As shown in Fig.3, $e = (u, v)$ denotes an edge on $P_G(s, t)$, with u closer to s than v; $M_t(u)$ denotes the set of nodes reachable in $S_G(t)$ from t without passing through edge (u, v), the length from t to the nodes in $M_t(u)$ doesn't change after deleting the edge e; $N_t(u) = V - M_t(u)$ denotes the remaining nodes (i,e., the subtree of $S_G(t)$ rooted at u), the length from t to the nodes in $N_t(u)$ may increase as a consequence of deleting e.

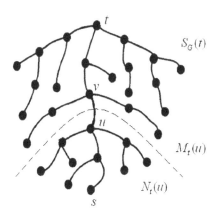

Fig. 3. $M_t(u)$ and $N_t(u)$

Since the replacement path $P_{G-e}(u, t)$ joining u and t must contain an edge in $E_t(u) = \{(x, y) \in E - (u, v) | (x \in N_t(u)) \wedge (y \in M_t(u))\}$, it follows that it corresponds to the set of edges whose weights satisfy the following condition

$$d_{G-e}(u, t) = \min_{x, y \in E_t(u)} \{d_{G-e}(u, x) + w(x, y) + d_{G-e}(y, t)\}.$$

Fig.4 illustrates the situation.

Since $x \in N_t(u)$, then $d_{G-e}(u, x) = d_G(u, x) = d_G(t, x) - d_G(t, u)$ and since $y \in M_t(u)$, then $d_{G-e}(y, t) = d_G(y, t)$.

Hence, $d_{G-e}(u, t) = \min_{x \in N_t(u), y \in M_t(u)} \{d_G(t, x) - d_G(t, u) + w(x, y) + d_G(y, t)\}.$

4.1 Algorithm* for Computing the Anti-block Vital Edge

The algorithm* for obtaining the anti-block vital edge is based on the results mentioned above.

Step 1. Compute the shortest path tree $S_G(t)$ rooted at t by using the algorithm of Dijkstra and record $P_G(u, t)$, $d_G(u, t)$ and k (the number of edges along $P_G(s, t)$).

Step 2. Set $i = 0$.

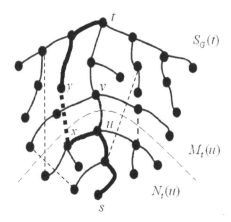

Fig. 4. Edge (u, v) is removed, dashed lines represent the linking edges. In bold, the replacement path at u with its linking edge (x, y).

Step 3. Remove the edge $e_i = (u, v)$ from $P_G(s, t)$ and compute $S_{G-e_i}(t)$, $M_t(u)$ and $N_t(u)$.

Step 4. Compute $d_{G-e_i}(u, t) = \min\limits_{x \in N_t(u), y \in M_t(u)} \{d_G(t, x) - d_G(t, u) + w(x, y) + d_G(y, t)\}$ and $c_{ut} = \frac{d_{G-e_i}(u,t)}{d_G(u,t)}$.

Step 5. Set $i = i + 1$. If $i < k$, then turn to step 3; otherwise turn to step 6.

Step 6. Compute c_{ut}, $c_{u't} = \max \{c_{ut}\}$ and the anti-block vital edge e'.

4.2   Analysis of the Time Complexity on the Algorithm*

For the shortest path tree $S_G(t)$ can be computed in $O(n^2)$ time by using the algorithm of Dijkstra, the computation time of $S_{G-e_i}(t)$, $M_t(u)$ and $N_t(u)$ for each e_i is $O(1)$, and the computation time of step 4 is $O(m)$. Since step 2-5 are loop computation and its repeat times is k, then the total time for step 2-5 is $O(mn)$ for $k \leq n - 1$. The computation time of step 6 is $O(n)$. It is known that the time complexity of the algorithm* is $O(mn)$.

From the above analysis, the following theorem holds.

Theorem 2. In a graph with n nodes and m edges, the anti-block vital edge problem on a shortest path between two nodes s and t can be solved in $O(mn)$ total time.

5   Applications of the Anti-block Vital Edge in Urban Traffic Networks

The grid-type network and the circular-type network are two examples of urban traffic networks. For two special cases, the anti-block vital edge have some properties. We will discuss them in details.

As shown in Fig.5, let $G = (V, E)$ denote an undirected planar grid-type network with $(m + 1)(n + 1)$ nodes, there are $m + 1$ rows of nodes in horizontal direction and $n + 1$ columns of nodes in vertical direction. $E = \{(v_{ij}, v_{i,j+1})\} \cup \{(v_{ij}, v_{i+1,j})\}$, $i = 0, 1, 2, \cdots, m, j = 0, 1, 2, \cdots, n$ denotes the set of the edges in G. The weight of each edge is 1.

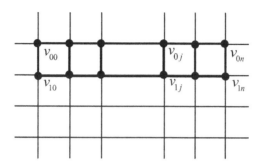

Fig. 5. Anti-block coefficient analysis in a grid-type network

Theorem 3. In an undirected planar grid-type network, if a source node and a destination node locate on a line, then the anti-block vital edge of the shortest path between the two nodes is the edge adjacent destination node.

Proof. Let v_{00} denote a source node and v_{0n} a destination node. If there is no any blockage happening in G, the shortest path from v_{00} to v_{0n} is $P_G(v_{00}, v_{0n}) = (v_{00}, v_{01}, \cdots, v_{0j}, \cdots, v_{0n})$. Let $P_G(v_{0j}, v_{0n})$ denote the shortest path from v_{0j} to v_{0n}, $P_{G-e}(v_{0j}, v_{0n})$ denote the shortest path from v_{0j} to v_{0n} which does not make use of $e = (v_{0j}, v_{0,j+1})$, $j = 0, 1, 2, \cdots, n-1$, $d_G(v_{0j}, v_{0n})$ denote the length of $P_G(v_{0j}, v_{0n})$, $d_{G-e}(v_{0j}, v_{0n})$ denote the length of $P_{G-e}(v_{0j}, v_{0n})$.

Since $c_{v_{0j}, v_{0n}} = \frac{d_{G-e}(v_{0j}, v_{0n})}{d_G(v_{0j}, v_{0n})} = \frac{n-j+2}{n-j}$, $c_{v_{0,j+1}, v_{0n}} = \frac{d_{G-e}(v_{0,j+1}, v_{0n})}{d_G(v_{0,j+1}, v_{0n})} = \frac{n-j-1+2}{n-j-1}$ and $\frac{n-j+1}{n-j-1} > \frac{n-j+2}{n-j}$, then $c_{v_{0j}, v_{0n}} < c_{v_{0,j+1}, v_{0n}}$.

Similarly, the following inequality holds: $c_{v_{00}, v_{0n}} < c_{v_{01}, v_{0n}} < \cdots < c_{v_{0,n-1}, v_{0n}}$.

It is known that the anti-block vital edge is the edge adjacent destination node in the shortest path under the assumption of source node and destination node locating on a line in a grid-type network. This ends the proof.

As shown in Fig.6, $G = (V, E)$ denotes an undirected planar circular-type network with $(m+1)(n+1)+1$ nodes, $E = \{(s, v_{0j})\} \cup \{(v_{ij}, v_{i,j+1})\} \cup \{(v_{ij}, v_{i+1,j})\}$, $i = 0, 1, 2, \cdots, m, j = 0, 1, 2, \cdots, n$ denotes the set of the edges in G. Let the polar angle θ of the edge $e = (s, v_{00})$ and the edge $e = (s, v_{01})$ satisfy $\theta < \frac{\pi}{2}$ and the weight of edge $e = (v_{ij}, v_{i,j+1})$ be 1. Therefore, the weight of $e = (v_{ij}, v_{i,j+1})$ is $(i + 1)\theta$, $i = 0, 1, 2, \cdots, m$.

Theorem 4. In an undirected planar circular-type network, if v_{m0} is a source node and s is a destination node, then the anti-block vital edge is every edge e of the shortest path $P_G(v_{m0}, s)$ from v_{m0} to s.

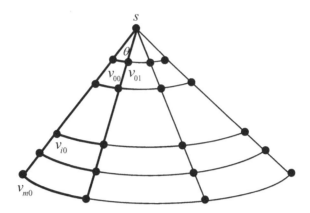

Fig. 6. Anti-block coefficient analysis in a circular-type network

Proof. If there is no any blockage happening in G, the shortest path from v_{m0} to s is $P_G(v_{m0}, s) = (v_{m0}, \cdots, v_{i0}, \cdots, v_{10}, v_{00}, s)$. Let $P_G(v_{i0}, s)$ denote the shortest path from v_{i0} to s, $P_{G-e}(v_{i0}, s)$ denote the shortest path from v_{i0} to s which does not make use of $e = (v_{i0}, v_{i-1,0})$ or $e = (v_{00}, s)$, where $i = 1, 2, \cdots, m$. Let $d_G(v_{i0}, s)$ denote the length of $P_G(v_{i0}, s)$ and $d_{G-e}(v_{i0}, s)$ denote the length of $P_{G-e}(v_{i0}, s)$.

Since $c_{v_{i0},s} = \frac{d_{G-e}(v_{i0},s)}{d_G(v_{i0},s)} = \frac{i \cdot \theta + i}{i} = \theta + 1$, $i = 0, 1, 2, \cdots, m$, then the following equality holds: $c_{v_{m0},s} = c_{v_{m-1,0},s} = \cdots = c_{v_{i0},s} = \cdots = c_{v_{00},s} = \theta + 1$. It is known that every edge along the shortest path $P_G(v_{m0}, s)$ is the anti-block vital edge of $P_G(v_{m0}, s)$. This ends the proof.

6   Conclusions

From the transportation network management point of view, it is valuable to identify the result by the failure of a component. In this paper, under the assumption that a sudden blockage of an edge is possible in a transportation network, we define a different parameter - anti-block vital edge (AVE for short) for measuring the vitality of an edge along the shortest path $P_G(s, t)$ between the source s and destination t in G. Our approach of building all the replacement paths along $P_G(s, t)$ and the ratio of the length of the replacement path $P_{G-e}(u, t)$ to the length of $P_G(u, t)$ for each edge $e = (u, v) \in P_G(s, t)$ reveals their importance in network applications. The maximum ratio is also a key parameter for measuring the competitive ratio of a strategy for the online blockage problems. We show that the problem of finding the AVE can be solved in $O(mn)$ time in a general network, where n and m denote the number of nodes and edges in the graph, respectively. Some applications of the AVE for two special traffic networks are shown. There are some further directions to work, such as how to improve the algorithm and its time complexity for computing the AVE and find the anti-block vital node when blockage happens at a node.

References

1. Corley, H.W., Sha, D.Y.: Most vital links and nodes in weighted networks. Operation Research Letters 1, 157–161 (1982)
2. Malik, K., Mittal, A.K., Gupta, S.K.: The k most vital arcs in the shortest path problem. Operation Research Letters 8, 223–227 (1989)
3. Nardelli, E., Proietti, G., Widmayer, P.: A faster computation of the most vital edge of a shortest path between two nodes. Information Processing Letters 79(2), 81–85 (2001)
4. Nardelli, E., Proietti, G., Widmayer, P.: Finding the detour critical edge of a shortest path between two nodes. Information Processing Letters 67(1), 51–54 (1998)
5. Hershberger, J., Suri, S.: Vickrey prices and shortest paths: What is an edge worth? In: Proc. 42nd Annu. IEEE Symp.Found. Comput. Sci., pp. 252–259 (2001)
6. Hershberger, J., Suri, S., Bhosle, A.: On the difficulty of some shortest path problems. In: Proc. 20th Sympos. Theoret. Aspects Comput. Sci. LNCS (2003)
7. Nardelli, E., Proietti, G., Widmayer, P.: Finding the most vital node of a shortest path. Theoretical Computer Science 296, 167–177 (2003)
8. Li, Y., Guo, Y.: Study on vital edges of shortest paths in traffic and transportation networks. Chinese Journal of Management Science 12(4), 69–73 (2004)
9. Bhosle, A.M.: Improved algorithms for replacement paths problems in restricted graphs [J]. Operations Research Letters 33, 459–466 (2005)
10. Papadimitriou, C.H., Yannakakis, M.: Shortest paths without a map. In: Ronchi Della Rocca, S., Ausiello, G., Dezani-Ciancaglini, M. (eds.) Automata, Languages and Programming. LNCS, vol. 372, pp. 610–620. Springer, Heidelberg (1989)
11. Bar-Noy, A., Schieber, B.: The canadian traveler problem. In: Proceedings of the Second Annual ACM-SIAM Symposium on Discrete Algorithms, pp. 261–270 (1991)
12. Su, B.: Research on strategy for sequential unexpected blockages during the transportation process. PHD. Dissertation, 1 (2005)
13. Dijkstra, E.W.: A note on two problems in connexion with graphs. Numer. Math. pp. 269–271 (1959)

K-Connected Target Coverage Problem in Wireless Sensor Networks

Deying Li[1,2], Jiannong Cao[1], Ming Liu[1], and Yuan Zheng[1]

[1] Internet and Mobile Computing Lab, Department of Computing
Hong Kong Polytechtic University,
Hung Hom, Kowloon, Hong Kong
[2] School of Information, Renmin University of China
100872, Beijing, China
deyingli@ruc.edu.cn,
{csjcao,csmliu,csyuazheng}@comp.polyu.edu.hk

Abstract. An important issue in deploying a WSN is to provide target coverage with high energy efficiency and fault-tolerance. Sensor nodes in a wireless sensor network are resource-constrained, especially in energy supply, and prone to failure. In this paper, we study the problem of constructing energy efficient and fault-tolerant target coverage. More specifically, we propose solutions to forming *k*-connected coverage of targets with the minimal number of active nodes. We first address the *k*-connected augmentation problem, and then show that the *k*-connected target coverage problem is NP-hard. We propose two heuristic algorithms to solve the problem. We have carried out extensive simulations to study the performance of the proposed algorithms. The evaluation results have demonstrated their desirable efficiency.

Keywords: NP-hard problems, heuristic algorithms, *k*-connected target coverage, *k*-connected augmentation, wireless sensor networks.

1 Introduction

In wireless sensor networks (WSN), an important task is to monitor and collect the relevant data in a geographical region or a set of targets. Since sensor nodes in a WSN are often deployed in an arbitrary manner, one of the fundamental issues in the task of target monitoring is *target coverage* which aims at covering the specified targets by a subset of the deployed sensor nodes with minimum resource consumption. Recent research has reported that significant energy savings can be achieved by elaborately managing the duty cycle of the nodes in a WSN with high node density. In this approach, some nodes are scheduled to sleep (or enter a power saving mode) while the remaining active nodes keep working. However, the excessive number of sleeing nodes may cause the WSN to be disconnected, i.e. the set of active nodes will be isolated. In addition, the overloading of the active nodes may cause them to be easily exhausted and even to fail. These will consequently invalidate the data collection and transmission.

A. Dress, Y. Xu, and B. Zhu (Eds.): COCOA 2007, LNCS 4616, pp. 20–31, 2007.

Previous research has been reported on determining a minimal number of active sensors required to maintain the initial coverage area as well as the connectivity. Using a minimal set of active nodes reduces power consumption and prolongs network lifetime. However, the existing works have concentrated on providing only single connectivity. In practice, connectivity affects the robustness and achievable throughput of communication in a WSN, and single connectivity often is not sufficient for many WSN applications because a single failed node could disconnect the network. Therefore, it is desirable for have sufficient connectivity in a wireless sensor network to ensure the successful data operations.

In this paper, we study the problem of constructing energy efficient and fault-tolerant target coverage. More specifically, we propose solutions to forming *k*-connected coverage of targets with the minimal number of active nodes. Our goal is to design algorithms to construct a k-connected target coverage graph that uses the minimum number of sensors such that (1) the communication graph formed by these sensors is *k*-connected, and (2) any target can be covered by at least one of these sensors. We first consider the *k*-connected augmentation problem, i.e., for a given graph $G=(V, E)$ and a subset of V, adding the minimum number of nodes such that the resulting subgraph is *k*-connected. We show that the *k*-connected augmentation problem is NP-hard and then give the heuristic algorithms accordingly. Then, we show that the problem of *k*-connected target coverage is also NP-hard and then propose two heuristic algorithms to solve the problem of *k*-connected target coverage. To the best of our knowledge, our algorithms are the first ones proposed for solving the *k*-connected augmentation problem.

The rest of paper is organized as follows. Section 2 briefly reviews the related work; section 3 describes the network model and defines the problem to be studied in this paper; section 4 studies the *k*-connected augmentation problem and the *k*-connected coverage problem. Solutions to the problems are proposed. Section 5 presents the simulations results and section 6 concludes the paper.

2 Related Work

There are many studies in literature on the coverage problem ([1-5 etc.]) in WSNs. Different formulations of the coverage problem have been proposed, depending on the subject to be covered (area versus discrete points) [4,5], the sensor deployment mechanism (random versus deterministic [6]), as well as other wireless sensor network properties (e.g. network connectivity and minimum energy consumption). For energy efficient area coverage, the works in [7,8] consider a large population of sensors, deployed randomly for area monitoring. The goal is to achieve an energy-efficient design that maintains area coverage. Slijepcevic et al [7] model the sensed area as a collection of fields. The proposed most-constrained least-constraining algorithm computes the disjoint covers successively, selecting sensors that cover the critical element (field covered by a minimal number of sensors). Cardei et al [2] model the disjoint sets as disjoint dominating sets in an undirected graph. The maximum disjoint dominating sets computation is NP-complete, and any polynomial-time approximation algorithm has a lower bound of 1.5. A graph-coloring mechanism is proposed for computing the disjoint dominating sets. Huang and Tseng [8] proposed solutions to solve two versions of the k-coverage problem, namely k-UC and k-NC. Authors modeled the coverage problem as a decision problem, whose goal is to determine whether each location of the target sensing area is sufficiently covered or not. Their solutions were based on checking the perimeter of each sensor's sensing range.

Zhang and Hou [9] proved an important, but intuitive result that if the communication range Rc is at least twice the sensing range Rs, a complete coverage of a convex area implies connectivity of the working nodes. They further discussed the case $Rc > Rs$. Wang et al [10] generalized the result in [9] by showing that, when the communication range Rc is at least twice the sensing range Rs, a k-covered network will result in a k-connected network. Wu and Yang [11] proposed two density control models for designing energy conserving protocols in sensor networks, using the adjustable sensing range of several levels.

Zhou et al [12] addressed the problem of selecting a minimum size connected k-cover, which is defined as a set of sensors M such that each point in query region is covered by at least k, and the communication graph induced by M is connected. They present a centralized $O(\log n)$-approximation algorithm, and also present a distributed algorithm. Zhou et al [13] also address k_1-connected k_2-cover problem, in which a distributed and localized Voronoi-based algorithm.

The energy-efficient target coverage problem deals with the coverage of a set of targets with minimum energy cost [1,6,18]. Cardei and Du [1] addressed the target coverage problem where the disjoint sets are modeled as disjoint set covers, such that every cover completely monitors all the target points. Disjoint set coverage problem was proved to be NP-complete and a lower approximation bound of 2 for any polynomial-time approximation algorithm was indicated. The disjoint set cover problem is reduced to a maximum flow problem, which is modeled as a mixed integer programming. Simulation shows better results in terms of numbers of disjoint set covers computed, compared with most-constrained least-constraining algorithm [7], when every target is modeled as a field. Cardei et al [18] proposed an approach which differs from [1] by not requiring the sensor sets to be disjoint and by allowing sensors to participate in multiple sets,. They proposed two heuristics that efficiently compute the sets, using linear programming and the greedy approach.

Several schemes [14, 15, 16] have been proposed to maintain k-connectivity in topology control which involve the use of spanning graph.

In this paper, we consider the k-connected target coverage problem. Our objective is to find a minimum number of active nodes such that they cover all targets and the graph induced by them is k-connected. Our problem is different from the existing problems in the following ways.

(1) For k_1-connected k_2-cover problem, the existing papers [10, 13] address the area coverage, but we address target coverage.

(2) For target coverage problem, our objective is different from the existing works [1,18], which didn't consider connectivity. The work in [19] considers only 1-connected coverage.

(3) For connectivity, the existing works [14,15,16] aims to maintain a spanning subgraph to be k-connected, but in our work, we select minimum number of nodes to form k-connected.

3   Network Model and Problem Definition

In this section, we formulate the *k*-connected target coverage problem.

Assume that *n* sensors $v_1, v_2, \ldots, v_n$ are deployed in a region to monitor *m* targets I_1, $I_2, \ldots, I_m$. Each node v_i has a sensing region $S(v_i)$ and a communication range *R*. Any target inside $S(v_i)$ is covered by v_i. v_i can directly communicate with v_j if their Euclidian distance is less than communication range *R*, i.e., v_j lies in the communication range of v_i.

Now, we formally define the *k*-connected target coverage problem. First, we describe some necessary definitions.

Definition 1 (Communication graph). Given a set of sensors *V* in a sensor network and *R* being transmission range, the communication graph of *V* is a undirected graph with *V* as the set of nodes and edges between any two nodes if they can directly communication with each other, i.e. the distance between them is at most *R*.

Definition 2 (*k*-connectivity). The communication graph $G=(V, E)$ is *k*-connected if for any two nodes v_i and v_j in *V*, there are *k* node-disjoint paths between v_i and v_j.

Definition 3 (target coverage). Given a set of *m* targets *I* and a set of sensors *V*, a subset *C* of *V* is said to be coverage for the targets, if any target can be covered by at least one sensor in *C*.

Definition 4 (*k*-connected target coverage). Consider a sensor network consisting of a set *V* of sensors and a target set *I*, where each sensor v_i has a sensing range $S(v_i)$ and communication range *R*. A subset *C* of *V* is said to be *k*-connected target coverage for targets set *I* if the following two conditions hold:

  1) Each target I_j in *I* is covered by at least one sensor in *C*.
  2) The communication graph induced by *C* is *k*-connected.

In order to reduce the energy consumption, our objective is to use a minimum number of sensor nodes to completely cover all the targets and the graph induced by these sensor nodes is *k*-connected.

The problem studied in this paper can be now formally defined as follows:

The *k*-connected target coverage problem: Given a sensor network and a set of targets *I* over the network, the *k*-connected target coverage problem is to find a *k*-connected target coverage *C* such that the number of sensor nodes for *C* is minimized. This problem is NP-hard as it is a generalization of set cover problem, which is well-known NP-hard.

4   Algorithms to *k*-Connected Target Coverage

In this section, before studying the *k*-connected target coverage problem, we first study an introductory problem, namely the *k*-connected augmentation problem. We prove that the problem is NP-hard and propose the corresponding heuristic algorithms for the cases of *k*=2 and *k*>2, respectively. Based on the discussion of the *k*-connected augmentation problem, we further investigate the problem of k-connected target coverage problem. We prove that the k-connected target coverage problem is also NP-hard and propose two heuristic to solve this problem.

4.1 k-Connected Augmentation Problem

In this subsection, we address the k-connected augmentation problem. We first give a definition.

Definition 5 (k-connected augmentation problem). Given a graph $G=(V, E)$ which is k-connected, a subset C of V, and a integer k, the problem is to find a subset X such that the subgraph induced by $C \cup X$ is a k-connected and $|X|$ is minimized.

To the best of our knowledge, there is no algorithm for the k-connected augmentation problem ($k>2$). In the following two subsections, we will give some results for $k=2$ and general k respectively.

We know that when $k=1$, 1-connected augmentation problem is exactly the Steiner tree problem, which is NP-hard, therefore k-connected augmentation problem is also NP-hard.

A. 2-connected augmentation problem

We propose an algorithm for the 2-connected augmentation problem. Without loss of generality, we assume the subgraph $G[C]$ induced by C is a connected graph. Based on the existing approximation algorithms for the Steiner tree problem, we can propose an algorithm for solving this problem. We first construct an auxiliary graph-weighted block cut tree-which based on subgraph $G[C]$.

Definition 6: Subgraph $G[C]$ induced by C in $G=(V, E)$ is a graph $G[C]=(C, E_1)$, where any two nodes u and v in C, $(u, v) \in E$ if and only if $(u, v) \in E_1$.

Definition 7: The maximal 2-connected components of a graph are called blocks. Specially, a bridge is also called a block (an edge is called a bridge if the graph induced by removing this edge becomes disconnected.) A node u is called as cut node of $G=(V, E)$, if when it is removed from the graph, the resulted subgraph $G[V-\{u\}]$ will be disconnected.

Main idea for constructing block cut tree is constructing a tree among cut nodes and blocks of $G[C]$. In this established tree, a cut node has an edge to the block if this cut node is in the block. The algorithm of constructing a weighted block cut tree is as following:

Algorithm 1 Construct block tree
Input: Connected graph $G[C]$
Output: Weighted block tree BCT of $G[C]$
Step 1: Let a_1, a_2,... and B_1, B_2..., be the cut nodes and blocks of $G[C]$, respectively. The node set $V(BCT)$ is the union of V_a and V_b, where $V_a=\{ a_1, a_2,...\}$ and $V_b=\{ b_i| B_i$ is a block of $G[C]\}$. Associated with each node in $V(BCT)$ is a set, for $a_i \in V_a$, $X_i=\{a_i\}$. For $b_j \in V_b$, $Y_j=\{v_t| v_t \in B_j$ and is not a cut node in $G[C]\}$.
Step 2: The edge set $E(BCT)$ of edges (a_i, b_j), a_i is a cut node that belongs to B_j.

In constructing the weighted block cut tree BCT of $G[C]$, the leaves of BCT should be nodes b_j, $b_j \in V_b$. Let $L=\{ b_j$ is a leaf of $BCT\}$. We randomly select only one node v_{j_i} from each Y_j, where Y_j corresponds to $b_j \in L$. Let $V_L=\{ v_{j_i} \mid v_{j_i} \in Y_j , b_j \in L \}$.

It is easy to know that if $G[V_L \cup X]$ is a Steiner tree spanning all the nodes of V_L, $G[V_L \cup X \cup C]$ is a 2-connected subgraph. In addition, $V_L \subseteq V_1 \subseteq C$, a Steiner tree spanning all nodes of V_1 in $G[(V-C) \cup V_1]$ must be a Steiner tree spanning all nodes of V_L.

In the 2-connected augmentation problem, we need to find a subset X of V-C, such that $G[X \cup C]$ is 2-connected and $|X|$ is minimized. Therefore, the 2-connected augmentation problem can be transformed to the following problem:

Steiner tree problem in Subgraph (STS): Given a graph $G=(V, E)$ and a subset C of V, find a Steiner tree $G[V_L \cup X]$ that spans all nodes of V_L such that $|X|$ is minimized, where V_L is reduced from the weighted block cut tree BCT of $G[C]$.

We know that $G[V_L \cup X \cup C]$ is an approximate solution for 2-connected subgraph problem if $G[V_L \cup X]$ is an approximation solution for STS. We can use the algorithm proposed in [17] to get an approximation solution for STS.

Our proposed algorithm for the 2-connected subgraph problem as follows:

Algorithm 2: Construct 2-connteced subgraph including C
Input: Graph $G=(V, E)$ and a subset C of V
Output: 2-connteced subgraph including C
Step 1: Construct a weighted block cut tree BCT of $G[C]$, and get a subset V_L of C
Step 2: Construct a Steiner tree T spanning all nodes of V_L in $G[V_L \cup (V-C)]$ such that the number of Steiner nodes is minimized; therefore $G[V(T) \cup C]$ is 2-connected subgraph

B. k-connected Augmentation Problem

Because the k-connected augmentation problem is NP-hard, we cannot have an efficient optimal algorithm. We therefore design heuristic for solving the k-connected augmentation problem. To demonstrate the correctness of our algorithm, we first give a Lemma

Definition 8. u is k-connected to v in G if there are k node-disjoint paths between u and v in G.

Lemma 1. If G is k-connected and $\forall u_1, u_2 \in N(v)$, where $N(v)$ is a neighbor set of v in G, u_1 is k-connected to u_2 in $G[V$-$v]$, then G-$\{v\}$ is k-connected.

Proof. We need to prove that $G'=G-v$ is connected after removal of any k-1 nodes in G'. Given any two nodes v_1 and v_2 in G', without loss of generality, we assume $\{v_1,v_2\} \cap N(v) = \Phi$. We now prove that v_1 is still connected to v_2 after removal of the set of any k-1 nodes $W = \{w_1, w_2, ... w_{k-1}\}$, where $w_i \in V(G) - \{v, v_1, v_2\}$. Since G is k-connected, there are at least k disjoint node paths. We denote the set of disjoint node paths from v_1 and v_2 in G by $S_{v_1 v_2}(G)$, therefore, $\mid S_{v_1 v_2}(G) \mid \geq k$. If $v \notin S_{v_1 v_2}(G)$, it is obvious that v_1 is still connected to v_2 after removal of the set of any k-1 nodes $W = \{w_1, w_2, ... w_{k-1}\}$. Therefore, we only consider $v \in S_{v_1 v_2}(G)$. Let G'' be the resulting graph after v and W are removed from G, and let s_1 be the number of paths in $S_{v_1 v_2}(G')$ that are broken due to the removal of nodes in W, i.e. $s_1 = |\{p \in S_{v_1 v_2}(G') | \exists w \in W, w \in p\}|$. Since the paths in $S_{v_1 v_2}(G')$ are pair-wise-internally-node-disjoint, the removal of any one node in W breaks at most one path in the set, given $|W|=k-1$, we have $s_1 \leq k-1$.

If $|S_{v_1 v_2}(G')| \geq k$, then $|S_{v_1 v_2}(G'')| \geq |S_{v_1 v_2}(G')| - s_1 \geq 1$, i.e. v_1 is still connected v_2 in G''. Now we consider the case where $|S_{v_1 v_2}(G')| < k$. This occurs only when the removal of v breaks one path $p^0 \in S_{v_1 v_2}(G)$, without loss of generality, let the order of nodes on p^0 be v_1, u_1, v, u_2, v_2. Since the removal of v reduces the number of pair-wise-internally-node-disjoint paths between v_1 and v_2 by at most one, $|S_{v_1 v_2}(G) - \{p^0\}| \geq k-1$. Hence $|S_{v_1 v_2}(G')| = k-1$.

Now we consider two cases:

1) $s_1 < k-1$: $|S_{v_1 v_2}(G'')| \geq |S_{v_1 v_2}(G')| - s_1 \geq 1$, i.e. v_1 is still connected to v_2 in G''.

2) $s_1 = k-1$: hence every node in W belongs to some path in $S_{v_1 v_2}(G')$. Since p^0 is internally-disjoint with all paths in $S_{v_1 v_2}(G')$, we have $p^0 \cap W = \Phi$. Thus v_1 is connected to u_1 and u_2 is connected to v_2 in G''. Let s_2 be the number of paths in $S_{u_1 u_2}(G')$ that are broken due to the removal of nodes in W, i.e., $s_2 = |\{p \in S_{v_1 v_2}(G') | \exists w \in W, w \in p\}|$. Since $|S_{u_1 u_2}(G')| \geq k$ and $s_2 \leq k-1$, $|S_{u_1 u_2}(G'')| \geq 1$, i.e. u_1 is still connected to u_2 in G''. Therefore, v_1 is still connected to v_2 in G''.

We have proved that for any two nodes $v_1, v_2 \in G'$, v_1 is connected to v_2 after the removal of any k-1 nodes from $G'-\{v_1, v_2\}$. Therefore G' is k-connected.

Using Lemma 1, we can design a heuristic algorithm, Algorithm 3 shown below, for the k-connected augmentation problem. The algorithm has a run time of $O(n^5)$, because in the For-loop, each if-loop needs to run the maximum flow algorithm which needs time $O(n^3)$ for at most n times.

Algorithm 3: Construct a k-connected subgraph

Input: $G(V, E)$, a k-connected graph, and $C \subseteq V$

Output: $G[V_k]$, a k-connected subgraph of G induced by V_k, $C \subseteq V_k \subseteq V$

Step 1: $V_k := V$;

Step 2: Sort all nodes in $V-C$ in an increasing order of degree in G as $v_1, v_2, ... v_m$ such that $d(v_1) \leq d(v_2) \leq ... \leq d(v_m)$;

Step 3: For $i=1$ to m,

if $\forall u_1, u_2 \in N(v_i)$, u_1 is k-connected to u_2 in $G[V_k - \{v_i\}]$, then

$V_k = V_k - \{v_i\}$; $i := i+1$

Step 4: Output a k-connected subgraph of G

4.2 *k*-Connected Target Coverage Problem and Its Algorithm

As mentioned before, the *k*-connected target coverage problem is NP-hard. We design two heuristics for solving the problem. First, we propose the TS algorithm based on the algorithms proposed in subsections 4.A and 4.B for the *k*-connected coverage problem. After that, directly applying Lemma 1, we propose another heuristic algorithm, namely RA algorithm.

The main idea of the TS algorithm is that the algorithm includes two steps. The first step is to construct a coverage of the targets using the set cover algorithm, and the second step is to add some nodes to this coverage such that the subgraph composed by both the newly added nodes and the nodes already existing in the coverage is *k*-connected. The TS algorithm is formally presented as follows:

TS Algorithm: Construct an approximate solution for *k*-connected coverage
Input: Given $G=(V, E)$, a set I of targets.
Output: *k*-connected subgraph which its node set forms coverage for I
Step 1: Construct a set cover $C \subseteq V$ for I such that $|C|$ is minimized.
Step 2: Connect set C into *k*-connected subgraph, i.e. finding a subset X of V-C to C such that $G[C \cup X]$ is *k*-connected subgraph and $|X|$ is minimized.

The problem in Step1 is a NP-hard and there is a typical algorithm to find an approximate solution C. Then we use the algorithms in subsections 4.A and 4.B to get an solution $G[V_k]$ which is *k*-connected ($k \geq 2$) augmentation and $C \subseteq V_k$, then V_k is a solution for the *k*-connected ($k \geq 2$) target coverage problem.

Another algorithm is called reverse algorithm (RA). The main idea of the reverse algorithm is that, initially, each sensor node in the sensor network is active, and then active nodes change to inactive one at a time if it satisfies two conditions: (1) after deleting this node, the remain nodes also form a coverage, and (2) any two neighbors of the node has *k* node-disjoint paths in remaining graph after deleting this node.

Reverse algorithm: Construct an solution for *k*-connected target coverage
Input: Given $G=(V, E)$, a set I of targets, and $I_v, \forall v \in V$, a subset of I
Output: *k*-connected subgraph whose node set forms a coverage for I
Step 1: $V_k := V$;
Step 2: Sort all the nodes in V in increasing order of degree in I as $v_1, v_2, ... v_n$ such that $D_I(v_1) \leq D_I(v_2) \leq ... \leq D_I(v_n)$, where $D_I(v) = |\{I_j | I_j \text{ is covered by } v\}|$
Step 3: For i=1 to n,
    if $\forall u_1, u_2 \in N(v_i)$, u_1 is *k*-connected to u_2 in $G[V_k - \{v_i\}]$, and $V_k - \{v_i\}$ is a coverage for I , then
        $V_k = V_k - \{v_i\}$; $i := i + 1$
Step 4: Output the *k*-connected target coverage.

Theorem 2. The reverse algorithm can produce an approximation solution for the *k*-connected target coverage problem with $O(n^5)$ time complexity.

Proof: It is easy to know that the algorithm can produce an approximation solution for the k-connected coverage problem.

In Step 3, in for-loop: each if-loop needs to use the maximum flow algorithm which need time $O(n^3)$ for $|N(v_i)|$, it takes total times at most $O(n^4)$. Therefore it takes time $O(n^5)$ for Step 3. Therefore the algorithm has time complexity of $O(n^5)$.

5 Performance Evaluation

In this section we evaluate the performance of the proposed algorithms by carrying out extensive simulations. In our simulation, N sensor nodes and M targets are independently and randomly distributed in a 500×500 square region. We assume the sensing range, S, of all sensor nodes are identical i.e., if the distance between any sensor node and a target is no more than S, the target is covered by the sensor node. In addition, the communicating range, R, of all sensor nodes are also assumed to be identical, i.e., if the distance between any pair of nodes is no more than R, there exists a direct communication between the two nodes. In the simulation we consider the following tunable parameters:

- N, the number of randomly deployed sensor nodes. We vary N between 25 and 70.
- M, the number of targets to be covered. We vary M between 10 and 50
- R, the communicating range. We R between 120 to 200.
- S, the sensing range. We vary S between 30 and 110

We simulate the proposed TS algorithm (TS) and RA algorithm (RA) and compared their performances in terms of the number of active sensor nodes. We test the performance of the two proposed algorithms in k-connected networks (k=2, 3). We present the averages of 100 separate runs for each result shown in the figures. In the simulations, any randomly generated topology that is not connected or targets that are not covered by all the sensor nodes are discarded.

In the first experiment, we consider 50 sensor nodes and 10 targets randomly distributed, and we vary communicating range between 120 and 200 with an increment of 5, while the sensing range is set to 70.

In Fig. 1, we present the number of active nodes obtained by using the TS and RA heuristics, depending on the communicating range. The numbers of active nodes returned by the two heuristics are close and they decrease with the network density. When the communicating range is larger, each sensor node can communicate with more sensors, thus fewer active nodes are needed.

In Fig. 2, we measure the number of active nodes when the number of sensor nodes varies between 25 and 70, and the communicating range is set to 150, sensing range is 70. We consider 10 targets randomly deployed. The number of active sensors increases with the number of sensors, as more sensors need to participate so that each active pairs communicate with k-disjoint paths.

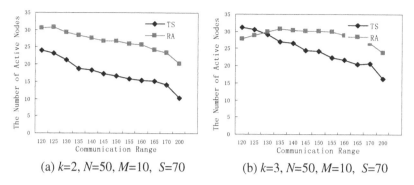

(a) *k*=2, *N*=50, *M*=10, *S*=70   (b) *k*=3, *N*=50, *M*=10, *S*=70

Fig. 1. The number of active nodes with communicating range

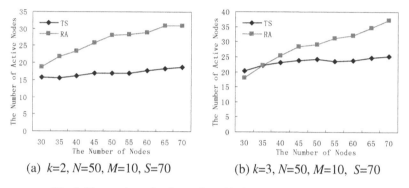

(a) *k*=2, *N*=50, *M*=10, *S*=70   (b) *k*=3, *N*=50, *M*=10, *S*=70

Fig. 2. The number of active nodes with the number of sensor nodes

In Fig. 3, we measure the number of active nodes when the sensing range varies between 70 and 110, and the communicating range is set to 150. We consider 50 sensors and 10 targets randomly deployed. The number of active sensor nodes is not increased with increasing sensing range, because when the sensing range is larger, each target is covered by more sensors.

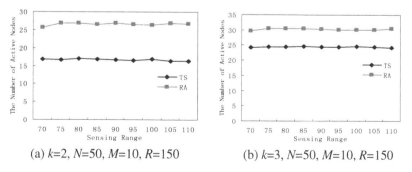

(a) *k*=2, *N*=50, *M*=10, *R*=150   (b) *k*=3, *N*=50, *M*=10, *R*=150

Fig. 3. The number of active nodes with sensing range

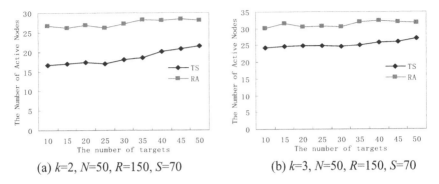

(a) k=2, N=50, R=150, S=70   (b) k=3, N=50, R=150, S=70

Fig. 4. The number of active nodes with the number of targets

In Fig. 4, we measure the number of active nodes when the number of targets varies between 10 and 500, the communicating range is set to 150, and the sensing range is set to 70. We consider 50 sensors randomly deployed. The number of active sensor nodes increased with the number of targets, as more targets needs to be covered.

6 Conclusion

In this paper, we have studied the problem of how to construct k-connected target coverage with the minimized number of active nodes in wireless sensor networks. We first discuss the k-connected augmentation problem in WSNs. Then, based on the result of the k-connected augmentation problem, we show that the k-connected coverage problem is NP-hard, and propose two heuristics to construct k-connected traget coverage. We have carried out extensive simulations to evaluate the performance of the two heuretic algorithms. The simulation results demonstrated the high effectiveness and efficiency of our algorithms

Acknowledgments. This research is partially supported by the Hong Kong Polytechnic University under the research grants G-YE57 and A-PF77 and the National Natural Science Foudation of China under grant 10671208 , and Key Lab of Data Enginerring and Knowledge Enginerring, MOE.

References

1. Cardei, M., Du, D.-Z.: Improving Wireless Sensor Network Lifetime through Power Aware Organization. ACM Wireless Networks 11, 333–340 (2005)
2. Cardei, D. MacCallum, X. Cheng, M. Min, X. Jia, D.Y. Li, and D.-Z. Du: Wireless Sensor Networks with Energy Efficient Organization. Journal of Interconnection Networks, 3(2002) 213-229
3. Cardei, M., Wu, J.: Energy-Efficient Coverage Problems in Wireless Ad Hoc Sensor Networks. to appear in Journal of Computer Communications on Sensor Networks

4. Li, X., Wan, P., Wang, Y., Frieder, O.: Coverage in Wireless Ad-hoc Sensor Networks. IEEE Transactions on Computers 52, 753–763 (2003)
5. Meguerdichian, S., Koushanfar, F., Potkonjak, M., Srivastava, M.B.: Coverage Problems in Wireless Ad-hoc Sensor Networks. In: Proc. of INFOCOM'01, pp. 1380–1387
6. Kar, K., Banerjee, S.: Node Placement for Connected Coverage in Sensor Networks. In: Proc. of WiOpt 2003: Modeling and Optimization in Mobile, Ad Hoc and Wireless Networks
7. Slijepcevic, S., Potkonjak, M.: Power Efficient Organization of Wireless Sensor Networks. In: Proc. of IEEE Int. Conference on Communications, vol. 2, pp. 472–476 (2001)
8. Huang, C., Tseng, Y.: The Coverage Problem in a Wireless Sensor Network. WSNA'03, San Diego, California, USA (September 19, 2003)
9. Zhang, H., Hou, J.C.: Maintaining Sensing Coverage and Connectivity in Large Sensor Networks. NSF International workshop on Theoretical and algorithmic Aspects of sensor, Ad Hoc Wireless and Peer-to-Peer Networks (2004)
10. Wang, X., Xing, G., Zhang, Y., Lu, C., Pless, R., Gill, C.D.: Integrated Coverage and Connectivity Configuration in Wireless Sensor Networks. The First ACM Conference on Embedded Networked Sensor Systems. ACM Press, New York (2003)
11. Wu, J., Yang, S.: Energy-Efficient Node Scheduling Models in Sensor Networks with Adjustable Ranges. Intel Journal of Foundations of Computer Science (2005)
12. Zhou, Z., Das, S., Gupta, H.: Connected K-Coverage Problem in Sensor Networks. In: ICCCN 2004, pp. 373–378 (2004)
13. Zhou, Z., Das, S., Gupta, H.: Fault Tolerant Connected Sensor Cover with Variable Sensing and Transmission Ranges. In: IEEE SECON 2005 (2005)
14. Bahramgiri, M., Hajiaghayi, M., Mirrokni, V.S.: Fault-tolerant and 3-dimensional Distributed Topology Control Algorithms in Wireless Multi-hop Networks. In: Proc. of ICCCN 2002, pp. 392–397 (2002)
15. Li, N., Hou, J.C.: FLSS: A Fault-tolerant Topology Control Algoritjm for Wireless Networks. In: Proc. MobiCom, pp. 275–286 (2005)
16. Li, X.-Y., Wan, P.-J., Wang, Y., Yi, C.-W.: Fault Tolerant Deployment and Topology Control in Wireless Networks. In: Proc. of MobiHoc, pp. 117–128 (2003)
17. Guha, S., Khuller, S.: Improved Methods for Approximating Node Weighted Steiner Trees and Connected Dominating Sets. Inf. Comput. 150, 57–74 (1999)
18. Cardei, M., Thai, M., Li, Y.S., Wu, W.L.: Energy-Efficient Target Coverage in Wireless Sensor Networks. In: INFOCOM 2005
19. Jaggi, N., Abouzeid, A.A.: Energy-efficient Connected Coverage in Wireless Sensor Networks. In: AMOC, pp. 85–100 (2006)

Searching Cycle-Disjoint Graphs

Boting Yang, Runtao Zhang, and Yi Cao

Department of Computer Science, University of Regina
{boting,zhang23r,caoyi200}@cs.uregina.ca

Abstract. In this paper, we study the problem of computing the minimum number of searchers who can capture an intruder hiding in a graph. We propose a linear time algorithm for computing the vertex separation and the optimal layout for a unicyclic graph. The best algorithm known so far is given by Ellis et al. (2004) and needs $O(n \log n)$ time, where n is the number of vertices in the graph. By a linear-time transformation, we can compute the search number and the optimal search strategy for a unicyclic graph in linear time. We show how to compute the search number for a k-ary cycle-disjoint graph. We also present some results on approximation algorithms.

1 Introduction

Given a graph in which an intruder is hiding on vertices or edges, the searching problem is to find the minimum number of searchers to capture the intruder. The graph searching problem has many applications [4,5,7].

Let G be a graph without loops and multiple edges. Initially, all vertices and edges of G are *contaminated*, which means an intruder can hide on any vertices or anywhere along edges. There are three actions for searchers: (1) place a searcher on a vertex; (2) remove a searcher from a vertex; and (3) slide a searcher along an edge from one end vertex to the other. A *search strategy* is a sequence of actions designed so that the final action leaves all edges of G cleared. An edge uv in G can be *cleared* in one of two ways by a sliding action: (1) two searchers are located on vertex u, and one of them slides along uv from u to v; or (2) a searcher is located on vertex u, where all edges incident with u, other than uv, are already cleared, and the searcher slides from u to v. The intruder can move along a path that contains no searcher at a great speed at any time. For a graph G, the minimum number of searchers required to clear G is called the *search number*, denoted by $s(G)$. A search strategy for a graph G is *optimal* if this strategy can clear G using $s(G)$ searchers. Let $E(i)$ be the set of cleared edges just after action i. A search strategy is said to be *monotonic* if $E(i) \subseteq E(i+1)$ for every i. LaPaugh [8] and Bienstock and Seymour [1] proved that for any connected graph G, allowing recontamination cannot reduce the search number. Thus, we only need to consider monotonic search strategies. Megiddo et al. [9] showed that determining the search number of a graph G is NP-hard. They also gave a linear time algorithm to compute the search number of a tree and an $O(n \log n)$ time algorithm to find the optimal search strategy, where n is the

A. Dress, Y. Xu, and B. Zhu (Eds.): COCOA 2007, LNCS 4616, pp. 32–43, 2007.

number of vertices in the tree. Peng et al. [10] proposed a linear time algorithm to compute the optimal search strategy of trees.

Search numbers are closely related to several other important graph parameters, such as vertex separation and pathwidth. A *layout* of a connected graph $G(V, E)$ is a one to one mapping $L: V \rightarrow \{1, 2, \ldots, |V|\}$. Let $V_L(i) = \{x : x \in V(G),$ and there exists $y \in V(G)$ such that the edge $xy \in E(G)$, $L(x) \leq i$ and $L(y) > i\}$. The *vertex separation of G with respect to L*, denoted by $\mathrm{vs}_L(G)$, is defined as $\mathrm{vs}_L(G) = \max\{|V_L(i)| : 1 \leq i \leq |V(G)|\}$. The *vertex separation of G* is defined as $\mathrm{vs}(G) = \min\{\mathrm{vs}_L(G) : L$ is a layout of $G\}$. We say that L is an *optimal layout* if $\mathrm{vs}_L(G) = \mathrm{vs}(G)$. Kinnersley [6] showed that $\mathrm{vs}(G)$ equals the pathwidth of G. Ellis et al. [2] proved that $\mathrm{vs}(G) \leq s(G) \leq \mathrm{vs}(G) + 2$ for any connected undirected graph G. They gave a transformation called 2-expansion from G to G' such that $\mathrm{vs}(G') = s(G)$. They also described an algorithm for trees to compute the vertex separation in linear time. Based on this algorithm, Ellis and Markov [3] gave an $O(n \log n)$ algorithm for computing the vertex separation and the optimal layout of a unicyclic graph.

The rest of this paper is organized as follows. In Section 2, we give definitions and notation. In Section 3, we present a linear time algorithm for computing the search number and the optimal search strategy for a tree by applying the labeling method. In Section 4, we improve Ellis and Markov's algorithm from $O(n \log n)$ to $O(n)$ for computing the vertex separation and the optimal layout of a unicyclic graph. In Section 5, we show how to compute the search number of a k-ary cycle-disjoint graph. In Section 6, we investigate approximation algorithms for computing the search number of a cycle-disjoint graph.

2 Preliminaries

All graphs in this paper are finite without loops and multiple edges. A *rooted tree* is a tree with one vertex designated as the root of the tree. We use $T[r]$ to denote a rooted tree T with root r. For any two vertices v_1 and v_2 in $T[r]$, if there is a path from r to v_2 that contains v_1, then we say v_2 is a *descendant* of v_1; specifically, if v_2 is adjacent to v_1, we say v_2 is a *child* of v_1. Each vertex of $T[r]$ except r is a descendant of r. For any edge with end vertices u and v, if v is the child of u, then we orient this edge with the direction from u to v. This edge is denoted by (u, v). After this orientation, we obtain a directed rooted tree $T[r]$ such that the in-degree of r is 0 and the in-degree of any other vertex is 1. For any vertex v of $T[r]$, the subtree induced by v and all its descendant vertices is called the *vertex-branch* at v, denoted by $T[v]$. $T[r]$ can be considered as a vertex-branch at r. For any directed edge (u, v), the graph $T[v] + (u, v)$ is called the *edge-branch* of (u, v), denoted by $T[uv]$. $T[uv]$ is also called an *edge-branch* at u.

A vertex-branch $T[x]$ is said to be *k-critical* if $s(T[x]) = k$ and there are exactly two edge-disjoint edge-branches in $T[x]$ such that they share a common vertex and each has search number k. This common vertex is called a *k-critical vertex*. An edge-branch $T[xy]$ is *k-critical* if $s(T[xy]) = k$ and $T[xy]$ contains a

k-critical vertex-branch. We use k^+ to denote a *critical element*, where k is a positive integer. The value of k^+, denoted as $|k^+|$, is equal to k.

Let $T[r]$ be a rooted tree and v be a vertex in $T[r]$. If $s(T[v]) = s_1$ and $T[v]$ is s_1-critical, let v_1 be the s_1-critical vertex in $T[v]$ and let $T[v, v_1]$ denote the subtree obtained by deleting all edges and vertices (except v_1) of $T[v_1]$ from $T[v]$. If $s(T[v, v_1]) = s_2$ and $T[v, v_1]$ is s_2-critical, let v_2 be the s_2-critical vertex in $T[v, v_1]$ and let $T[v, v_1, v_2]$ denote the subtree obtained by deleting all edges and vertices (except v_2) of $T[v_2]$ from $T[v, v_1]$. Repeat this process until we first encounter a subtree $T[v, v_1, \ldots, v_k]$ that is a single vertex v or whose search number is equal to s_{k+1} and which is not s_{k+1}-critical. If $T[v, v_1, \ldots, v_k]$ is a single vertex, then the label of v, denoted by $L(v)$, is defined by $(s_1^+, s_2^+, \ldots, s_k^+)$; otherwise, the label $L(v)$ is defined by $(s_1^+, s_2^+, \ldots, s_k^+, s_{k+1})$. Specifically, if $s(T[v]) = s_1 > 0$ and $T[v]$ is not s_1-critical, then the label $L(v)$ is defined by (s_1). Let (u, v) be an edge in $T[r]$. If $s(T[uv]) = s_1$ and $T[uv]$ is s_1-critical, let v_1 be the s_1-critical vertex in $T[uv]$ and let $T[uv, v_1]$ denote the subtree obtained by deleting all edges and vertices (except v_1) of $T[v_1]$ from $T[uv]$. If $s(T[uv, v_1]) = s_2$ and $T[uv, v_1]$ is s_2-critical, let v_2 be the s_2-critical vertex in $T[uv, v_1]$ and let $T[uv, v_1, v_2]$ denote the subtree obtained by deleting all edges and vertices (except v_2) of $T[v_2]$ from $T[uv, v_1]$. Repeat this process until we first encounter a subtree $T[uv, v_1, \ldots, v_k]$ whose search number is equal to s_{k+1} and which is not s_{k+1}-critical. The label of uv, denoted by $L(uv)$, is defined by $(s_1^+, s_2^+, \ldots, s_k^+, s_{k+1})$. Specifically, if $s(T[uv]) = s_1 > 0$ and $T[uv]$ is not s_1-critical, then the label of uv is defined by (s_1). Both vertex labels and edge labels have the following property.

Lemma 1. *For a labeled tree, each vertex label or edge label consists of a sequence of strictly decreasing elements such that each element except the last one must be a critical element.*

3   Labeling Method for Trees

From [2], we know that the search number of a tree can be found in linear time by computing the vertex separation of the 2-expansion of the tree. From [9], we know that the search number of a tree can also be found in linear time by using a hub-avenue method. However, in order to apply the labeling method proposed in [2] to compute search numbers of other special graphs (refer to the full version [11] of this paper), we modify this method so that it can compute the search number of a tree directly.

For a tree T, if we know the search number of all edge-branches at a vertex in T, then $s(T)$ can be computed from combining these branches' information.

Algorithm. SEARCHNUMBER($T[r]$)
Input: A rooted tree $T[r]$.
Output: $s(T[r])$.

1. Assign label (0) to each leaf (except r if r is also a leaf) and assign label (1) to each pendant edge in $T[r]$.

2. **if** there is an unlabeled vertex v whose all out-going edges have been labeled,
   then compute the label $L(v)$;
       if v has an in-coming edge (u, v), **then**
           computer the label $L(uv)$;
       else stop and **output** the value of the first element in the label $L(v)$.
   repeat Step 2.

Because a rooted tree $T[r]$ has a unique root r, every vertex except r has a unique in-coming edge and r has no in-coming edge. Thus, if v has no in-coming edge at line 3 of Step 2, then v must be the root r and the value of the first element in its label is equal to the search number of $T[r]$. We can prove the following results for rooted trees.

Lemma 2. *For a rooted tree $T[r]$ and a vertex v in $T[r]$, let $v_1, v_2, \ldots, v_k$ be all the children of v. Let $a = \max\{s(T[vv_i]) \mid 1 \leq i \leq k\}$ and b be the number of edge-branches with search number a.*
(i) If $b \geq 3$, then $s(T[v]) = a + 1$.
(ii) If $b = 2$ and no edge-branch at v is a-critical, then $s(T[v]) = a$ and $T[v]$ is a-critical.
(iii) If $b = 2$ and at least one branch at v is a-critical, then $s(T[v]) = a + 1$.
(iv) If $b = 1$ and no branch at v is a-critical, then $s(T[v]) = a$.
(v) If $b = 1$ and $T[vv_j]$ is a-critical, let u be the a-critical vertex in $T[vv_j]$, and let $T[v, u]$ be the subtree formed by deleting all edges of $T[u]$ from $T[v]$. If $s(T[v, u]) = a$, then $s(T[v]) = a + 1$; and if $s(T[v, u]) < a$, then $s(T[v]) = a$ and $T[v]$ is a-critical.

In Step 2 of the algorithm SEARCHNUMBER($T[r]$), let $v_1, v_2, \ldots, v_k$ be all the children of v. Each label $L(vv_i)$ contains the structure information of the edge-branch $T[vv_i]$. For example, if $L(vv_i) = (s_1^+, s_2^+, \ldots, s_m^+, s_{m+1})$, it means $T[vv_i]$ has a s_1-critical vertex u_1, $T[vv_i, u_1]$ has a s_2-critical vertex u_2, $\ldots$, $T[vv_i, u_1, \ldots, u_{m-1}]$ has a s_m-critical vertex u_m, and $T[vv_i, u_1, \ldots, u_m]$ has search number s_{m+1} and it is not s_{m+1}-critical. From Lemma 2, we can compute $L(v)$ that contains the structure information of the vertex-branch $T[v]$ by using the structure information of all edge-branches at v. Since the label of an edge (x, y) contains the information of the edge-branch $T[y] + (x, y)$, we can compute $L[xy]$ from $L[y]$. By using appropriate data structures for storing labels, each loop in Step 2 can be performed in $O(s(T[vv']) + k)$ time, where $T[vv']$ is the edge-branch that has the second largest search number among all edge-branches at v and k is the number of children of v. By using a recursion to implement Step 2 of SEARCHNUMBER($T[r]$), we can prove the following result.

Theorem 1. *If n is the number of vertices in a tree T, then the running time of computing $s(T)$ is $O(n)$.*

After we find the search number, we can use the information obtained in Algorithm SEARCH-NUMBER($T[r]$) to compute an optimal monotonic search strategy in linear time.

4   Unicyclic Graphs

Ellis and Markov [3] proposed an $O(n \log n)$ algorithm to compute the vertex separation of a unicyclic graph. In this section we will give an improved algorithm that can do the same work in $O(n)$ time. All definitions and notation in this section are from [3]. Their algorithm consists of three functions: main, vs_uni and vs_reduced_uni (see Fig. 28, 29 and 30 in [3] for their descriptions).

Let U be a unicyclic graph and e be a cycle edge of U. In function main, it first computes the vertex separation of the tree $U - e$, and then invokes function vs_uni to decide whether vs(U) = vs($U - e$). vs_uni is a recursive function that has $O(\log n)$ depth, and in each iteration it computes the vertex separation of a reduced tree $U' - e$ and this takes $O(n)$ time. Thus, the running time of vs_uni is $O(n \log n)$. vs_uni invokes the function vs_reduced_uni to decide whether a unicyclic graph U is k-conforming. vs_reduced_uni is also a recursive function that has $O(\log n)$ depth, and in each iteration it computes the vertex separation of $T_1[a]$ and $T_1[b]$ and this takes $O(n)$ time. Thus, the running time of vs_reduced_uni is also $O(n \log n)$.

We will modify all three functions. The main improvements of our algorithm are to preprocess the input of both vs_uni and vs_reduced_uni so that we can achieve $O(n)$ running time. The following is our improved algorithm, which computes the vertex separation and the corresponding layout for a unicyclic graph U.

program main_modified
1   For each constituent tree, compute its vertex separation, optimal layout and type.
2   Arbitrarily select a cycle edge e and a cycle vertex r. Let $T[r]$ denote $U - e$ with root r. Compute vs($T[r]$) and the corresponding layout X.
3   Let L be the label of r in $T[r]$. Set $\alpha \leftarrow$ vs($T[r]$), $k \leftarrow$ vs($T[r]$).
4   **while** the first element of L is a k-critical element and the corresponding k-critical vertex v is not a cycle vertex in U, **do**
            Update U by deleting $T[v]$ and update L by deleting its first element;
            Update the constituent tree $T[u]$ that contains v by deleting $T[v]$
                and update the label of u in $T[u]$ by deleting its first element;
            $k \leftarrow k - 1$;
5   **if** (vs_uni_modified(U, k))
            then output(α, the layout created by vs_uni_modified);
            else output($\alpha + 1$, X);

function vs_uni_modified(U, k): Boolean
Case 1: U has one k-critical constituent tree;
            compute vs(T');
            if vs(T') = k, **then return** (false) **else return** (true);
Case 2: U has three or more non-critical k-trees;
            return (false);
Case 3: U has exactly two non-critical k-trees T_i and T_j;
            compute vs($T_1[a]$), vs($T_1[b]$), vs($T_2[c]$) and vs($T_2[d]$);
            /* Assume that vs(T_1) $\geq$ vs(T_2). */
            /* Let L_a be the label of a in $T_1[a]$, and L_b be the label of b in $T_1[b]$. */

/* Let L_c be the label of c in $T_2[c]$, and L_d be the label of d in $T_2[d]$. */
/* Let U' be U minus the bodies of T_i and T_j. */
return (vs_reduced_uni_modified($U', L_a, L_b, L_c, L_d, k$));

Case 4: U has exactly one non-critical k-tree T_i;
/* let q be the number of $(k-1)$-trees that is not type NC. */

Case 4.1: $0 \le q \le 1$;
      return (true);

Case 4.2: $q = 2$;
      for each tree T_j from among the two $(k-1)$-trees, **do**
        compute the corresponding vs($T_1[a]$), vs($T_1[b]$), vs($T_2[c]$) and vs($T_2[d]$);
        if (vs_reduced_uni_modified($U', L_a, L_b, L_c, L_d, k$)) **then return** (true);
        /* U' is equal to U minus the bodies of T_i and T_j. */
      return (false);

Case 4.3: $q = 3$;
      for each tree T_j from among the three $(k-1)$-trees, **do**
        compute the corresponding vs($T_1[a]$), vs($T_1[b]$), vs($T_2[c]$) and vs($T_2[d]$);
        if (vs_reduced_uni_modified($U', L_a, L_b, L_c, L_d, k$)) **then return** (true);
        /* U' is equal to U minus the bodies of T_i and T_j. */
      return (false);

Case 4.4: $q \ge 4$;
      return (false);

Case 5: U has no k-trees;
/* let q be the number of $(k-1)$-trees that is not type NC. */

Case 5.1: $0 \le q \le 2$;
      return (true);

Case 5.2: $q = 3$;
      for each choice of two trees T_i and T_j from the three $(k-1)$-trees, **do**
        compute the corresponding vs($T_1[a]$), vs($T_1[b]$), vs($T_2[c]$) and vs($T_2[d]$);
        if (vs_reduced_uni_modified($U', L_a, L_b, L_c, L_d, k$)) **then return** (true);
        /* U' is equal to U minus the bodies of T_i and T_j. */
      return (false);

Case 5.3: $q = 4$;
      for each choice of two trees T_i and T_j from the four $(k-1)$-trees, **do**
        compute the corresponding vs($T_1[a]$), vs($T_1[b]$), vs($T_2[c]$) and vs($T_2[d]$);
        if (vs_reduced_uni_modified($U', L_a, L_b, L_c, L_d, k$)) **then return** (true);
        /* U' is equal to U minus the bodies of T_i and T_j. */
      return (false);

Case 5.4: $q \ge 5$;
      return (false).

function vs_reduced_uni_modified(U, L_a, L_b, L_c, L_d, k)): Boolean
      /* Let a_1, b_1, c_1, d_1 be the first elements of L_a, L_b, L_c, L_d respectively. */
      /* Let $|a_1|, |b_1|, |c_1|, |d_1|$ be the value of a_1, b_1, c_1, d_1 respectively. */
      /* We assume that $|a_1| \ge |c_1|$. */

Case 1: $|a_1| = k$;
      return (false).

Case 2: $|a_1| < k - 1$;
      return (true).

Case 3: $|a_1| = k - 1$;
      if both a_1 and b_1 are $(k-1)$-critical elements, **then**

/* Let u be the $(k-1)$-critical vertex in $T_1[a]$
    and let v be the $(k-1)$-critical vertex in $T_1[b]$. */
if $u = v$ and u is not a cycle vertex, **then**
    update L_a and L_b by deleting their first elements;
    update U by deleting $T[u]$;
    update the label of the root of the constituent tree containing u
        by deleting its first element;
    if $|c_1|$ is greater than the value of the first element in current L_a,
        then return (vs_reduced_uni_modified$(U, L_c, L_d, L_a, L_b, k-1)$.
        else return (vs_reduced_uni_modified$(U, L_a, L_b, L_c, L_d, k-1)$.
    else /* ($u = v$ and u is a cycle vertex) or $(u \neq v)$ */
        return (T_2 contains no $k-1$ types other than NC constituents);
else return ((neither a_1 nor d_1 is $(k-1)$-critical element)
    or (neither b_1 nor c_1 is $(k-1)$-critical element)).

Lemma 3. *Let U be a unicyclic graph, e be a cycle edge and r be a cycle vertex in U. Let $T[r]$ denote the tree $U - e$ with root r. If $\text{vs}(T[r]) = k$, then U has a k-constituent tree of type Cb if and only if the first element in the label of r in $T[r]$ is a k-critical element and the corresponding k-critical vertex is not a cycle vertex.*

The correctness of the modified algorithm follows from the analysis in Sections 4 and 5 in [3]. We now compare the two algorithms. In our main_modified function, if the condition of the *while-loop* is satisfied, then by Lemma 3, U has a k-constituent tree of type Cb that contains v. Let $T'[u]$ be this constituent tree and u be the only cycle vertex in $T'[u]$. The first element in the label of u in $T'[u]$ must be k-critical element. Let $L(r)$ be the label of r in $T[r]$ and $L(u)$ be the label of u in $T'[u]$. We can obtain the label of r in $T[r] - T[v]$ and the label of u in $T'[u] - T'[v]$ by deleting the first element of each label, according to the definition of labels [3]. This work can be done in constant time. However, without choosing a cycle vertex as the root of T, their algorithm needs $O(n)$ time to compute these two labels. Function vs_uni in [3] can only invoke itself in Case 1 when U has a k-constituent tree of type Cb. Our main_modified function invokes function vs_uni_modified only when the condition of the *while-loop* is not satisfied. By Lemma 3, in this case, U does not have a k-constituent tree of type Cb. Thus in Case 1 of vs_uni_modified, the tree must be of type C, and recursion is avoided. In their function vs_reduced_uni, $vs(T_1)$ and $vs(T_2)$ are computed using $O(n)$ time. However, we compute them before invoking vs_reduced_uni_modified. Let L_a, L_b, L_c and L_d be the label of a in $T_1[a]$, b in $T_1[b]$, c in $T_2[c]$ and d in $T_2[d]$ respectively. All the information needed by vs_reduced_uni_modified is these four labels. While recursion occurs, we can obtain new labels by simply deleting the first elements from the old ones, which requires only constant time. Hence, the time complexity of vs_reduced_uni_modified can be reduced to $O(1)$ if we do not count the recursive iterations.

We now analyze the running time of our modified algorithm. Since function vs_reduced_uni_modified only ever invokes itself and the depth of the recursion is $O(\log n)$, its running time is $O(\log n)$. In function vs_uni_modified, Case 1 needs $O(n)$; Cases 3, 4.2, 4.3, 5.2 and 5.3 need $O(n) + O(\log n)$; and other cases can be done in $O(1)$. Thus, the running time of vs_uni_modified is $O(n) + O(\log n)$. In the main_modified function, all the work before invoking vs_uni_modified can be done in $O(n) + O(\log n)$. Hence, the total running time of the modified algorithm is $O(n)$. Therefore, we have the following theorem.

Theorem 2. *For a unicyclic graph G, the vertex separation and the optimal layout of G can be computed in linear time.*

For a graph G, the 2-expansion of G is the graph obtained by replacing each edge of G by a path of length three. By Theorem 2.2 in [2], the search number of G is equal to the vertex separation of the 2-expansion of G. From Theorem 2, we have the following result.

Corollary 1. *For a unicyclic graph G, the search number and the optimal search strategy of G can be computed in linear time.*

5 k-Ary Cycle-Disjoint Graphs

A graph G is called a *cycle-disjoint graph (CDG)* if it is connected and no pair of cycles in G share a vertex. A *complete k-ary tree T* is a rooted k-ary tree in which all leaves have the same depth and every internal vertex has k children. If we replace each vertex of T with a $(k+1)$-cycle such that each vertex of internal cycle has degree at most 3, then we obtain a cycle-disjoint graph G, which we call a *k-ary cycle-disjoint graph (k-ary CDG)*. In T, we define the level of the root be 1 and the level of a leaf be the number of vertices in the path from the root to that leaf. We use T_k^h to denote a complete k-ary tree with level h and G_k^h to denote the k-ary CDG obtained from T_k^h. In this section, we will show how to compute the search numbers of k-ary CDGs. Similar to [3], we have the following lemmas.

Lemma 4. *Let G be a graph containing three connected subgraphs G_1, G_2 and G_3, whose vertex sets are pairwise disjoint, such that for every pair G_i and G_j there exists a path in G between G_i and G_j that contains no vertex in the third subgraph. If $s(G_1) = s(G_2) = s(G_3) = k$, then $s(G) \geq k+1$.*

Lemma 5. *For a connected graph G, let $C = v_1 v_2 \ldots v_m v_1$ be a cycle in G such that each v_i $(1 \leq i \leq m)$ connects to a connected subgraph X_i by a bridge. If $s(X_i) \leq k, 1 \leq i \leq m$, then $s(G) \leq k+2$.*

Lemma 6. *For a connected graph G, let v_1, v_2, v_3, v_4 and v_5 be five vertices on a cycle C in G such that each v_i $(1 \leq i \leq 5)$ connects to a connected subgraph X_i by a bridge. If $s(X_i) \geq k, 1 \leq i \leq 5$, then $s(G) \geq k+2$.*

Lemma 7. *For a connected graph G, let $C = v_1 v_2 v_3 v_4 v_1$ be a 4-cycle in G such that each v_i $(1 \leq i \leq 4)$ connects to a connected subgraph X_i by a separation edge. If $s(G) = k+1$ and $s(X_i) = k, 1 \leq i \leq 4$, then for any optimal monotonic search strategy of G, the first cleared vertex and the last cleared vertex must be in two distinct graphs $X_i + v_i$, $1 \leq i \leq 4$.*

Lemma 8. *For a CDG G with search number k, let S be an optimal monotonic search strategy of G in which the first cleared vertex is a and the last cleared vertex is b. If there are two cut-vertices a' and b' in G such that an edge-branch $G_{a'}$ of a' contains a and an edge-branch $G_{b'}$ of b' contains b and the graph G' obtained by removing $G_{a'}$ and $G_{b'}$ from G is connected, then we can use k searchers to clear G' starting from a' and ending at b'.*

For a vertex v in G, if $s(G) = k$ and there is no monotonic search strategy to clear G starting from or ending at v using k searchers, then we say that v is a *bad vertex* of G.

Lemma 9. *Let G be a connected graph and C be a cycle of length at least four in G, and v_1 and v_2 be two vertices on C such that each v_i $(1 \leq i \leq 2)$ connects to a connected subgraph X_i by a bridge $v_i v_i'$. If $s(X_1) = s(X_2) = k$ and v_1' is a bad vertex of X_1 or v_2' is a bad vertex of X_2, then we need at least $k+2$ searchers to clear G starting from v_3 and ending at v_4, where v_3 and v_4 are any two vertices on C other than v_1 and v_2.*

Lemma 10. *For a connected graph G, let v_1, v_2, v_3 and v_4 be four vertices on a cycle C in G such that each v_i $(1 \leq i \leq 4)$ connects to a connected subgraph X_i by a bridge $v_i v_i'$. If $s(X_i) = k$, and v_i' is a bad vertex of X_i, $1 \leq i \leq 4$, then $s(G) \geq k + 2$.*

From the above lemmas, we can prove the major result of this section.

Theorem 3. *Let T_k^h be a complete k-ary tree with level h and G_k^h be the corresponding k-ary CDG.*
  (i) If $k = 2$ and $h \geq 3$, then $s(T_2^h) = \lfloor \frac{h}{2} \rfloor + 1$ and $s(G_2^h) = \lfloor \frac{h}{2} \rfloor + 2$.
  (ii) If $k = 3$ and $h \geq 2$, then $s(T_3^h) = h$ and $s(G_3^h) = h + 1$.
  (iii) If $k = 4$ and $h \geq 2$, then $s(T_4^h) = h$ and $s(G_4^h) = h + \lceil \frac{h}{2} \rceil$.
  (iv) If $k \geq 5$ and $h \geq 2$, then $s(T_k^h) = h$ and $s(G_k^h) = 2h$.

Proof. The search numbers of complete k-ary trees can be verified directly by the algorithm SEARCHNUMBER($T[r]$). Thus, we will only consider the search numbers of k-ary CDGs.

  (i) The search number of G_2^h can be verified by a search strategy based on SEARCHSTRATEGY($T[r]$).

  (ii) We now prove $s(G_3^h) = h + 1$ by induction on h. Let $R = r_0 r_1 r_2 r_3 r_0$ be the cycle in G_3^h that corresponds to the root of T_3^h. Suppose r_0 is the vertex without any outgoing edges. When $h = 2$, it is easy to see that $s(G_3^2) = 3$ and all four vertices of R are not bad vertices in G_3^2. Suppose $s(G_3^h) = h + 1$ holds when $h < n$ and all four vertices of R are not bad vertices in G_3^h. When $h = n$, R has three edge-branches with search number n. It follows from Lemma 4 that $s(G_3^n) \geq n + 1$. We will show how to use $n + 1$ searchers to clear the graph by the following strategy: use n searchers to clear $G[r_1]$ ending at r_1; keep one searcher on r_1 and use n searchers to clear $G[r_2]$ ending at r_2; use one searcher to clear the edge $r_1 r_2$; slide the searcher on r_1 to r_0 and slide the searcher on r_2 to r_3; use one searcher to clear the edge $r_0 r_3$; then clear $G[r_3]$ with n searchers starting from r_3. This strategy never needs more than $n + 1$ searchers. Thus, $s(G_3^n) = n + 1$. From this strategy, it is easy to see that all four vertices of R are not bad vertices in G_3^n.

(iii) We will prove $s(G_4^h) = h + \lceil \frac{h}{2} \rceil$ by induction on h. Let $R = r_0 r_1 r_2 r_3 r_4 r_0$ be the cycle in G_4^h that corresponds to the root of T_4^h. Suppose r_0 is the vertex without any outgoing edges. We want to show that if h is odd, then no bad vertex is on R, and if h is even, then r_0 is a bad vertex of G_4^h.

When $h = 2$, it is easy to see that $s(G_4^2) = 3$ and r_0 is a bad vertex in G_4^2. When $h = 3$, by Lemma 10, $s(G_4^3) \geq 5$ and it is easy to verify that 5 searchers can clear G_4^3 starting from any one of the five vertices on R. Suppose these results hold for G_4^h when $h < n$. We now consider the two cases when $h = n$.

If n is odd, $G[r_i]$ has search number $n - 1 + (n - 1)/2$ and r_i is a bad vertex in $G[r_i]$, $1 \leq i \leq 4$. By Lemma 10, we have $s(G_4^n) \geq n - 1 + (n - 1)/2 + 2 = n + (n + 1)/2$. We will show how to use $n + (n + 1)/2$ searchers to clear the graph by the following strategy. Let v be any one of the cycle vertex of R. We first place two searchers α and β on v and then slide β along R starting from v and ending at v. Each time when β arrives a vertex of R, we clear the subgraph attached to this vertex using $n - 1 + (n - 1)/2$ searchers. This strategy never needs more than $n + (n + 1)/2$ searchers. Thus, $s(G_4^n) = n + (n + 1)/2$. It is also easy to see that all five vertices of R are not bad vertices in G_4^n.

If n is even, $G[r_i]$ has search number $n - 1 + n/2$ and r_i is not a bad vertex in $G[r_i]$, $1 \leq i \leq 4$. By Lemma 4, we have $s(G_4^n) \geq n + n/2$. We will show how to use $n + n/2$ searchers to clear the graph by the following strategy: use $n - 1 + n/2$ searchers to clear $G[r_1]$ ending at r_1; use $n - 1 + n/2$ searchers to clear $G[r_2]$ ending at r_2; use one searcher to clear the edge $r_1 r_2$; slide the searcher on r_1 along the path $r_1 r_0 r_4$ to r_4; slide the searcher on r_2 to r_3 along the edge $r_2 r_3$; use one searcher to clear the edge $r_3 r_4$; clear $G[r_3]$ with $n - 1 + n/2$ searchers starting from r_3 and finally clear $G[r_4]$ with $n - 1 + n/2$ searchers starting from r_4. This strategy never needs more than $n + n/2$ searchers. Thus, $s(G_4^n) = n + n/2$ and, by Lemma 7, r_0 is a bad vertex in G_4^n.

(iv) The search number of G_k^h, $k \geq 5$, can be verified directly from Lemmas 5 and 6.

6   Approximation Algorithms

Megiddo et al. [9] introduced the concept of the hub and the avenue of a tree. Given a tree T with $s(T) = k$, only one of the following two cases must happen: (1) T has a vertex v such that all edge-branches of v have search number less than k, this vertex is called a *hub* of T; and (2) T has a unique path $v_1 v_2 \ldots v_t$, $t > 1$, such that v_1 and v_t each has exactly one edge-branch with search number k and each v_i, $1 < i < t$, has exactly two edge-branches with search number k, this unique path is called an *avenue* of T.

Theorem 4. *Given a CDG G, if T is a tree obtained by contracting each cycle of G into a vertex, then $s(T) \leq s(G) \leq 2s(T)$.*

Corollary 2. *For any CDG, there is a linear time approximation algorithm with approximation ratio 2.*

Lemma 11. *Let G be a CDG in which every cycle has at most three vertices with degree more than two. Let T be the tree obtained from G by contracting every cycle of G into a vertex. If the degree of each cycle vertex in G is at most three, then $s(G) \leq s(T) + 1$.*

Let $S = (a_1, \ldots, a_k)$ be an optimal monotonic search strategy for a graph. The *reversal* of S, denoted as S^R, is defined by $S^R = (\bar{a}_k, \bar{a}_{k-1}, \ldots, \bar{a}_1)$, where each $\bar{a}_i$, $1 \leq i \leq k$, is the *converse* of a_i, which is defined as follows: the action "place a searcher on vertex v" and the action "remove a searcher from vertex v" are converse with each other; and the action "move the searcher from v to u along the edge vu" and the action "move the searcher from u to v along the edge uv" are converse with each other.

Lemma 12. *If S is an optimal monotonic search strategy of a graph G, then S^R is also an optimal monotonic search strategy of G.*

Lemma 13. *Given a graph G, for any two vertices a and b of G, there is a search strategy that uses at most $s(G) + 1$ searchers to clear G starting from a and ending at b.*

Theorem 5. *Let G be a connected graph and $X_1, X_2, \ldots, X_m$ be an edge partition of G such that each X_i is a connected subgraph and each pair of X_i share at most one vertex. Let G^* be a graph of m vertices such that each vertex of G^* corresponds to a X_i and there is an edge between two vertices of G^* if and only if the corresponding two X_i share a common vertex. If G^* is a tree, then there is a search strategy that uses at most $\max_{1 \leq i \leq m} s(X_i) + \lceil \Delta(G^*)/2 \rceil s(G^*)$ searchers to clear G, where $\Delta(G^*)$ is the maximum degree of G^*.*

Proof. We prove the result by induction on $s(G^*)$. If $s(G^*) = 1$, then G^* is a single vertex or a path, and $\lceil \Delta(G^*)/2 \rceil = 1$. Suppose that G^* is the path $v_1 v_2 \ldots v_m$ and v_i corresponds to X_i, $1 \leq i \leq m$. Let a_i be the vertex shared by X_i and X_{i+1}, $1 \leq i \leq m-1$ and let a_0 be a vertex in X_1 and a_m be a vertex in X_m. By Lemma 13, we can use $s(X_i) + 1$ searchers to clear each X_i starting from a_{i-1} and ending at a_i, for $X_1, X_2, \ldots, X_m$. Therefore, there is a search strategy uses at most $\max_i s(X_i) + 1$ searchers to clear G. Suppose that this result holds for $s(G^*) \leq n$, $n \geq 2$. When $s(G^*) = n + 1$, we consider the following two cases.

CASE 1. G^* has a hub v. Let $X(v)$ be the subgraph of G that corresponds to v and S be an optimal search strategy of $X(v)$. Each subgraph that corresponds to a neighbor of v in G^* shares a vertex with $X(v)$ in G. Divide these shared vertices into $\lceil \deg(v)/2 \rceil$ pairs such that for each pair of vertices a_i and a_i', a_i is cleared before a_i' is cleared in S, $1 \leq i \leq \lceil \deg(v)/2 \rceil$. Let v_i (resp. v_i') be the neighbor of v such that its corresponding subgraph of G, denoted by $X(v_i)$ (resp. $X(v_i')$), shares a_i (resp. a_i') with $X(v)$. Let v be the root of G^*, let T_i (resp. T_i') be the vertex-branch of v_i (resp. v_i') and let $X(T_i)$ (resp. $X(T_i')$) be the subgraph of G that is the union of the subgraphs that correspond to all vertices in T_i (resp. T_i'). Obviously a_i (resp. a_i') is the only vertex shared by $X(v)$ and $X(T_i)$ (resp. $X(T_i')$). Since v is a hub of G^*, we know that $s(T_i) \leq n$. Thus, $s(X(T_i)) \leq \max_i s(X_i) + \lceil \Delta(T_i)/2 \rceil n \leq \max_i s(X_i) + \lceil \Delta(G^*)/2 \rceil n$. First, we place a searcher on each a_i, $1 \leq i \leq \lceil \deg(v)/2 \rceil$. Then use $\max_i s(X_i) + \lceil \Delta(G^*)/2 \rceil n$ searchers to clear each subgraph $X(T_i)$ separately. After that, we perform S to clear $X(v)$. Each time after some a_i is cleared by S, we remove the searcher on a_i and place it on a_i', $1 \leq i \leq \lceil \deg(v)/2 \rceil$. Finally, after $X(v)$ is cleared, we again use $\max_i s(X_i) + \lceil \Delta(G^*)/2 \rceil n$ searchers to clear each subgraph $X(T_i')$ separately. Therefore,we can clear G with no more than $\max_i s(X_i) + \lceil \Delta(G^*)/2 \rceil n + \lceil \deg(v)/2 \rceil \leq \max_i s(X_i) + \lceil \Delta(G^*)/2 \rceil (n + 1)$ searchers.

CASE 2. G^* has an avenue $v_1 v_2 \ldots v_t$, $t > 1$. Let v_0 be a neighbor of v_1 other than v_2 and let v_{t+1} be a neighbor of v_t other than v_{t-1}. Let $X(v_i)$, $0 \leq i \leq t+1$, be the subgraph of G that corresponds to v_i. For $0 \leq i \leq t$, let b_i be the vertex shared by $X(v_i)$ and $X(v_{i+1})$. For $1 \leq i \leq t$, let S_i be an optimal search strategy of $X(v_i)$ such that b_{i-1} is cleared before b_i is cleared. Thus, we can use a similar search strategy described in CASE 1 to clear each $X(v_i)$ and all the subgraphs that correspond to the edge-branches of v_i. Note that when we clear $X(v_i)$, b_{i-1} and b_i form a pair as a_i and a_i' in CASE 1. In such a strategy we never need more than $\max_i s(X_i) + \lceil \Delta(G^*)/2 \rceil (n + 1)$ searchers.

In Theorem 5, if each X_i is a unicyclic graph, then we have a linear time approximation algorithm for cycle-disjoint graphs. We can design a linear time approximation algorithm when each $s(X_i)$ can be found in linear time.

References

1. Bienstock, D., Seymour, P.: Monotonicity in graph searching. Journal of Algorithms 12, 239–245 (1991)
2. Ellis, J., Sudborough, I., Turner, J.: The vertex separation and search number of a graph. Information and Computation 113, 50–79 (1994)
3. Ellis, J., Markov, M.: Computing the vertex separation of unicyclic graphs. Information and Computation 192, 123–161 (2004)
4. Fellows, M., Langston, M.: On search, decision and the efficiency of polynomial time algorithm. In: 21st ACM Symp. on Theory of Computing, pp. 501–512. ACM Press, New York (1989)
5. Frankling, M., Galil, Z., Yung, M.: Eavesdropping games: A graph-theoretic approach to privacy in distributed systems. Journal of ACM 47, 225–243 (2000)
6. Kinnersley, N.: The vertex separation number of a graph equals its path-width. Information Processing Letters 42, 345–350 (1992)
7. Kirousis, L.M., Papadimitriou, C.H.: Searching and pebbling. Theoretical Computer Science 47, 205–218 (1986)
8. LaPaugh, A.S.: Recontamination does not help to search a graph. Journal of ACM 40, 224–245 (1993)
9. Megiddo, N., Hakimi, S.L., Garey, M., Johnson, D., Papadimitriou, C.H.: The complexity of searching a graph. Journal of ACM 35, 18–44 (1988)
10. Peng, S., Ho, C., Hsu, T., Ko, M., Tang, C.: Edge and node searching problems on trees. Theoretical Computer Science 240, 429–446 (2000)
11. Yang, B., Zhang, R., Cao, Y.: Searching cycle-disjoint graphs. Technical report CS-2006-05, Department of Computer Science, University of Regina (2006)

An Asymptotic PTAS for Batch Scheduling with Nonidentical Job Sizes to Minimize Makespan[⋆]

Yuzhong Zhang and Zhigang Cao[⋆⋆]

School of Operations Research and Management Science, Qufu Normal University,
Rizhao, Shandong, China
Tel(Fax): (86)0633-8109019
cullencao@eyou.com

Abstract. Motivated by the existence of an APTAS(Asymptotic PTAS) for bin packing problem, we consider the batch scheduling problem with nonidentical job sizes to minimize makespan. For the proportional special version, i.e., there exists a fixed number α such that $p_j = \alpha s_j$ for every $1 \leq j \leq n$, we first present a lower bound of $3/2$ for the approximation ratio and then design an APTAS for it. Our basic idea is quite simple: we first enumerate all the partial schedules of relatively large jobs; Then for every partial schedule we insert the small jobs, *split* them if necessary; Further then, we choose the best of all the obtained schedules; Finally, we collect the split small jobs and put them into new batches. As we can round the large jobs into only a constant number of different kinds at a reasonable expense of accuracy, the running time can be bounded. When the optimal objective value of instances in our consideration can not be arbitrarily small, $\inf\limits_{I}\{P_{\max} : P_{\max}$ is the largest processing time in $I\} \neq 0$ for instance, our result is perfect in the sense of worst-case performance.

1 Introduction and Our Contributions

In this paper, we study the problem of batch scheduling with nonidentical job sizes to minimize makespan. We are given a list of jobs $\{J_1, J_2, \cdots, J_n\}$, each job J_j is characterized by a double of real numbers (p_j, s_j), where p_j is the processing time and s_j the job size. A number of jobs can be processed as a batch simultaneously, as long as the total size does not exceed the machine capacity, and the processing time of a batch is given by the longest job in this batch. No preemption is allowed. Our goal is to batch the given jobs and schedule the batches in some sequence such that makespan is minimized. Throughout this paper(except in Theorem 1), we assume that the machine capacity is 1, and thus $s_j \in (0, 1]$, as is always assumed. Without loss of generality we further assume that $\alpha = 1$, i.e., $p_j = s_j (1 \leq j \leq n)$.

[⋆] Supported by NNSF of China(NO.10671108).
[⋆⋆] Corresponding author.

A. Dress, Y. Xu, and B. Zhu (Eds.): COCOA 2007, LNCS 4616, pp. 44–51, 2007.

Batch scheduling problems are encountered in many environments and our research is motivated by the burn-in model in semiconductor industry. In the industry of semiconductor manufacturing, the last stage is the final testing (called the burn-in operation). In this stage, chips are loaded onto boards which are then placed in an oven and exposed to high temperature. Each chip has a pre-specified minimum burn-in time, and a number of chips can be baked in an oven simultaneously as long as the oven can hold them, and the baking process is not allowed to be preempted, that is, once the processing of a batch is started, the oven is occupied until the process is completed. To ensure that no defective chips will pass to the customer, the processing time of a batch is that of the longest one among these chips. As the baking process in burn-in operations can be long compared to other testing operations(e.g.,120 hours as opposed to 4-5 hours for other operations), an efficient algorithm for batching and scheduling is highly non-trivial.

Since the early 1990s, due to its deep root in the real world, the batch scheduling problem has attracted a lot of attention and many variants have been discussed(see [2,4,5,10]). However, up to now, most of the discussions have been restricted on the model with identical job sizes. As to the more general and more practical case with nonidentical job sizes, since it was proposed by R. Uzsoy([11]), relatively few cases have been studied.

In Uzsoy's paper, he considered the problem of minimizing makespan and the problem of minimizing total completion time, both problems are proven to be NP-hard. For the problem of minimizing makespan, the author gave four heuristics, all of which have beautiful computational results. However, no theoretical analysis is provided. Later, G. Zhang et al.([13]) analyzed the four heuristics. They proved that the worst-case ratio of Algorithm LPT-FF is no greater than 2 but each of the other three can behave badly enough, i.e., the worst-case ratios of them tend to infinity. Then, they provided a highly non-trivial algorithm MLPT-FF with worst-case ratio 7/4. Several researchers also studied this problem by simulated annealing or branch-and-bound([1,3,7]). Recently, S. Li et al.([8]) considered the case with non-identical job arrivals. In G.Zhang et al.'s paper, they also considered the special case with no processing time of a *large* job less than the processing time of a small job, where a job is called a large job if its job size is greater than 1/2 and a small job otherwise, they presented an algorithm with worst-case ratio 3/2 and proved this is the best possible unless P=NP, in the sense of worst-case performance.

In our paper, we consider a further special case, $1|B, p_j \equiv \alpha s_j|C_{\max}$ in the three-tuple denotation of R.Graham et al.([6]), where $p_j \equiv \alpha s_j$ means there exists a constant α such that for arbitrary job $J_j(1 \le j \le n)$, $p_j = \alpha s_j$ holds. This restriction is not severe since in the semiconductor industry requiring that a bigger chip should be baked longer than a smaller one is quite reasonable, and many times we can assume that the ratio $p_j/s_j(1 \le j \le n)$ is a constant regardless of j. We design an Asymptotic PTAS for this problem, which can work as good as an PTAS as long as the optimal objective values of instances in our consideration can not be arbitrarily small. Adding up with the lower bound of $3/2$ we give for the worst-case ratio, our result is perfect in the sense of worst-case performance.

Our basic idea follows from W. Fernandez de la Vega and G.S. Lueker([12]), who present an APTAS for the bin packing problem. The differences between the two problems, however, force us to find new techniques. Moreover, we use more delicate rules in treating the processing times.

The rest of the paper is organized as follows. Section 2 gives some necessary preliminary knowledge and notations. In Section 3 we describe the main result, and in Section 4 we draw a conclusion and direct the further researches.

2   Preliminaries and Notations

It is well known that there does not exist any algorithm for bin packing problem with worst-case ratio better than $3/2$, unless P=NP([9]), therefore, it is impossible for us to design a PTAS for it. However, it does admit an APTAS(*Asymptotic PTAS*), which is almost as good as a PTAS when the instances in our consideration is *large* enough([12]). So first of all, let's introduce the concept of APTAS.

In the rest of this paper, we will denote by $Opt(I, P)$ the optimal objective value of an instance I for a given problem P, and $\mathcal{A}(I, P)$ the objective value obtained by algorithm $\mathcal{A}$. Without causing any trouble, we simply write them as $Opt(I)$ and $\mathcal{A}(I)$. We will denote by I both an instance of a problem and a family of jobs.

Definition. *A family of algorithms* $\{\mathcal{A}_\varepsilon\}_\varepsilon$ *for a given problem is said to be an APTAS iff, for arbitrary* $\varepsilon > 0$, *there exists a positive number* $N(\varepsilon)$ *such that* $\sup_I\{\frac{\mathcal{A}_\varepsilon(I)}{Opt(I)} : Opt(I) \geq N(\varepsilon)\} \leq 1 + \varepsilon$ *and the running time of* $\mathcal{A}_\varepsilon$ *is bounded by a polynomial in the input size of the problem while* ε *is regarded as a constant.*

The following simple combinatorial result will be applied in the next section and we present it as a lemma.

Lemma 1. *The number of nonnegative integral solutions to the n-variable equation* $x_1 + x_2 + \cdots + x_n = m$ *is* C_{m+n-1}^{n-1}, *where* m *and* n *are positive integers. And that to* $x_1 + x_2 + \cdots + x_n \leq m$ *is* C_{m+n}^n.   □

The NP-Completeness of EQUALPARTITION will be used later, we describe it as follows.

EQUALPARTITION: *Given a set of* $2m$ *nonnegative integers* $X = \{a_1, a_2, \cdots, a_{2m}\}$, *is it possible to partition* X *into two parts* X_1 *and* X_2 *such that* $\sum_{a_i \in X_1} a_i = \sum_{a_i \in X_2} a_i = (a_1 + a_2 + \cdots + a_{2m})/2$ *and* $|X_1| = |X_2| = m$? *Where* $|X_1|$ *and* $|X_2|$ *denote the cardinality of them, respectively.*

Lemma 2. EQUALPARTITION *belongs to NP-C.*   □

For simplicity, we denote by BSM the batch scheduling problem with nonidentical job sizes to minimize makespan. We also denote by PBSM the proportional BSM, the main problem in our consideration.

3   The Main Result

3.1   Lower Bound for PBSM

It is trivially true that $3/2$ is a lower bound for BSM as it includes BINPACKING as a special case. While PBSM does not include BINPACKING as a special case, we can still show that it takes $3/2$ as a lower bound.

Theorem 1. *There does not exist any algorithm for PBSM with worst-case ratio less than 3/2, unless P=NP.*

Proof. If not, suppose that $\mathcal{A}$ is an exception whose worst-case ratio is $3/2$-β, where $0 < \beta \leq 1/2$, next we will show that EQUALPARTITION can be solved in polynomial time of its input size, which contradicts the fact that EQUAL PARTITION is NP-C.

Given any instance of EQUALPARTITION $I_1 = \{a_1, a_2, \cdots, a_{2m}\}$, construct an instance of PBSM I_2 as follows: There are $2m$ jobs with $p_1 = s_1 = (M + 2a_1), p_2 = s_2 = (M + 2a_2), \cdots, p_{2m} = s_{2m} = (M + 2a_{2m})$, and the machine capacity is $B = a_1 + a_2 + \cdots + a_{2m} + mM$ and M is large enough, say $M = ((3 - 2\beta)/\beta)a_{max} + 1$.

It is easy to see that answering 'yes' for I_1 is equivalent to $Opt(I_2) < 3M$. Next, we will show that $Opt(I_2) < 3M$ if and only if $\mathcal{A}(I_2) < 3M$ which will complete the theorem. 'If' is obvious, so it remains to show the converse.

$Opt(I_2) < 3M$ implies that there are exactly two batches in the optimal schedule and we can assume that $Opt(I_2) = 2(M + a_i + a_j)$ for some $1 \leq i < j \leq 2m$. Thus:

$$\mathcal{A}(I_2) \leq (3/2 - \beta)Opt(I_2) = 3M + (3 - 2\beta)(a_i + a_j) - 2\beta M < 3M \qquad \square$$

3.2   A Polynomially Solvable Special Case of BSM

In this subsection we will show that a special case of BSM with fixed number of processing times, fixed number of job sizes and no job size smaller than ε, which will be denoted by BSM$'$, is polynomially solvable.

Theorem 2. *The number of substantially different feasible schedules for BSM$'$ is a polynomial in the input size and therefore BSM$'$ is polynomially solvable.*

Proof. We say two jobs are of the same class iff they have the same sizes and processing times. By hypothesis of the theorem, there are a fixed number of, say k different classes of jobs. we say two batches are of the same kind iff they include the same number of jobs from every class.

We first claim that there are at most $r = C^{1/\lfloor \varepsilon \rfloor}_{k+1/\lfloor \varepsilon \rfloor}$ different kinds of batches, where $\lfloor x \rfloor$ means the largest integer less than or equal to x. In fact, it is bounded by the number of positive integral solutions to $x_1 + x_2 + \cdots + x_k \leq \lfloor 1/\varepsilon \rfloor$, by lemma 1 we have the claim.

As our objective is to minimize makespan, we can view two schedules with the same batches but different sequencing of these batches as identical. Since for any schedule there are at most n batches, the number of substantially different schedules is at most C_{r+n-1}^{n-1}(Lemma 1 is applied once more),which is a polynomial in n (but not in $1/\varepsilon$, $O((n + (1/\varepsilon)^k)^{(1/\varepsilon)^k})$ in fact).

As for the second part of the theorem, we merely have to design an algorithm to enumerate all the different schedules and choose the optimal one. The detailed proof is left to the readers.

Similar to Theorem 2, we have the following result as a byproduct.   □

Corollary. *For any scheduling problem (without batching), if it is polynomially solvable, then its corresponding batch scheduling problem with fixed number of distinct processing times, fixed number of job sizes and no job size smaller than a constant, is polynomially solvable.*   □

3.3   The Main Procedure

We are given a set of jobs $I = \{J_1, J_2, \cdots, J_n\}$ and remember that $p_j = s_j \in (0, 1]$. For any given error parameter $\varepsilon > 0$, let $\varepsilon_0 = \varepsilon/(1 + \varepsilon)$. We say a job is *long* if its processing time(and its size as well) is greater than or equal to ε_0, and *short* otherwise. Denote by $\mathcal{L}$ and $\mathcal{S}$ the set of long jobs and short jobs, respectively. Before discussion, we further assume that all the jobs have been reindexed such that $p_1 \leq p_2 \leq \cdots \leq p_n$.

Suppose that there are m long jobs, let $Q = \lceil m \rceil \varepsilon_0^3$, where $\lceil x \rceil$ means the smallest integer greater than or equal to x. Without loss of generality(which will be seen later), we assume that $m > \lceil 2/\varepsilon_0^3 \rceil$, i.e. $Q > 2$.

First of all, we define two partitions of $\mathcal{L}$:

$PT1 = \{\overline{\mathcal{J}}_0, \overline{\mathcal{J}}_1, \cdots, \overline{\mathcal{J}}_{k_1}\}$, where $\overline{\mathcal{J}}_0$ contains the first $Q - 1$ jobs, and $\overline{\mathcal{J}}_1, \cdots, \overline{\mathcal{J}}_{k_1}$ in turn all take Q jobs except possibly the last one.

$PT2 = \{\tilde{\mathcal{J}}_1, \tilde{\mathcal{J}}_2, \cdots, \tilde{\mathcal{J}}_{k_2}\}$ is similar to $PT1$ except that the first $k_2 - 1$ sets all have Q jobs.

It is easy to calculate that $k_1 = \lceil (m - Q + 1)/Q \rceil$ and $k_2 = \lceil m/Q \rceil$, both of which are bounded by a constant $\lceil 1/\varepsilon_0^3 \rceil$. More detailed yet still easy analysis gives the following proposition of k_1 and k_2.

Lemma 3. *If $k_1 = k_2$ does not hold, then $k_1 = k_2 - 1$. And $k_1 = k_2$ implies that $\overline{\mathcal{J}}_{k_1} = \{J_m\}$.*   □

Based on the two partitions, we construct two auxiliary instances:

$$\overline{I} = (\bigcup_{i=0}^{k_1} \overline{\mathcal{J}}_i') \bigcup \mathcal{S}, \quad \tilde{I} = (\bigcup_{i=1}^{k_2} \tilde{\mathcal{J}}_i') \bigcup \mathcal{S}$$

Where $\overline{\mathcal{J}}_i'$ ($0 \leq i \leq k_1$) is constructed by letting all the jobs in $\overline{\mathcal{J}}_i$ be the longest one, and $\tilde{\mathcal{J}}_i'$ by letting all the jobs in $\tilde{\mathcal{J}}_i$ be the shortest one. Let $\overline{I}_0 = \overline{I}/\overline{\mathcal{J}}_0'$, by Proposition 1, it's not hard to have:

Lemma 4. $Opt(\overline{I}_0) \leq Opt(\tilde{I})$ □

Next, we will show that $\overline{I}$ is a perfect approximation of I.

Lemma 5. $Opt(I) \leq Opt(\overline{I}) \leq (1 + \varepsilon_0)Opt(I)$

Proof. We only have to show the latter part as the former one is trivially true. It's trivial that $Opt(\widetilde{I}) \leq Opt(I)$, thus by Lemma 3 we have:

$$Opt(\overline{I}) \leq Opt(\overline{I}_0) + Opt(\overline{\mathcal{J}}_0)$$
$$\leq Opt(\widetilde{I}) + (Q - 1) \cdot 1$$
$$\leq Opt(I) + m\varepsilon_0^3$$

For any optimal schedule of I, the total size of long jobs is at least $m\varepsilon_0$, which will forms at least $\lceil m\varepsilon_0 \rceil$ batches with processing time greater than or equal to ε_0, so $Opt(I) \geq (m\varepsilon_0) \cdot \varepsilon_0 = m\varepsilon_0^2$, therefore:

$$Opt(\overline{I}) \leq Opt(I) + \varepsilon_0 Opt(I) = (1 + \varepsilon_0)Opt(I)$$

which completes the proof.   □

Now we are about to describe the final Algorithm *EIR* (Enumerate long jobs, Insert short jobs, Rearrange split jobs) for PBSM. We denote by $\overline{I}'$ exactly the same instance as I but of a slightly different problem in which short jobs can be *split*, i.e., a short job $J_j = (p_j, s_j)$ can be split into two jobs $J_{j1} = (p_j, s_{j1})$ and $J_{j2} = (p_j, s_{j2})$, where $s_{j1} + s_{j2} = s_j$ and J_{j1} and J_{j2} can be placed into two different batches. This technic is explored by G. Zhang et al.. Obviously, $Opt(\overline{I}') \leq Opt(\overline{I})$.

Algorithm EIR

Step 1. Compute the optimal schedule of $\overline{I}'$ as follows:

1.1 Compute all the feasible schedules of long jobs in $\overline{I}'$;
1.2 For any partial schedule of long jobs, insert short jobs in the original order, open a new batch or split them if necessary;
1.3 Select the optimal entire schedule $\overline{\pi}'$ from all the obtained schedules.
Step 2. Collect all the split short jobs in $\overline{\pi}'$ and put $\lfloor 1/\varepsilon_0 \rfloor$ of them in a new batch to obtain a feasible schedule $\overline{\pi}$ of $\overline{I}$.
Step 3. Obtain the corresponding original schedule π from $\overline{\pi}$, output π as the final solution to I.

Lemma 6. *The schedule obtained in step 1 is an optimal one for* $\overline{I}'$.

Proof. For any given partial schedule of long jobs, we insert the remaining short ones just as G. Zhang et al. do in Algorithm A_1, so the optimality can be verified quite similarly by an interchanging strategy. Hence, the enumerating of all the possible partial schedules gives the lemma.   □

Theorem 3. *Algorithm EIR is an APTAS for PBSM.*

Proof. As for the accuracy, suppose that $\overline{\pi}'$ has l batches, whose processing times are $p^1, p^2, \cdots, p^l$, respectively. Denote by s the number of batches in $\overline{\pi}$ formed by split jobs. It's easy to calculate that $s \leq \lceil (l-1)/\lfloor 1/\varepsilon_0 \rfloor \rceil$, and we assume that s is exactly $\lceil (l-1)/\lfloor 1/\varepsilon_0 \rfloor \rceil$ for simplicity (otherwise, we can add some *dummy* jobs). Denote the processing times of these batches by $p^{[1]}, p^{[2]}, \cdots, p^{[s]}$, respectively. Then $p^{[1]} \leq \varepsilon_0, p^{[2]} \leq p^{2+\lfloor 1/\varepsilon_0 \rfloor}, \cdots, p^{[s]} \leq p^{2+(s-1)\lfloor 1/\varepsilon_0 \rfloor}$. Therefore:

$$
\begin{aligned}
C_{max}(\overline{\pi}) &= (p^1 + p^2 + \cdots + p^l) + (p^{[1]} + p^{[2]} + \cdots + p^{[s]}) \\
&\leq Opt(\overline{I}') + \frac{1}{\lfloor 1/\varepsilon_0 \rfloor}[(1 + p^1 + p^2) + (p^3 + \cdots + p^{2+\lfloor 1/\varepsilon_0 \rfloor}) \\
&\quad + \cdots + (p^{3+(s-2)\lfloor 1/\varepsilon_0 \rfloor} + \cdots + ^{2+(s-1)\lfloor 1/\varepsilon_0 \rfloor}) + \cdots + p^l] \\
&\leq Opt(\overline{I}') + \varepsilon_0[1 + Opt(\overline{I}')] \\
&= (1 + \varepsilon_0)Opt(\overline{I}') + \varepsilon_0 \\
&\leq (1 + \varepsilon_0)Opt(\overline{I}) + \varepsilon_0
\end{aligned}
$$

Thus by Lemma 4: $EIR(I) = C_{max}(\pi) \leq C_{max}(\overline{\pi}) \leq (1 + \varepsilon_0)Opt(\overline{I}) + \varepsilon_0 \leq (1 + \varepsilon_0)^2 Opt(I) + \varepsilon_0 \leq (1 + \varepsilon)Opt(I) + \varepsilon$

As for the running time, by Theorem 2, there are $O((n + (1/\varepsilon)^{(1/\varepsilon)^3})(1/\varepsilon)^{(1/\varepsilon)^3})$ iterations in step 1. Since each iteration takes $O(n)$ time, the running time when Algorithm *EIR* terminates step 1.2 is $O(n \cdot (n + (1/\varepsilon)^{(1/\varepsilon)^3})(1/\varepsilon)^{(1/\varepsilon)^3})$. While selecting the optimal schedule in step 1.3 takes $O((n + (1/\varepsilon)^{(1/\varepsilon)^3})(1/\varepsilon)^{(1/\varepsilon)^3})$ time, taking out the split jobs and forming new batches in step 2 takes $O(n)$ time, and step 3 takes $O(n)$ time, the total running time of Algorithm *EIR* can be bounded by $O(n \cdot (n + (1/\varepsilon)^{(1/\varepsilon)^3})(1/\varepsilon)^{(1/\varepsilon)^3})$, which is a polynomial in n.   □

4   Conclusion and Remarks

In this paper, we consider the proportional batch scheduling problem with non-identical job sizes to minimize makespan. We first give a lower bound of $3/2$ and then present an APTAS, with the restraint the α is a constant. Noticing that when the largest processing time can not be arbitrarily small, i.e., $\inf_{I}\{P_{max} : P_{max} \text{ is the largest processing time in } I\} \neq 0$, our algorithm is as good as a PTAS , which is the best possible in the sense of worst-case performance and doesn't exist by the lower bound.

Pity that the running time in our algorithm is quite huge(although polynomial in n), as for further discussions, can we still reduce it? This is of great interest.

References

1. Azizoglu, M., Webster, S.: Scheduling a batch processing machine with non-identical job sizes. International Journal of Production Research 38(10) (2000)
2. Brucker, P., Gladky, A., Hoogeveen, H., Kovalyow, M.Y., Potts, C.N., Tautenhahn, T., van de Velde, S.L.: Scheduling a batching machine. Journal of Scheduling 1, 31–54 (1998)

3. Dupont, L., Dhaenens-Flipo, C.: Minimizing the makespan on a batch machine with non-identical job sizes: an exact procedure. Computer &Operations Research 29, 807–819 (2002)
4. Deng, X.T., Feng, H.D., Zhang, P.X., Zhang, Y.Z., Zhu, H.: Minimizing mean completion time in batch processing system. Algorithmica 38(4), 513–528 (2004)
5. Deng, X.T., Poon, C.K., Zhang, Y.Z.: Approximation Algorithms in batch scheduling. Journal of Combinational Optimization 7, 247–257 (2003)
6. Graham, R.L., Lawler, E.L., Lenstra, J.K., Rinnooy Kan, A.H.G.: Optimization and approximation in deterministic sequencing and scheduling. Annals of Discrete Mathematics 5, 287–326 (1979)
7. Jolai, F., Ghazvini, Pupont, L.: Minimizing mean flow times criteria on a single batch processing machine with nonidentical job sizes. International Journal of Production Econonmics 55, 273–280 (1998)
8. Li, S.G., Li, G.J., Wang, X.L., Liu, Q.M.: Minimizing makespan on a single batching machine with release times and non-identical jog sizes. Operations Research Letters 33, 157–164 (2005)
9. Lenstra, J.K., Shmoys, D.R.: Computing near-optimal schedules. In: Chétienne, P., et al. (eds.) Scheduling theory and its applications, Wiley, New York (1995)
10. Potts, C.N., Lovalyov, M.Y.: Scheduling with batching: a review. European Journal of Operational Research 120, 228–249 (2000)
11. Uzsoy, R.: Scheduling a single batch processing machine with non-identical job sizes. International Journal of Production Research 32(7), 1615–1635 (1994)
12. de la Vega, W.F., Lueker, G.S.: Bin packing can be solved within in linear time. Combinatorica 1(4), 349–355 (1981)
13. Zhang, G.C., Cai, X.Q., Lee, C.Y., Wong, C.K.: Minimizing makespan on a single batch processing machine with non-identical job sizes. Naval Research Logistics 48, 226–247 (2001)

A New Dynamic Programming Algorithm for Multiple Sequence Alignment

Jean-Michel Richer, Vincent Derrien, and Jin-Kao Hao

LERIA - University of Angers, 2 bd Lavoisier, 49045 Angers Cedex 01, France
{richer,derrien,hao}@info.univ-angers.fr

Abstract. Multiple sequence alignment (MSA) is one of the most basic and central tasks for many studies in modern biology. In this paper, we present a new progressive alignment algorithm for this very difficult problem. Given two groups A and B of aligned sequences, this algorithm uses Dynamic Programming and the sum-of-pairs objective function to determine an optimal alignment C of A and B. The proposed algorithm has a much lower time complexity compared with a previously published algorithm for the same task [11]. Its performance is extensively assessed on the well-known BAliBase benchmarks and compared with several state-of-the-art MSA tools.

Keywords: multiple alignment, dynamic programming.

1 Introduction

In biology, the *Multiple Sequence Alignment* of nucleic acids or proteins is one of the most basic and central tasks which is a prior to phylogeny reconstruction, protein structure modeling or gene annotation. The goal of the alignment operation is to identify similarities at the primary sequence level which usually implies structural and functional similarity.

A multiple alignment of a set of sequences helps visualize conserved regions of residues by organizing them into a matrix where similar residues ideally appear in the same column. In order to obtain this matrix it is necessary to use *edit operations* which consist of a match, a substitution or an insertion. A match puts two equal residues in the same column while a substitution uses two different residues. An insertion consists in inserting a special character, called a gap, whenever characters of a sequence have to be shifted from one column to be aligned with similar residues of other sequences.

To align two sequences, a simple polynomial algorithm based on dynamic programming (DP) has been designed with linear gap penalty [14]. This algorithm is based on a scoring scheme of edit operations and is influenced by two parameters: a substitution matrix and a model of gaps. A substitution matrix (PAM [1], BLOSUM [9]) assigns a score to a match or a substitution and the gap model helps score the insertions.

Obtaining an accurate alignment is a difficult task which requires to design scoring schemes which are biologically sound. In practice, the sum-of-pairs [10] is the most widely used for its simplicity although some other models have been developed [18,15].

We can compute the optimal alignment of a set of k sequences of length n by extending [14] to a k-dimension DP algorithm [13], but its complexity in $\mathcal{O}(n^k 2^k)$ is too

A. Dress, Y. Xu, and B. Zhu (Eds.): COCOA 2007, LNCS 4616, pp. 52–61, 2007.

time consuming to tackle the alignments problems that biologists encounter everyday. In fact, the problem of aligning $k > 2$ sequences is known to be NP-hard [24]. For this reason various heuristic methods have been designed to decrease the complexity of the standard algorithm and obtain sub-optimal alignments. These heuristic methods fall into two categories.

Progressive methods (PM) [5] are the most widely used optimization techniques. They consist in iteratively aligning the most closely related sequences or groups of sequences until all sequences are aligned. The most famous progressive methods are *CLUSTALW* [21] and *T-Coffee* [17], MUSCLE [4] and MAFFT [12]. The PM approach has the advantage to be simple, efficient and provides good results. Nevertheless this approach suffers from its greedy nature: mistakes made in the alignment of previous sequences can not be corrected as more sequences are added. The order in which the sequences are aligned is determined by an efficient clustering method such as neighbor-joining [19]. Progressive Methods therefore automatically construct a phylogenetic tree as well as an alignment.

The iterative methods (IM) start from an initial alignment (or a set of initial alignments) and iteratively improve it following some objective function (SAGA [16], Prob-Cons [3], DiAlign-T [20], PRRN/PRRP [8]). Many of these methods in fact combine iterative and progressive optimization and can lead to an alignment of better quality but generally require more computational effort [4,12].

The group-to-group (also called alignment of alignments) algorithm represents a natural simplification of the k-dimension DP algorithm and is the core of progressive and iterative methods [26]. Aligning alignments (AA) is the problem of finding an optimal alignment of two alignments under the sum-of-pairs objective function. An approximate version of AA widely used is based on profiles. A profile is a table that lists the frequencies of each amino acid for each column of an alignment. To improve the quality of the overall alignment it is interesting to compute the exact SP score of two alignments.

Recently Kececioglu and Starrett [11] gave the outline of an algorithm to exactly align two alignments with affine gap cost using shapes. We believe that this algorithm requires more computational effort than needed and can be described in a more elegant manner. We have designed a generic framework which is a generalization of the algorithm of Gotoh [6] that can be instanciated in order to perform an exact pairwise alignment or to align two alignments using linear or affine gap penalties. The method has been implemented in the software MALINBA and we give here some of the results obtained on the BAliBASE benchmarks.

The rest of the paper is organized as follows. In the next section we will give a formal description of some important notions used for the alignment of sequences. Section 3 presents the generic framework that we have defined to align alignments. The next section provides some results from the BAliBase database of alignments compared to other MSA softwares.

2   Formal Definition of the Problem

Let us consider that an alphabet is a set of distinct letters for which we identify a special symbol called a gap generally represented by the character ' – '. A sequence is expressed over an alphabet and is a string of characters where each character stands for a residue, i.e. a nucleic acid (DNA) or an amino acid (protein). Aligning two sequences or two sets of sequences can be performed by using edit operations and the result in a matrix called an alignment:

Definition 1. - Alignment - *Let $S = \{S_1, \ldots, S_k\}$ be a set of sequences defined over an alphabet $\Sigma : \forall u \in \{1, \ldots, k\}, S_u = \langle x_1^u, \ldots, x_{|S_u|}^u \rangle$ where $|S_u|$ is the length of S_u. An alignment A^S is a matrix:*

$$A^S = \begin{bmatrix} a_1^1 & \cdots & a_q^1 \\ \vdots & & \vdots \\ a_1^k & & q_q^k \end{bmatrix}$$

such that $\forall u \in \{1, \ldots, k\}, \forall v \in \{1, \ldots, q\}, \quad a_v^u \in \Sigma$. The matrix A^S verifies the following properties:

- *$\forall u \in \{1, \ldots, k\}, \quad max(|S_u|) \leq q \leq \sum_{u=1}^{u=k} |S_u|$,*
- *$\nexists j \in \{1, \ldots, q\} \quad such\ that \quad \forall u \in \{1, \ldots, k\}, \quad a_j^u = -$,*
- *$\forall u \in \{1, \ldots, k\}$, there exists an isomorphism $f_u : \{1, \ldots, |S_u|\} \rightarrow \{1, \ldots, q\}$ such that $\langle a_{f_u(1)}^u, a_{f_u(2)}^u, \ldots, a_{f_u(|S_u|)}^u \rangle = S_u$*

Example 1. For example, the set of sequences S could be aligned as follows:

S	an alignment of S
ACCT	AC-CT
AC	AC---
ACT	AC--T
CAAT	-CAAT
CT	-C--T
CAT	-CA-T

2.1 Sum-of-Pairs

As previously mentioned, to establish the quality of an alignment we use an objective function called the sum-of-pairs which depends on a substitution matrix w and a model of gap $g(n)$. In the reminder of this paper we will consider that the substitution matrix corresponds to a measure of similarity which means that similar residues will be rewarded by a positive score and dissimilar residues will get a negative score. There exist two widely used models of gaps called linear and affine[1].

Definition 2. - Gap model - *A gap model is an application $g : \mathbb{N} \rightarrow \mathbb{R}$ which assigns a score, also called a penalty, to a set of consecutive gaps. This penalty is generally negative.*

---

[1] The linear gap model is sometimes refered as a *constant model* and the affine gap model is sometimes refered as *linear* which is confusing.

Definition 3. - Linear gap model - *For this model, the penalty is proportional to the length of the gap and is given by* $g(n) = n \times g_o$ *where* $g_o < 0$ *is the* opening *penalty of a gap and* n *the number of consecutive gaps.*

Definition 4. - Affine gap model - *For this model the insertion of a new gap has a more important penalty than the extension of an existing gap, this can be stated by the following formula:*

$$g(n) = \begin{cases} 0 & \text{if } n = 0 \\ g_o + (n-1) \times g_e & \text{if } n \geq 1 \end{cases}$$

where $g_o < 0$ *is the* gap opening *penalty and* $g_e < 0$ gap extension *penalty and are such that* $|g_e| < |g_o|$.

Definition 5. - Sum-of-pairs of an alignment - *Let* A^S *be an alignment of a set of sequences* $S = \{S_1, \ldots, S_k\}$. *The sum-of-pairs is given by the following formula:*

$$sop(A^S) = \sum_{c=1}^{q} sop^c(A_c^S)$$

where $sop^c(A_c^S)$ *is the score of the c column of the alignment given by:*

$$sop^c(A_c^S) = \sum_{r=1}^{k-1} \sum_{s=r+1}^{k} \delta_{r,s} \times w(a_c^r, a_c^s) \times \lambda \begin{pmatrix} a_{c-1}^r & a_c^r \\ a_{c-1}^s & a_c^s \end{pmatrix}$$

with:

- $0 < \delta_{r,s} \leq 1$ *is a weighting coefficient that allows to remedy problems arising from biased sequences, in order for example to avoid over-represented sequences to dominate the alignment. For simplicity's sake we will chose* $\delta_{r,s} = 1$ *in the remainder of this paper . When* $\delta_{r,s} \neq 1$ *the sum-of-pairs is called the weighted sum-of-pairs [7].*
- *we introduce here* λ *which is the key feature of our work and is an application* $\Sigma^4 \to \mathbb{R}$, *induced by the gap model.* λ *takes into account the previous edit operation used to obtain the current column of the alignment.*

Definition 6. - λ for a linear gap model - *For a linear gap model, a gap has always the same cost wherever it is placed in the alignment.* $\forall c \in \{1, \ldots, q\}$:

$$\lambda \begin{pmatrix} a_{c-1}^r & a_c^r \\ a_{c-1}^s & a_c^s \end{pmatrix} = \begin{cases} 0 & \text{if } c - 1 = 0 \\ 1 & \text{if } a_c^r \neq - \text{ and } a_c^s \neq - \\ g_{op} & \text{if } a_c^r = - \text{ or } a_c^s = - \end{cases}$$

Definition 7. - λ for an affine gap model - *For the affine gap model, the previous edit operation and especially insertions will influence the cost of the penalty.* $\forall c \in \{1, \ldots, q\}$:

$$\lambda \begin{pmatrix} a_{c-1}^r & a_c^r \\ a_{c-1}^s & a_c^s \end{pmatrix} = \begin{cases} 0 & \text{if } c - 1 = 0 \\ 1 & \text{if } a_c^r \neq - \text{ and } a_c^s \neq - \\ g_{op} & \text{if } (a_c^r = - \text{ and } a_{c-1}^r \neq -) \text{ or if } (a_c^s = - \text{ et } a_{c-1}^s \neq -) \\ g_{ext} & \text{if } (a_c^r = - \text{ and } a_{c-1}^r = -) \text{ or if } (a_c^s = - \text{ et } a_{c-1}^s = -) \end{cases}$$

3　A Generic Framework for Aligning Alignments

The problem of aligning alignments can be stated as follows :

Definition 8. - **Aligning alignment** - *Given two multiple alignments A_v and A_h, find an optimal alignment of A_v and A_h for the sum-of-pairs objective function for a given substitution matrix w and gap model $g(n)$.*

The framework that we now define is a generalization of the algorithm of [6] based on a measure of similarity. We refer the reader to [25] for a better understanding of the computation process which is based on two steps. The first step is the initialization of the first row and first column of the matrix and the second step is the recursive relation used to compute each remaining cell of the matrix. To decrease the complexity we introduce three auxillary matrices called D, V and H. D is used to record the cost of a match or a substitution, V and H are used to record the cost of an insertion respectively in A_v and A_h. Table 1 represents the possible moves in the dynamic programming matrix. For example, DH means a match or substituion between A_v and A_h followed by an insertion in A_h. The M matrix records the optimal cost of the global alignment. To obtain the optimal alignment we use the *traceback* technique (see [25]). We consider that A_v and A_h are defined as follows :

$$A_h = \begin{bmatrix} x_1^1 & \cdots & x_{q_h}^1 \\ \vdots & & \vdots \\ x_1^{k_h} & \cdots & x_{q_h}^{k_h} \end{bmatrix} \quad A_v = \begin{bmatrix} y_1^1 & \cdots & y_{q_v}^1 \\ \vdots & & \vdots \\ y_1^{k_v} & \cdots & y_{q_v}^{k_v} \end{bmatrix}$$

Table 1. possible moves for the dynamic programming algorithm

↘ ↓ ↘ ↓	DD DH VD VH	
→ ↘ → ↓	DH **D** VH **V**	
↘ ↓	HD HV	
→ →	HH **H**	

1. initialization, $\forall i \in \{1, \ldots, k_v\}, \forall j \in \{1, \ldots, k_h\}$:

$$M_{0,0} = D_{0,0} = H_{0,0} = V_{0,0} = 0$$
$$D_{i,0} = H_{i,0} = -\infty$$
$$D_{0,j} = V_{0,j} = -\infty$$
$$H_{0,j} = H_{0,j-1} + sop^j(A_h) + \sum_{e=1}^{k_v} \sum_{f=1}^{k_h} w(x_j^f, -) \times (g(j) - g(j-1))$$
$$V_{i,0} = V_{i-1,0} + sop^i(A_v) + \sum_{e=1}^{k_v} \sum_{f=1}^{k_h} w(-, y_i^e) \times (g(i) - g(i-1))$$

2. recursive relation, $\forall i \in \{1, \ldots, k_v\}, \forall j \in \{1, \ldots, k_h\}$:

$$M_{i,j} = max\{D_{i,j}, H_{i,j}, V_{i,j}\}$$

with

$$
D_{i,j} = max \begin{cases}
D_{i-1,j-1} + \displaystyle\sum_{e=1}^{k_v}\sum_{f=1}^{k_h} w(x_j^f, y_i^e) \times \lambda \begin{pmatrix} x_{j-1}^f & x_j^f \\ y_{i-1}^e & y_i^e \end{pmatrix} \\[4mm]
+ \displaystyle\sum_{r=1}^{k_v-1}\sum_{s=r+1}^{k_v} w(y_i^r, y_i^s) \times \lambda \begin{pmatrix} y_{i-1}^r & y_i^r \\ y_{i-1}^s & y_i^s \end{pmatrix} + \displaystyle\sum_{r=1}^{k_h-1}\sum_{s=r+1}^{k_h} w(x_j^r, x_j^s) \times \lambda \begin{pmatrix} x_{j-1}^r & x_j^r \\ x_{j-1}^s & x_j^s \end{pmatrix} \\[4mm]
H_{i-1,j-1} + \displaystyle\sum_{e=1}^{k_v}\sum_{f=1}^{k_h} w(x_j^f, y_i^e) \times \lambda \begin{pmatrix} x_{j-1}^f & x_j^f \\ - & y_i^e \end{pmatrix} \\[4mm]
+ \displaystyle\sum_{r=1}^{k_v-1}\sum_{s=r+1}^{k_v} w(y_i^r, y_i^s) \times \lambda \begin{pmatrix} - & y_i^r \\ - & y_i^s \end{pmatrix} + \displaystyle\sum_{r=1}^{k_h-1}\sum_{s=r+1}^{k_h} w(x_j^r, x_j^s) \times \lambda \begin{pmatrix} x_{j-1}^r & x_j^r \\ x_{j-1}^s & x_j^s \end{pmatrix} \\[4mm]
V_{i-1,j-1} + \displaystyle\sum_{e=1}^{k_v}\sum_{f=1}^{k_h} w(x_j^f, y_i^e) \times \lambda \begin{pmatrix} - & x_j^f \\ y_{i-1}^e & y_i^e \end{pmatrix} \\[4mm]
+ \displaystyle\sum_{r=1}^{k_v-1}\sum_{s=r+1}^{k_v} w(y_i^r, y_i^s) \times \lambda \begin{pmatrix} y_{i-1}^r & y_i^r \\ y_{i-1}^s & y_i^s \end{pmatrix} + \displaystyle\sum_{r=1}^{k_h-1}\sum_{s=r+1}^{k_h} w(x_j^r, x_j^s) \times \lambda \begin{pmatrix} - & x_j^r \\ - & x_j^s \end{pmatrix}
\end{cases}
$$

$$
H_{i,j} = max \begin{cases}
D_{i,j-1} + \displaystyle\sum_{e=1}^{k_v}\sum_{f=1}^{k_h} w(x_j^f, -) \times \lambda \begin{pmatrix} x_{j-1}^f & x_j^f \\ y_i^e & - \end{pmatrix} \\[4mm]
+ \displaystyle\sum_{r=1}^{k_v-1}\sum_{s=r+1}^{k_v} w(-, -) \times \lambda \begin{pmatrix} y_{i-1}^r & - \\ y_{i-1}^s & - \end{pmatrix} + \displaystyle\sum_{r=1}^{k_h-1}\sum_{s=r+1}^{k_h} w(x_j^r, x_j^s) \times \lambda \begin{pmatrix} x_{j-1}^r & x_j^r \\ x_{j-1}^s & x_j^s \end{pmatrix} \\[4mm]
H_{i,j-1} + \displaystyle\sum_{e=1}^{k_v}\sum_{f=1}^{k_h} w(x_j^f, -) \times \lambda \begin{pmatrix} - & x_j^f \\ - & - \end{pmatrix} \\[4mm]
+ \displaystyle\sum_{r=1}^{k_v-1}\sum_{s=r+1}^{k_v} w(-, -) \times \lambda \begin{pmatrix} - & - \\ - & - \end{pmatrix} + \displaystyle\sum_{r=1}^{k_h-1}\sum_{s=r+1}^{k_h} w(x_j^r, x_j^s) \times \lambda \begin{pmatrix} x_{j-1}^r & x_j^r \\ x_{j-1}^s & x_j^s \end{pmatrix} \\[4mm]
V_{i,j-1} + \displaystyle\sum_{e=1}^{k_v}\sum_{f=1}^{k_h} w(x_j^f, -) \times \lambda \begin{pmatrix} - & x_j^f \\ y_i^e & - \end{pmatrix} \\[4mm]
+ \displaystyle\sum_{r=1}^{k_v-1}\sum_{s=r+1}^{k_v} w(-, -) \times \lambda \begin{pmatrix} y_i^r & - \\ y_i^s & - \end{pmatrix} + \displaystyle\sum_{r=1}^{k_h-1}\sum_{s=r+1}^{k_h} w(x_j^r, x_j^s) \times \lambda \begin{pmatrix} - & x_j^r \\ - & x_j^s \end{pmatrix}
\end{cases}
$$

$$
V_{i,j} = max \begin{cases}
D_{i-1,j} + \displaystyle\sum_{e=1}^{k_v}\sum_{f=1}^{k_h} w(-, y_i^e) \times \lambda \begin{pmatrix} x_j^f & - \\ y_{i-1}^e & y_i^e \end{pmatrix} \\[4mm]
+ \displaystyle\sum_{r=1}^{k_v-1}\sum_{s=r+1}^{k_v} w(y_i^r, y_i^s) \times \lambda \begin{pmatrix} y_{i-1}^r & y_i^r \\ y_{i-1}^s & y_i^s \end{pmatrix} + \displaystyle\sum_{r=1}^{k_h-1}\sum_{s=r+1}^{k_h} w(x_j^r, x_j^s) \times \lambda \begin{pmatrix} x_{j-1}^r & x_j^r \\ x_{j-1}^s & x_j^s \end{pmatrix} \\[4mm]
H_{i-1,j} + \displaystyle\sum_{e=1}^{k_v}\sum_{f=1}^{k_h} w(-, y_i^e) \times \lambda \begin{pmatrix} x_j^f & - \\ - & y_i^e \end{pmatrix} \\[4mm]
+ \displaystyle\sum_{r=1}^{k_v-1}\sum_{s=r+1}^{k_v} w(y_i^r, y_i^s) \times \lambda \begin{pmatrix} - & y_i^r \\ - & y_i^s \end{pmatrix} + \displaystyle\sum_{r=1}^{k_h-1}\sum_{s=r+1}^{k_h} w(-, -) \times \lambda \begin{pmatrix} x_j^r & - \\ x_j^s & - \end{pmatrix} \\[4mm]
V_{i-1,j} + \displaystyle\sum_{e=1}^{k_v}\sum_{f=1}^{k_h} w(-, y_i^e) \times \lambda \begin{pmatrix} - & - \\ - & y_i^e \end{pmatrix} \\[4mm]
+ \displaystyle\sum_{r=1}^{k_v-1}\sum_{s=r+1}^{k_v} w(y_i^r, y_i^s) \times \lambda \begin{pmatrix} y_{i-1}^r & y_i^r \\ y_{i-1}^s & y_i^s \end{pmatrix} + \displaystyle\sum_{r=1}^{k_h-1}\sum_{s=r+1}^{k_h} w(-, -) \times \lambda \begin{pmatrix} - & - \\ - & - \end{pmatrix}
\end{cases}
$$

The score of each edit operation depends on 4 different factors that we call α, β, γ and δ. For example, to obtain $DD_{i,j}$, we need to compute :

- the α factor which corresponds to the former edit operation $D_{i-1,j-1}$
- the β factor is the sum-of-pairs score of column j of A_h :

$$\sum_{r=1}^{k_h-1} \sum_{s=r+1}^{k_h} w(x_j^r, x_j^s) \times \lambda \begin{pmatrix} x_{j-1}^r & x_j^r \\ x_{j-1}^s & x_j^s \end{pmatrix}$$

- the γ factor is the sum-of-pairs score of the column i of A_v :

$$\sum_{r=1}^{k_v-1} \sum_{s=r+1}^{k_v} w(y_i^r, y_i^s) \times \lambda \begin{pmatrix} y_{i-1}^r & y_i^r \\ y_{i-1}^s & y_i^s \end{pmatrix}$$

- the δ factor results from the interaction of column j of A_h with column i of A_v :

$$\sum_{e=1}^{k_v} \sum_{f=1}^{k_h} w(x_j^f, y_i^e) \times \lambda \begin{pmatrix} x_{j-1}^f & x_j^f \\ y_{i-1}^e & y_i^e \end{pmatrix}$$

Proposition 1. - **Complexity of aligning alignment** - *The complexity of the computation of the alignment of two alignments composed of k sequences of length n is $\mathcal{O}(n^2 k^2)$.*

Proof : for each $M_{i,j}$, we need to compute 9 values for which we need :

$$\underbrace{k_v \times k_h}_{\delta} + \underbrace{\frac{1}{2} \times (k_v - 1) \times k_v}_{\gamma} + \underbrace{\frac{1}{2} \times (k_h - 1) \times k_h}_{\beta}$$

computations. If we consider that $k_v = k_h = k$ and that $q_v = q_h = n$, we then have to perform : $(n+1) \times (n+1) \times 9 \times (2k-1) \times k \approx n^2 \times 9 \times 2k^2$ computations. This value is to be compared with the complexity of [11] which is $\mathcal{O}((3+\sqrt{2})^k (n-k)^2 k^{3/2})$. For example, to align 2 alignments of 10 sequences of 100 residues, [11] would normally have to perform 7.2×10^{11} computations while we would require only 1.7×10^7.

3.1   Instanciation of the Framework

In the case of a pairwise alignment with a linear gap penalty, the β and γ factors are not involved because there is only one sequence in each alignment. The extension penalty is equal to the opening penalty. The simplification of formulas show that $\forall i, j \; D_{i,j} = H_{i,j} = V_{i,j}$. It is then not necessary to use the D, V and H matrices and the simplified formula is equal to the Needleman and Wunsch formula: $M_{i,j} = max\{M_{i-1,j-1} + w(x_j, y_i), M_{i,j-1} + g_o, M_{i-1,j} + g_o\}$.

4   Experimentations

The generic framework presented so far has been implemented in the software MA-LINBA (Multiple Affine or LINear Block Alignment) which is a modified version of

PLASMA [2] written in C++. Compared to PLASMA, MALINBA uses an *exact* version of the DP algorithm to align two columns of 2 alignments while PLASMA relies on the insertion of columns of gaps in one of the two subalignments and local rearrangements of residues in blocks.

To evaluate the quality of the alignments obtained by MALINBA and other MSA programs we have performed some tests on the BAliBase 2.0 database of benchmarks.

4.1 BAliBase

BAliBase is designed for assessing MSA algorithms [22] and is divided into five reference sets which were designed to test different aspects of alignment softwares. Set 1 is composed of approximately equidistant sequences. Set 2 is made of families whith orphan sequences while Set 3 contains divergent families. Set 4 has sequences with large N/C terminal insertions and sequences of Set 5 contain large internal insertions. All reference alignments were refined manually by BAliBase authors.

To assess alignment accuracy we use the bali_score program which helps compute two scores : the *BAliBase sum-of-pairs score* (SPS) which in fact is a ratio between the number of correctly aligned residue pairs found in the test alignment and the total number of aligned residue pairs in core blocks of the reference alignment [23]. We also report the column score (CS) defined as the number of correctly aligned columns found in the test alignment divided by the total number of columns in core blocks of the reference alignment. The closer to 1.0 these scores are, the better the alignment is.

4.2 Results

We have compared our results obtained with MALINBA to five widely known MSA systems: (1) CLUSTALW 1.83, the most popular progressive alignment software; (2) the nwnsi version of MAFFT 5.86 using iterative refinement techniques; (3) MUSCLE 3.6; (4) PROBCONS 1.11; (5) T-COFFEE 4.96. All sotwares were run using default parameters on an Intel Core 2 Duo E6400 with 1 Gb of RAM. For this test MALINBA used 7 specific sets of parameters and we kept the alignments that provided the best SPS scores. Given the results of table 2 which reports the average SPS end CS scores, we can rank the softwares as follows using average SPS or CS scores : CLUSTAL < MALINBA, T-COFFEE < MUSCLE < MAFFT < PROBCONS.

Table 2. Results of the SPS and CS score for MSA Softwares and overall execution time

Softwares	Set 1		Set 2		Set 3		Set 4		Set 5		Time
	SPS	CS	SPS	CS	SPS	CS	SPS	CS	SPS	CS	(in s)
CLUSTAL	0.809	0.707	0.932	0.592	0.723	0.481	0.834	0.623	0.858	0.634	120
MAFFT	0.829	0.736	0.931	0.525	0.812	0.595	**0.947**	0.822	**0.978**	**0.911**	98
MUSCLE	0.821	0.725	0.935	0.593	0.784	0.543	0.841	0.593	0.972	0.901	75
PROBCONS	**0.849**	**0.765**	**0.943**	**0.623**	**0.817**	**0.631**	0.939	**0.828**	0.974	0.892	711
TCOFFEE	0.814	0.712	0.928	0.524	0.739	0.480	0.852	0.644	0.943	0.863	1653
MALINBA	0.811	0.705	0.911	0.522	0.752	0.346	0.899	0.734	0.942	0.842	343

Better alignments have been obtained with MALINBA by fine tuning some parameters. However these results are not reported here. Finally, let us say that despite its complexity our method is quite fast (see the time column in table 2).

5 Conclusion

We have designed a generic framework to align alignments with the sum-of-pairs objective function. This framework is exact and can be used to align two sequences or two alignments using linear of affine gap penalties. This framework was implemented in MALINBA and tested on the BAliBase benchmark and proves to be efficient. Although quite simple, we believe that many improvements can be considered in order to increase the quality of alignments obtained on the BAliBase dataset. For example, local rearrangements on misaligned regions after each progressive step using secondary structure information could probably help improve the column score.

References

1. Dayhoff, M.O., Schwartz, R.M., Orcutt, B.C.: A model of evolutionary change in proteins. In: Dayhoff, M.O. (ed.) Atlas of Protein Sequence and Structure. National Biomedical Research Foundation, vol. 5, chapter 22, pp. 345–352 (1978)
2. Derrien, V., Richer, J-M., Hao, J-K.: Plasma: un nouvel algorithme progressif pour l'alignement multiple de sÃ©quences. Actes des PremiÃ¨res JournÃ©es Francophones de Programmation par Contraintes (JFPC'05), Lens, France (2005)
3. Do, C.B., Mahabhashyam, S.P., Brudno, M., Batzoglou, S.: Probabilistic consistency-based multiple sequence alignment. Genome Research 15, 330–340 (2005)
4. Edgar, R.C.: Muscle: multiple sequence alignment with high accuracy and high throughput. Nucleic Acid Research 32, 1792–1797 (1994)
5. Feng, D.-F., Doolitle, R.F.: Progressive sequence alignment as a prerequisite to correct phylogenetic trees. Journal of Molecular Evolution 25, 351–360 (1987)
6. Gotoh, O.: An improved algorithm for matching biological sequences. Journal of Molecular Biology 162, 705–708 (1982)
7. Gotoh, O.: A weighting system and algorithm for aligning many phylogenetically related sequences. Computer Applications in the Biosciences (CABIOS) 11, 543–551 (1995)
8. Gotoh, O.: Multiple sequence alignment: algorithms and applications. Adv. Biopys. 36, 159–206 (1999)
9. Henikoff, S., Henikoff, J.G.: Amino acid substitution matrices from protein blocks. In: Proceedings of the National Academy of Science, vol. 89, pp. 10915–10919 (1992)
10. Humberto, C., Lipman, D.: The multiple sequence alignment problem in biology. SIAM Journal on Applied Mathematics 48(5), 1073–1082 (1988)
11. John, K., Dean, S.: Aligning alignments exactly. In: Proceedings of RECOMB 04, San Diego, March 27-31, 2004, pp. 27–31 (2004)
12. Katoh, Misawa, Kuma, Miyata: Mafft: a novel method for rapid mulitple sequence alignment based on fast fourier transform. Nucleic Acid Research 30, 3059–3066 (2002)
13. Lipman, D.J., Altschul, S.F., Kececioglu, J.D.: A tool for multiple sequence alignment. In: Proc. Natl. Acad Sci., pp. 4412–4415 (1989)
14. Needleman, S.B., Wunsch, C.D.: A general method applicable to the search for similarities in the amino acid sequence of two proteins. JMB 3(48), 443–453 (1970)

15. Notredame, C., Higgins, D., Heringa, J.: T-coffee: A novel method for multiple sequence alignments. Journal of Molecular Biology 302, 205–217 (2000)
16. Notredame, C., Higgins, D.G.: Saga: Sequence alignment by genetic algorithm. Nuc. Acids Res. 8, 1515–1524 (1996)
17. Notredame, C., Higgins, D.G., Heringa, J.: T-coffee: A novel method for fast and accurate multiple sequence alignment. Journal of Molecular Biology, 205–217 (2000)
18. Notredame, C., Holme, L., Higgins, D.G.: Coffee: A new objective function for multiple sequence alignmnent. Bioinformatics 14(5), 407–422 (1998)
19. Saitou, N., Nei, M.: The neighbor-joining method: a new method for reconstructing phylogenetic trees. Mol. Biol. Evol. 4, 406–425 (1987)
20. Subramanian, A.R., Weyer-Menkhoff, J., Kaufmann, M., Morgenstern, B.: Dialign-t: An improved algorithm for segment-based multiple sequence alignment. BMC Bioinformatics 6, 66 (2005)
21. Thompson, J.D., Higgins, D.G., Gibson, T.J.: Clustalw: improving the sensitivity of progressive multiple sequence alignment through sequence weighting, position-specific gap penalties and weight matrix choice. Nucl. Acids Res. 22, 4673–4690 (1994)
22. Thompson, J.D., Plewniak, F., Poch, O.: Balibase: A benchmark alignments database for the evaluation of multiple sequence alignment programs. Bioinformatics 15, 87–88 (1999)
23. Thompson, J.D., Plewniak, F., Poch, O.: A comprehensive comparison of multiple sequence alignment programs. Nucleic Acid Research 27, 2682–2690 (1999)
24. Wang, L., Jiang, T.: On the complexity of multiple sequence alignment. Journal of Computational Biology 1(4), 337–348 (1994)
25. Waterman, M.S.: Introduction to Computational Biology. Chapman and Hall/CRC, Boca Raton (2000)
26. Yamada, S., Gotoh, O., Yamana, H.: Improvement in accuracy of multiple sequence alignment using novel group-to-group sequence alignment algorithm with piecewise linear gap cost. BMC Bioinformatics 7(1), 524 (2006)

Energy Minimizing Vehicle Routing Problem

İmdat Kara[1], Bahar Y. Kara[2], and M. Kadri Yetis[3]

[1] Başkent University, Dept. Ind. Eng., Ankara, Turkey
`ikara@baskent.edu.tr`
[2] Bilkent University, Dept. Ind. Eng., Ankara, Turkey
`bkara@bilkent.edu.tr`
[3] Havelsan A.S., Ankara, Turkey
`kyetis@yahoo.com`

Abstract. This paper proposes a new cost function based on distance and load of the vehicle for the Capacitated Vehicle Routing Problem. The vehicle-routing problem with this new load-based cost objective is called the Energy Minimizing Vehicle Routing Problem (EMVRP). Integer linear programming formulations with $O(n^2)$ binary variables and $O(n^2)$ constraints are developed for the collection and delivery cases, separately. The proposed models are tested and illustrated by classical Capacitated Vehicle Routing Problem (CVRP) instances from the literature using CPLEX 8.0.

Keywords: Capacitated vehicle routing problem, Energy minimizing vehicle routing problem, Integer programming.

1 Introduction

One of the most important and widely studied combinatorial problem is the Travelling Salesman Problem (TSP) and its variants, which is NP-hard. The problems of finding optimal routes for vehicles from one or several depots to a set of locations/customers are the variants of the multiple Travelling Salesman Problem (m-TSP) and known as Vehicle Routing Problems (VRPs). Vehicle routing problems have many practical applications, especially in transportation and distribution logistics. An extensive literature exists on these problems and their variations (e.g. Golden and Assad [8], Bodin [4], Laporte [12], Laporte and Osman [13], Ball et al. [2], Toth and Vigo [16] [17]).

The Capacitated Vehicle Routing Problem (CVRP) is defined on a graph $G = (V, A)$ where $V = \{0, 1, 2, \ldots, n\}$ is the set of nodes (vertices), 0 is the *depot (origin, home city)*, and the remaining nodes are *customers*. The set $A = \{(i, j) : i, j \in V, i \neq j\}$ is an arc (or edge) set. Each customer $i \in V \backslash \{0\}$ is associated with a positive integer demand q_i and each arc (i, j) is associated a travel cost c_{ij} (which may be symmetric, asymmetric, deterministic, random, etc.). There are m vehicles with identical capacity Q. The CVRP consists of determining a set of m vehicle routes with minimum cost in such a way that; each route starts and ends at the depot, each customer is visited by exactly one route, and the total demand of each route does not exceed the vehicle capacity Q.

A. Dress, Y. Xu, and B. Zhu (Eds.): COCOA 2007, LNCS 4616, pp. 62–71, 2007.

CVRP was first defined by Dantzig and Ramser in 1959 [5]. In that study, the authors used distance as a surrogate for the cost function. Since then, the cost of traveling from node i to node j, i.e., c_{ij}, has usually been taken as the distance between those nodes (for recent publications, see e.g. Baldacci et al. [1], Letchford and Salazar-Gonzalez [14], Yaman [21]).

The real cost of a vehicle traveling between two nodes depends on many variables: the load of the vehicle, fuel consumption per mile (kilometer), fuel price, time spent or distance traveled up to a given node, depreciation of the tires and the vehicle, maintenance, driver wages, time spent in visiting all customers, total distance traveled, etc. (Baldacci et al.[1], Toth and Vigo [17], Desrochers et al. [6]). Most of the attributes are actually distance or time based and can be approximated by the distance. However, some variables either cannot be represented by the distance between nodes or involve travel costs that may not be taken as constant. Examples of such variables are vehicle load, fuel consumption per mile (kilometer), fuel price or time spent up to a given node. Most of these types of variables may be represented as a function of the flow, especially , as a function of the load of vehicles on the corresponding arc. Thus, for some cases, in addition to the distance traveled, we need to include the load of the vehicle as additional indicator of the cost.

We observe that, some researches with different objectives have been conducted on TSP (see e.g. Bianco [3], Gouveia and VoB [10], Lucena [15], Tsitsiklis [19]). To the best of our knowledge, the vehicle routing literature dealing with single criteria optimization has not previously included the flow on the arcs to the traveling cost, which is the main motivation of this research. In this study, we propose a new cost function which is a product of the distance traveled and the weight of the vehicle on that arc. The contributions of this paper may be summarized as:

- Define a new cost function for vehicle routing problems as a multiple of length of the arc traveled and the total load of the vehicle on this arc. Name this problem as Energy Minimizing Vehicle Routing Problem (EMVRP).
- Present polynomial size integer programming formulations for EMVRP for collection and delivery cases.

We briefly show the relation between the energy used and load of a vehicle and define the new cost function in Section 2. Problem identification and integer programming formulations of the EMVRP for both collection and delivery cases are presented in Section 3. The proposed models are tested and illustrated by standard CVRP instances obtained from the literature and the results are given in Section 4. Concluding remarks are in Section 5.

2   New Cost Function

For vehicle routing problems where vehicles carry goods from an origin (center, factory and/or warehouse) to the customer, or from the customer to the origin, the traveling cost between two nodes can be written as,

$$\text{Cost} = f(\text{load}, \text{distance traveled}, \text{others})$$

where $f(.)$ is any function. We derive a cost function that mainly focuses on the total energy consumption of the vehicles. Recall from mechanics that,

$$\text{Work} = \text{force} * \text{distance}$$

In the CVRP, the movement of the vehicles can be considered as an impending motion where the force causing the movement is equal to the friction force (see for example Walker [20]). Remember also that,

$$\text{Friction force} = \text{Coefficient of friction} * \text{weight}.$$

Thus, we have

$$\text{Work} = \text{Friction force} * \text{distance}$$
$$\text{Work} = \text{Coefficient of friction} * \text{weight} * \text{distance}$$

The coefficient of friction can be considered as constant on roads of the same type. Then, the work done by a vehicle over a link (i, j) will be:

$$\text{Work} = \text{weight of the vehicle(over link}(i, j)) * \text{distance(of link}(i, j)).$$

Since work is energy, minimizing the total work done is equivalent to minimizing the total energy used (at least in terms of fuel consumption). Obviously, the weight of the vehicle equals the weight of the empty vehicle (tare) plus the load of the vehicle. Thus, if one wants to minimize the work done by each vehicle, or to minimize the energy used, one needs to use the cost as,

$$\text{Cost of}(i, j) = [\text{Load of the vehicle over}(i, j) + \text{Tare}] * \text{distance of}(i, j), \qquad (1)$$

There seems to be no such definition and objective cost function in the vehicle routing literature. However, there are references on the Internet as shown in Figure 1, (Goodyear website, [22]) indicating that fuel consumption changes with vehicle load.

Figure 1 shows that, miles per gallon decrease with increased vehicle weight. Thus for a CVRP in which goods are carried and fuel prices are relatively more important than the drivers wages, considering the load of the vehicle as well as the distances will produce a more realistic cost of traveling from one customer to another. This analysis shows that for such CVRPs we may define a more realistic cost of traveling from one customer to another by considering the load of the vehicle as well as the distances. We refer the CVRP in which cost is defined as in expression (1) the Energy Minimizing Vehicle Routing Problem (EMVRP).

In the CVRP, vehicles collect and/or deliver the items and/or goods from/to each customer on the route. Thus, the load of a vehicle changes throughout the tour. They show an increasing step function in the case of collection and a decreasing step function in the case of delivery. Thus, load of a vehicle cumulate or accumulate along the tour. For this reason, one must consider the collection and delivery situations, carefully.

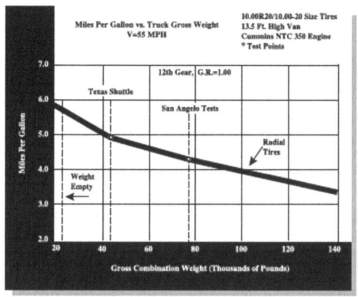

Fig. 1. Miles per Gallon versus vehicle weight [22]

3   Integer Programming Formulations

3.1   Problem Identification

Consider a vehicle routing problem defined over a network $G = (V, A)$ where $V = \{0, 1, 2, \ldots, n\}$ is the node set, 0 is the depot and $A = \{(i, j) : i, j \in V, i \neq j\}$ is the set of arcs, and, components of which are given as:

d_{ij} is the distance from node i to node j,
q_i is the nonnegative weight (e.g. demand or supply) of node i,
m is the number of identical vehicles,
Q_0 is the tare of a vehicle (truck),
Q is the capacity of a vehicle.

We define Energy Minimizing Vehicle Routing Problem (EMVRP) as the problem of constructing vehicle routes such that:

- Each node is served exactly one vehicle,
- Each route starts and ends at the depot,
- The load on the arcs cumulate as much as preceding nodes supply in the case of collection or accumulate as much as preceding nodes demand in the case of delivery,
- The load of a vehicle does not exceed its capacity Q,
- The objective is to find a set of m vehicle routes of minimum total cost, i.e., minimum total energy.

We use the following decision variables in formulating this problem:

$x_{ij} = 1$ if the arc (i, j) is on the tour, and zero otherwise;

y_{ij} is the weight of a vehicle if it goes from i to j, and zero otherwise.

Definition of the y_{ij} is the core of this approach. The weight on the first arc of any tour must take a predetermined value, i.e., tare and then must always increase (or decrease) by q_i units just after node i. In the case of collection, the flow variable shows an increasing step function; for delivery, it shows a decreasing step function. Therefore a model constructed for one case may not be suitable for the other case. The following observation states very important relationship between these situations.

Observation 1. *When the distance matrix is symmetric, the optimal route of the delivery (collection) case equals the optimal route of the collection (delivery) case traversed in the reverse order.*

Proof. Consider a route which consist of k nodes: $n_0 - n_1 - n_2 - \ldots - n_k - n_0$, where n_0 is the depot. For the collection case, the cost of this tour (i.e., total energy used) is:

$$Q_0 \, d_{01} + \sum_{j=1}^{k-1} \left(Q_0 + \sum_{i=1}^{j} q_i \right) d_{j,j+1} + \left(Q_0 + \sum_{i=1}^{k} q_i \right) d_{k0} \tag{2}$$

For the delivery case, the cost of the reverse route $n_0 - n_k - n_{k-1} - \ldots - n_1 - n_0$ is:

$$\left(Q_0 + \sum_{i=1}^{k} q_i \right) d_{0k} + \sum_{j=1}^{k-1} \left(Q_0 + \sum_{i=1}^{j} q_i \right) d_{j+1,j} + Q_0 \, d_{10} \tag{3}$$

Observe that (2) and (3) are the same for symmetric $D = [d_{ij}]$ matrices. □

3.2 Formulations

For the symmetric-distance case, one does not need to differentiate between collection and delivery since the solution of one will determine the solution of the other. For the case of an asymmetric distance matrix, due to the structure of the problem, we present decision models for collection and delivery cases, separetely. The model for the collection case is:

$$F_1 : Min \sum_{i=0}^{n} \sum_{j=0}^{n} d_{ij} \, y_{ij} \tag{4}$$

s.t.

$$\sum_{i=1}^{n} x_{0i} = m \tag{5}$$

$$\sum_{i=1}^{n} x_{i0} = m \tag{6}$$

$$\sum_{i=0}^{n} x_{ij} = 1, \qquad j = 1, 2, \ldots, n \tag{7}$$

$$\sum_{j=0}^{n} x_{ij} = 1, \qquad i = 1, 2, \ldots, n \tag{8}$$

$$\sum_{\substack{j=0 \\ j \neq i}}^{n} y_{ij} - \sum_{\substack{j=0 \\ j \neq i}}^{n} y_{ji} = q_i, \qquad i = 1, 2, \ldots, n \tag{9}$$

$$y_{0i} = Q_0 x_{0i}, \qquad i = 1, 2, \ldots, n \tag{10}$$

$$y_{ij} \leq (Q + Q_0 - q_j) x_{ij}, \qquad (i, j) \in A \tag{11}$$

$$y_{ij} \geq (Q_0 + q_i) x_{ij}, \qquad \forall (i, j) \in A \tag{12}$$

$$x_{ij} = 0 \text{ or } 1, \qquad (i, j) \in A \tag{13}$$

where $q_0 = 0$.

The cost of traversing an arc (i, j) is the product of the distance between the nodes i and j and weight on this arc and this is satisfied by the objective function given in (4). Constraints (5) and (6) ensure that m vehicles are used. Constraints (7) and (8) are the degree constraints for each node. Constraint (9) is the classical conservation of flow equation balancing inflow and outflow of each node, which also prohibits any illegal subtours. Constraint (10) initialize the flow on the first arc of each route, cost structure of the problem necessitates such an initialization. Constraints (11) take care of the capacity restrictions and forces y_{ij} to zero when the arc (i, j) is not on any route, and constraint (12) produce lower bounds for the flow on any arc. Integrality constraints are given in (13). We do not need nonnegativity constraints since we have constraints given in (12). Let us call constraints (10), (11) and (12) as the *bounding constraints* of the formulation. Validity of them is shown in proposition 1 below.

Proposition 1. *In the case of collection, the constraints given in (10), (11) and (12) are valid for EMVRP.*

Proof. As it is explained before, we need initialization value of y_{ij}s for each tour that constraints (10) do it, otherwise y_{ij}s may not be the actual weight on the arcs. Constraints (12) is valid since going from i to j the flow must be at least the initial value plus the weight of the node i (unless node i is the depot, in which case $q_0 = 0$). Similarly, since the vehicle is destined for node j, it will also collect the q_j units at node j (unless j is the depot). In that case, the flow on the arc upon arriving at node j should be enough to take the weight of node

$j(q_j)$, i.e., $y_{ij} + q_j x_{ij} \leq (Q + Q_0)x_{ij}$, which produce constraints (11). Similar constraints for classical CVRP may be seen in (Gouveia [9], Baldacci et al.[1], Letchford and Salazar-Gonzalez [14], Yaman [21]).   □

Due to Observation 1, the delivery problem for the symmetric case need not be discussed. For the asymmetric case, the delivery problem will be modeled by replacing constraints (9) ,(10), (11) and (12) with the following given below.

$$\sum_{\substack{j=0\\j\neq i}}^{n} y_{ij} - \sum_{\substack{j=0\\j\neq i}}^{n} y_{ji} = q_i, \qquad i = 1, 2, \ldots, n \qquad (14)$$

$$y_{i0} = Q_0 x_{i0}, \qquad i = 1, 2, \ldots, n \qquad (15)$$

$$y_{ij} \leq (Q + Q_0 - q_i)\, x_{ij}, \qquad \forall (i, j) \in A \qquad (16)$$

$$y_{ij} \geq (Q_0 + q_j)\, x_{ij}, \qquad \forall (i, j) \in A \qquad (17)$$

Thus the model for the delivery case is:

$$F_2 : Min \sum_{i=0}^{n} \sum_{j=0}^{n} d_{ij}\, y_{ij}$$

$$\text{s.t. } (5)\text{-}(8),(13) \text{ - } (17).$$

where $q_0 = 0$.

Both of the proposed models have $n^2 + n$ binary and $n^2 + n$ continuous variables, and $2n^2 + 6n + 2$ constraints, thus proposed formulations contain $O(n^2)$ binary variables and $O(n^2)$ constraints.

3.3   Extension to Distance Constraints

In certain applications of the CVRP, there is an additional restriction on the total distance traveled by each vehicle (or cost, or time). This problem is known as the Distance-Constrained VRP (abbreviated as DVRP). In the case of the EMVRP, if such a side condition is imposed, we may easily put the necessary constraints into the proposed models. We need to define additional decision variables as:

z_{ij} the total distance traveled by a vehicle (or cost, or time) from the origin to node j when it goes from i to j.

Note that if the arc (i, j) is not on the optimal route, then z_{ij} must be equal to zero. The distance-constrained EMVRP can be modeled by including constraints (18)-(21) in both collection and delivery cases.

$$\sum_{\substack{j=0\\j\neq i}}^{n} z_{ij} - \sum_{\substack{j=0\\j\neq i}}^{n} z_{ji} = \sum_{j=0}^{n} d_{ij}x_{ij} \qquad i = 1, 2, \ldots, n \qquad (18)$$

$$z_{ij} \leq (T - d_{j0})\, x_{ij} \qquad j \neq 0, (i, j) \in A \qquad (19)$$

$$z_{0i} = d_{0i}x_{0i} \qquad i = 1, 2, \ldots, n \qquad (20)$$

$$z_{ij} \geq (d_{0i} + d_{ij})\, x_{ij} \qquad i \neq 0 (i, j) \in A \qquad (21)$$

where T is the maximum distance that a vehicle can travel. These constraints are taken from Kara [11]. Constraints (18) sum the value of the z_{ij}s and eliminate all illegal subtours. Constraints (19), (20)and (21) are the distance bounding constraints ensuring that the total distance of each route cannot exceed the predetermined value T.

4   Illustrative Examples

In this section, we conduct some numerical examples of EMVRP formulation focusing on the collection case. We use two CVRP instances from the literature and we solve the instances via CPLEX 8.0 on an Intel Pentium III 1400 MHz computer. We want to test the effect of the new objective function on the optimal routes (i.e. distance-based routes versus energy-based routes). For that purpose we define two scenarios. Scenario 1 is the EMVRP and Scenario 2 is the standard CVRP (distance minimizing CVRP, which tries to minimize the total distances without considering the vehicle loads). We choose 2 symmetric instances, eil3 and gr17 from the literature [23]. For eil3 $m = 4$ and $Q = 6000$, and for gr-17 $m = 3$ and $Q = 6$. For each problem, we assume $Q_0 = 15\%$ of Q. Table 1 summarizes the results. The second and third columns of Table 1 provide the solutions of Scenario 1 and 2 of eil3 and the 4th and 5th columns provide those of gr-17.

Table 1. Computational Results for eil3 and gr-17 Problems [23]

	eil3		gr-17	
	Scenario 1 EMVRP	Scenario 2 CVRP	Scenario 1 EMVRP	Scenario 2 CVRP
Energy Min.	779.400	810.700	7331	8810
Distance Min.	277	247	3088	2685
Selected Routes	0-4-7-10-6-0 0-8-5-2-0 0-9-12-0 0-11-3-1-0	0-1-0 0-8-5-3-0 0-9-12-10-6-0 0-11-4-7-2-2	0-1-4-10-2-5-16-0 0-9-14-13-7-6-0 0-15-11-8-3-12-0	0-12-16-13-5-7-6-0 0-14-9-1-4-10-2-0 0-15-11-8-3-0

As Table 1 demonstrates, there is a considerable difference between energy-minimizing and distance-minimizing solutions. The cost of the route that minimizes total distance may be 13% less than the solution which minimizes energy. Counter intuitively for both examples, energy usage increases as total distance decreases. Observe from Table 1 that the routes selected under two scenarios are completely different.

Even though we propose a model with $O(n^2)$ binary variables and $O(n^2)$ constraints for the EMVRP, the CPU times of CPLEX over moderate sized problems were not promising. It is therefore necessary to develop efficient solution procedures for the EMVRP like heuristics proposed for CVRP (Gendreau, et al. [7], Toth and Vigo [18]). However, these modifications are beyond the scope of this paper.

5   Conclusion

This paper proposes a new objective function for the vehicle routing problem in which goods are carried and fuel prices are relatively more important than the drivers wages. For such CVRPs we define a more realistic cost of traveling from one customer to another by considering the load of the vehicle as well as the distances. We refer the CVRP as the Energy Minimizing Vehicle Routing Problem (EMVRP), where cost is defined as a multiple of length of the arc traveled and the total load of the vehicle on this arc.

Integer programming formulations with $O(n^2)$ binary variables and $O(n^2)$ constraints are developed for both collection and delivery cases of EMVRP. The adaptability of the formulations to the distance-constrained case is also demonstrated. The proposed models are tested and demonstrated by using CPLEX 8.0 on two instances taken from the literature.

The adaptability and usability of the proposed models to the other network design problems, such as multiple traveling repairman problem and school-bus routing problems, are under consideration.

References

1. Baldacci, R., Hadjiconstantinou, E., Mingozzi, A.: An exact algorithm for the capacitated vehicle routing problem based on a two-commodity network flow formulation. Operations Research 52, 723–738 (2004)
2. Ball, M.O., Magnanti, T.L., Monma, C.L., Nemhauser, G.L.: Handbooks in Operations Research and Management Science: Network routing. North-Holland, Amsterdam (1995)
3. Bianco, L., Mingozzi, A., Ricciardelli, S.: The traveling salesman problem with cumulative costs. Networks 23, 81–91 (1993)
4. Bodin, L.D.: Twenty years of routing and scheduling. Operations Research 38(4), 571–579 (1990)
5. Dantzig, G.B., Ramser, J.H.: The truck dispatching problem. Management Science 6, 80–91 (1959)
6. Desrochers, M., Lenstra, J.K., Savelsbergh, M.W.P.: A classification scheme for vehicle routing and scheduling problems. European Journal of Operational Research 46, 322–332 (1990)
7. Gendreau, M., Hertz, A., Laporte, G.: A tabu search heuristic for the vehicle routing problem. Management Science 40, 1276–1290 (1994)
8. Golden, B.L., Assad, A.A.: Vehicle Routing: Methods and Studies. North-Holland, Amsterdam (1988)
9. Gouveia, L.: A result on projection for the vehicle routing problem. European Journal of Operational Research 85, 610–624 (1995)
10. Gouveia, L., VoB, S.: A classification of formulations for the (time-dependent) traveling salesman problem. European Journal of Operational Research 83, 69–82 (1995)

11. Kara, İ.: Flow based Integer programming formulations for the distance constrained vehicle routing problem. Technical report 2006/01, Başkent University Faculty of Engineering, Ankara, Turkey (2006)
12. Laporte, G.: The vehicle routing problem: An overview of exact and approximate algorithms. European Journal of Operational Research 59, 345–358 (1992)
13. Laporte, G., Osman, I.H.: Routing problems: A bibliography. Annals of Operations Research 61, 227–262 (1995)
14. Letchford, A.N., Salazar-Gonzalez, J-J.: Projection results for vehicle routing. Mathematical Programming, Ser. B 105, 251–274 (2006)
15. Lucena, A.: Time-dependent traveling salesman problem-The deliveryman case. Networks 20, 753–763 (1990)
16. Toth, P., Vigo, D. (eds.): The Vehicle Routing Problem. SIAM Monographs on Discrete Mathematics and Applications. SIAM (2002)
17. Toth, P., Vigo, D.: An overview of vehicle routing problems. In: Toth, P., Vigo, D. (eds.) The Vehicle Routing Problem. SIAM Monographs on Discrete Mathematics and Applications, pp. 1–26. SIAM (2002)
18. Toth, P., Vigo, D.: The Granular Tabu Search and its application to the vehicle routing problem. INFORMS Journal on Computing 15(4), 333–346 (2003)
19. Tsitsiklis, J.N.: Special cases of traveling salesman and repairman problems with time windows. Networks 22, 263–282 (1992)
20. Walker, K.M.: Applied Mechanics for Engineering Technology, 6th edn. Prentice Hall, Englewood Cliffs (2000)
21. Yaman, H.: Formulations and valid inequalities for the heterogeneous vehicle routing problem. Mathematical Programming, Ser. A 106, 365–390 (2006)
22. Goodyear web site:
http://www.goodyear.com/truck/pdf/commercialtiresystems/FuelEcon.pdf
23. VRPLIB:
http://www.or.deis.unibo.it/research_pages/ORinstances/VRPLIB/VRPLIB.html

On the On-Line k-Taxi Problem with Limited Look Ahead

Weimin Ma[1,2], Ting Gao[1], and Ke Wang[1]

[1] School of Economics and Management
Beijng University of Aeronautics and Astronautics, Beijing, 100083, P.R. China
mawm@buaa.edu.cn, gt1986sl@163.com, wangke@sem.buaa.edu.cn
[2] School of Economics and Management, Zhongyuan University of Technology,
Zhengzhou, Henan Province, 450007, P.R. China

Abstract. Based on some new results concerning k-taxi problem, a new variant of the problem, namely the on-line k-taxi problem with limited look ahead (OTLLA) is proposed by our team. Compared with the traditional k-taxi problem, in which only the start and end points of the current service request are known at each step, the OTLLA has a basic realistic consideration: the start point of the first service request is given before the whole service sequence plays, and at each step the end of the current service request and the start point of the next service request are known. In this paper, after the formulation of the model of the OTLLA, some results concerning the competitive analysis for some special cases of OTLLA are given: the competitive algorithm so called Partial Greedy Algorithm (PGA) is designed and the competitive ratios are obtained. Furthermore, some comparisons between some on-line algorithms are developed. Finally, some conclusions are given and some future research directions are discussed.

1 Introduction

In real life, many ongoing decision-making activities, such as currency exchange, stock transactions or mortgage financing, must be carried out in an on-line fashion, with no secure knowledge of future events. However, that knowledge often influences the decision result in a fatal way. Faced with this lack of knowledge, players of these decision-making games often have two choices. One is to use models based on assumptions about the future distribution of relevant quantities. The other is to analyze the worst case and then make some decision to let the worst case be better. Unfortunately, these two approaches may give some on-line solutions that are far from the relevant optimal solutions. An alternate approach in such situations is to use competitive analysis (first applied to on-line algorithms by Sleator and Tarjian in [1]). In this approach, the performance of an on-line strategy is measured against that of an optimal off-line strategy having full knowledge of future events. An advantage of this performance measure over the traditional average-case measure is that for most nontrivial decision-making activities it is extremely difficult to come up with an accurate probabilistic model.

A. Dress, Y. Xu, and B. Zhu (Eds.): COCOA 2007, LNCS 4616, pp. 72–80, 2007.
© Springer-Verlag Berlin Heidelberg 2007

Over the past two decades, on-line problems and their competitive analysis have received considerable interest. On-line problems had been investigated already in the seventies and early eighties of last century but an extensive, systematic study started only when Sleator and Tarjian [1] suggested comparing an on-line algorithm to an optimal off-line algorithm and Karlin, Manasse, Rudolph and Sleator [2] coined the term competitive analysis. In the late eighties and early nineties of last century, three basic on-line problems were studied extensively, namely paging, the k-server problem and metrical task systems. The k-server problem, introduced by Manasse et al. [3], generalizes paging as well as more general caching problems. The problem consists of scheduling the motion of k mobile servers that reside on the points of a metric space S. The metrical task system, introduced by Borodin et al. [4], can model a wide class of on- line problems. An on-line algorithm deals with events that require an immediate response. Future events are unknown when the current event is dealt with. The task system [4], the k-server problem [5], and on-line/off-line games [6] all attempt to model on-line problems and algorithms. During the past few years, apart from the three basic problems, many on-line problems have been investigated in application areas such as data structures, distributed data management, scheduling and load balancing, routing, robotics, financial games, graph theory, and a quantity of problems arising in computer systems. The adversary method for deriving lower bounds on the competitive ratio has been implicitly used by Woodall [7] in the analysis of the so-called Bay Restaurant Problem. Kierstead and Trotter [8] use the adversary method in their investigation of on-line interval graph coloring. Yao [9] formulates a theorem that starts with the words "For any on-line algorithm..." and which proves the impossibility of an on-line bin-packing algorithm with a competitive ratio strictly better than $3/2$. More traditional results concerning on-line competitive analysis can be found in Albers and Leonardi [10] and first chapter of Fiat and Woeginger [11].

Recent years, some researchers attempt to apply the theories of the on-line competitive analysis to some realistic problem in the domain of Economics and Management decision-making and some useful results are obtained. Based on the traditional Ski-Rental Problem [12], Fleischer [13] recently initiated the algorithmic study of the Bahncard problem. Karlin et al. [14] researched this problem for finite expiration periods by a new randomized online algorithm. In paper [15], Fujinware and Iwama reconsidered the classical online Ski-Rental problems in the way of average-case analysis technique. Xu and Xu [16] employed the competitive analysis to well study the on-line leasing problem. Ma et al. [17,18,19,20] successfully delt with several kind of problems, including k-truck scheduling, snacks problem and transportation problem, with the on-line algorithm theory.

In this paper, we originally proposed the on-line k-taxi problem with limited look ahead (OTLLA). The problem based on the traditional k-taxi problem [21]. After the model of OTLLA is formulated, some on-line algorithms are designed and some comparisons between on-line algorithms the relevant lower bounds of competitive ratio are also discussed. The rest of paper is organized as follows: in section 2, the model of OTLLA is established. Section 3 presents some results concerning the on-line algorithms for the OTLLA. In section 4, some comparisons between on-line algorithms developed. Finally, in section 5, we conclude the paper and discuss the future research directions.

2   The Model of the OTLLA

Let $G = (V, E)$ denote an edge weighted graph with n points and the weights of edges satisfy the triangle inequality, where V is a metric space consisting of n points, and E is the set of all weighted edges. For $u, v, w \in V$, the weight of edge meet triangle inequality: $d(u, v) + d(v, w) \geq d(u, w)$ where $d(x, y)$ indicates the distance of the shortest path between points of the x and y. Let $d_{max} = \max d(v_i, v_j)$ and $d_{min} = \min d(v_i, v_j)$ where $i \neq j$ and $v_i, v_j \in V$. Define the following parameter,

$$\lambda = \frac{d_{max}}{d_{min}} \tag{1}$$

obviously $\lambda \geq 1$.

We assume that k taxis occupy a k-points which is a subset of V. A service request $r = (a, b)$, $a, b \in V$, implies there are some customs on point a that must be moved to point b. A service request sequence R consists of some service requests in turn, namely $R = (r_1, r_2, \cdots, r_m)$, where $r = (a_i, b_i)$, $a_i, b_i \in V$. The start point a_1 of the first request is given before the whole service sequence plays. In the whole process, when the ith request r_i arrives, the end point of the ith request and start point of the $(i + 1)$th request r_{i+1} are known by the player. The problem is to decide to move which taxi when a new service request occurs on the basis that we have no information about future possible requests. All discussion is based on the following essential assumptions:

 i) Graph G is connected.
 ii) When a new service request occurs, k taxis are all free.

For any request sequence $R = (r_1, r_2, \cdots, r_m)$, let $C_{OPT}(R)$ denote the optimal (minimum) cost to complete the whole sequences with the off-line algorithm OPT who knows the whole request sequence before its playing; Let $C_{ON}(R)$ denote the cost to complete the whole sequences with the on-line algorithm ON who has not any knowledge about the future requests, namely only part of sequence known. If there exist some constants α and β to satisfy the following inequality,

$$C_{ON}(R) \leq \alpha \cdot C_{OPT}(R) + \beta \tag{2}$$

the on-line algorithm ON is called α-competitive algorithm and α is called competitive ratio. Obviously we get $\alpha \geq 1$ and our objective to design some on-line algorithms with competitive ratio as small as possible. The constant β is related with the original locations of the k taxis. We also note that a competitive algorithm must perform well on all kinds of input sequence.

Based on above model, the OTLLA aims to obtain some on-line algorithms with competitive ratio as small as possible so as to minimize the whole distance of the transportation of all k taxis in the on-line fashion, namely without any knowledge about the future requests.

3   Results of Competitive Analysis

In this section, we will present some results of competitive analysis for OTLLA.

Assume that k taxis are at k different points before the first service request come, and there is a car at a_1. (As from the previous formulation, the OTLLA knows that the first request will occur at point a_1 before the whole service starts). Otherwise, we can make it through by finite movement of k taxis at different points also ensure that there is a car at a_1. And the moving cost must less than or equal to a constant $(k-1) \cdot d_{max}$. This constant makes no influence to the discussion of competitive ratio.

An on-line algorithm so called Partial Greedy Algorithm is designed as follows:

Partial Greedy Algorithm (PGA): *For ith service request $r_i = (a_i, b_i)$:*

1) *If there are taxis at both of a_i and b_i, the taxi at a_i take the passenger from a_i to b_i as well as the taxi at b_i move to a_i to at the same time. The cost of complete the service request is $C_{\mathrm{PGA}}(r_i) = 2 \cdot d(a_i, b_i)$ and there no point which have more than one taxi. The move are $a_i \rightarrow b_i$ and $b_i \rightarrow a_i$.*
2) *If there is a taxi at a_i but no taxi at b_i, the taxi at a_i take the passenger from a_i to b_i. The cost of complete the service request is $C_{\mathrm{PGA}}(r_i) = d(a_i, b_i)$ and there no point which have more than one taxi. The move is $a_i \rightarrow b_i$.*
3) *If there is a taxi at b_i but no taxi at a_i, the taxi at b_i move to a_i first and then take the passenger from a_i to b_i. The cost of complete the service request is $C_{\mathrm{PGA}}(r_i) = 2 \cdot d(a_i, b_i)$ and there no point which have more than one taxi. The move is $b_i \rightarrow a_i \rightarrow b_i$.*
4) *If there is a taxi at neither a_i point nor b_i point, the following cases concerning the relationship between a_i and a_{i-1} need to be consider:*
   i) *If $a_i = a_{i-1}$, the taxi at b_{i-1} (because the $(i-1)$th service request is $r_{i-1} = d(a_{i-1}, b_{i-1})$ there must be one taxi at b_{i-1}) move to a_i, then take the passenger from a_i to b_i the cost of complete the service request is $C_{\mathrm{PGA}}(r_i) = d(a_{i-1}, b_{i-1}) + d(a_i, b_i)$ and there no point which have more than one taxi. The move are $b_{i-1} \rightarrow a_i \rightarrow b_i$.*
   ii) *If $a_i \neq a_{i-1}$, schedule the nearest taxi on the point c_i, where $c_i \neq a_i$ to a_i and then take the passenger from a_i to b_i the cost of complete the service request is $C_{\mathrm{PGA}}(r_i) = d(c_i, a_i) + d(a_i, b_i)$ and there no point which have more than one taxi. The move are $c_i \rightarrow a_i \rightarrow b_i$.*

For the algorithm PGA, we have the following theorem.

Theorem 1. *For OTLLA problem with k taxis, if $k = n$ or $k = n - 1$ hold, the PGA is an on-line algorithm with competitive ratio 2, where $n = |V|$ indicates the number of points of Graph G.*

Proof. If $k = n$ or $k = n - 1$ hold, the 4)th case of the algorithm PGA will never happen because the number of taxi is too many to let more than one point without taxi. Then the algorithm need not to make full use of the information looked ahead about the start and end point of the next request. The OTLLA then degenerates to the traditional k-taxi problem. The competitive ratio of PGA is 2 for the moment. Please refer to detailed proof of reference [21].   □

In order to prove the competitive ratio for another special case of OTLLA, namely the case with $k = n - 2$, we need to prove some lemmas first.

Lemma 1. *For OTLLA problem with k taxis, if $k = n - 2$ holds, according to Algorithm PGA, at least one of the following inequalities holds:*

$$C_{\text{PGA}}(r_i) + C_{\text{PGA}}(r_{i+1}) \leq 2 \cdot d(a_i, b_i) + d(a_{i+1}, b_{i+1}) + d_{max} \qquad (3)$$

$$C_{\text{PGA}}(r_i) + C_{\text{PGA}}(r_{i+1}) \leq d(a_i, b_i) + 2 \cdot d(a_{i+1}, b_{i+1}) + d_{max} \qquad (4)$$

Proof. For any request $r_i = (a_i, b_i)$ according to the algorithm PGA, we have:

$$C_{\text{PGA}}(r_i) \leq d(a_i, b_i) + d_{max} \qquad (5)$$

Because in case 1) and 3) of the PGA, the cost $C_{\text{PGA}}(r_i) = 2 \cdot d(a_i, b_i) \leq d(a_i, b_i) + d_{max}$. Similarly, the equation (5) holds for the other cases of PGA. At the beginning of the game, according to the formulation of the OTLLA, there must exist a taxi at the point a_i. And then according to all cases of PGA, we can easily get,

$$C_{\text{PGA}}(r_1) \leq 2 \cdot d(a_1, b_1) \qquad (6)$$

Combining the inequalities (5) and (6), apparently the following inequality holds,

$$C_{\text{PGA}}(r_1) + C_{\text{PGA}}(r_2) \leq 2 \cdot d(a_1, b_1) + d(a_2, b_2) + d_{max} \qquad (7)$$

So for the case $i = 1$ the Lemma 1 holds.

The following proof will consider the general case of ith and $(i+1)$th service request. For $r_i = d(a_i, b_i)$,

I) The cases of 1), 2) and 3) of PGA occur. The $C_{\text{PGA}}(r_i) = 2 \cdot d(a_i, b_i)$ holds. Combining the equation (5), we get,

$$C_{\text{PGA}}(r_i) + C_{\text{PGA}}(r_{i+1}) \leq 2 \cdot d(a_i, b_i) + d(a_{i+1}, b_{i+1}) + d_{max} \qquad (8)$$

The Lemma 1 holds.

II) The case of 4) of PGA occurs. While there is taxi neither at point a_i nor at point b_i, the cost to complete request satisfy $C_{\text{PGA}}(r_i) \leq d(a_i, b_i) + d_{max}$. For the cost of next request $r_{i+1} = d(a_{i+1}, b_{i+1})$, we need to consider the following three cases.

a) $a_{i+1} \neq a_i$ and $a_{i+1} \neq b_i$.

Under this condition there must be a taxi at a_{i+1} because for completing the last request r_i, a taxi is moved from a point exclude a_{i+1}. Thus to complete the request r_{i+1}, only the cases 1) or 2) of PGA could occur and the cost satisfies $C_{\mathrm{PGA}}(r_{i+1}) \leq 2 \cdot d(a_{i+1}, b_{i+1})$. Then easily to get,

$$C_{\mathrm{PGA}}(r_i) + C_{\mathrm{PGA}}(r_{i+1}) \leq d(a_i, b_i) + 2 \cdot d(a_{i+1}, b_{i+1}) + d_{max} \quad (9)$$

and Lemma 1 holds.

b) $a_{i+1} = a_i$.

(i) $b_{i+1} = c_i$. According to algorithm of PGA, after satisfying the request $r_i = d(a_i, b_i)$ there must be taxi neither at a_i nor c_i. So move the taxi at b_i to a_{i+1} and then take the passenger from a_{i+1} to b_{i+1}. The cost of complete r_{i+1} is $C_{\mathrm{PGA}}(r_{i+1}) \leq d(a_{i+1}, b_{i+1}) + d(a_i, b_i)$. Then easily to get, the Lemma 1 holds.

(ii) $b_{i+1} \neq c_i$. According to algorithm of PGA, after satisfying the request $r_i = d(a_i, b_i)$ there must be taxi neither at a_i nor c_i and thus there must be a taxi at b_{i+1}. Then we easily to get the cost of complete r_{i+1} satisfy $C_{\mathrm{PGA}}(r_{i+1}) \leq 2 \cdot d(a_{i+1}, b_{i+1}) + d(a_i, b_i)$. The Lemma 1 is obtained at moment.

c) $a_{i+1} = b_i$.

Under this condition there must be a taxi at a_{i+1}. Similar with the case a), we easily to know the Lemma 1 holds.

Combining above cases, we know Lemma 1 holds.

The proof is completed.   □

Lemma 2. *For OTLLA problem with k taxis, if $k = n - 2$ holds, according to algorithm PGA, the following inequality holds:*

$$C_{\mathrm{PGA}}(r_i) + C_{\mathrm{PGA}}(r_{i+1}) \leq \frac{3 + \lambda}{2} \cdot [d(a_i, b_i) + d(a_{i+1}, b_{i+1})] \quad (10)$$

Proof. For any request $r_i = (a_i, b_i)$according to the algorithm PGA and Lemma 1, we can prove the Lemma 2 with the following two case:

i). The inequality $C_{\mathrm{PGA}}(r_i) + C_{\mathrm{PGA}}(r_{i+1}) \leq 2 \cdot d(a_i, b_i) + d(a_{i+1}, b_{i+1}) + d_{max}$ holds. Then we get,

$$\frac{C_{\mathrm{PGA}}(r_i) + C_{\mathrm{PGA}}(r_{i+1})}{d(a_i, b_i) + d(a_{i+1}, b_{i+1})} \leq \frac{2 \cdot d(a_i, b_i) + d(a_{i+1}, b_{i+1}) + d_{max}}{d(a_i, b_i) + d(a_{i+1}, b_{i+1})} \quad (11)$$

$$= 2 + \frac{d_{max} - d(a_{i+1}, b_{i+1})}{d(a_i, b_i) + d(a_{i+1}, b_{i+1})}$$

$$\leq 2 + \frac{d_{max} - d_{min}}{2 \cdot d_{min}}$$

$$= \frac{3 + \lambda}{2}$$

Above second inequality holds for $d_{min} \leq d(a_i, b_i), d(a_{i+1}, b_{i+1}) \leq d_{max}$.

ii). The inequality $C_{\text{PGA}}(r_i) + C_{\text{PGA}}(r_{i+1}) \leq d(a_i, b_i) + 2 \cdot d(a_{i+1}, b_{i+1}) + d_{max}$ holds. Then we similarly get,

$$\frac{C_{\text{PGA}}(r_i) + C_{\text{PGA}}(r_{i+1})}{d(a_i, b_i) + d(a_{i+1}, b_{i+1})} \leq \frac{d(a_i, b_i) + 2 \cdot d(a_{i+1}, b_{i+1}) + d_{max}}{d(a_i, b_i) + d(a_{i+1}, b_{i+1})} \quad (12)$$

$$= 2 + \frac{d_{max} - d(a_i, b_i)}{d(a_i, b_i) + d(a_{i+1}, b_{i+1})}$$

$$\leq 2 + \frac{d_{max} - d_{min}}{2 \cdot d_{min}}$$

$$= \frac{3 + \lambda}{2}$$

The proof is completed. □

Theorem 2. *For OTLLA problem with k taxis, if $k = n - 2$ holds, the competitive ratio of algorithm PGA is $\frac{3+\lambda}{2}$, where $n = |V|$ indicates the number of points of Graph G and $\lambda = \frac{d_{max}}{d_{min}}$.*

Proof. For any service request sequence $R = (r_i, r_2, \cdots, r_m)$, the following inequality holds,

$$C_{\text{OPT}}(R) \geq \sum_{i=1}^{m} d(a_i, b_i) \quad (13)$$

The above inequality hold because $\sum_{i=1}^{m} d(a_i, b_i)$ is the least mileage with passengers. This is the most minimum cost which need to pay to complete service request sequence R.

For algorithm PGAthe cost to complete service request sequence R satisfy the following inequality,

$$C_{\text{PGA}}(R) = \sum_{i=1}^{m} [C_{\text{PGA}}(r_i)] + \beta \quad (14)$$

$$= [C_{\text{PGA}}(r_1) + C_{\text{PGA}}(r_2)] + [C_{\text{PGA}}(r_3) + C_{\text{PGA}}(r_4)] + \cdots$$

$$+ [C_{\text{PGA}}(r_{m-1}) + C_{\text{PGA}}(r_m)] + \beta$$

$$\leq \frac{3+\lambda}{2} \cdot [d(a_1, b_1) + d(a_2, b_2)] + \frac{3+\lambda}{2} \cdot [d(a_3, b_3) + d(a_4, b_4)] + \cdots$$

$$+ \frac{3+\lambda}{2} \cdot [d(a_{m-1}, b_{m-1}) + d(a_m, b_m)] + \beta$$

$$= \frac{3+\lambda}{2} \cdot \sum_{i=1}^{m} d(a_i, b_i) + \beta$$

$$\leq \frac{3+\lambda}{2} \cdot C_{\text{OPT}}(R) + \beta$$

The first inequality of above formula holds for Lemma 2. Considering the parity of m, if m is a odd number there is a constant difference in above formula. But this constant can not influence the competitive ratio, e.g., the case with $m \to \infty$.

The proof is completed. □

4   Comparison Between Competitive Algorithms

The Quality standards of on-line algorithms are the corresponding competi-
tive ratios. In paper [21], the authors propose the algorithm so called Position
Maintaining Strategy (PMS) to handle the traditional k-taxi problem. Because
OTLLA is a degenerated variant of the traditional one the competitive ratio of
PMS holds for the special case,e.g., $k = n - 2$ of OTLLA. According to PMS
[21] and our result, for the special case $k = n - 2$ of OTLLA, we have,

$$\alpha_{\mathrm{PMS}} = 1 + \lambda \qquad (15)$$

$$\alpha_{\mathrm{PGA}} = \frac{3 + \lambda}{2} \qquad (16)$$

and then,

$$\alpha_{\mathrm{PGA}} - \alpha_{\mathrm{PMS}} = \frac{3 + \lambda}{2} - (1 + \lambda) = \frac{1 - \lambda}{2} \leq 0 \qquad (17)$$

The above inequality holds for $\lambda \geq 1$. We have the following Corollary.

Corollary 1. *For OTLLA problem with k taxis, if $k = n - 2$ holds, the competitive ratio of algorithm PGA is better than algorithm PMS.*

5   Conclusion and Discussion

In this paper, the on-line taxi problem with limited look ahead (OTLLA) is
proposed and studied. For some special cases, e.g., $k = n$, $k = n-1$ and $k = n-2$,
the competitive algorithm Partial Greedy Algorithm (PGA) is designed and
some good competitive ratios are obtained. The rigorous proofs are given. Some
comparisons between PGA and PMS are developed.

   Although we get some results for OTLLA, the general case, e.g., only consider
the number taxis k, is still open. To design some good algorithm to deal with
the general case maybe is good direction for the future research. Following that,
in the modle of OTLLA, the depth of look ahead is just knowing the start point
of next request. If this condition vary what will happen? Thirdly, for problem
OTLLA, we failed to get any result concerning the lower bound of competitive
ratio. Any further research about it is useful.

Acknowledgements. The work was partly supported by the National Natural
Science Foundation of China (70671004, 70401006, 70521001), Beijing Natural
Science Foundation (9073018), Program for New Century Excellent Talents in
University (NCET-06-0172) and A Foundation for the Author of National Ex-
cellent Doctoral Dissertation of PR China.

References

1. Sleator, D.D., Tarjan, R.E.: Amortized efficiency of list update and paging rules. Communication of the ACM 28, 202–208 (1985)
2. Karlin, R., Manasse, M., Rudlph, L., Sleator, D.D.: Competitive snoopy caching. Algorithmica 3, 79–119 (1988)
3. Manasse, M.S., McGeoch, L.A., Sleator, D.D.: Competitive algorithms for on-line problems. In: Proc. 20th Annual ACM Symp. on Theory of Computing, pp. 322–333. ACM Press, New York (1988)
4. Borodin, A., Linial, N., Sake, M.: An optimal on-line algorithm for metrical task systems. Journal of ACM 39, 745–763 (1992)
5. Manasse, M.S., McGeoch, L.A., Sleator, D.D.: Competitive algorithms for server problems. Journal of Algorithms (11), 208–230 (1990)
6. Ben-david, S., Borodin, A., Karp, R.M., ardos, G.T, Wigderson, A.: On the power if randomization in on-line algorithms. In: Proc. 22nd Annual ACM Symp. on Theory of Computing, pp. 379–386. ACM Press, New York (1990)
7. Woodall, D.R.: The bay restaurant- a linear storage problem. American Mathematical Monthly 81, 240–246 (1974)
8. Kierstead, H.A., Trotter, W.T.: An extremal problem in recursive combinatorics. Congressus Numerantium 33, 143–153 (1981)
9. Yao, A.C.C: New algorithm for bin packing. J. Assoc. Comput. Mach. 27, 207–227 (1980)
10. Albers, S.: Online algorithms: a survey. Mathematical Programming 97(1-2), 3–26 (2003)
11. Fiat, A., Woeginger, G.J.: Online algorithms: The state of the art. LNCS, vol. 1442. Springer, Heidelberg (1998)
12. Karp, R.: On-line algorithms versus offline algorithms: How Much is it Worth to Know the Future? In: Proc. IFIP 12th World Computer Congress, vol. 1, pp. 416–429 (1992)
13. Fleischer, R.: On the Bahncard Problem. In: Proc. TCS'01, pp. 161–174 (2001)
14. Karlin, A.R., Kenyon, C., Randall, D.: Dynamic TCP Acknowledgement and Other Stories about $e/(e-1)$. In: Proc. STOC '01, pp. 502–509 (2001)
15. Fujiwara, H., Iwama, K.: Average-case Competitive Analyses for Ski-rental Problems. In: Bose, P., Morin, P. (eds.) ISAAC 2002. LNCS, vol. 2518, pp. 476–488. Springer, Heidelberg (2002)
16. Xu, Y.F., Xu, W.J.: Competitive Algorithms for Online Leasing Problem in Probabilistic Environments. In: ISNN'01, pp. 725–730 (2004)
17. Ma, W.M., Xu, Y.F., Wang, K.L.: On-line k-truck problem and its competitive algorithm. Journal of Global Optimization 21(1), 15–25 (2001)
18. Ma, W.M., Xu, Y.F., You, J., Liu, J., Wang, K.L.: On the k-Truck Scheduling Problem. International Journal of Foundations of Computer Science 15(1), 127–141 (2004)
19. Ma, W.M., You, J., Xu, Y.F., Liu, J., Wang, K.L.: On the on-line number of snacks problem. Journal of Global Optimization 24(4), 449–462 (2002)
20. Ma, W.M., Liu, J., Chen, G.Q., You, J.: Competitive analysis for the on-line truck transportation problem. Journal of global optimization 34(4), 489–502 (2006)
21. Xu, Y.F., Wang, K.L., Zhu, B.: On the k-taxi problem. Information 1, 2–4 (1999)

The Minimum Risk Spanning Tree Problem[*]

Xujin Chen, Jie Hu, and Xiaodong Hu

Institute of Applied Mathematics
Chinese Academy of Sciences
P.O. Box 2734, Beijing 100080, China
{xchen,hujie,xdhu}@amss.ac.cn

Abstract. This paper studies a spanning tree problem with interval data that finds diverse applications in network design. Given an underlying network $G = (V, E)$, each link $e \in E$ can be established by paying a cost $c_e \in [\underline{c}_e, \overline{c}_e]$, and accordingly takes a risk $\frac{\overline{c}_e - c_e}{\overline{c}_e - \underline{c}_e}$ of link failure. The *minimum risk spanning tree* (MRST) problem is to establish a spanning tree in G of total cost no more than a given constant so that the risk sum over the links on the spanning tree is minimized. In this paper, we propose an exact algorithm for the MRST problem that has time-complexity of $O(m^2 \log m \log n(m + n \log n))$, where $m = |E|$ and $n = |V|$.

1 Introduction

In contrast to classical discrete optimization problems, a large number of recent network designs involve interval data – each network link is associated with an interval modeling the uncertainty about the real value of the link-cost which can take any value in the interval [13,4].

Continuing the diverse research efforts on network optimization with interval data [12,3,7,6], we study in the paper the minimum risk spanning tree problem which arises in a variety of applications. For example, nodes in a communication network wish to communicate with one another, which is usually realized by establishing network links to form a spanning tree in the network. The cost c_e of establishing link e can be any real value in the interval $[\underline{c}_e, \overline{c}_e]$. The higher the cost/payment c_e, the better the link e established, and the lower the risk $\frac{\overline{c}_e - c_e}{\overline{c}_e - \underline{c}_e}$ of link failure at e. Particularly, when paying the lowest possible $c_e = \underline{c}_e$, the established e, due to its poor quality, is prone to malfunction constantly, and suffers from a full risk of link failure; when paying high enough $c_e = \overline{c}_e$, the established e keeps functioning properly for a long period of time, and runs no risk of link failure. In practice, on the one hand, the total cost that can be paid is typically budgeted, meaning that some links in the spanning tree have to accept low payments for establishments, and take high risks of failures; on the other hand, the more links fail, the longer it takes to repair them, and the worse the communication becomes. To deal with this dilemma, the *minimum risk*

---

[*] Supported in part by the NSF of China under Grant No. 10531070, 10671199, and 70221001.

A. Dress, Y. Xu, and B. Zhu (Eds.): COCOA 2007, LNCS 4616, pp. 81–90, 2007.

spanning tree (MRST) problem consists in finding a spanning tree of minimum (total) risk under given budget constraint. Other real-life applications of the MRST problem arise, for example, in the design of transportation network under uncertain construction time or uncertain expenses [9].

Related work. Despite the practical importance of the MRST problem, few previous works have provided the exact model for this problem, in which *not only* a spanning tree *but also* a cost allocation to its links should be determined (aiming for minimization on the total risk). Most literature has focus on finding *only* spanning tree(s) with various objectives.

Among vast literature on spanning tree problems, the work on the *robust spanning tree problem with interval data* (RSTID) and the *constrained minimum spanning tree problem* (CMST) are most related to the present study on the MRST problem.

The RSTID problem has been popularly studied under the *robust deviation (minimax regret) criterion*, which, given an undirected graph with edge-cost intervals, asks for a spanning tree that minimizes the maximum deviation of its cost from the costs of the minimum spanning trees obtained for all possible realizations of the edge costs within the given intervals. In spite of considerable research attention the RSTID problem [19,18] has attracted, the NP-completeness of the problem [2] stems from the graph topology and the structures of cost intervals, and explains the reason for the lack of efficient algorithms [15,16].

As one of the most extensively studied *bicriteria network design problems* [14], the CMST problem is to find a spanning tree of minimum weight whose length is no greater than a given bound, where weight and length are real functions defined on the edge set of the underlying graph. Aggarwal et al. [1] showed that the CMST problem is NP-hard. Ravi and Goemans [17] devised a polynomial approximation scheme based on their $(1, 2)$-approximation algorithm for the CMST problem. Recently, an exact pseudo-polynomial algorithm and a fully polynomial bicriteria approximation scheme were proposed by Hong et al [11].

Our contributions. In this paper we establish the mathematical model for the MRST problem, and develop a $O(m^2 \log m \log n(m + n \log n))$-time algorithm that obtain the optimal solution of the problem on networks with n nodes and m edges.

The polynomial-time solvability of the MRST problem established exhibits the essential difference between the MRST problem and the NP-hard spanning tree problems under interval uncertainty mentioned above. Moreover, in real world, the network designer may have his own preferences of money to risk depending on varying trade-offs between them. Our model and algorithm are quite flexible in the sense that with different budget levels, they are usually able to produce a couple of candidates (spanning trees and associated cost allocations) for selections by the network designer, who is willing to take some risk to save some amount of budget (say, for future use) at his most preferred trade-off between money and risk.

Our approach is based on a key observation which enables us to reduce the MRST problem to two special cases of the CMST problem. It is worth noting that this reduction does not imply benefits from the aforementioned pseudo-polynomial time algorithm for the general CMST problem [11] since the algorithm is derived from a two-variable extension of the matrix-tree theorem [5], and therefore not polynomial for the two special cases to which the MRST problem reduces. We overcome this difficulty and solve these two special cases in polynomial time by borrowing Ravi and Goeman's elegant method for approximating the general CMST problem [17].

Organization of the paper. In section 2, we provide the mathematical model for the MRST problem, and make our key observation on a *nice property* of some optimal solutions to the MRST problem. In Section 3, we first solve in polynomial time two special cases of the CMST problem, which then leads us to an efficient algorithm for finding an optimal solution to the MRST problem that enjoys the nice property. In Section 4, we conclude this paper with remarks on future research.

2   Mathematical Model

We model the network as an undirected graph $G = (V, E)$ with vertex set $V = V(G)$ of size $n = |V|$ and edge set $E = E(G)$ of size $m = |E|$, where vertex (resp. edge) corresponds to network node (resp. link), and G has no loop[1]. Each edge $e \in E$ is associated with an interval $[\underline{c}_e, \overline{c}_e]$ indicating the lowest cost $\underline{c}_e \in \mathbb{R}_+$ and highest cost $\overline{c}_e \in \mathbb{R}_+$ of establishing e. We consider $\underline{c} = (\underline{c}_e : e \in E)$ and $\overline{c} = (\overline{c}_e : e \in E)$ as rational vectors in $\mathbb{R}_+^E$ with $\underline{c} \leq \overline{c}$. For ease of description, we make the notational convention that $\frac{0}{0} = 0$, and quantify the *risk* at e as $\frac{\overline{c}_e - c_e}{\overline{c}_e - \underline{c}_e}$ for any $c_e \in [\underline{c}_e, \overline{c}_e]$. Let $\mathscr{T}$ denote the set of all spanning trees T's in G, and let $\mathsf{C} \in \mathbb{R}_+$ (with $\mathsf{C} \geq \min_{T \in \mathscr{T}} \sum_{e \in E(T)} \underline{c}_e$) be the (cost) budget. The *minimum risk spanning tree* (MRST) problem can be formulated as the following mathematical programming:

$$
\begin{aligned}
\min \ & \tau(T, c) = \sum_{e \in E(T)} \frac{\overline{c}_e - c_e}{\overline{c}_e - \underline{c}_e} \\
\text{s.t. } & T \in \mathscr{T} \\
& c \in \mathbb{R}_+^{E(T)}, \text{ and } c_e \in [\underline{c}_e, \overline{c}_e] \text{ for all } e \in E(T) \\
& \sum_{e \in E(T)} c_e \leq \mathsf{C}
\end{aligned}
$$

An instance of the MRST problem on G with intervals $[\underline{c}_e, \overline{c}_e]$, $e \in E$, and budget C is denoted by $(G, \underline{c}, \overline{c}, \mathsf{C})$, and its (feasible) *solution* refers to a pair (T, c) satisfying the three constraints in the above programming. In this way, a solution (T, c) gives not only a spanning tree T, but also a cost allocation c to the edges of T under which the risk of T is $\tau(T, c)$.

The following lemma exhibits a nice property of some optimal solution to the MRST problem which plays a key role in establishing the polynomial-time solvability of the MRST problem. We present here a brief justification.

---

[1] A loop is an edge with its both ends identical.

Lemma 1. *There exists an optimal solution* (T^*, c^*) *to the MRST problem for which we have an edge* $f \in E(T^*)$ *such that* $c_f \in [\underline{c}_f, \bar{c}_f]$, *and* $c_e \in \{\underline{c}_e, \bar{c}_e\}$ *for all* $e \in E(T^*) - \{f\}$.

Proof. Let (T^*, c^*) be an optimal solution to the MRST problem on $(G, \underline{c}, \bar{c}, \mathsf{C})$ such that the set $S_{(T^*,c^*)} := \{g : g \in E(T^*) \text{ and } c^*_g \in (\underline{c}_g, \bar{c}_g)\}$ contains as few edges as possible. If there exist two different edges $e, f \in S_{(T^*,c^*)}$, and $\bar{c}_e - \underline{c}_e \leq \bar{c}_f - \underline{c}_f$, then take $\delta := \min\{\bar{c}_e - c^*_e, c^*_f - \underline{c}_f\}$ and define $c' \in \mathbb{R}^{E(T^*)}_+$ by $c'_e := c^*_e + \delta$, $c'_f := c^*_f - \delta$, and $c'_g := c^*_g$ for all $g \in E(T^*) - \{e, f\}$; it is easily shown that (T^*, c') is an optimal solution to the MRST problem such that $S_{(T^*,c')}$ is a proper subset of $S_{(T^*,c^*)}$, which contradicts the minimality of $S_{(T^*,c^*)}$. Therefore $|S_{(T^*,c^*)}| \leq 1$, implying that the optimal solution (T^*, c^*) enjoys the property stated in the lemma. □

For our easy reference, we call every optimal solution of the property in Lemma 1 as to a *simple optimal solution* to the MRST problem.

3   Efficient Algorithms

In this section, we first present two polynomial-time algorithms each solving a special case of the CMST problem; then we use these two algorithms as subroutines to design an exact $O(m^2 \log m \log n(m + n \log n))$-time algorithm for the MRST problem.

3.1   Polynomially Solvable Cases of the CMST Problem

In an undirected graph with real function π defined on its edge set, for a tree T in the graph, $\pi(T)$ represents the summation $\sum_e \pi(e)$ over edges e of T.

The constrained minimum spanning tree problem. Given an undirected graph $H = (\tilde{V}, \tilde{E})$ with nonnegative real functions, *weight (function)* w and *length (function)* l, both defined on its edge set $\tilde{E} = \tilde{E}(H)$, the *constrained minimum spanning tree* (CMST) problem consists of finding a spanning tree in H of minimum (total) weight under the constraint that its (total) length does not exceed a prespecified *bound* L, i.e., solving the the following combinatorial optimization problem (CMST):

$$\mathsf{W} = \min w(\tilde{T})$$
$$\text{s.t. } \tilde{T} \text{ is a spanning tree in } H$$
$$l(\tilde{T}) \leq \mathsf{L}$$

We use abbreviation "the (CMST) on (H, w, l, L)" to mean "an instance of the CMST problem on H with weight w, length l, and (length) bound L". To make the (CMST) on (H, w, l, L) nontrivial, it can be assumed that there exist spanning tree(s) $\tilde{T}, \tilde{T}'$ in H satisfying

$$l(\tilde{T}) \geq \mathsf{L} \geq l(\tilde{T}'). \tag{3.1}$$

The standard Lagrangian relaxation dualizes the length constraint in (CMST) by considering for any $z \geq 0$ the problem: $v(z) = \min\{\varsigma(\tilde{T}) : \tilde{T}$ is a spanning tree in $H\} - z\mathsf{L}$, where function ς on $\tilde{E}$ is given by $\varsigma(e) = w(e) + zl(e)$, $e \in \tilde{E}$. Observe that the plot of $v(z)$ as z varying is the lower envelope of the lines with interception $w(\tilde{T})$ and slope $l(\tilde{T}) - \mathsf{L}$ for all spanning trees $\tilde{T}$ in H. It follows from (3.1) that $\mathsf{LR} = \max_{z \geq 0} v(z)$ is finite, and upper bounded by W. This enables us to apply the following result of Ravi and Goemans [17] in our algorithm design.

Theorem 1. [17] *If* $\mathsf{LR} < \infty$, *then there is a* $O(|\tilde{E}| \log |\tilde{E}|(|\tilde{E}| + |\tilde{V}| \log |\tilde{V}|))$-*time serial algorithm (resp. a* $O(\log^2 |\tilde{E}|(|\tilde{E}| + |\tilde{V}| \log |\tilde{V}|))$-*time parallel algorithm) which outputs a spanning tree* $\tilde{T}$ *in* H *such that for some spanning tree* $\tilde{T}'$ *in* H *the following hold:*

(i) $l(\tilde{T}) \geq \mathsf{L} \geq l(\tilde{T}')$, $w(\tilde{T}) \leq \mathsf{W}$, *and* $w(\tilde{T}') \leq \mathsf{W}$ *if* $l(\tilde{T}') = \mathsf{L}$;
(ii) $\tilde{T} = \tilde{T}' \cup \{e\} \setminus \{e'\}$ *for some* $e, e' \in \tilde{E}$, *and* $e = e'$ *if and only if* $l(\tilde{T}') = \mathsf{L}$.

Let us call the algorithm in Theorem 1 the *RG algorithm*. We remark that if $|\tilde{E}| \leq |\tilde{V}|^p$ for some constant p, then the RG algorithm can be implemented in $O(|\tilde{E}| \log^2 |\tilde{V}| + |\tilde{V}| \log^3 |\tilde{V}|)$ as mentioned in Section 4 of [17].

Solving the CMST problem with $\{0, 1\}$-*length or* $\{0, 1\}$-*weight.* Despite the NP-hardness of the general CMST problem [1], the next two algorithms find optimal solutions for the special cases when edge lengths or edge weights are 0 or 1.

ALGORITHM CMST$\{0,1\}$L
Input : instance of the CMST problem on (H, w, l, L) in which $H = (\tilde{V}, \tilde{E})$ and
        length l is a $\{0, 1\}$-function on $\tilde{E}$.
Output : an optimal solution (a spanning tree $\tilde{T}^*$ in H) to the (CMST) on
        (H, w, l, L), or a declaration of infeasibility.

1. $\tilde{T}^* \leftarrow$ a spanning tree in H of minimum cost w.r.t. l
2. **if** $l(\tilde{T}^*) > \mathsf{L}$ **then** report "the (CMST) has no feasible solution", and stop
3. $\tilde{T}^* \leftarrow$ a spanning tree in H of minimum cost w.r.t. w
4. **if** $l(\tilde{T}^*) \leq \mathsf{L}$ **then** go to Step 6
5. $\tilde{T}^* \leftarrow \tilde{T}$ returned by the RG algorithm on $(H, w, l, \lfloor \mathsf{L} \rfloor)$
6. Output $\tilde{T}^*$

Corollary 1. ALGORITHM CMST$\{0,1\}$L *returns an optimal solution to the CMST problem with* $\{0, 1\}$-*length function in* $O(|\tilde{E}| \log |\tilde{E}|(|\tilde{E}| + |\tilde{V}| \log |\tilde{V}|))$ *time.*

Proof. If Step 5 was not executed, then the algorithm did find an optimal solution, else Steps 1 – 4 assume that (3.1) holds. Suppose that ALGORITHM CMST$\{0,1\}$L outputs a spanning tree T^*, which is returned in Step 5 by the RG algorithm.
   It is clear that for $\{0, 1\}$-length function l, the (CMST) on (H, w, l, L) is equivalent to the (CMST) on $(H, w, l, \lfloor \mathsf{L} \rfloor)$. Moreover (3.1) implies $\mathsf{LR} < \infty$. Let $\tilde{T}$ and $\tilde{T}'$ be as stated in Theorem 1. Then $\tilde{T}^* = \tilde{T}$ and $w(\tilde{T}^*) \leq \mathsf{W}$ by Theorem 1(i). To justify the optimality of $\tilde{T}^*$, it suffices to show $l(\tilde{T}) \leq \lfloor \mathsf{L} \rfloor$.

Notice from Theorem 1 that either $\lfloor L \rfloor = l(\tilde{T}') = l(\tilde{T})$ or $l(\tilde{T}) \geq \lfloor L \rfloor > l(\tilde{T}')$. In the former case, we are done. In the latter case, we deduce from Theorem 1(ii) that there exist $e, e' \in \tilde{E}$ with $l(e) = 1$ and $l(e') = 0$ such that $l(\tilde{T}) = l(\tilde{T}') + l(e) - l(e') = l(\tilde{T}') + 1 < \lfloor L \rfloor + 1$, which implies $l(\tilde{T}) \leq \lfloor L \rfloor$ as $l(\tilde{T})$ is an integer.

Using Fibonacci heap priority queues, the spanning trees in Step 1 and 3 can be found in $O(|\tilde{E}| + |\tilde{V}| \log |\tilde{V}|)$ time [8]. It is instant from Theorem 1 that the running time of ALGORITHM CMST$\{0,1\}$L is dominated by that of Step 5, and the result follows.   □

In the case of weight w being a $\{0, 1\}$-function, we have $0 \leq W \leq |\tilde{V}|$, which suggests the utilization of ALGORITHM CMST$\{0,1\}$L in the following binary search to guess W and the corresponding optimal spanning tree. In the pseudo-code below variables *"left"* and *"right"* hold the left and right endpoints of the current search interval, respectively.

ALGORITHM CMST$\{0,1\}$W

Input : instance of the CMST problem on (H, w, l, L) in which $H = (\tilde{V}, \tilde{E})$ and
        weight w is a $\{0, 1\}$-function on $\tilde{E}$.

Output : an optimal solution $\tilde{T}^*$ to the (CMST) on (H, w, l, L), or a declaration
        of infeasibility.

1. $\tilde{T}^* \leftarrow$ a spanning tree in H of minimum cost w.r.t. l
2. **if** $l(\tilde{T}^*) > L$ **then** report "the (CMST) has no feasible solution" and stop
3. $left \leftarrow 0$, $right \leftarrow |\tilde{V}|$
4. **repeat**
5. $middle \leftarrow \lceil (left + right)/2 \rceil$
6.   Apply ALGORITHM CMST$\{0,1\}$L to (CMST) on $(H, l, w, middle)$
7.   **if** ALGORITHM CMST$\{0,1\}$L returns tree $\tilde{T}$ and $l(\tilde{T}) \leq L$
8.   **then** $right \leftarrow middle$, $\tilde{T}^* \leftarrow \tilde{T}$
9.   **else** $left \leftarrow middle$
10. **until** $right - left \leq 1$
11. Apply ALGORITHM CMST$\{0,1\}$L to (CMST) on $(H, l, w, left)$
12. **if** ALGORITHM CMST$\{0,1\}$L returns tree $\tilde{T}$ and $l(\tilde{T}) \leq L$ **then** $\tilde{T}^* \leftarrow \tilde{T}$
13. Output $\tilde{T}^*$

Corollary 2. ALGORITHM CMST$\{0,1\}$W *returns an optimal solution for the CMST problem with $\{0, 1\}$-weight function in* $O(|\tilde{E}| \log |\tilde{E}| \log |\tilde{V}|(|\tilde{E}| + |\tilde{V}| \log |\tilde{V}|))$ *time.*

Proof. The correctness of ALGORITHM CMST$\{0,1\}$W follows from the correctness of ALGORITHM CMST$\{0,1\}$L (Corollary 1) and the fact that the condition in Step 7 is satisfied if and only if $W \leq middle$, which guarantees $right \geq W$ throughout the algorithm. The time complexity is implied by Corollary 1 since only $O(\log |\tilde{V}|)$ repetitions (Steps 4 – 10) are executed.   □

For brevity, we use "ALG_CMST$\{0,1\}$L" (resp. "ALG_CMST$\{0,1\}$W") as shorthand for "ALGORITHM CMST$\{0,1\}$L" (resp. "ALGORITHM CMST$\{0,1\}$W").

3.2   An Efficient Algorithm for the MRST Problem

As is customary, for vector/function $\pi \in \mathbb{R}_+^J$ and set $J' \subseteq J$, the restriction of π on J' is written as $\pi|_{J'}$.

$O(m)$-time construction. In view of Lemma 1, we make use of a construction in [7] as a step to establish the close connection between the MRST problem and the CMST problem. *Based on* the triple $(G, \underline{c}, \overline{c})$ with $G = (V, E)$, $\underline{c}, \overline{c} \in \mathbb{R}_+^E$ and $\underline{c} \leq \overline{c}$, we *construct* in $O(m)$ time an undirected graph $\overline{G} = (\overline{V}, \overline{E})$, $\{0, 1\}$-function $w \in \{0, 1\}^{\overline{E}}$, and real function $l \in \mathbb{R}_+^{\overline{E}}$ on $\overline{E}$ such that $\overline{V} := V$; $\overline{E} := \{\underline{e}, \overline{e} : e \in E\}$ with every $e \in E$ corresponding to *two distinct* $\underline{e}, \overline{e} \in \overline{E}$ both having ends the same as e; and $w(\underline{e}) := 1$, $w(\overline{e}) := 0$, $l(\underline{e}) := \underline{c}_e$, $l(\overline{e}) := \overline{c}_e$, for every $e \in E$. So

$$|\overline{V}| = n \text{ and } |\overline{E}| = 2m. \tag{3.2}$$

Next we present a 1-1 correspondence between the spanning trees in $\overline{G}$ and the pairs (T, c) in which T is a spanning tree in G, $c \in \mathbb{R}_+^{E(T)}$, and

$$c_e \in \{\underline{c}_e, \overline{c}_e\} \text{ for all } e \in E(T). \tag{3.3}$$

Given a spanning tree $\overline{T}$ in $\overline{G}$, we use $\text{PAIR}(\overline{T})$ to denote the unique pair (T, c) in which spanning tree T in G and $c \in \mathbb{R}_+^{E(T)}$ satisfy (i) for any $e \in E$, $e \in E(T)$ and $c_e = \underline{c}_e$ if and only if $\underline{e} \in \overline{E}(\overline{T})$, and (ii) for any $e \in E$, $e \in E(T)$ and $c_e = \overline{c}_e$ if and only if $\overline{e} \in \overline{E}(\overline{T})$. (Clearly, $\overline{T}$ and (i), (ii) imply (3.3).) Conversely, given a spanning tree T in G, and $c \in \mathbb{R}_+^{E(T)}$ which satisfy (3.3), we use $\text{TREE}(T, c)$ to denote the unique spanning tree $\overline{T}$ in $\overline{G}$ such that (i) and (ii) hold. Thus $\overline{T} = \text{TREE}(T, c)$ if and only if $(T, c) = \text{PAIR}(\overline{T})$. Moreover the construction from $\overline{T}$ to $\text{PAIR}(\overline{T})$, and from (T, c) to $\text{TREE}(T, c)$ can be completed in $O(m)$ time. The following immediate corollary of the 1-1 correspondence serves as a technical preparation for proving the correctness of our algorithm for the MRST problem.

Lemma 2. *If $\overline{T} = \text{TREE}(T, c)$, or equivalently $(T, c) = \text{PAIR}(\overline{T})$, then $w(\overline{T}) = \sum_{e \in E(T)} \frac{\overline{c}_e - c_e}{\overline{c}_e - \underline{c}_e}$ and $l(\overline{T}) = \sum_{e \in E(T)} c_e$.*

$O(m^2 \log m \log n(m + n \log n))$-time exact algorithm. Recalling Lemma 1, we now develop an efficient algorithm for the MRST problem that finds a *simple optimal solution*. The basic idea behind our algorithm is testing all m edges to find an edge f such that f and some simple optimal solution (T^*, c^*) satisfy Lemma 1. The testing is realized via applications of $\text{ALG\_CMST}\{0,1\}\text{L}$ and $\text{ALG\_CMST}\{0,1\}\text{W}$ to instances involving constructions described above.

In the following algorithm, the set $\mathcal{S}$ holds feasible solutions to the MRST problem found by the algorithm. By *contracting* an edge in a graph, we mean the operation of identifying the ends of the edge and deleting the resulting loop.

ALGORITHM MRST

Input : instance of the MRST problem on $(G, \underline{c}, \overline{c}, \mathsf{C})$ with $G = (V, E)$; $\underline{c}, \overline{c} \in \mathbb{R}_+^E$, $\underline{c} \leq \overline{c}$; $\min_{T \in \mathcal{T}} \sum_{e \in E(T)} \underline{c}_e \leq \mathsf{C} \in \mathbb{R}_+$.

Output : an optimal solution (T, c) to the MRST problem on $(G, \underline{c}, \overline{c}, \mathsf{C})$.

1. $\mathcal{S} \leftarrow \emptyset$
2. Construct $\overline{G} = (\overline{V}, \overline{E})$, $w \in \{0,1\}^{\overline{E}}$, $l \in \mathbb{R}_+^{\overline{E}}$ based on $(G, \underline{c}, \overline{c})$
3. **for** every $f \in E$ **do**
4. $\mathcal{T} \leftarrow \emptyset$
5.   Obtain graph $H = (\tilde{V}, \tilde{E})$ from $\overline{G}$ by contracting $\underline{f}, \overline{f}$ [2]
6.   Apply ALG_CMST$\{0,1\}$W to (CMST) on $(H, w|_{\tilde{E}}, l|_{\tilde{E}}, \mathsf{C} - \underline{c}_f)$
7.   **if** ALG_CMST$\{0,1\}$W returns tree $\tilde{T}_1$
8.   **then** $\mathcal{T} \leftarrow \{\tilde{T}_1\}$
9.   Apply ALG_CMST$\{0,1\}$L to (CMST) on $(H, l|_{\tilde{E}}, w|_{\tilde{E}}, w(\tilde{T}_1))$
10.   **if** ALG_CMST$\{0,1\}$L returns tree $\tilde{T}_2$ and $l(\tilde{T}_2) \leq \mathsf{C} - \underline{c}_f$
11.   **then** $\mathcal{T} \leftarrow \mathcal{T} \cup \{\tilde{T}_2\}$
12.   **if** $\mathcal{T} \neq \emptyset$ **then for** every $\tilde{T} \in \mathcal{T}$ **do**
13. $(T', c') \leftarrow$ PAIR$(\tilde{T})$ [3], $E' \leftarrow$ the edge set of T'
14. $T \leftarrow$ a spanning tree in G with edge set $E' \cup \{f\}$
15. $c|_{E(T)\setminus\{f\}} \leftarrow c'$, $c_f \leftarrow \min\{\overline{c}_f, \mathsf{C} - l(\tilde{T})\}$
16. $\mathcal{S} \leftarrow \mathcal{S} \cup \{(T, c)\}$
17.   **end-for**
18. **end-for**
19. Take $(T, c) \in \mathcal{S}$ with minimum $\tau(T, c)$
20. Output (T, c)

Theorem 2. ALGORITHM MRST *solves the MRST problem and obtain the optimal solution in time* $O(m^2 \log m \log n(m + n \log n))$.

Proof. Recalling (3.2), it is straightforward from Corollary 1 and Corollary 2 that ALGORITHM MRST runs in time $O(m^2 \log m \log n(m+n \log n))$ as claimed.

It can be seen from Steps 6 – 8 and Step 10 that $l(\tilde{T}) \leq \mathsf{C} - \underline{c}_f$ for any $\tilde{T} \in \mathcal{T}$. Therefore in Step 15 we have $\underline{c}_f \leq c_f \leq \overline{c}_f$. Furthermore Lemma 2 assures that any (T, x) produced and putted into $\mathcal{S}$ by Steps 12 – 17 is a feasible solution to the MRST problem on $(G, \underline{c}, \overline{c}, \mathsf{C})$.

Let (T^*, c^*) be a simple optimal solution to the MRST problem whose existence is guaranteed by Lemma 1. By Step 19, it suffices to show that the final $\mathcal{S}$ contains an element (T, c) with $\tau(T, c) \leq \tau(T^*, c^*)$. To this end, by Lemma 1, hereafter symbol f denotes the edge $f \in E(T^*)$ such that

(1) $c_e^* \in \{\underline{c}_e, \overline{c}_e\}$ for all $e \in E(T^*) \setminus \{f\}$.

Let T_* (resp. $G_* = (V_*, E_*)$) be the tree (resp. graph) obtained from T^* (resp. G) by contracting f. Then T_* is a spanning tree in G_*, $E_* = E \setminus \{f\}$, and a construction based on the triple $(G_*, \underline{c}|_{E\setminus\{f\}}, \overline{c}|_{E\setminus\{f\}})$ would yield graph $\overline{G}_* = H = (\tilde{V}, \tilde{E})$, $\{0,1\}$-function $w|_{\tilde{E}} \in \{0,1\}^{\tilde{E}}$ and real function $l|_{\tilde{E}} \in \mathbb{R}_+^{\tilde{E}}$ on $\tilde{E} = \overline{E} \setminus \{\underline{f}, \overline{f}\}$. Let $c_* := c^*|_{E(T^*)\setminus\{f\}}$. Then (1) implies that with (T_*, c^*) in place of (T, c), (3.3) is satisfied. Hence we deduce from Lemma 2 that $\overline{T}_* :=$ TREE(T_*, c_*) is a spanning tree in H such that

---

[2] Notice that $\tilde{E} = \overline{E} \setminus \{\underline{f}, \overline{f}\}$.
[3] T' is a spanning tree in the graph obtained from G by contracting f.

(2) $w(\overline{T}_*) = \sum_{e \in E(T^*) \setminus \{f\}} \frac{\overline{c}_e - c_e^*}{\overline{c}_e - \underline{c}_e} = \tau(T^*, c^*) - \frac{\overline{c}_f - c_f^*}{\overline{c}_f - \underline{c}_f} \leq \tau(T^*, c^*)$ and

(3) $l(\overline{T}_*) = \sum_{e \in E(T^*) \setminus \{f\}} c_e^* \leq \mathsf{C} - c_f^* \leq \mathsf{C} - \underline{c}_f$.

It follows (3) that, in Step 6, a spanning tree $\tilde{T}_1$ in H with $l(\tilde{T}_1) \leq \mathsf{C} - \underline{c}_f$ and $w(\tilde{T}_1) \leq w(\overline{T}_*)$ is returned by $\mathrm{ALG\_CMST\{0,1\}W}$, and in Steps 7 – 8, this $\tilde{T}_1$ is putted into $\mathcal{T}$. If $w(\tilde{T}_1) < w(\overline{T}_*)$, then $w(\tilde{T}_1) \leq w(\overline{T}_*) - 1$ as w is a $\{0,1\}$-function, and later in Steps 12 – 17 when taking $\tilde{T} = \tilde{T}_1$, the pair $(T', c') = \mathrm{PAIR}(\tilde{T}_1)$ consists of a spanning tree T' in G_* and $c' \in \mathbb{R}_+^{E_*(T)} = \mathbb{R}_+^{E'}$, and a solution (T, c) of the MRST problem on $(G, \underline{c}, \overline{c}, \mathsf{C})$ is putted into $\mathcal{S}$ such that $\tau(T, c) = \sum_{e \in E(T)} \frac{\overline{c}_e - c_e}{\overline{c}_e - \underline{c}_e} = \left(\sum_{e \in E_*(T')} \frac{\overline{c}_e - c_e'}{\overline{c}_e - \underline{c}_e} \right) + \frac{\overline{c}_f - c_f}{\overline{c}_f - \underline{c}_f} = w(\tilde{T}_1) + \frac{\overline{c}_f - c_f}{\overline{c}_f - \underline{c}_f}$

(by Lemma 2); and therefore $\tau(T, c) \leq w(\tilde{T}_1) + 1 \leq w(\overline{T}_*) \overset{\mathrm{by}(2)}{\leq} \tau(T^*, c^*)$ as desired. Hence we may assume that

(4) $w(\tilde{T}_1) = w(\overline{T}_*)$.

So the spanning tree $\overline{T}_*$ in H is a feasible solution to the CMST problme on $(H, l|_{\tilde{E}}, w|_{\tilde{E}}, w(\tilde{T}_1))$, and in Step 9, $\mathrm{ALG\_CMST\{0,1\}L}$ returns a spanning tree $\tilde{T}_2$ in H with

(5) $w(\tilde{T}_2) \leq w(\tilde{T}_1) \overset{\mathrm{by}(4)}{=} w(\overline{T}_*) \overset{\mathrm{by}(2)}{=} \tau(T^*, c^*) - \frac{\overline{c}_f - c_f^*}{\overline{c}_f - \underline{c}_f}$, and

(6) $l(\tilde{T}_2) \leq l(\overline{T}_*) \overset{\mathrm{by}(3)}{\leq} \mathsf{C} - \underline{c}_f$.

From (6) we see that in Steps 10 – 11, tree $\tilde{T}_2$ is put into $\mathcal{T}$. Subsequently, in Steps 12 – 17 when taking $\tilde{T} = \tilde{T}_2$, a solution (T, c) of the MRST problem on $(G, \underline{c}, \overline{c}, \mathsf{C})$ is put into $\mathcal{S}$ such that $c_f = \min\{\overline{c}_f, \mathsf{C} - l(\tilde{T}_2)\} \overset{\mathrm{by}(6)}{\geq} \min\{\overline{c}_f, \mathsf{C} - l(\overline{T}_*)\} \overset{\mathrm{by}(3)}{\geq} c_f^*$ and $\tau(T, c) = w(\tilde{T}_2) + \frac{\overline{c}_f - c_f}{\overline{c}_f - \underline{c}_f} \overset{\mathrm{by}(5)}{\leq} \tau(T^*, c^*) - \frac{\overline{c}_f - c_f^*}{\overline{c}_f - \underline{c}_f} + \frac{\overline{c}_f - c_f}{\overline{c}_f - \underline{c}_f} \leq \tau(T^*, c^*)$, completing the proof. $\square$

4   Concluding Remark

In this paper we have studied the minimum risk spanning tree (MRST) problem, which finds practical applications in network design with interval data. It comes as a surprise that our efficient algorithm proves the tractability of the MRST problem, in comparison with the fact that most previously studied spanning tree problems under interval uncertainty are NP-hard. As a byproduct, we show that the constrained spanning tree problem with $\{0, 1\}$ weight or length is polynomial-time solvable.

Similar phenomenon appears in the shortest path planning with interval data. In contrast to the intractability in robustness optimization and stochastic applications, the so called *minimum risk-sum path problem* [7] admits polynomial-time algorithms which use as subroutines the dynamic programming procedures for the the *constrained shortest path problem* [10].

As future research, it would be interesting to know if similar results could be obtained using our methodology for other combinatorial optimization problems whose classic versions are polynomial-time solvable (e.g., network flow problems).

References

1. Aggarwal, V., Aneja, Y.P., Nair, K.P.: Minimal spanning tree subject to a side constraint. Computers and Operations Research 9, 287–296 (1982)
2. Aron, I.D., Van Hentenryck, P.: On the complexity of the robust spanning tree problem with interval data. Operations Research Letters 32, 36–40 (2004)
3. Averbakh, I., Lebedev, V.: Interval data minmax regret network optimization problems. Discrete Applied Mathematics 138, 289–301 (2004)
4. Averbakh, I.: Computing and minimizing the relative regret in combinatorial optimization with interval data. Discrete Optimization 2, 273–287 (2005)
5. Barahona, F., Pulleyblank, W.R.: Exact arborescences, matchings and cycles. Discrete Applied Mathematics 16, 91–99 (1987)
6. Chanas, S., Zieliński, P.: On the hardness of evaluating criticality of activities in a planar network with duration intervals. Operations Research Letters 31, 53–59 (2003)
7. Chen, X., Hu, J., Hu, X.: On the minimum risk-sum path problem. In: ESCAPE 2007. Proceedings of The International Symposium on Combinatorics, Algorithms, Probabilistic and Experimental methodologies (to appear, 2007)
8. Cormen, T.H., Leiserson, C.E., Rivest, R.L.: Introduction to Algorithms. McGraw Hill, New York (1990)
9. Demchenko, A.I.: Design of transportation networks under uncertainty. In: Proceedings of the Seminar on Interval Mathematics, Saratov, pp. 10–16 (1990)
10. Hassin, R.: Approximation schemes for the restricted shortest path problem. Mathematics of Operations Research 17, 36–42 (1992)
11. Hong, S-. P, Chung, S-.J., Park, B.H.: A fully polynomial bicriteria approximation scheme for the constrained spanning tree problem. Operation Research Letters 32, 233–239 (2004)
12. Kasperski, A., Zieliński, P.: An approximation algorithm for interval data minmax regret combinatorial optimization problems. Information Processing Letters 97, 177–180 (2006)
13. Kouvelis, P., Yu, G.: Robust Discrete Optimization and its Applications. Kluwer Academic Publishers, Boston (1997)
14. Marathe, M.V., Ravi, R., Sundaram, R., Ravi, S.S., Rosenkrantz, D.J., Hunt III, H.B.: Bicriteria network design problems. Journal of Algorithms 28, 142–171 (1998)
15. Montemanni, R.: A Benders decomposition approach for the robust spanning tree problem with interval data. European Journal of Operational Research 174, 1479–1490 (2006)
16. Montemanni, R., Gambardella, L.M.: A branch and bound algorithm for the robust spanning tree problem with interval data. European Journal of Operational Research 161, 771–779 (2005)
17. Ravi, R., Goemans, M.X.: The constrained spanning tree problem. In: Karlsson, R., Lingas, A. (eds.) SWAT 1996. LNCS, vol. 1097, pp. 66–75. Springer, Heidelberg (1996)
18. Salazar-Neumann, M.: The robust minimum spanning tree problem: compact and convex uncertainty. Operations Research Letters 35, 17–22 (2007)
19. Yaman, H., Karasan, O.E., Pinar, M.C.: The robust spanning tree problem with interval data. Operations Research Letters 29, 31–40 (2001)

The Size of a Minimum Critically
m-Neighbor-Scattered Graph

Fengwei Li and Qingfang Ye

Department of Mathematics, Shaoxing University, Zhejiang,
Shaoxing 312000, P.R. China
`fengwei.li@eyou.com`

Abstract. It seems reasonable that for a connected representing graph
of a spy network, the more edges it has, the more jeopardy the spy
network is in. So, a spy network which has the minimum number of edges
is the comparatively reliable network we want. As a special kind of graph,
a critically m-neighbor-scattered graph is important and interesting in
applications in communication networks. In this paper, we obtain some
upper bounds and a lower bound for the size of a minimum critically
m-neighbor-scattered graph with given order p and $4 - p \leq m \leq -1$.
Moreover, we construct a $(1 + \epsilon)$-approximate graph for the minimum
critically m-neighbor-scattered graph of order p for sufficiently small m
and sufficiently large p.

1   Introduction

Throughout this paper, a graph $G = (V, E)$ always means a finite simple connected graph with vertex set V and edge set E. We shall use $\lfloor x \rfloor$ to denote the
largest integer not larger than x, and $\lceil x \rceil$ the smallest integer not smaller than
x. $d_G(v)$ denotes the degree of a vertex v of G and δ_G denotes the minimum
degree of G. $\emptyset$ denotes the *nullset*. We use Bondy and Murty [1] for terminology
and notations not defined here.

The scattering number of a graph G was introduced by Jung [4] as an alternative measure of the vulnerability of G to disruption caused by the removal of
some vertices. The concept of scattering number is defined as follows:

Definition 1.1. (Jung [4]). The (vertex) scattering number of a graph G is
defined as $s(G) = max\{\omega(G - X) - |X| : \omega(G - X) > 1\}$, where $\omega(G - X)$ is
the number of connected components in the graph $G - X$.

Definition 1.2 A vertex cut-set X is called an s-set of G if it satisfies that
$s(G) = \omega(G - X) - |X|$.

Definition 1.3 A graph G is said to be an *m-scattered* graph if $s(G) = m$.

In [3] Gunther and Hartnell introduced the idea of modelling a spy network
by a graph whose vertices represent the agents and whose edges represent lines
of communication. Clearly, if a spy is discovered, the espionage agency can no

A. Dress, Y. Xu, and B. Zhu (Eds.): COCOA 2007, LNCS 4616, pp. 91–101, 2007.

longer trust any of the spies with whom he or she was in direct communication, and so the betrayed agents become effectively useless to the network as a whole. Such betrayals are clearly equivalent, in the modelling graph, to the removal of the closed neighborhood of v, where v is the vertex representing the particular agent who has been subverted.

Therefore instead of considering the scattering number of a communication network, we discuss the neighbor-scattering number for the disruption of graphs caused by the removal of some vertices and their adjacent vertices.

Let u be a vertex in G. The *open neighborhood* of u is defined as $N(u) = \{v \in V(G)|(u,v) \in E(G)\}$; whereas the *closed neighborhood* of u is defined as $N[u] = \{v\} \cup N(u)$. We define analogously the open neighborhood $N(S) = \cup_{u \in S} N(u)$ for any $S \subseteq V(G)$ and the closed neighborhood $N[S] = \cup_{u \in S} N[u]$. A vertex $u \in V(G)$ is said to be *subverted* when the closed neighborhood $N[u]$ is deleted from G. A *vertex subversion strategy* of G, X, is a set of vertices whose closed neighborhood is deleted from G. The survival subgraph, G/X, is defined to be the subgraph left after the subversion strategy X is applied to G, i.e., $G/X = G - N[X]$. X is called a *cut-strategy* of G if the survival subgraph G/X is disconnected, or is a clique, or is a $\emptyset$.

Definition 1.4 (Gunther and Hartnell [2]). The *(vertex) neighbor-connectivity* of a graph G is defined as $K(G) = min\{|X| : X$ is a cut-strategy of $G\}$, where the minimum is taken over all the cut-strategies X of G.

Definition 1.5 (Wei and Li [6]). The *(vertex) neighbor-scattering number* of a graph G is defined as $S(G) = max\{\omega(G/X) - |X| : X$ is a cut-strategy of $G\}$, where $\omega(G/X)$ is the number of connected components in the graph G/X.

Definition 1.6 A cut-strategy X of G is called an S-set of G if it satisfies that $S(G) = \omega(G/X) - |X|$.

Definition 1.7 If the neighbor-scattering number of a graph G, $S(G) = m$, then G is called an *m-neighbor-scattered* graph. A graph G is said to be *critically m-neighbor-scattered* if $S(G) = m$, and for any vertex v in G, $S(G/\{v\}) > S(G)$.

For a connected representing graph of a spy network, the more edges it has, the more jeopardy the spy network is in. So, a spy network which has the minimum number of edges is comparatively reliable network we want, and hence we are interested in constructing the minimum critically m-neighbor-scattered graphs.

Definition 1.8 A graph is called *minimum critically m-neighbor-scattered* if no critically m-neighbor-scattered graph with the same number of vertices has fewer edges than G.

From the definition we know that, in general, the less the neighbor-scattering number of a graph is, the more stable the graph is. So, in this paper we always assume that $S(G) = m$ and $4 - |V(G)| \leq m \leq -1$. In Section 2, we give a class of critically m-neighbor-scattered graphs. In Section 3, some upper and lower bounds for the size of a minimum critically m-neighbor-scattered graph are given, and moreover, for sufficiently small m and sufficiently large p, a $(1 + \epsilon)$-approximate graph for the minimum critically m-neighbor-scattered graph of order p is constructed.

2 A Class of Critically m-Neighbor-Scattered Graphs

In 1990, Wu and Cozzens [7] introduced an operation, E, to construct a class of k-neighbor-connected, $k \geq 1$, graphs from a given k-connected graph. The operation E is defined as follows:

E is an operation on a graph G to create a collection of graphs, say G^E.

A new graph $G^e \in G^E$ is created as follows:

(1) Each vertex v of G is replaced by a clique C_v of order $\geq deg(v) + 1$.

(2) C_{v_1} and C_{v_2} are joined by, at most, one edge and they are joined by an edge if, and only if, vertices v_1 and v_2 are adjacent in G.

(3) Each vertex in C_v is incident with, at most, one edge not entirely contained in C_v.

Using the same method, we can construct a class of s-neighbor-scattered, $(4 - |V(G)|) \leq s \leq (|V(G)| - 2)$, graphs from a given s-scattered graph G.

Theorem 2.1. *Let G be a connected noncomplete s-scattered graph. Apply operation E to G to obtain a graph G^e, then G^e is a connected noncomplete s-neighbor-scattered graph.*

Proof. For $s(G) = s$, let X be an s-set of the graph G, i.e., $\omega(G - X) > 1$, $s = s(G) = \omega(G - X) - |X|$. We use X' in G^e to denote the vertex set corresponding to X in G. It is obvious that deleting X from G is equivalent to deleting the neighborhoods of the corresponding vertices of X' in G^e and $\omega(G - X) = \omega(G^e/X') \geq 2$. Hence, by the definition of neighbor-scattering number, we have $S(G^e) \geq \omega(G^e/X') - |X'| = \omega(G - X) - |X| = s$.

We can prove that $S(G^e) \leq s$. Otherwise, if $S(G^e) > s$, by the construction of G^e, there must exist an S-set X' of G^e such that $\omega(G^e/X') \geq 2$. Correspondingly, there must exist a vertex cut-set X of G with $|X| = |X'|$ and $\omega(G - X) = \omega(G^e/X')$. So we have $s(G) = max\{\omega(G - X) - |X| : \omega(G - X) > s\} \geq \omega(G - X) - |X| = \omega(G^e/X') - |X'| > s$, a contradiction to the fact that $s(G) = s$. Hence we know that $S(G^e) = s$ when $s(G) = s$.   ∎

Example 1. In Figure 1, we give a graph G and a new graph G^e under the operation E.

As we known, in 1962 Harary investigated a problem on reliable communication networks. For any fixed integers n and p such that $p \geq n + 1$, Harary constructed a class of graphs $H_{n,p}$ which are n-connected with the minimum number of edges on p vertices. Thus, Harary graphs are examples of graphs which in some sense have the maximum possible connectivity and hence are of interests as possibly having good stability properties.

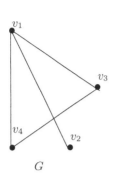 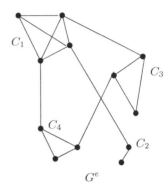

Fig. 1.

In order to give our main result, we introduce a new concept, *generalized Harary graphs*, denoted by $G_{n,p}$ and defined as follows:

Case 1. If n is even, let $n = 2r$, then $G_{n,p}$ has vertices $0, 1, 2, \cdots, p-1$, and two vertices i and j are adjacent if and only if $|i - j| \leq r$, where the addition is taken modulo p.

Case 2. If n is odd $(n > 1, n \neq 3)$ and p is even, let $n = 2r + 1$ $(r > 0)$, then $G_{n,p}$ is constructed by first drawing $G_{2r,p}$, and then adding edges joining vertex i to vertex $i + \frac{p}{2}$ for $1 \leq i \leq \frac{p}{2}$.

Case 3. If $n = 3$ and p is even, then $G_{n,p}$ is constructed by first drawing a cycle C_p with vertices $0, 1, 2, \cdots, p-1$, and two vertices 0 and $\frac{p}{2}$ are adjacent for $1 \leq i \leq \frac{p}{2}$, then adding edges joining the vertex i to $p - i$ for $1 \leq i \leq \frac{p}{2}$.

Case 4. If n is odd $(n > 1)$ and $p(p \neq 7)$ is odd, let $n = 2r + 1(r > 0)$, then $G_{2r+1,p}$ is constructed by first drawing $G_{2r,p}$, and then adding edges joining the vertex i to $i + \frac{p+1}{2}$ for $0 \leq i \leq \frac{p-1}{2}$.

Case 5. If $n = 3$ and $p = 7$, then $G_{n,p}$ is constructed by first drawing a cycle C_p with vertices $0, 1, 2, \cdots, p-1$, and then adding edges joining the vertex 0 to $\frac{p+1}{2}$ and $\frac{p-1}{2}$, and then adding edges joining the vertex i to $p - i$ for $1 \leq i \leq \frac{p-1}{2}$.

Example 2. We give some examples of generalized Harary graphs in Figure 2.

It is easy to see that the graph $G_{3,p}$ is not isomorphic to $H_{3,p}$, but it has the same order, size and connectivity as those of $H_{3,p}$, where $p = 7$ or even.

Lemma 2.2.([5]) *Let $H_{n,p}$ be a noncomplete Harary graph, then $s(H_{n,p}) = 2 - n$ except for that $s(H_{3,7}) = s(H_{3,n}) = 0$, if n is even.*

Corollary 2.3 *Let $G_{n,p}$ be a noncomplete generalized Harary graph, then $s(G_{n,p}) = 2 - n$.*

Proof. It is easy to prove that $s(G_{3,7}) = s(G_{3,p}) = 2 - 3 = -1$. ∎

Lemma 2.4. *Let $G_{n,p}$ be a noncomplete generalized Harary graph, then $s(G_{n,p} - \{v\}) > s(G_{n,p})$ for any vertex v in $G_{n,p}$.*

Proof. It is easily checked that Lemma 2.4 to be true.   ∎

Theorem 2.5 *For any positive integer n and negative integer m, such that $4 - n \leq m \leq -1$, there exists a class of critically m-neighbor-scattered graphs each of which has n cliques.*

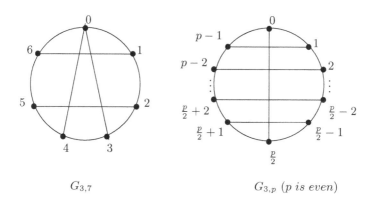

$G_{3,7}$ $G_{3,p}$ (p is even)

Fig. 2.

Proof. For any positive integer n and negative integer m, such that $4 - n \leq m \leq -1$, we can construct a generalized Harary graph $G_{2-m,n}$. From Corollary 2.3 we know that it is an m-scattered graph. From Theorem 2.1 and Lemma 2.4, we know that applying the operation of E to $G_{2-m,n}$ and $G_{2-m,n} - \{v\}$ for any vertex v in $G_{2-m,n}$, we obtain a class of critically m-neighbor-scattered graphs $G_{2-m,n}^{E}$ each of which has n cliques.   ∎

Example 3. We give a critically (-3)-neighbor-scattered graph $G_{5,6}^{e}$ with 6 cliques in Figure 3.

3   The Size of a Minimum Critically m-Neighbor-Scattered Graph

From the above theorem, we know that for any positive integers p, n and negative integer m, such that $4-n \leq m \leq -1$, we can construct a class of graphs $G_{2-m,n}^{E}$, each of which is a critically m-neighbor-scattered graph of order p. For brevity, we use $G(2 - m, n)$ to denote this class of graphs.

Let the vertices in $G_{2-m,n}$ be $v_0, v_1, v_2, \cdots, v_{n-1}$, and the corresponding cliques in each of $G(2 - m, n)$ be $C_0, C_1, C_2, \cdots, C_{n-1}$. Set the number of vertices of cliques $C_0, C_1, C_2, \cdots, C_{n-1}$ be $x_0, x_1, x_2, \cdots, x_{n-1}$, respectively, where $x_i \geq deg(v_i) + 1 = 3 - m$, for all $i = 1, 2, 3, \cdots, n - 1$, $x_0 \geq deg(v_0) + 1 = 3 - m$, if and only if at least one of m and n is even, and $x_0 \geq deg(v_0) = 4 - m$ if both

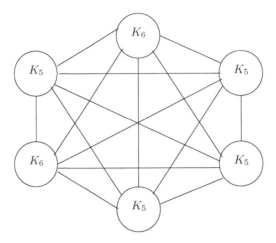

Fig. 3.

of m and n are odd. So, we get that $\sum_{i=0}^{n-1} x_i = p$. Thus, the number of edges in each of $G(2 - m, n)$ is

$$|E(G(2-m,n))| = \begin{cases} \frac{1}{2}(\sum_{i=0}^{n-1} x_i(x_i - 1) + (3 - m)n), \\ if\ at\ least\ one\ of\ m\ and\ n\ is\ even \\ \frac{1}{2}(\sum_{i=0}^{n-1} x_i(x_i - 1) + (3 - m)n + 1), \\ if\ both\ of\ m\ and\ n\ are\ odd \end{cases}$$

In order to discuss the size of a minimum critically m-neighbor-scattered graph, we minimize $|E(G(2 - m, n))|$, the size of $G(2 - m, n)$, under the condition $\sum_{i=0}^{n-1} x_i = p$, and we let $\widetilde{G}(2 - m, n)$ be a subclass of $G(2 - m, n)$ having the smallest number of edges, which is denoted as $\tilde{g}(2 - m, n)$. Set $x = (x_0, x_1, x_2, \cdots, x_{n-1})$, and let $f(x)$ denote the number of edges of $G(2 - m, n)$. It is easy to see that $f(x) \leq \tilde{g}(2 - m, n)$.

Since m and $2 - m$ have the same parity, we distinguish two cases:

Case 1. At least one of m and n is even ($n \geq 3 - m$). We have the following nonlinear integer programming:

$$min_{x_i} f(x) = \sum_{i=0}^{n-1} x_i^2 + (3 - m)n - p$$

$$s.t \begin{cases} h(x) = \sum_{i=0}^{n-1} x_i = p, \\ g_i(x) = x_i \geq 3 - m, \quad for\ all\ i = 0, 1, 2, \cdots, n - 1 \\ x_i \in Z^+, \qquad\qquad\quad for\ all\ i = 0, 1, 2, \cdots, n - 1 \end{cases} \quad (2.1)$$

where Z^+ denotes the set of positive integers.

Since $x_i \geq 3 - m$, $p = \sum_{i=0}^{n-1} x_i \geq \sum_{i=0}^{n-1}(3 - m) = (3 - m)n$, we have $n \leq \frac{p}{3-m}$. Since n is an integer, we have $n \leq \lfloor \frac{p}{3-m} \rfloor$

Case 2. Both m and n are odd, $(n \geq 3 - m)$. We have the following nonlinear integer programming:
$$min_{x_i} f(x) = \sum_{i=0}^{n-1} x_i^2 + (3-m)n - p + 1$$

$$s.t \begin{cases} h(x) = \sum_{i=0}^{n-1} x_i = p, \\ g_0(x) = x_0 \geq 4 - m, \\ g_i(x) = x_i \geq 3 - m, \quad for \ all \ i = 1, 2, \cdots, n-1 \\ x_i \in Z^+, \quad\quad\quad\quad for \ all \ i = 0, 1, \cdots, n-1 \end{cases} \quad (2.2)$$

where Z^+ denotes the set of positive integers.

Since $x_0 \geq 4 - m$, $x_i \geq 3 - m(i = 1, 2, \cdots, n-1)$, $p = \sum_{i=0}^{n-1} x_i \geq 4 - m + \sum_{i=1}^{n-1}(3-m) = (3-m)n + 1$, we have $n \leq \frac{p-1}{3-m}$. Since n is an integer, we have $n \leq \lfloor \frac{p-1}{3-m} \rfloor \leq \lfloor \frac{p}{3-m} \rfloor$.

Now let us solve the above two nonlinear integer programming. We first pay our attention to problem (2.1)

We use the well-known *Lagrangian Method* to solve this nonlinear integer programming and we know that $x_i = \lfloor \frac{p}{n} \rfloor$ or $\lfloor \frac{p}{n} \rfloor + 1$ is the optimal solution to problem (2.1).

For problem (2.2), we can use similar method to solve it and obtain the same solution as that of problem (2.1).

For convenience, we denote $\lfloor \frac{p}{n} \rfloor$ by Q, rearrange the sequence of x_i, such that

$$x_i = \begin{cases} Q + 1, \ if \ 0 \leq i \leq R - 1 \\ Q, \quad\quad if \ R \leq i \leq n - 1 \end{cases}$$

where $R = p - nQ$. Therefore,

$$\widetilde{g}(2-m,n) = \begin{cases} \frac{1}{2}((Q+1)^2 R + Q^2(n-R) - p + (3-m)n), \\ if \ at \ least \ one \ of \ m \ and \ n \ is \ even \\ \frac{1}{2}((Q+1)^2 R + Q^2(n-R) - p + (3-m)n + 1), \\ if \ both \ of \ m \ and \ n \ are \ odd \end{cases}$$

$$= \begin{cases} \frac{1}{2}(2QR + R + Q^2 n - p + (3-m)n), \\ if \ at \ least \ one \ of \ m \ and \ n \ is \ even \\ \frac{1}{2}(2QR + R + Q^2 n - p + (3-m)n + 1), \\ if \ both \ of \ m \ and \ n \ are \ odd \end{cases}$$

Next we find a lower bound of the size of a minimum critically m-neighbor-scattered graph with order p.

Lemma 3.1. ([8]) *For any graph G, $S(G) \geq 1 - K(G)$.*
where $K(G)$ denotes the neighbor-connectivity of G.

Lemma 3.2. ([2]) *If G is a K-neighbor-connected graph, then for any vertex v of G, $deg(v) \geq K(G)$.*

From the above two Lemmas, we get that

Theorem 3.3. *For any connected S-neighbor-scattered graph G, $\delta(G) \geq 1 - S(G)$.*

Theorem 3.4. *Let m be a negative integer. If G is a minimum critically m-neighbor-scattered graph of order p, then $|E(G)| \geq \lceil \frac{1}{2}(1-m)p \rceil$.*

Proof. By Theorem 3.3 we know that $\delta(G) \geq 1 - S(G)$. Since G is a critically m-neighbor-scattered graph of order p, we have $S(G) = m$ and $\delta(G) \geq 1 - m$. Thus, $|E(G)| \geq \lceil \frac{1}{2}(1-m)p \rceil$.   ∎

Next we find an upper bound for the size of a minimum critically m-neighbor-scattered graph of order p. In the following nonlinear integer programming, we regard n as a variable, p and m as fixed integers,

$$min_n f(n)$$

$$s.t \begin{cases} n \geq 3 - m, \\ x_i - (3 - m) \geq 0, i = 0, 1, \cdots, n-1 \\ x_i \in Z^+, i = 0, 1, \cdots, n-1 \end{cases} \quad (2.5)$$

where

$$f(n) = \begin{cases} 2QR + R + Q^2 n - p + (3-m)n, \\ \text{if at least one of } m \text{ and } n \text{ is even} \\ 2QR + R + Q^2 n - p + (3-m)n + 1, \\ \text{if both of } m \text{ and } n \text{ are odd} \end{cases}$$

Before giving our main result, we give some lemmas first.

Lemma 3.5.([7]) *For any fixed positive integers s, t, if $s + 1 \leq n \leq \lfloor \frac{t}{n} \rfloor$, $Q = \lfloor \frac{t}{n} \rfloor$, $R = t - nQ$, then the function $f(n)$ is decreasing.*

Lemma 3.6.([7]) *Let t, n, s be three integers, $n \geq s + 1$. If $n = \lfloor \frac{t}{s} \rfloor$, then $s = \lfloor \frac{t}{n} \rfloor$.*

Theorem 3.7. *Let m be a negative integer. If G is a minimum critically m-neighbor-scattered graph of order p, then $|E(G)| \leq \lceil \frac{1}{2}(3-m)p + \frac{1}{2}(3-m)R \rceil$, where $R = p - \lfloor \frac{p}{3-m} \rfloor (3-m)$, the remainder of the order p divided by $3 - m$.*

Proof. Let n be an integer such that $n \geq 3 - m$. We use p to denote the order of each of the graphs $\widetilde{G}(2 - m, n)$. Hence, $\widetilde{g}(2 - m, n) = \lceil \frac{1}{2}(2QR + R + Q^2 n - p + (3-m)n) \rceil$, where $R = p - \lfloor \frac{p}{3-m} \rfloor (3-m)$, and $R = p - nQ$. By Theorem 2.5, $\widetilde{G}(2 - m, n)$ is a class of critically m-neighbor-scattered graphs, and hence $|E(G)| \leq \widetilde{g}(2 - m, n)$.

In the following we will prove that $n \leq \lfloor \frac{p}{3-m} \rfloor$. Otherwise, if $n > \lfloor \frac{p}{3-n} \rfloor$, since n is an integer, then we have $n > \frac{p}{3-m}$. Therefore, $n(3 - m) \geq p$. On the other hand, by the construction of the graphs $\widetilde{G}(2 - m, n)$, $x_i = |C_i| \geq 3 - m$, for all $i = 0, 1, \cdots, n - 1$. Thus $p = \sum_{i=0}^{n-1} x_i \geq (3-m)n$, i.e., $n \leq \lfloor \frac{p}{3-m} \rfloor$, a contradiction. Hence, $n \leq \lfloor \frac{p}{3-m} \rfloor$. By Lemma 3.5, the function $f(n)$ is decreasing, for $3 - m \leq n \leq \lfloor \frac{p}{3-m} \rfloor$. Hence $f(n)$ attains its minimum value when $n = \lfloor \frac{p}{3-m} \rfloor$.

By Lemma 3.6, when $n = \lfloor \frac{p}{3-m} \rfloor$ and $n \geq 3 - m$, we have $3 - m = \lfloor \frac{p}{n} \rfloor$. Hence $Q = \lfloor \frac{p}{n} \rfloor = 3 - m$ and $R = p - nQ = p - (3-m)n = p - \lfloor \frac{p}{3-m} \rfloor (3-m)$. So, the minimum value of $f(n)$ is $f(\lfloor \frac{p}{3-m} \rfloor)$, i.e.,

$$minf(n) = f(\lfloor \tfrac{p}{3-m} \rfloor)$$

$$= \begin{cases} 2QR + R + (3-m)^2 n - p + (3-m)n, \\ if\ at\ least\ one\ of\ m\ and\ n\ is\ even \\ 2QR + R + (3-m)^2 n - p + (3-m)n + 1, \\ if\ both\ of\ m\ and\ n\ are\ odd \end{cases}$$

$$= \begin{cases} (3-m)p + (3-m)R, \\ if\ at\ least\ one\ of\ m\ and\ n\ is\ even \\ (3-m)p + (3-m)R + 1, \\ if\ both\ of\ m\ and\ n\ are\ odd \end{cases}$$

Therefore, when $n = \lfloor \frac{p}{3-m} \rfloor$, $\widetilde{g}(2-m,n) = \frac{1}{2}f(n) = \lceil \frac{1}{2}((3-m)p + (3-m)R) \rceil$, where $R = p - \lfloor \frac{p}{3-m} \rfloor(3-m)$. So, $|E(G)| \leq \lceil \frac{1}{2}(3-m)p + \frac{1}{2}(3-m)R \rceil$. ∎

Since $0 \leq R \leq 1 - m$, we have $\frac{1}{2}(3-m)p + \frac{1}{2}(3-m)R \leq \frac{1}{2}(3-m)p + \frac{1}{2}(3-m)(1-m) = \frac{1}{2}(3-m)(1+p-m)$. So we have the following corollary.

Corollary 3.8. *Let m be a negative integer. If G is a minimum critically m-neighbor-scattered graph with order p, then $\lceil \frac{1}{2}(1-m)p \rceil \leq |E(G)| \leq \lceil \frac{1}{2}(3-m)(1+p-m) \rceil$.*

Corollary 3.9. *Let m be a negative integer. If the order of a minimum critically m-neighbor-scattered graph G, p, is a multiple of $3-m$, and $\delta(G) = 3-m$, then $|E(G)| = \lceil \frac{1}{2}(3-m)p \rceil$.*

Proof. When the order of a minimum critically m-neighbor-scattered graph G, p, is a multiple of $3-m$, then, $R = p - \lfloor \frac{p}{3-m} \rfloor(3-m) = 0$. So, by Theorem 3.7, $|E(G)| \leq \lceil \frac{1}{2}(3-m)p \rceil$. From the fact that $\delta(G) = 3-m$, we have $|E(G)| \geq \lceil \frac{1}{2}(3-m)p \rceil$. Thus the proof is complete. ∎

From above we know that Theorem 3.7 gives an upper and lower bounds for the size of a minimum critically m-neighbor-scattered graph G of order p. Let $f \triangleq f(m,p)$ denote the number of edges in a minimum critically m-neighbor-scattered graph among the graphs $G' = G^E_{2-m,\lfloor \frac{p}{3-m} \rfloor}$ of order p. It is obvious that $\delta_{G'} \geq 2 - m$. Then we have

$$\lceil \tfrac{1}{2}(2-m)p \rceil \leq f \leq \lceil \tfrac{1}{2}(3-m)(p+R) \rceil.$$

On the other hand, let $f_{opt} \triangleq f_{opt}(m,p)$ denote the number of edges in a minimum critically m-neighbor-scattered graph among all graphs G of order p. So, by Theorem 3.7 we have

$$f_{opt} \geq \lceil \tfrac{1}{2}(1-m)p \rceil \geq \tfrac{1}{2}(1-m)p.$$

Since $R \leq 1 - m$, we have

$$f - f_{opt} \leq p + \tfrac{1}{2}(3-m)R \leq p + \tfrac{1}{2}(3-m)(1-m).$$

Then
$$\frac{f}{f_{opt}} \leq 1 + \frac{2}{1-m} + \frac{3-m}{p}.$$

From the construction of graph $G' = G^E_{2-m,\lfloor \frac{p}{3-m} \rfloor}$, we know that $\lfloor \frac{p}{3-m} \rfloor \geq 3-m$, and so $p \geq (3-m)(3-m)$. Then
$$\frac{f}{f_{opt}} \leq 1 + \frac{1}{1-m} + \frac{3-m}{(3-m)(3-m)} \leq 1 + \frac{1}{1-m} + \frac{1}{3-m} < 1 + \frac{2}{3-m}.$$

Let $\epsilon = \frac{2}{3-m}$ and we regard m and p as variables. When $m \leq -1$ becomes smaller and smaller, the critically m-neighbor-scattered graph G' becomes more and more reliable. This means that $f(G')$ is a $(1+\epsilon)$-approximate size (graph) of a minimum critically m-neighbor-scattered graph among all graphs of order p. So we get a method to construct a $(1+\epsilon)$-approximate minimum critically m-neighbor-scattered graph G of order p. The construction is given as follows:

Step 1. *Construct a generalized Harary graph $G_{2-m,\lfloor \frac{p}{3-m} \rfloor}$.*

Step 2. *Construct graph $G^E_{2-m,\lfloor \frac{p}{3-m} \rfloor}$ such that this graph has $R = p - \lfloor \frac{p}{3-m} \rfloor (3 - m)$ cliques of order $4-m$ and $\lfloor \frac{p}{3-m} \rfloor - R$ cliques of order $3-m$.*
where E is the operation given before. It is easily seen that the size of the above graph is
$$f_{app} = |E(G)| = R\binom{4-m}{2} + (\lfloor \frac{p}{3-m} \rfloor - R)\binom{3-m}{2} + \lfloor \frac{(2-m)\lfloor \frac{p}{3-m} \rfloor}{2} \rfloor.$$

Example 4. When $p = 31, m = -2$ are given, we can construct an approximate minimum critically (-2)-neighbor-scattered graph G with order $p = 31$, $\lfloor \frac{p}{3-m} \rfloor = \lfloor \frac{31}{5} \rfloor = 6$, $R = p - \lfloor \frac{p}{3-m} \rfloor (3-m) = 1$, and $|E(G)| = R\binom{4-m}{2} + (\lfloor \frac{p}{3-m} \rfloor - R)\binom{3-m}{2} + \lfloor \frac{(2-m)\lfloor \frac{p}{3-m} \rfloor}{2} \rfloor = 77$. See Figure 4.

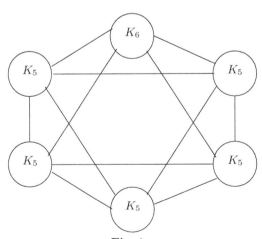

Fig. 4.

References

1. Bondy, J.A., Murty, U.S.R.: Graph Theory with Applications. Macmillan/Elsevier, London/New York (1976)
2. Gunther, G.: Neighbor-connectivity in regular graphs. Discrete Applied Math. 11, 233–243 (1985)
3. Gunther, G., Hartnell, B.L.: On minimizing the effects of betrayals in a resistance movement. In: Proc. Eighth Manitoba Conference on Numerical Mathematics and Computing, pp. 285–306 (1978)
4. Jung, H.A.: On a class of posets and the corresponding comparability graphs. J. Combin. Theory Ser. B 24, 125–133 (1978)
5. Ouyang, K.Z., Ouyang, K.Y.: Relative breaktivity of graphs. J. Lanzhou University (Natural Science) 29(3), 43–49 (1993)
6. Wei, Z.T. (Li, X.L., supervisor): On the reliability parameters of networks, Master's Thesis, Northwesten Polytechnical University, pp. 30–40 (2003)
7. Wu, S.-S.Y., Cozzens, M.: Cozzens, The minimum size of critically m-neighbor-conneced graphs. Ars Combinatoria 29, 149–160 (1990)
8. Li, F.W., Li, X.L.: Computational complexity and bounds for neighbor-scattering number of graphs. In: Proceedings of the 8th International Symposium on Parallel Architectures, Algorithms and Networks, Washington, DC, pp. 478–483. IEEE Computer Society, Los Alamitos (2005)

A New Hybrid Algorithm for Feature Selection and Its Application to Customer Recognition*

Luo Yan[1] and Yu Changrui[2]

[1] Institute of System Engineering, Shanghai Jiao Tong University, 200052 Shanghai, China
[2] School of Information Management and Engineering,
Shanghai University of Finance and Economics, Shanghai, 200433, China
{yanluo,yucr}@sjtu.edu.cn

Abstract. This paper proposes a novel hybrid algorithm for feature selection. This algorithm combines a global optimization algorithm called the simulated annealing algorithm based nested partitions (NP/SA). The resulting hybrid algorithm NP/SA retains the global perspective of the nested partitions algorithm and the local search capabilities of the simulated annealing method. We also present a detailed application of the new algorithm to a customer feature selection problem in customer recognition of a life insurance company and it is found to have great computation efficiency and convergence speed.

1 Introduction

For a complex decision problem, data collected in the world are so large that it becomes more and more difficult for the decision-maker to access them. This has lead to an increased industry and academic interest in knowledge discovery in databases. The process of discovering useful information in large databases consists of numerous steps, which may include integration of data from numerous databases, manipulation of the data to account for missing and incorrect data, and induction of a model with a learning algorithm [1].Traditionally, data mining draws heavily on both statistics and artificial intelligence, but numerous problems in data mining and knowledge discovery can also be formulated as optimization problems [2] [3].

An important problem in knowledge discovery is analyzing the relevance of the features, usually called feature or attribute subset selection. It is the problem of selecting a subset of d features from a set of D features based on some optimization criterion [4]. The primary purpose of feature selection is to design a more compact classifier with as little performance degradation as possible.

The literature on feature selection is extensive within the machine-learning and knowledge-discovery communities. Some of the methods applied to this problem in the past include genetic algorithms [5], correlation-based algorithms [6], evolutionary search [7], rough sets theory [8], randomized search [9], and the nested partitions method [10]. In this paper we propose a new alternative for feature selection based on a combinatorial optimization method called NP/SA which retains the global perspective of the NP algorithm and the local search capabilities of the SA method.

* This research work is supported by the Natural Science Fund of China (# **70501022**) and China Postdoctoral Science Foundation (# **20060400169**).

A. Dress, Y. Xu, and B. Zhu (Eds.): COCOA 2007, LNCS 4616, pp. 102–111, 2007.
© Springer-Verlag Berlin Heidelberg 2007

The feature-selection problem is generally difficult to solve. The number of possible feature subsets is 2^n, where n is the number of features, and evaluating every possible subset is therefore prohibitively expensive unless n is very small. In this paper we focus on data mining where the data are nominal, i.e. each feature can take only finitely many values.

The remainder of the paper is organized as follows. In Section 2 we review the general procedure of the NP algorithm and the SA method. Then we present a combined NP/SA algorithm, i.e. an improved NP algorithm based on simulated annealing. In Section 3 we apply the hybrid NP/SA algorithm to the customer feature selection problem of a life insurance company. Section 4 contains some concluding remarks and future research directions.

2 Algorithm Development

2.1 The Nested Partitions Method

The NP method, an optimization algorithm proposed by Shi and Ólafsson [10], may be described as an adaptive sampling method that uses partitioning to concentrate the sampling effort in those subsets of the feasible region that are considered the most promising. It combines global search through global sampling of the feasible region, and local search that is used to guide where the search should be concentrated. This method has been found to be promising for difficult combinatorial optimization problems such as the traveling salesman problem [11], buffer allocation problem [12], product design problem [13] [14], and so on.

Suppose the finite feasible region of a complex decision problem is Θ. Our objective is to optimize the objective performance function $f: \Theta \to \mathbf{R}$, that is, to solve:

$$\max_{\theta \in \Theta} f(\theta),$$

where $|\Theta| < \infty$. Also, to simplify the analysis, we assume that there exists a unique solution $\theta_{opt} \in \Theta$ to the above problem, which satisfies $f(\theta_{opt}) > f(\theta)$ for all $\theta \in \Theta \setminus \{\theta_{opt}\}$.

Definition 1. A region partitioned using a fixed scheme is called a valid region. In a discrete system a partitioned region with a singleton is called a singleton region. The collection of all valid regions is denoted by Σ. Singleton regions are of special interest in the process of optimization, and $\Sigma_0 \subset \Sigma$ denotes the collection of all such valid regions.

The optimization process of the NP method is a sequence of set partitions using a fixed partitioning scheme, with each partition nested within the last. The partitioning is continued until eventually all the points in the feasible region correspond to a singleton region.

Definition 2. The singleton regions in Σ_0 are called regions of maximum depth. More generally, we define the depth, $dep: \Sigma \to N_0$, of any valid region iteratively with Θ having depth zero, subregions of Θ having depth one, and so forth. Since they cannot be partitioned further, we call the singleton regions in Σ_0 regions of maximum depth.

Definition 3. If a valid region $\sigma \in \Sigma$ is formed by partitioning a valid region $\eta \in \Sigma$, then σ is called a subregion of region η, and region η is called a superregion of region σ. We define the superregion function $s : \Sigma \to \Sigma$ as follows. Let $\sigma \in \Sigma \backslash \Theta$. Define $s(\sigma) = \eta \in \Sigma$, if and only if $\sigma \subset \eta$ and if $\sigma \subseteq \xi \subseteq \eta$ then $\xi = \eta$ or $\xi = \sigma$. For completeness we define $s(\Theta) = \Theta$.

A set performance function $I : \Sigma \to \mathbf{R}$ is defined and used to select the most promising region and is therefore called the promising index of the region.

In the k-th iteration of the NP method there is always a region $\sigma(k) \subseteq \Theta$ that is considered the most promising, and as nothing is assumed to be known about location of good solutions before the search is started, $\sigma(0) = \Theta$. The most promising region is then partitioned into $M_{\sigma(k)}$ subregions, and what remains of the feasible region $\sigma(k)$ is aggregated into one region called the surrounding region. Therefore, in the k-th iteration $M_{\sigma(k)} + 1$ disjoint subsets that cover the feasible region are considered.

Each of these regions is sampled using some random sampling scheme, and the samples used to estimate the promising index for each region. This index is a set performance function that determines which region becomes the most promising region in the next iteration. If one of the subregions is found to be best, this region becomes the most promising region. If the surrounding region is found to be best, the method backtracks to a larger region. The new most promising region is partitioned and sampled in a similar fashion. This generates a sequence of set partitions, with each partition nested within the last. The partitioning is continued until eventually all the points in the feasible region correspond to a singleton region. For detailed description on a generic implementation of the NP algorithm, we refer the interested reader to [10].

2.2 Hybrid NP/SA Algorithm

We now describe the hybrid NP/SA algorithm in detail. The simulated annealing algorithm (SA) is essentially a heuristic algorithm. The technique has been widely applied to a variety of problems including many complex decision problems. Often the solution space of a decision problem has many local minima. The SA method, though by itself it is a local search algorithm, avoids getting trapped in a local minimum by accepting cost increasing neighbors with some probability [15]. Applying the ideas of SA to the random sampling of the NP algorithm will greatly improve the ability of global optimization of the NP algorithm and the ability of local optimization of the SA method. Merging the SA method into the NP algorithm, we get the Simulated Annealing algorithm based Nested Partitions (NP/SA). Note that NP/SA is not simply merging the whole SA into the random sampling of the NP algorithm, but combining the basic idea of SA with the complete optimization process of the NP algorithm properly so as to improve the optimization efficiency of the NP algorithm.

Similar to the preparatory work of SA implementation, firstly we need to set the initial annealing temperature T_0, the final annealing temperature T_f, and the number N of random samples at each annealing temperature. NP/SA is an improvement of the NP algorithm. It has the same operations in partitioning, calculation of promising indices and backtracking. The random sampling of NP/SA is improved. Actually, NP/SA does not implement a complete annealing process in every sampled region to obtain an optimal solution over the region. Instead, NP/SA carry out the optimization according to the same annealing temperature over the feasible regions at the same depth. According to the maximum depth $dep(\sigma)$ ($\sigma \in \Sigma_0$) of singleton region in the feasible region, the annealing speed $\Delta T = (T_0 - T_f)/dep(\sigma)$ is set.

Respectively optimize the uncrossed $M_{\sigma(k)} + 1$ feasible regions obtained through the k-th partitioning at the annealing temperature $T_k = T_0 - dep(\sigma(k)) \cdot \Delta T$ according to the SA method. That is to say, starting from a certain initial point $X^{(0)}$, randomly sample the feasible regions. If $f(X^{(k)}) \geq f(X^{(0)})$, where $f(X^{(k)})$ is the function value of the sampled point $X^{(k)}$, $X^{(k)}$ is accepted and taken as the initial point $X^{(0)}$ to continue the optimization; otherwise, if $f(X^{(k)}) < f(X^{(0)})$, $X^{(k)}$ is accepted with a probability of $\exp((f(X^{(k)}) - f(X^{(0)}))/T)$ and taken as the initial point $X^{(0)}$ to continue the optimization. When N points are sampled, the function value $f(X^{(0)})$ at the optimal point is used as the promising index function of each feasible region to fix the next most feasible region. The pseudo-code of the optimization process is following.

```
σ(k)=Θ;
d(σ(k))=0;
Repeat
Partition the current promising regionσ(k) into M_σ(k)
subregions.
T(k)=T(0)-dep(σ(k))*ΔT
For i=1 to M_σ(k)+1 do
For j=1 to N do
Generate_state_x(j);
δ=f(x(j))-f(x(k));
ifδ>0 then k=j
    else if random(0,1)<exp(-δ/T(k)) then
        k=j;
        Promising(i)=f(x(k));
End
If promising(i)>promising(m) then m=i;
if m<=M_σ(k) thenσ(k+1)=subregion(m);
        dep(σ(k))= dep(σ(k))+1;
    else backtrack(σ(k-1));
        dep(σ(k))= dep(σ(k))-1;
until it reaches the maximum depth and stabilizes.
```

2.3 Feasibility Analysis of NP/SA

The NP algorithm allows for the introduction of other algorithm and thoughts. It implicitly contains a requirement: the modifications to the operators of the NP algorithm are allowed so long as two conditions are satisfied. They are: (*a*) the probability of each point in the feasible region being sampled is larger than 0, and (*b*) the promising index corresponds with the performance function of the singleton region.

Although NP/SA is different from the pure NP algorithm in fixing the optima in the partitioned regions, its essential sampling method is still random sampling. This ensures that the probability of each point in the feasible region being sampled is larger than 0. Therefore, NP/SA completely satisfies condition (*a*) of the NP algorithm.

When the partitioning process of the NP/SA algorithm moves on to singleton, there is only one feasible point in the feasible region and only one point is obtained through sampling. The promising index at this point is the function value of this point; hence it corresponds with the performance function over the singleton. Thus, NP/SA satisfies the condition (*b*) of the NP algorithm.

In all, the introduction of SA into the NP algorithm satisfies the openness of the latter one, which ensures that NP/SA converges to the global optimal solution with a probability of 1.

2.4 Convergence of NP/SA

Due to the systematic partitioning of the feasible region the hybrid NP/SA algorithm converges to a global optimum. As the NP algorithm evolves, the sequence of most promising regions $\{\sigma(k)\}_{k=1}^{\infty}$ forms a Markov chain with state space Σ. The singleton regions with the global optima are denoted as the absorbing states. In literature [10], L. Shi and S. Ólafsson proved that, the expected number of nested partitioning when the NP algorithm converges to the optimal solution is given by the following equation:

$$E[Y] = 1 + \sum_{\eta \in \Sigma_1} \frac{1}{P_\eta[T_{\sigma_{opt}} < T_\eta]} + \sum_{\eta \in \Sigma_2} \frac{P_\Theta[T_\eta < \min\{T_\Theta, T_{\sigma_{opt}}\}]}{P_\eta[T_\Theta < T_\eta] \cdot P_\Theta[T_{\sigma_{opt}} < \min\{T_\Theta, T_\eta\}]},$$

where T_η is the hitting time of state $\eta \in \Sigma$, i.e. the first time that the Markov chain visits the state, $P_\eta[\cdot]$ denotes the probability of an event given that the chain starts in state $\eta \in \Sigma$, σ_{opt} is the region corresponding to the unique global optimum, and $\Sigma_1 = \{\eta \in \Sigma \setminus \{\sigma_{opt}\} | \sigma_{opt} \subseteq \eta\}$, $\Sigma_2 = \{\eta \in \Sigma | \sigma_{opt} \not\subseteq \eta\}$ and $\Sigma = \{\sigma_{opt}\} \cup \Sigma_1 \cup \Sigma_2$ are disjoint state spaces.

NP/SA introduces SA into the NP algorithm, which increases the probability of obtaining the global optima in the sampled regions and further increases the probability of the state of the Markov chain changes in the correct direction. Consequently, probability $P_\eta[T_{\sigma_{opt}} < T_\eta]$ at time $\eta \in \Sigma_1$, probability $P_\eta[T_\Theta < T_\eta]$ at time $\eta \in \Sigma_2$, and $P_\Theta[T_{\sigma_{opt}} < \min\{T_\Theta, T_\eta\}]$ are increased while probability $P_\Theta[T_\eta < \min\{T_\Theta, T_{\sigma_{opt}}\}]$ at time $\eta \in \Sigma_2$ is decreased. The combined effect of these factors reduces the expected number of nested partitioning when the NP algorithm converges to the global optimal solution, and thus speeds up the convergence of the algorithm.

3 Application to the Customer Feature Selection Problem in Customer Recognition

In this section we apply the hybrid NP/SA algorithm to the customer feature selection problem in customer recognition of a life insurance company where the objective is to select the most valid customer features that has the properties of combinatorial optimization, i.e., to recognize the customers who have high customer lifecycle values (CLV) and can establish or keep customer relationship with the company.

3.1 The Separability Criterion for Customer Feature Selection

Different classes of sample customers are separable because they locate at different domains of the feature space. The greater the distances among these domains are, the larger the separability of these customer classes is. Therefore, we use the separability criterion based on inter-class distance. Let $\vec{x}_k^{(i)}$ and $\vec{x}_l^{(j)}$ denote respectively the customer feature vectors with dimension d in sample classes w_i and w_j. The distance between the two vectors is denoted by $\delta(\vec{x}_k^{(i)}, \vec{x}_l^{(j)})$. Let c denote the number of the classes of sample customers, n_i and n_j denote the sample number of w_i and w_j respectively, and P_i and P_j denote their prior probabilities respectively. The average distance between different classes of customer feature vectors can be expressed as

$$ J_d(\vec{x}) = \frac{1}{2} \sum_{i=1}^{c} P_i \sum_{j=1}^{c} P_j \frac{1}{n_i n_j} \sum_{k=1}^{n_i} \sum_{l=1}^{n_j} \delta(\vec{x}_k^{(i)}, \vec{x}_l^{(j)}), $$

where P_i is the prior probability of customer class w_i and can be estimated using training customer sample numbers, that is, $\bar{P}_i = \frac{n_i}{n}$.

When the distance between two vectors takes the form of Euclidean distance,

$$ \delta(\vec{x}_k^{(i)}, \vec{x}_l^{(j)}) = (\vec{x}_k^{(i)} - \vec{x}_l^{(j)})^T (\vec{x}_k^{(i)} - \vec{x}_l^{(j)}). $$

Then the average distance between different classes of customer feature vectors can be denoted as

$$ J_d(\vec{x}) = tr(S_w + S_b), $$

where the intra-class dispersion matrix

$$ S_w = \sum_{i=1}^{c} P_i \frac{1}{n_i} \sum_{k=1}^{n_i} (\vec{x}_k^{(i)} - \vec{m}_i)(\vec{x}_k^{(i)} - \vec{m}_i)^T, $$

and the inter-class dispersion matrix

$$S_b = \sum_{i=1}^{c} P_i(\bar{m}_i - \bar{m})(\bar{m}_i - \bar{m})^T .$$

$\bar{m}_i$ is the mean vector of the sample customer class w_i, and $\bar{m}$ is the total mean vector of the customer sample sets. Moreover, we have

$$\bar{m}_i = \frac{1}{n_i} \sum_{k=1}^{n_i} \bar{x}_k^{(i)} , \; \bar{m} = \sum_{i=1}^{c} P_i \bar{m}_i .$$

As it is better that the intra-class separability of the customer recognition result is low while the inter-class separability is high, we can use the following criterion:

$$J = tr(S_w^{-1} S_b) .$$

3.2 Hybrid NP/SA Algorithm for the Customer Feature Selection Problem

Mathematically, the customer attribute selection problem can be stated as follows. The customer feature selection is the problem of selecting a subset of d features from a set of D original features obtained from the sample customers so as to separate various classes of sample customers. Suppose the original feature vector of customers is $\bar{x}_D$. Then, the optimization problem of customer feature selection is denoted as

$$\max z = J(\bar{x})$$

$$s.t. \begin{cases} |\bar{x}| = d \\ \bar{x} \subset \bar{x}_D \end{cases} .$$

The possible combinatorial number of selecting d features from D original customer features is $C_D^d = \dfrac{D!}{(D-d)!d!}$. When the dimension of the original feature vectors is high, the exhaustive attack method can not be applied to solve this problem because of its enormous computation work. Therefore, a feasible algorithm that satisfies the validity criterion of customer recognition is necessary. We now describe how the hybrid NP/SA algorithm applies to the customer feature selection problem of a life insurance company.

After an initial selection of customer data in the information system of the company, we obtain the original customer features and the corresponding customer information including twelve features: ID numbers, marriage conditions, carieers, sex, the earliest ages when they bought the life insurances, the maximum payment periods, the sum of paid insurance, the number of insurance types they have bought, the sum of payment times, the sum of insurance shares, the percentage of the valid insurance policies, and the average payment for one time. Among them, as ID number is only a mark of a customer, it is not considered a feature for customer recognition. The other eleven features are used for customer recognition and are correspondingly denoted as oldfea[1], oldfea[2],···, oldfea[11]. The objective of our optimization is to select five most effective customer features for customer recognition from the eleven features, i.e., $D=11$ and $d=5$.

Meanwhile, through analysis and investigation, the experts in the company got the information of a customer in three dimensions: customer lifecycle value (CLV), customer perception value (CPV), and available competitiveness (AC). For the information, we use 1 to denote a high value of a customer dimension and −1 to denote a low value of a customer dimension.

We initialize different types of customer features. The initialization method includes valuation of the discrete feature variables. For example, for the feature of marriage, the value of 'single' is 0 while that of 'married' is 1. After normalizing the customer features, we obtain the customer feature values on $[0,1]$.

The initial annealing temperature $T_0 = 100$. The final annealing temperature $T_f = 0.001$. The maximum depth of partitioning is $dep(\Sigma_0) = D - d = 6$. Thus, the annealing temperature in the feasible region of depth dep is

$$T_{dep} = 100 - dep \cdot (100 - 0.001)/6 = 100 - 16.6665dep.$$

Table 1. The result of customer feature selection

Arrays	The corresponding features	Features in the dimension of CLV	Features in the dimension of CPV	Features in the dimension of AC
Oldfea[1]	Marriag conditions	0	0	0
Oldfea[2]	Carieers	0	0	0
Oldfea[3]	Sex	0	0	0
Oldfea[4]	The earliest ages when they bought life insurances	0	0	0
Oldfea[5]	The maximum payment periods	0	0	0
Oldfea[6]	The sum of paid insurance	1	1	1
Oldfea[7]	The number of insurance types they have bought	1	1	0
Oldfea[8]	The sum of payment times	1	1	1
Oldfea[9]	The sum of insurance shares	0	1	1
Oldfea[10]	The percentage of the valid insurance policies	1	1	1
Oldfea[11]	The average payment for one time	1	0	1
Times of Backtracking		0	0	0

Let the arrays oldfea[i] ($i=1,\ldots,11$) denote the current most promising regions. The features corresponding to the factors valued 1 are the combination of features comprised in the feasible region. 30 points in each subregion is randomly sampled. $30 \cdot dep$ points are sampled in the entire region except the current most promising region.

The NP/SA algorithm is then implemented for selecting the effective feature combinations that can identify the values of the three dimensions. The algorithm in C programming language is realized. The result of customer feature selection is shown in Table 1, where 1 stands for selection and 0 stands for non-selection.

In the optimization process using the hybrid NP/SA method, backtracking implicates that the last time partitioning, sampling, and promising indices are invalid. The algorithm should backtrack to the last iteration and continue with a new optimization process. Therefore, backtracking is a significant symbol of convergence speed. In Table 1 the times that backtracking occurs is zero, which strongly indicates that the hybrid NP/SA method is very effective for solving customer feature selection problems in customer recognition.

4 Conclusions and Future Research

This paper proposes a new hybrid algorithm for feature selection. It combines the nested partitions (NP) algorithm and simulated annealing algorithm (SA) in a novel way. The resulting algorithm retains the global perspective of the nested partitions algorithm and the local search capabilities of the simulated annealing method. This new optimization algorithm was shown empirically to be very efficient for a difficult customer feature selection problem in customer recognition.

However, further theoretical and empirical development is needed for the algorithm. Customer feature selection problems provide a rich class of NP-hard problems that will be used for extensive empirical testing of the algorithm. We will also consider different objective functions. The theoretical development of the hybrid NP/SA algorithm will focus on using the Markov structure to obtain results about the time until convergence and to develop stopping rules.

References

1. Olafsson, S., Yang, J.: Intelligent Partitioning for Feature Selection. INFORMS Journal on Computing 17(3), 339–355 (2005)
2. Basu, A.: Perspectives on operations research in data and knowledge management. Eur. J. Oper. Res. 111, 1–14 (1998)
3. Bradley, P.S., Fayyad, U.M., Mangasarian, O.L.: Mathematical programming for data mining: Formulations and challenges. INFORMS J. Comput. 11, 217–238 (1999)
4. Oh, I.-S., Lee, J.-S., Moon, B.-R.: Hybrid genetic algorithms for feature selection. IEEE Trans. on Pattern Analysis and Machine Intelligence 26(11), 1424–1437 (2004)
5. Yang, J., Honavar, V.: Feature subset selection using a genetic algorithm. In: Motada, H., Liu, H. (eds.) Feature Selection, Construction, and Subset Selection: A Data Mining Perspective, pp. 117–136. Kluwer, Dordrecht (1998)

6. Hall, M.A.: Correlation-based feature selection for discrete and numeric class machine learning. In: Proc. 17th Internat. Conf. Mach. Learn, Stanford University, pp. 359–366. Morgan Kaufmann, San Francisco (2000)
7. Kim, Y.S., Street, W.N., Menczer, F.: Feature selection in unsupervised learning via evolutionary search. In: Proc. 6th ACM SIGKDD Internat. Conf. Knowledge Discovery Data Mining, Boston, MA, pp. 365–369. ACM, New York (2000)
8. Modrzejewski, M.: Feature selection using rough sets theory. In: Brazdil, P.B. (ed.) Proc. Eur. Conf. Mach. Learn., Vienna, Austria, pp. 213–226. Springer, Heidelberg (1993)
9. Skalak, D.: Prototype and feature selection by sampling and random mutation hill climbing algorithms. In: Proc. 11th Internat. Conf. Mach. Learn., New Brunswick, NJ, pp. 293–301. Morgan Kaufmann, San Francisco (1994)
10. Shi, L., Olafsson, S.: Nested partitions method for global optimization. Oper. Res. 48, 390–407 (2000)
11. Shi, L., Olafsson, S., Sun, N.: New parallel randomized algorithms for the traveling salesman problem. Computers & Operations Research 26, 371–394 (1999)
12. Shi, L., Men, S.: Optimal buffer allocation in production lines. IIE Transactions 35, 1–10 (2003)
13. Shi, L., Olafsson, S., Chen, Q.: A new hybrid optimization algorithm. Computers & Industrial Engineering 36, 409–426 (1999)
14. Shi, L., Olafsson, S., Chen, Q.: An optimization framework for product design. Management Science 47(2), 1681–1692 (2001)
15. Ahmed, M.A., Alkhamis, T.M.: Simulation-based optimization using Simulated Annealing with ranking and selection. Computers & Operations Research 29, 387–402 (2002)

Steiner Forests on Stochastic Metric Graphs*

Vangelis Th. Paschos[1,**], Orestis A. Telelis[2], and Vassilis Zissimopoulos[2]

[1] LAMSADE, CNRS UMR 7024, Université Paris-Dauphine, France
paschos@lamsade.dauphine.fr
[2] Department of Informatics and Telecommunications, University of Athens, Greece
{telelis,vassilis}@di.uoa.gr

Abstract. We consider the problem of connecting given vertex pairs over a stochastic metric graph, each vertex of which has a probability of presence independently of all other vertices. Vertex pairs requiring connection are always present with probability 1. Our objective is to satisfy the connectivity requirements for every possibly materializable subgraph of the given metric graph, so as to optimize the expected total cost of edges used. This is a natural problem model for cost-efficient Steiner Forests on stochastic metric graphs, where uncertain availability of intermediate nodes requires fast adjustments of traffic forwarding. For this problem we allow a priori design decisions to be taken, that can be modified efficiently when an actual subgraph of the input graph materializes. We design a fast (almost linear time in the number of vertices) modification algorithm whose outcome we analyze probabilistically, and show that depending on the a priori decisions this algorithm yields 2 or 4 approximation factors of the optimum expected cost. We also show that our analysis of the algorithm is tight.

1 Introduction

We consider the problem of laying out routes that connect simultaneously given source-destination vertex pairs over a metric graph $G_0(V_0, E_0)$. Vertices of the metric graph G_0 other than the sources and destinations may be used, but we are uncertain of their availability, in that each such vertex is present with some probability independently of all other vertices. Sources and destinations are present with probability 1. Our objective is to take some a priori decisions regarding the layout of required routes, so as to be able to come up with feasible routes for every possibly materializable subgraph $G_1(V_1, E_1)$, $V_1 \subseteq V_0$, of G_0, and minimize the expected total cost of edges used over the distribution of all such subgraphs. This is the well known *Steiner Forest* problem, defined over a *stochastic* metric graph G_0.

---

* Research partially supported by the Greek Ministry of Education and Research under the project PYTHAGORAS II.
** Part of this work was carried out while the author was visiting the Department of Informatics and Telecommunications, University of Athens. Financial support of the Greek Ministry of Education and Research and hospitality of members of the Department of Informatics and Telecommunications are gratefully acknowledged.

A. Dress, Y. Xu, and B. Zhu (Eds.): COCOA 2007, LNCS 4616, pp. 112–123, 2007.
© Springer-Verlag Berlin Heidelberg 2007

A brute-force way to cope with this problem is to precompute a feasible and approximate (or maybe optimum) solution for every possible subgraph of G_0 that may materialize, and apply an appropriate solution when the subgraph actually appears. In principle there need not be a constraint on the computational effort applied for taking a priori decisions, as long as they support fast response to the actually materialized data. In this light however, we require that such a response should be of strictly lower complexity compared to the a priori computational effort. A straightforward pattern for implementing this setting is for example to compute an optimum a priori solution over G_0, and if this solution is not feasible for the materialized subgraph G_1, use a polynomial-time approximation algorithm to obtain a completely different feasible solution for G_1. On the other hand, a natural challenge is to design such an efficient response strategy (algorithm), that can be supported by polynomial-time computable a priori decisions. In this paper we design and analyze such a strategy for *repairing* an a priori polynomial-time computable feasible solution for G_0, so as to render it feasible for G_1. We show that this strategy also approximates the optimum expected cost over all materializable subgraphs G_1.

The problem model we consider finds natural application in networks, where uncertain availability of intermediate nodes requires fast adjustments of traffic forwarding. The Steiner Forest problem is a well-known NP-hard multicommodity network design problem (even in metric graphs), generalizing the Steiner tree problem, and the only known approximation algorithm yields approximation factor 2 and was analyzed in [1,8] (see also [18]). Recent years have seen a detailed study and sensitivity analysis of this algorithm, mainly in the context of *Stochastic Network Design*, which owes its roots to *Stochastic Programming* [5,4], where some elements of the input data set to an optimization problem are associated to a distribution describing their probability of occurence. Stochastic Programming was introduced by the seminal work of Dantzig [5] and thereafter has evolved into an independent discipline of Operations Research that handles uncertainty in optimization problems by usage of probabilities, statistics and mathematical programming (see [4] for a description of the field). We refer the reader to [11,7] for approximation results on Stochastic Steiner Forest models and to [13,12,10,6,9] for additional recent approximation results on stochastic network design problems in general.

Our work is mostly related to the framework of *Probabilistic Combinatorial Optimization*, introduced in [2,14], where repairing strategies as the one described previously are analyzed probabilistically, so that the expected cost of their outcome can be computed efficiently (this ensures that the problem of taking a priori decisions for a particular strategy belongs to class NPO). Several network design problems have been treated in the probabilistic combinatorial optimization framework, including minimum coloring [17], maximum independent set [16], longest path [15], and minimum spanning tree [3]. Apart from probabilistic analysis of repairing strategies, results in [17,16] also include derivation of approximability properties.

The article is structured as follows. At first we introduce notation. In section 2 we present a repairing strategy (algorithm) and derive the expected cost of the repaired feasible solution for the actually materialized subgraph. Approximation properties of the proposed strategy with respect to a priori decisions (solutions) are analyzed in section 3. We show that our approximation results are tight in section 4, and conclude in section 5.

Notation. In what follows we denote by $G_0(V_0, E_0)$ the input metric graph and let $\langle s_r, t_r \rangle$, $r = 1 \ldots k$, denote the k pairs that we have to connect for the Steiner Forest instance. Each vertex $v_i \in V_0 \setminus \{s_r, t_r | r = 1 \ldots k\}$ is accociated to a probability p_i of survival in the actually materialized graph $G_1(V_1, E_1)$, $V_1 \subseteq V_0$. Vertices s_r, t_r, $r = 1 \ldots k$ are assumed to be always present in G_1. G_1 emerges as the complete metric subgraph of G_0, by an independent random Bernoulli trial for each vertex $v_i \in V_0$.

We will elaborate on a feasible a priori solution that is a forest. We denote it as an edge subset $F_0 \subseteq E_0$, consisting of f_0 trees and write $F_0 = \cup_{l=1}^{f_0} T_{0,l}$. A feasible (possibly repaired) solution over the actually materialized subgraph G_1 will be a forest $F_1 = \cup_{l=1}^{f_1} T_{1,l}$, with $T_{1,l} \subseteq E_1$, $l = 1 \ldots f_1$. The subset of F_0 that remains valid for G_1 is denoted with F_0' and we refer with $T_{0,l}'$ to the subset of the tree $T_{0,l}$ that remains valid in G_1. Thus it is $F_0' = \cup_{l=1}^{f_0} T_{0,l}'$. Given two vertices v_i and v_j of some tree T, with $[v_i \cdots v_j]_T$ we denote the set of edges of the unique path connecting v_i and v_j on T.

2   A Repairing Strategy

In this paragraph we design and analyze probabilistically a repairing algorithm for an a priori feasible solution F_0. When the subgraph G_1 materializes, the algorithm identifies the trees of F_0 that become disconnected in G_1 (due to abscense of some edges incident to missing vertices), and reconnects each tree separately by using additional edges from E_1. Clearly this procedure generates a Steiner Forest that is feasible for the originally given pairs $\langle s_r, t_r \rangle$, $r = 1 \ldots k$, and ensures $f_1 = f_0$ i.e., both the a priori and repaired forests have the same number of trees. We explain the procedure followed by the algorithm for reconnecting a particular tree of F_0 that has been disconnected in G_1. This same procedure is followed for every such disconnected tree separately.

Consider the tree $T_{0,l}$, such that $T_{0,l}' \subset T_{0,l}$. The algorithm orders the vertices of $T_{0,l}$ using a Depth-First-Search, starting from an arbitrary leaf-vertex of $T_{0,l}$. Vertices of $T_{0,l}$ are inserted in an ordered list $\mathcal{L}$ in order of visitation by DFS in the following way: if v_i and v_{i+1} are two distinct vertices visited by DFS consecutively for the first time, but no (v_i, v_{i+1}) edge exists in $T_{0,l}$, then they are appended to $\mathcal{L}$ along with the parent vertex u of v_{i+1}, in the order v_i, u, v_{i+1}. Thus $\mathcal{L}$ may contain some vertices more than once (in fact, as many times as their children in $T_{0,l}$). However, $|\mathcal{L}| = O(|T_{0,l}|)$. We note that a different ordered list is produced for each tree of the a priori solution that needs to be repaired.

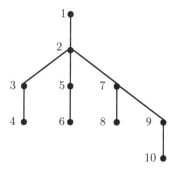

(a) A tree of the a priori solution.

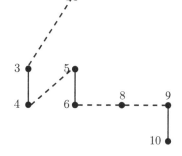

(b) The repaired tree in absence of vertices v_2, v_7. Dashed edges were added by the repairing algorithm.

Fig. 1. Functionality of the repairing algorithm over a particular tree of an a priori forest

When the actual subgraph G_1 materializes, the algorithm sets $T_{1,l} = T'_{0,l}$. Then it removes from $\mathcal{L}$ every copy of vertex $v \in V_0 \setminus V_1$ thus producing the list $\mathcal{L}'$. It scans $\mathcal{L}'$ in order and for every two consecutive vertices v_i, v_j it inserts in $T_{1,l}$ an edge (v_i, v_j) if $i < j$ and v_i, v_j are not already connected in $T_{1,l}$. We illustrate the functionality of the repairing algorithm over a particular tree by an example.

Example. Fig. 1(a) depicts a tree of an a priori solution numbered according to DFS visitation starting from a leaf-vertex. The corresponding ordered list produced in this way is $\mathcal{L} = \{1, 2, 3, 4, 2, 5, 6, 2, 7, 8, 7, 9, 10\}$. Assuming that vertices 2 and 7 are absent from the vertex set of the actually materialized subgraph, all occurences of these vertices are dropped from $\mathcal{L}$ and $\mathcal{L}' = \{1, 3, 4, 5, 6, 8, 9, 10\}$ emerges. The repairing agorithm scans $\mathcal{L}'$ in order and adds edges $(1, 3)$, $(4, 5)$, $(6, 8)$, $(8, 9)$, so as to reconnect the remainders of the a priori tree, as shown in fig. 1(b).

We prove the following:

Proposition 1. *The repairing algorithm produces a connected tree $T_{1,l}$ out of tree $T_{0,l}$ of the a priori solution.*

Proof. For every vertex v_j in $\mathcal{L}'$ there is an appearance of v_j in $\mathcal{L}'$ after a vertex v_i with $i < j$, so that v_j is connected to v_i by the end of the repairing algorithm. This holds for all vertices, apart from the one appearing first in $\mathcal{L}'$. This implies that all vertices are connected into one component by the end of execution of the repairing algorithm for $T_{0,l}$. Furthermore the emerging construction cannot contain cycles for two reasons: $T_{0,l}$ did not have cycles and in order for a cycle to occur in the repaired solution $T_{1,l}$, insertion of at least one edge (v_i, v_j) is required while its endpoints have been already connected. This cannot happen by functionality of the repairing algorithm.   □

Since the repairing algorithm reconnects on G_1 every single tree of the a priori solution that was disconnected, the union of all such repaired trees along with trees that survived unaffected on G_1 yields a feasible Steiner Forest on G_1. These trees remain pairwise vertex-disjoint as they were in the a priori solution, because the repairing algorithm uses only edges to reconnect trees and no such edge connects vertices belonging in different trees. Thus $f_0 = f_1$ is guaranteed.

The complexity of the repairing algorithm is almost linear in the number of vertices of G_0. Indeed, a DFS over a tree $T_{0,l}$ is of $O(|T_{0,l}|)$ time, while by using UNION-FIND disjoint sets representation for maintaining connected components during the scan of $\mathcal{L}'$, an $O(|T_{0,l}|\alpha(|T_{0,l}|))$ time is spent. Summing over all trees of the a priori forest F_0, and because $|T_{0,l}| = O(n)$, we obtain $O(n\alpha(n))$ total time for producing the final feasible forest F_1.

Theorem 1. *Given an arbitrary feasible a priori solution F_0, the expected cost of a repaired solution F_1 is:*

$$E[c(F_1)] = \sum_{l=1}^{f_1}\left(\sum_{(v_i,v_j)\in T_{0,l}} p_ip_jc(v_i,v_j)+ \right.$$

$$+ \sum_{(v_i,v_j)\in E(V(T_{0,l}))\setminus T_{0,l}} c(v_i,v_j)p_ip_j \times \prod_{\substack{v_l\in[v_i,v_j]_{\mathcal{L}_l}:\\ i<j\ ,\ v_i,v_j\notin[v_i,v_j]_{\mathcal{L}_l}}} (1-p_l) \left. \right)$$

where $V(T_{0,l})$ is the set of vertices incident to edges of $T_{0,l}$, and $E(V(T_{0,l}))$ is the set of all edges induced by vertices in $V(T_{0,l})$. Furthermore, $\mathcal{L}_l$ is the ordered list for tree $T_{0,l}$ and $[v_i,v_j]_{\mathcal{L}_l}$ the sublist of $\mathcal{L}_l$ starting at v_i and ending in v_j not including these two vertices. For all sublists not satisfying the specified restrictions we define the product to be 0.

Proof. Each individual expression summed for tree $T_{0,l}$ consists of two terms, the first one expressing the expected cost of surviving edges in the materialized subgraph (that is the expected cost of $T'_{0,l}$), while the second expresses the expected cost of edges added to $T'_{0,l}$ by the repairing algorithm, so that $T'_{0,l}$ is augmented into a feasible tree $T_{1,l}$. The first term is justified by the fact that $(v_i,v_j) \in T_{0,l}$ survives in $T'_{0,l}$ if and only if both its endpoints survive. This happens with probability p_ip_j, since these two events are independent.

The second term emerges by inspection of the functionality of the repairing algorithm. When G_1 materializes, missing vertices (in $V_0 \setminus V_1$) are dropped from the ordered list encoding $\mathcal{L}_l$ and the modified list $\mathcal{L}'_l$ emerges. The repairing algorithm scans $\mathcal{L}'_l$ and for every pair of consecutive vertices v_i, v_j it connects them using an edge (v_i, v_j) if and only if $i < j$ and v_i is not connected to v_j already.

Vertices $v_i, v_j \in \mathcal{L}$ both survive in $\mathcal{L}'_l$ with probability p_ip_j. Vertices v_i and v_j are not connected to each other if all vertices between v_i and v_j in $\mathcal{L}_l$ are missing from $\mathcal{L}'_l$, and this happens with probability $\prod_{v_i \in [v_i, v_j]_{\mathcal{L}}} (1 - p_l)$. Furthermore, neither v_i nor v_j should appear as intermediates in the sublist $[v_i, v_j]_{\mathcal{L}_l}$, otherwise they should also be missing, and would not be encountered by the repairing algorithm. Finally, the sublist $[v_i, v_j]_{\mathcal{L}_l}$ should not be empty, otherwise a surviving edge (v_i, v_j) is implied, rendering v_j connected to v_i. $\square$

Clearly the expression given in theorem 1 is computable in polynomial-time. Thus:

Corollary 1. *The problem of a priori optimizing the steiner forest problem on stochastic metric graphs, for the proposed repairing algorithm, belongs to the class NPO.*

The problem is NP-hard: setting all survival probabilities of vertices of G_0 equal to 1, yields a deterministic steiner forest instance. Results of section 3 imply existence of a polynomial-time 4-approximation algorithm for a priori optimization of the expression of theorem 1.

3   Approximation

In this section we carry out appropriate analysis so as to show that the Steiner Forest problem over a stochastic metric graph can be approximated efficiently within a constant factor, by the repairing algorithm. The heart of our results is the following theorem:

Theorem 2. *If F_1 is a repaired feasible solution produced by the proposed repairing algorithm for the Steiner Forest problem on a stochastic metric graph, given an a priori feasible solution F_0, then $c(F_1) \leq 2c(F_0)$.*

The proof of this result is carried out by an analysis of the algorithm over each tree $T_{0,l}$ separately. The result emerges by summing over the trees of F_1. In the following we denote by $T_{r,l}$ the subset of edges added by the repairing algorithm to $T'_{0,l}$. We prove first some lemmas that will be combined towards the proof of the theorem.

Lemma 1. *For every edge $(v_i, v_j) \in T_{r,l}$ we have $c(v_i, v_j) \leq c([v_i \ldots v_j]_{T_{0,l}})$.*

Proof. Immediate by the triangle inequality holding for the cost function $c : E_0 \to \Re^+$. $\square$

According to lemma 1 we can express the cost of the repaired tree $T_{1,l}$ as follows:

$$c(T_{1,l}) = c(T'_{0,l}) + c(T_{r,l}) \leq \sum_{e \in T'_{0,l}} c(e) + \sum_{(v_i, v_j) \in T_{r,l}} c([v_i \ldots v_j]_{T_{0,l}}) \qquad (1)$$

Lemma 2. *For every three distinct edges (v_i, v_j), (v_k, v_l), (v_q, v_r) in $T_{r,l}$ the paths $[v_i \cdots v_j]_{T_{0,l}}$, $[v_k \cdots v_l]_{T_{0,l}}$, $[v_q \cdots v_r]_{T_{0,l}}$, do not share an edge in common.*

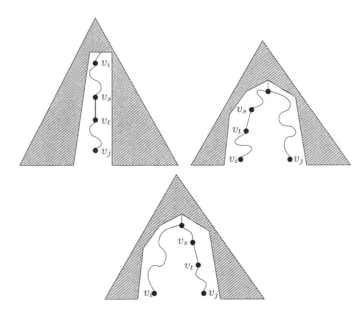

Fig. 2. Three cases that may happen for edge (v_s, v_t) with respect to v_i, v_j (proof of lemma 2)

Proof. By functionality of the repairing algorithm we have that $i < j$, $k < l$, $q < r$. Furthermore, if we assume without loss of generality that the vertex pairs were encountered in the order $\langle i, j \rangle$, $\langle k, l \rangle$, $\langle q, r \rangle$ during scanning of $\mathcal{L}'$, then we deduce that $j < l < r$. If the paths intersect in some common edge (v_s, v_t), then it must be $s, t \leq j$ (fig. 2 depicts all possible cases), thus $s, t < l$ and $s, t < r$. In this case edge (v_s, v_t) must have been scanned at least three times during DFS: once before visitation of each of the vertices v_j, v_l, v_r. But this contradicts the fact that a DFS scans each edge of a graph exactly twice.   □

The following lemma will help us to complete the proof of the theorem:

Lemma 3. *Consider two edges* (v_i, v_j), (v_k, v_l) *in* $T_{r,l}$. *For every edge* (v_s, v_t) *with* $(v_s, v_t) \in [v_i \cdots v_j]_{T_{0,l}} \cap [v_k \cdots v_l]_{T_{0,l}}$ *it holds* $(v_s, v_t) \notin T'_{0,l}$.

Proof. The proof is by contradiction. Suppose that $(v_s, v_t) \in [v_i \cdots v_j]_{T_{0,l}} \cap [v_k \cdots v_l]_{T_{0,l}}$ and $(v_s, v_t) \in T'_{0,l}$. Without loss of generality we assume that the repairing algorithm encountered first the pair $\langle v_i, v_j \rangle$ and afterwards the pair $\langle v_k, v_l \rangle$ in $\mathcal{L}'$. It must be $i < j$, $k < l$ and $j < l$ (v_j may coincide with v_k). Since $(v_s, v_t) \in [v_i \cdots v_j]_{T_{0,l}} \cap [v_k \cdots v_l]_{T_{0,l}}$, then we must have $s, t \leq j$ and, consequently, $s, t < l$. Furthermore, it must hold either that (i) $s, t > k$ or that (ii) $s, t > i$, otherwise it should be $s, t < i$ and, given that $s, t \leq j$ also, we would deduce that (v_s, v_t) would have been scanned twice during DFS, once before visitation of v_i and once before visitation of v_j. In this case it could not have been scanned again right before visitation of v_l. Now (i) cannot hold because

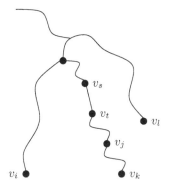

Fig. 3. Case $s, t > i$ examined in the proof of lemma 3: $(v_s, v_t) \in [v_i \cdots v_j]_{T_{0,l}} \cap [v_k \cdots v_l]_{T_{0,l}}$ and suppose that $(v_s, v_t) \in T'_{0,l}$. Then obviously v_s, v_t survive in $\mathcal{L}'$ and appear as intermediates in the two pairs $\langle v_i, v_j \rangle$ and $\langle v_k, v_l \rangle$.

$k \geq j$ and $s, t < j$. If (ii) holds, i.e. $s, t > i$, it is implied that the repairing algorithm did not encounter in $\mathcal{L}'$ vertices v_k, v_l and v_i, v_j consecutively (fig. 3), which is a contradiction. □

The proof of theorem 2 can now be completed as follows:

Proof. Relation (1) can be written:

$$
\begin{aligned}
c(T_{1,l}) &\leq \sum_{e \in T'_{0,l}} c(e) + \sum_{(v_i, v_j) \in T_{r,l}} c([v_i \dots v_j]_{T_{0,l}}) \\
&= \sum_{e \in T'_{0,l}} c(e) + \sum_{(v_i, v_j) \in T_{r,l}} \sum_{e \in [v_i \dots v_j]_{T_{0,l}}} c(e) \\
&= \sum_{e \in T'_{0,l}} c(e) + \sum_{(v_i, v_j) \in T_{r,l}} \left(\sum_{e \in [v_i \dots v_j]_{T_{0,l}} : e \in T'_{0,l}} c(e) + \sum_{e \in [v_i \dots v_j]_{T_{0,l}} : e \notin T'_{0,l}} c(e) \right)
\end{aligned}
$$

By lemmas 2 and 3 the following are implied:

$$
\sum_{(v_i, v_j) \in T_{r,l}} \sum_{e \in [v_i \dots v_j]_{T_{0,l}} : e \in T'_{0,l}} c(e) \leq \sum_{e \in T'_{0,l}} c(e) \tag{2}
$$

$$
\sum_{(v_i, v_j) \in T_{r,l}} \sum_{e \in [v_i \dots v_j]_{T_{0,l}} : e \notin T'_{0,l}} c(e) \leq 2 \sum_{e \in T_{0,l} \setminus T'_{0,l}} c(e) \tag{3}
$$

By replacing the relations (2) and (3) in the expression we obtain:

$$
c(T_{1,l}) \leq 2 \sum_{e \in T'_{0,l}} c(e) + 2 \sum_{e \in T_{0,l} \setminus T'_{0,l}} c(e) \leq 2c(T_{0,l}) \tag{4}
$$

Summing over all trees, since $f_0 = f_1$, we obtain that $c(F_1) \leq 2c(F_0)$. □

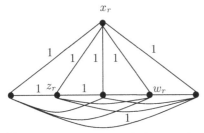

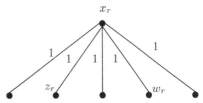

(a) The structure used in proving tightness of analysis for some S_r, with $|S_r| = 5$. Apart from edges with cost 1, all other edges have cost 2.

(b) An optimum star centered at x_r, with cost 5.

Fig. 4. Illustration of worst-case construction for showing tightness of analysis

Now we can state our main approximation result:

Theorem 3. *There is an $O(n\alpha(n))$ time repairing algorithm for the Steiner Forest problem on stochastic metric graphs that, when applied to an α-approximate a priori feasible solution, produces feasible solutions that are 2α-approximate to the optimum expected cost.*

Proof. By theorem 2 $c(F_1) \leq 2c(F_0)$. Let $OPT(G_0)$ and $OPT(G_1)$ be the costs of an optimum Steiner forest on G_0 and G_1 respectively for the given source-destination pairs, and $c(F_0) \leq \alpha OPT(G_0)$. It is $OPT(G_0) \leq OPT(G_1)$ for every possible subgraph G_1 of G_0, because every feasible solution for G_1 is also feasible for G_0. Thus $c(F_1) \leq 2\alpha OPT(G_1)$. Taking expectation over all possible subgraphs G_1 yields $E[c(F_1)] \leq 2\alpha E[OPT(G_1)]$.   □

Corollary 2. *There is an $O(n\alpha(n))$ time repairing algorithm for the Steiner Forest problem on stochastic metric graphs, that can be supported by a poly-nomial-time algorithm for taking a priori decisions [1,8], so as to yield factor 4 approximation of the optimum expected cost. The repairing algorithm is 2-approximate given an optimum feasible a priori solution.*

We note that in both cases mentioned in the corollary, the proposed repairing algorithm is faster than the algorithm used for a priori decisions, and is far more efficient than the trivial practices discussed in the introduction: in fact, any approximation algorithm used for taking a priori decisions (including the one of [1,8]) will incur $\Omega(n^2)$ complexity.

4   Tightness of Analysis

We show in this paragraph that the analysis of the repairing algorithm is tight for arbitrary a priori solution F_0. We construct a worst-case example.

We consider a metric graph $G_0(V_0, E_0)$. For some fixed constant k take k sets of vertices $S_1, \ldots, S_k$, along with a vertex $x_r \notin S_r$, $r = 1 \ldots k$, per subset. Let the input metric graph consist of the vertex set $V_0 = \left(\cup_{r=1}^k S_r \right) \cup \{x_r | r = 1 \ldots k\}$.

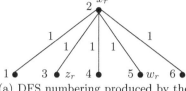
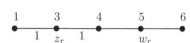

(a) DFS numbering produced by the repairing algorithm for star connection of S_r.

(b) A feasible tree produced over a disconnected star tree, when x_r is missing.

Fig. 5. DFS numbering and repaired tree assumed in showing tightness of analysis

We take $|S_r| = n$ so that $|V_0| = \Theta(n)$, because k is a fixed constant. We set $c(x_r, y) = 1$ for each $y \in S_r$, $r = 1 \ldots k$. For each S_r pick two distinct arbitrary vertices $w_r, z_r \in S_r$ and set $c(z_r, y) = 1$ for all $y \in S_r \setminus \{w_r, z_r\}$ and $c(z_r, w_r) = 2$. For all other edges of the graph we set their cost equal to 2. Fig. 4(a) shows the construction for a particular set S_r.

The Steiner Forest instance that we consider requires that each S_r is connected (it is trivial to express this requirement with source-destination pairs). We assume that the stochastic graph is defined by setting the survival probability of each x_r equal to p. An optimum a priori solution to this instance is defined as a forest consisting of an optimum connecting tree per vertex set S_r. We consider such an a priori solution that the corresponding tree for S_r is the star $T_r = \{(x, y) | y \in S_r\}$. Fig. 4(b) shows the construction for a particular vertex set S_r and the optimum star tree solution for this set.

Among the various cases that may occur in the actually materialized subgraph of G_0 we consider the one where all vertices x_r, $r = 1 \ldots k$ survive, and the case where all vertices x_r are missing. For the first case the a priori optimum solution remains feasible and has an optimum cost of $\sum_{r=1}^{k} |S_r|$, while in the second case, the repairing algorithm is executed for each tree T_r. Fig. 5 depicts the DFS numbering of a tree T_r by the repairing algorithm, and the corresponding repaired solution. It is easy to see that such a "chain" as the one appearing in fig. 5(b), must have a cost at least $2(|S_r| - 1) - 2 = 2(|S_r| - 2)$, because in this chain z_r may be incident to two vertices that are connected with two edges of cost 1 to it. However, the optimum cost for the materialized subgraph occurs if we connect per set S_r its vertices to z_r, and is equal to $|S_r|$. Clearly the optimum expected cost is at most $\sum_{r=1}^{k} |S_r| = k(n - 1)$, while the solution produced by the repairing algorithm has an expected cost of value at least:

$$p^k \sum_{r=1}^{k} |S_r| + 2(1 - p^k) \sum_{r=1}^{k} (|S_r| - 2) = kp^k(n - 1) + 2k(1 - p^k)(n - 2)$$

Hence, the approximation factor is asymptotically lower-bounded by:

$$\lim_{n \to \infty} \frac{kp^k(n - 1) + 2k(1 - p^k)(n - 2)}{k(n - 1)} = p^k + 2(1 - p^k)$$

which approaches arbitrarily close to 2 by choosing p arbitrarily close to 0.

5   Conclusions

We considered the Steiner Forest problem in stochastic metric graphs, where each vertex that is not a source or destination is present with a given probability independently of all other vertices. The problem amounts to coming up with a feasible Steiner Forest for every possible materializable subgraph of the given graph, so as to minimize the expected cost of the resulting solution taken over the distribution of these subgraphs. We designed an efficient algorithm that runs in almost linear time in the number of vertices that adjusts efficiently a priori taken decisions. Given that a priori decisions constitute a feasible forest on the original metric graph we were able to derive a polynomial time computable expression for the expected cost of a Steiner Forest produced by the proposed algorithm. Furthermore, we have shown that this algorithm at most doubles the cost of the a priori solution, and this leads to 2 approximation of the optimum expected cost given an optimum a priori solution, and 4 approximation given a 2 approximate solution. Our analysis of the proposed repairing algorithm was shown to be tight.

We note that for the more special case of the Steiner Tree problem in the same model, the well-known minimum spanning tree heuristic [18] that includes only vertices requiring connection, gives a feasible and 2-approximate a priori solution that trivially remains feasible and 2-approximate for the actually materialized subgraph. As a non-trivial aspect of future work we consider extending our results to the case of complete graphs with general cost functions. Simply using shortest-path distances on these graphs does not straightforwardly lead to efficient and approximate repairing algorithms.

References

1. Agrawal, A., Klein, P.N., Ravi, R.: When Trees Collide: An Approximation Algorithm for the Generalized Steiner Problem on Networks. SIAM Journal on Computing 24(3), 440–456 (1995)
2. Bertsimas, D.: Probabilistic Combinatorial Optimization. PhD thesis (1988)
3. Bertsimas, D.: The probabilistic minimum spanning tree problem. Networks 20, 245–275 (1990)
4. Birge, J.R., Louveaux, F.: Introduction to Stochastic Programming. LNCS. Springer, Heidelberg (1997)
5. Dantzig, G.W.: Linear programming under uncertainty. Management Science 1, 197–206 (1951)
6. Dhamdhere, K., Ravi, R., Singh, M.: On Two-Stage Stochastic Minimum Spanning Trees. In: Jünger, M., Kaibel, V. (eds.) IPCO 2005. LNCS, vol. 3509, pp. 321–334. Springer, Heidelberg (2005)
7. Fleischer, L., Koenemann, J., Leonardi, S., Schaefer, G.: Simple cost sharing schemes for multicommodity rent-or-buy and stochastic steiner tree. In: Proceedings of the ACM Symposium on Theory of Computing (STOC), pp. 663–670. ACM Press, New York (2006)
8. Goemans, M.X., Williamson, D.P.: A General Approximation Technique for Constrained Forest Problems. SIAM Journal on Computing 24(2), 296–317 (1995)

9. Gupta, A., Pál, M.: Stochastic Steiner Trees Without a Root. In: Caires, L., Italiano, G.F., Monteiro, L., Palamidessi, C., Yung, M. (eds.) ICALP 2005. LNCS, vol. 3580, pp. 1051–1063. Springer, Heidelberg (2005)

10. Gupta, A., Pál, M., Ravi, R., Sinha, A.: What About Wednesday? Approximation Algorithms for Multistage Stochastic Optimization. In: Proceedings of the International Workshop on Approximation and Randomized Algorithms (APPROX-RANDOM), pp. 86–98 (2005)

11. Gupta, A., Pál, M., Ravi, R., Sinha, A.: Boosted sampling: approximation algorithms for stochastic optimization. In: Gupta, A. (ed.) Proceedings of the ACM Symposium on Theory of Computing (STOC), pp. 417–426. ACM Press, New York (2004)

12. Gupta, A., Ravi, R., Sinha, A.: An Edge in Time Saves Nine: LP Rounding Approximation Algorithms for Stochastic Network Design. In: Proceedings of the IEEE Symposium on Foundations of Computer Science (FOCS), pp. 218–227. IEEE Computer Society Press, Los Alamitos (2004)

13. Immorlica, N., Karger, D.R., Minkoff, M., Mirrokni, V.S.: On the costs and benefits of procrastination: approximation algorithms for stochastic combinatorial optimization problems. In: Proceedings of the ACM-SIAM Symposium on Discrete Algorithms (SODA), pp. 691–700. ACM Press, New York (2004)

14. Jaillet, P.: Probabilistic Traveling Salesman Problems. PhD thesis (1985)

15. Murat, C., Paschos, V.Th.: The probabilistic longest path problem. Networks 33(3), 207–219 (1999)

16. Murat, C., Paschos, V.Th.: A priori optimization for the probabilistic maximum independent set problem. Theoretical Computer Science 270(1–2), 561–590 (2002)

17. Murat, C., Paschos, V.Th.: On the probabilistic minimum coloring and minimum k-coloring. Discrete Applied Mathematics 154(3), 564–586 (2006)

18. Vazirani, V.: Approximation Algorithms. LNCS. Springer, Heidelberg (2003)

On Threshold BDDs and the Optimal Variable Ordering Problem

Markus Behle

Max-Planck-Institut für Informatik, Stuhlsatzenhausweg 85, 66123 Saarbrücken,
Germany
behle@mpi-inf.mpg.de

Abstract. Many combinatorial optimization problems can be formulated as 0/1 integer programs (0/1 IPs). The investigation of the structure of these problems raises the following tasks: count or enumerate the feasible solutions and find an optimal solution according to a given linear objective function. All these tasks can be accomplished using binary decision diagrams (BDDs), a very popular and effective datastructure in computational logics and hardware verification.

We present a novel approach for these tasks which consists of an *output-sensitive* algorithm for building a BDD for a linear constraint (a so-called threshold BDD) and a *parallel* AND operation on threshold BDDs. In particular our algorithm is capable of solving knapsack problems, subset sum problems and multidimensional knapsack problems.

BDDs are represented as a directed acyclic graph. The size of a BDD is the number of nodes of its graph. It heavily depends on the chosen variable ordering. Finding the optimal variable ordering is an NP-hard problem. We derive a 0/1 IP for finding an optimal variable ordering of a threshold BDD. This 0/1 IP formulation provides the basis for the computation of the variable ordering spectrum of a threshold function.

We introduce our new tool azove 2.0 as an enhancement to azove 1.1 which is a tool for counting and enumerating 0/1 points. Computational results on benchmarks from the literature show the strength of our new method.

1 Introduction

For many problems in combinatorial optimization the underlying polytope is a 0/1 polytope, i.e. all feasible solutions are 0/1 points. These problems can be formulated as 0/1 integer programs. The investigation of the polyhedral structure often raises the following problem:

Given a set of inequalities $Ax \leq b$, $A \in \mathbb{Z}^{m \times d}$, $b \in \mathbb{Z}^m$, compute a list of all 0/1 points satisfying the system.

Binary decision diagrams (BDDs) are perfectly suited to compactly represent all 0/1 solutions. Once the BDD for a set of inequalities is built, counting the solutions and optimizing according to a linear objective function can be done in

A. Dress, Y. Xu, and B. Zhu (Eds.): COCOA 2007, LNCS 4616, pp. 124–135, 2007.

time linear in the size of the BDD, see e.g. [1,3]. Enumerating all solutions can be done by a traversal of the graph representing the BDD.

In section 2 of this paper we develop a new output-sensitive algorithm for building a QOBDD for a linear constraint (a so-called threshold BDD). More precisely, our algorithm constructs exactly as many nodes as the final QOBDD consists of and does not need any extra memory. In section 3 the synthesis of these QOBDDs is done by an AND operation on all QOBDDs *in parallel* which is also a novelty. Constructing the final BDD by sequential AND operations on pairs of BDDs (see e.g. [3]) may lead to explosion in size during computation even if the size of the final BDD is small. We overcome this problem by our parallel AND operation.

The size of a BDD heavily depends on the variable ordering. Finding a variable ordering for which the size of a BDD is minimal is a difficult task. Bollig and Wegener [4] showed that improving a given variable ordering of a general BDD is NP-complete. For the optimal variable ordering problem for a threshold BDD we present for the first time a 0/1 IP formulation in section 4. Its solution gives the optimal variable ordering and the number of minimal nodes needed. In contrast to all other exact BDD minimization techniques (see [7] for an overview) which are based on the classic method by Friedman and Supowit [8], our approach does not need to build a BDD explicitly. With the help of this 0/1 IP formulation and the techniques for counting 0/1 vertices described in [3] we are able to compute the variable ordering spectrum of a threshold function.

We present our new tool `azove` 2.0 [2] which is based on the algorithms developed in sections 2 and 3. Our tool `azove` is able to count and enumerate all 0/1 solutions of a given set of linear constraints, i.e. it is capable of constructing all solutions of the knapsack, the subset sum and the multidimensional knapsack problem. In section 5 we present computational results for counting the satisfiable solutions of SAT instances, matchings in graphs and 0/1 points of general 0/1 polytopes.

BDDs

BDDs were first proposed by Lee in 1959 [11]. Bryant [5] presented efficient algorithms for the synthesis of BDDs. After that, BDDs became very popular in the area of hardware verification and computational logics, see e.g. [12,16].

We provide a short definition of BDDs as they are used in this paper. A *BDD* for a set of variables $x_1, \ldots, x_d$ is a directed acyclic graph $G = (V, A)$, see figure **??**. All nodes associated with the variable x_i lie on the same level labeled with x_i, which means, we have an *ordered* BDD (OBDD). In this paper all BDDs are ordered. For the edges there is a parity function par: $A \to \{0, 1\}$. The graph has one node with in-degree zero, called the root and two nodes with out-degree zero, called leaf 0 resp. leaf 1. Apart from the leaves all nodes have two outgoing edges with different parity. A path $e_1, \ldots, e_d$ from the root to one of the leaves represents a variable assignment, where the level label x_i of the starting node of e_j is assigned to the value par(e_j). An edge crossing a level with nodes labeled x_i is called a *long* edge. In that case the assignment for x_i is free. All paths from

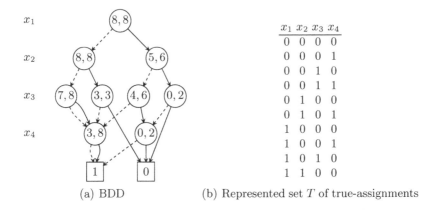

		x_1 x_2 x_3 x_4

<table>
<tr><td></td><td>(a) BDD</td><td>(b) Represented set T of true-assignments</td></tr>
</table>

Fig. 1. A threshold BDD representing the linear constraint $2x_1 + 5x_2 + 4x_3 + 3x_4 \leq 8$. Edges with parity 0 are dashed.

the root to leaf 1 represent the set $T \subseteq \{0,1\}^d$ of true-assignments. The *size* of a BDD is defined as the number of nodes $|V|$. Let w_l be the number of nodes in level l. The *width* of a BDD is the maximum of all number of nodes in a level $w = \max\{w_l \mid l \in 1, \ldots, d\}$.

Vertices $u, v \in V$ with the same label are *equivalent* if both of their edges with the same parity point to the same node respectively. If each path from root to leaf 1 contains exactly d edges the BDD is called *complete*. A complete and ordered BDD with no equivalent vertices is called a *quasi-reduced* ordered BDD (*QOBDD*). A vertex $v \in V$ is *redundant* if both outgoing edges point to the same node. If an ordered BDD does neither contain redundant nor equivalent vertices it is called *reduced* ordered BDD (*ROBDD*). For a fixed variable ordering both QOBDD and ROBDD are canonical representations.

A BDD representing the set $T = \{x \in \{0,1\}^d : a^T x \leq b\}$ of 0/1 solutions to the linear constraint $a^T x \leq b$ is called a *threshold BDD*. For each variable ordering the size of a threshold BDD is bounded by $O\left(d(|a_1|, \ldots, |a_d|)\right)$, i.e. if the weights $a_1, \ldots, a_d$ are polynomial bounded in d, the size of the BDD is polynomial bounded in d (see [16]). Hosaka et. al. [10] provided an example of an explicitly defined threshold function for which the size of the BDD is exponential for all variable orderings.

2   Output-Sensitive Building of a Threshold BDD

In this section we give a new output-sensitive algorithm for building a threshold QOBDD of a linear constraint $a^T x \leq b$ in dimension d. This problem is closely related to the knapsack problem. Our algorithm can easily be transformed to work for a given equality, i.e. it can also solve the subset sum problem.

A crucial point of BDD construction algorithms is the in advance detection of equivalent nodes [12]. If equivalent nodes are not fully detected this leads to isomorphic subgraphs. As the representation of QOBDDs and ROBDDs is canonical these isomorphic subgraphs will be detected and merged at a later stage which is a considerable overhead.

We now describe an algorithm that overcomes this drawback. Our detection of equivalent nodes is exact and complete so that only as many nodes will be built as the final QOBDD consists of. No nodes have to be merged later on. Be w the width of the BDD. The runtime of our algorithm is $O(dw \log(w))$

W.l.o.g. we assume $\forall i \in \{1, \ldots, d\}$ $a_i \geq 0$ (in case $a_i < 0$ substitute x_i with $1 - \bar{x}_i$). In order to exclude trivial cases let $b \geq 0$ and $\sum_{i=1}^{d} a_i > b$. For the sake of simplicity be the given variable ordering the canonical variable ordering $x_1, \ldots, x_d$. We assign weights to the edges depending on their parity and level. Edges with parity 1 in level l cost a_l and edges with parity 0 cost 0. The key to exact detection of equivalent nodes are two bounds that we introduce for each node, a lower bound lb and an upper bound ub. They describe the interval $[lb, ub]$. Let c_u be the costs of the path from the root to the node u. All nodes u in level l for which the value $b - c_u$ lies in the interval $[lb_v, ub_v]$ of a node v in level l are guaranteed to be equivalent with the node v. We call the value $b - c_u$ the slack. Figure 1(a) illustrates a threshold QOBDD with the intervals set in each node.

Algorithm 1.   Build QOBDD for the constraint $a^T x \leq b$

BuildQOBDD(slack, level)
1: **if** slack < 0 **then**
2:     **return** leaf 0
3: **if** slack $\geq \sum_{i=\text{level}}^{d} a_i$ **then**
4:     **return** leaf 1
5: **if** exists node v in level with $lb_v \leq$ slack $\leq ub_v$ **then**
6:     **return** v
7: **build** new node u in level
8: $l = $ level of node
9: 0-edge son $= $ BuildQOBDD(slack, $l + 1$)
10: 1-edge son $= $ BuildQOBDD(slack - a_l, $l + 1$)
11: **set lb** to max(lb of 0-edge son, lb of 1-edge son $+ a_l$)
12: **set ub** to min(ub of 0-edge son, ub of 1-edge son $+ a_l$)
13: **return** u

Algorithm 1 constructs the QOBDD top-down from a given node in a depth-first-search manner. We set the bounds for the leaves as follows: $lb_{\text{leaf } 0} = -\infty$, $ub_{\text{leaf } 0} = -1$, $lb_{\text{leaf } 1} = 0$ and $ub_{\text{leaf } 1} = \infty$. We start at the root with its slack set to b. While traversing downwards along an edge in step 9 and 10 we substract its costs. The sons of a node are built recursively. The slack always reflects the value of the right hand side b minus the costs c of the path from the root to the node. In step 5 a node is detected to be equivalent with an already built node v in that level if there exists a node v with slack $\in [lb_v, ub_v]$.

If both sons of a node have been built recursively at step 11 the lower bound is set to the costs of the longest path from the node to leaf 1. In case one of the sons is a long edge pointing from this level l to leaf 1 the value $lb_{\text{leaf 1}}$ has to be temporarily increased by $\sum_{i=l+1}^{d} a_i$ before. In step 12 the upper bound is set to the costs of the shortest path from the node to leaf 0 minus 1. For this reason the interval $[lb, ub]$ reflects the widest possible interval for equivalent nodes.

Lemma 1. *The detection of equivalent nodes in algorithm 1 is exact and complete.*

Proof. Assume to the contrary that in step 7 a new node u is built which is equivalent to an existing node v in the level. Again let c_u be the costs of the path from the root to the node u. Because of step 5 we have $b - c_u \notin [lb_v, ub_v]$.

Case $b - c_u < lb_v$:
In step 11 lb_v has been computed as the costs of the longest path from the node v to leaf 1. Let lb_u be the costs of the longest path from node u to leaf 1. Then there is a path from root to leaf 1 using node u with costs $c_u + lb_u \leq b$, so we have $lb_u < lb_v$. As the nodes u and v are equivalent they are the root of isomorphic subtrees, and thus $lb_u = lb_v$ holds.
Case $b - c_u > ub_v$:
With step 12 ub_v is the costs of the shortest path from v to leaf 0 minus 1. Be ub_u the costs of the shortest path from u to leaf 0 minus 1. Again the nodes u and v are equivalent so for both the costs we have $ub_u = ub_v$. Thus there is a path from root to leaf 0 using node u with costs $c_u + ub_u < b$ which is a contradiction.

Algorithm 1 can be modified to work for a given equality, i.e. it can also be used to solve the subset sum problem. The following replacements have to be made:

1: *replace* slack < 0 *with* slack $< 0 \vee$ slack $> \sum_{i=\text{level}}^{d} a_i$
3: *replace* slack $\geq \sum_{i=\text{level}}^{d} a_i$ *with* slack $= 0 \wedge$ slack $= \sum_{i=\text{level}}^{d} a_i$

3   Parallel AND Operation on Threshold BDDs

Given a set of inequalities $Ax \leq b$, $A \in \mathbb{Z}^{m \times d}$, $b \in \mathbb{Z}^m$, we want to build the ROBDD representing all $0/1$ points satisfying the system. This problem is closely related to the multidimensional knapsack problem. Our approach is the following. For each of the m linear constraints $a_i^T x \leq b_i$ we build the QOBDD with the method described in section 2. Then we build the final ROBDD by performing an AND operation on all QOBDDs in parallel. The space consumption for saving the nodes is exactly the number of nodes that the final ROBDD consists of plus d temporary nodes. Algorithm 2 describes our parallel *and*-synthesis of m QOBDDs.

We start at the root of all QOBDDs and construct the ROBDD from its root top-down in a depth-first-search manner. In steps 1 and 3 we check in parallel for trivial cases. Next we generate a signature for this temporary node of the ROBDD in step 5. This signature is a $1 + m$ dimensional vector consisting of the

---

Algorithm 2. Parallel conjunction of the QOBDDs $G_1, \ldots, G_m$

PARALLELANDBDDS($G_1, \ldots, G_m$)

1: **if** $\forall i \in \{1, \ldots, m\} : G_i = $ leaf 1 **then**
2:      **return** leaf 1
3: **if** $\exists i \in \{1, \ldots, m\} : G_i = $ leaf 0 **then**
4:      **return** leaf 0
5: **if** signature($G_1, \ldots, G_m$) $\in$ ComputedTable **then**
6:      **return** ComputedTable[signature($G_1, \ldots, G_m$)]
7: $x_i = $ NEXTVARIABLE($G_1, \ldots, G_m$)
8: 0-edge son = PARALLELANDBDDS($G_1|_{x_i=0}, \ldots, G_m|_{x_i=0}$)
9: 1-edge son = PARALLELANDBDDS($G_1|_{x_i=1}, \ldots, G_m|_{x_i=1}$)
10: **if** 0-edge son = 1-edge son **then**
11:      **return** 0-edge son
12: **if** $\exists$ node v in this level with same sons **then**
13:      **return** v
14: **build** node u with 0-edge and 1-edge son
15: ComputedTable[signature($G_1, \ldots, G_m$)] = u
16: **return** u

---

current level and the upper bounds saved in all current nodes of the QOBDDs. If there already exists a node in the ROBDD with the same signature we have found an equivalent node and return it. Otherwise we start building boths sons recursively from this temporary node in steps 8 and 9. From all starting nodes in the QOBDDs we traverse the edges with the same parity in parallel.

When both sons of a temporary node in the ROBDD were built we check its redundance in step 10. In step 12 we search for an already existing node in the current level which is equivalent to the temporary node. If neither is the case we build this node in the ROBDD and save its signature.

In practice the main problem of the parallel *and*-operation is the low hitrate of the ComputedTable. This is because equivalent nodes of the ROBDD can have different signatures and thus are not detected in step 5. In addition the space consumption for the ComputedTable is enormous and one is usually interested in restricting it. The space available for saving the signatures in the ComputedTable can be changed dynamically. This controls the runtime in the following way. The more space is granted for the ComputedTable the more likely equivalent node will be detected in advance which decreases the runtime. Note that because of the check for equivalence in step 12 the correctness of the algorithm does not depend on the use of the ComputedTable. If the use of the ComputedTable is little the algorithm naturally tends to exponential runtime.

4 Optimal Variable Ordering of a Threshold BDD Via 0/1 IP Formulation

Given a linear constraint $a^T x \leq b$ in dimension d we want to find an optimal variable ordering for building the threshold ROBDD. A variable ordering is called

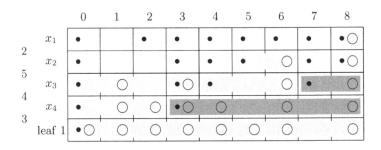

Fig. 2. Dynamic programming table for the linear constraint $2x_1+5x_2+4x_3+3x_4 \leq 8$. Variables U_{ln}, D_{ln} are shown as •, ○ resp. The light grey blocks represent the nodes in the ROBDD and the dark grey blocks represent the redundant nodes in the QOBDD.

optimal if it belongs to those variable orderings for which the size of the ROBDD is minimal. In the following we will derive a 0/1 integer programm whose solution gives the optimal variable ordering and the number of minimal nodes needed.

Building a threshold BDD is closely related to solving a knapsack problem. A knapsack problem can be solved with dynamic programming [13] using a table. We mimic this approach on a virtual table of size $(d+1) \times (b+1)$ which we fill with variables. Figure 2 shows an example of such a table for a fixed variable ordering. The corresponding BDD is shown in figure 1(a).

W.l.o.g. we assume $\forall i \in \{1, \ldots, d\}$ $a_i \geq 0$, and to exclude trivial cases, $b \geq 0$ and $\sum_{i=1}^{d} a_i > b$. Now we start setting up the 0/1 IP shown in figure 3. The 0/1 variables y_{li} (24) encode a variable ordering in the way that $y_{li} = 1$ iff the variable x_i lies on level l. To ensure a correct encoding of a variable ordering we need that each index is on exactly one level (2) and that on each level there is exactly one index (3).

We simulate a *down operation* in the dynamic programming table with the 0/1 variables D_{ln} (25). The variable D_{ln} is 1 iff there exists a path from the root to the level l such that b minus the costs of the path equals n. The variables in the first row (4) and the right column (5) are fixed. We have to set variable $D_{(l+1)n}$ to 1 if we followed the 0-edge starting from $D_{ln} = 1$

$$D_{ln} = 1 \rightarrow D_{(l+1)n} = 1 \ (12)$$

or according to the variable ordering given by the y_{li} variables, if we followed the 1-edge starting from $D_{l(n+a_i)} = 1$

$$y_{li} = 1 \wedge D_{l(n+a_i)} = 1 \rightarrow D_{(l+1)n} = 1 \ (15)$$

In all other cases we have to prevent $D_{(l+1)n}$ from being set to 1

$$y_{li} = 1 \wedge D_{ln} = 0 \rightarrow D_{(l+1)n} = 0 \ (16)$$
$$y_{li} = 1 \wedge D_{l(n+a_i)} = 0 \wedge D_{ln} = 0 \rightarrow D_{(l+1)n} = 0 \ (17)$$

In the same way, the *up operation* is represented by the 0/1 variables U_{ln} (26). The variable U_{ln} is 1 iff there exists a path upwards from the leaf 1 to the level l

$$\text{min} \qquad \sum_{\substack{l \in \{1,\ldots,d+1\} \\ n \in \{0,\ldots,b\}}} C_{ln} + 1 \tag{1}$$

s.t.

$$\forall i \in \{1,\ldots,d\} \qquad \sum_{l=1}^{d} y_{li} = 1 \tag{2}$$

$$\forall l \in \{1,\ldots,d\} \qquad \sum_{i=1}^{d} y_{li} = 1 \tag{3}$$

$$\forall n \in \{0,\ldots,b-1\} \qquad D_{1n} = 0 \tag{4}$$

$$\forall l \in \{1,\ldots,d+1\} \qquad D_{lb} = 1 \tag{5}$$

$$\forall n \in \{1,\ldots,b\} \qquad U_{(d+1)n} = 0 \tag{6}$$

$$\forall l \in \{1,\ldots,d+1\} \qquad U_{l0} = 1 \tag{7}$$

$$B_{(d+1)0} = 1 \tag{8}$$

$$\forall n \in \{1,\ldots,b\} \qquad B_{(d+1)n} = 0 \tag{9}$$

$$C_{(d+1)0} = 1 \tag{10}$$

$$\forall n \in \{1,\ldots,b\} \qquad C_{(d+1)n} = 0 \tag{11}$$

$$\forall l \in \{1,\ldots,d\}:$$

$$\forall n \in \{0,\ldots,b-1\} \qquad D_{ln} - D_{(l+1)n} \le 0 \tag{12}$$

$$\forall n \in \{1,\ldots,b\} \qquad U_{(l+1)n} - U_{ln} \le 0 \tag{13}$$

$$\forall n \in \{0,\ldots,b\}, j \in \{1,\ldots,n+1\} \quad D_{ln} + U_{l(j-1)} - \sum_{i=j}^{n} U_{li} - B_{l(j-1)} \le 1 \tag{14}$$

$$\forall l \in \{1,\ldots,d\}, i \in \{1,\ldots,d\}:$$

$$\forall n \in \{0,\ldots,b-a_i\} \qquad y_{li} + D_{l(n+a_i)} - D_{(l+1)n} \le 1 \tag{15}$$

$$\forall n \in \{b-a_i+1,\ldots,b-1\} \qquad y_{li} - D_{ln} + D_{(l+1)n} \le 1 \tag{16}$$

$$\forall n \in \{0,\ldots,b-a_i\} \qquad y_{li} - D_{l(n+a_i)} - D_{ln} + D_{(l+1)n} \le 1 \tag{17}$$

$$\forall n \in \{a_i,\ldots,b\} \qquad y_{li} + U_{(l+1)(n-a_i)} - U_{ln} \le 1 \tag{18}$$

$$\forall n \in \{1,\ldots,a_i-1\} \qquad y_{li} - U_{(l+1)n} + U_{ln} \le 1 \tag{19}$$

$$\forall n \in \{a_i,\ldots,b\} \quad y_{li} - U_{(l+1)(n-a_i)} - U_{(l+1)n} + U_{ln} \le 1 \tag{20}$$

$$\forall n \in \{0,\ldots,a_i-1\} \qquad y_{li} + B_{ln} - C_{ln} \le 1 \tag{21}$$

$$\forall n \in \{0,\ldots,a_i-1\} \qquad y_{li} - B_{ln} + C_{ln} \le 1 \tag{22}$$

$$\forall n \in \{a_i,\ldots,b\}, k \in \{n-a_i+1,\ldots,n\} \qquad y_{li} + B_{ln} + B_{(l+1)k} - C_{ln} \le 2 \tag{23}$$

$$\forall l \in \{1,\ldots,d\}, i \in \{1,\ldots,d\}: \qquad y_{li} \in \{0,1\} \tag{24}$$

$$\forall l \in \{1,\ldots,d+1\}, n \in \{0,\ldots,b\}: \qquad D_{ln} \in \{0,1\} \tag{25}$$

$$U_{ln} \in \{0,1\} \tag{26}$$

$$B_{ln} \in \{0,1\} \tag{27}$$

$$C_{ln} \in \{0,1\} \tag{28}$$

Fig. 3. 0/1 integer program for finding the optimal variable ordering of a threshold BDD for a linear constraint $a^T x \le b$ in dimension d

with costs n. The variables in the last row (6) and the left column (7) are fixed. We have to set $U_{ln} = 1$ if there is a 0-edge ending in $U_{(l+1)n} = 1$

$$U_{(l+1)n} = 1 \rightarrow U_{ln} = 1 \ (13)$$

or according to the variable ordering given by the y_{li} variables, if there is a 1-edge ending in $U_{(l+1)(n-a_i)} = 1$

$$y_{li} = 1 \land U_{(l+1)(n-a_i)} = 1 \rightarrow U_{ln} = 1 \ (18)$$

In all other cases we have to prevent U_{ln} from being set to 1

$$y_{li} = 1 \land U_{(l+1)n} = 0 \rightarrow U_{ln} = 0 \ (19)$$
$$y_{li} = 1 \land U_{(l+1)(n-a_i)} = 0 \land U_{(l+1)n} = 0 \rightarrow U_{ln} = 0 \ (20)$$

Next we introduce the 0/1 variables B_{ln} (27) which mark the beginning of the blocks in the dynamic programming table that correspond to the nodes in the QOBDD. These blocks can be identified as follows: start from a variable D_{ln} set to 1 and look to the left until a variable U_{ln} set to 1 is found

$$D_{ln} = 1 \land U_{l(j-1)} = 1 \land \bigwedge_{i=j}^{n} U_{li} = 0 \rightarrow B_{l(j-1)} = 1 \ (14)$$

We set the last row explicitly (8), (9).

At last we introduce the 0/1 variables C_{ln} (28) which indicate the beginning of the blocks that correspond to the nodes in the ROBDD. The variables C_{ln} only depend on the B_{ln} variables and exclude redundant nodes. The first blocks are never redundant

$$y_{li} = 1 \rightarrow B_{ln} = C_{ln} \ (21), (22)$$

If the 0-edge leads to a different block than the 1-edge, the block is not redundant

$$y_{li} = 1 \land B_{ln} = 1 \land \left(\bigvee_{k=n-a_i+1}^{n} B_{(l+1)k} = 1 \right) \rightarrow C_{ln} = 1 \ (23)$$

We set the last row explicitly (10), (11).

The objective function (1) is to minimize the number of variables C_{ln} set to 1 plus an offset of 1 for counting the leaf 0. An optimal solution to the IP then gives the minimal number of nodes needed for the ROBDD while the y_{li} variables encode the best variable ordering.

In practice solving this 0/1 IP is not faster than exact BDD minimization algorithms which are based on Friedman and Supowit's method [8] in combination with branch & bound (see [7] for an overview). Nevertheless it is of theoretical interest as the presented 0/1 IP formulation can be used for the computation of the variable ordering spectrum of a threshold function. The *variable ordering spectrum* of a linear constraint $a^T x \leq b$ is the function

$sp_{a^T x \leq b} \colon \mathbb{N} \to \mathbb{N}$, where $sp_{a^T x \leq b}(k)$ is the number of variable orderings leading to a ROBDD for the threshold function $a^T x \leq b$ of size k. In order to compute $sp_{a^T x \leq b}(k)$ we equate the objective function (1) with k and add it as the constraint $\sum_{\substack{l \in \{1,\ldots,d+1\} \\ n \in \{0,\ldots,b\}}} C_{ln} + 1 = k$ to the formulation given in figure 3. The number of $0/1$ vertices of the polytope corresponding to this formulation then equals $sp_{a^T x \leq b}(k)$. In [3] we provide a method for counting these $0/1$ vertices.

5   Computational Results

We developed the tool `azove` 2.0 which implements the output-sensitive building of QOBDDs and the parallel AND synthesis as described in sections 2 and 3. It can be downloaded from [2]. In contrast to version 1.1 which uses CUDD 2.4.1 [14] as BDD manager, the new version 2.0 does not need an external library for managing BDDs.

In the following we compare `azove` 2.0 to `azove` 1.1 which sequentially uses a pairwise AND operation [3]. We restrict our comparisson to these two tools since we are not aware of another software tool specialized in counting $0/1$ solutions for general type of problems. The main space consumption of `azove` 2.0 is due to the storage of the signatures of the ROBDD nodes. We restrict the number of stored signatures to a fixed number. In case more signatures need to be stored we start overwriting them from the beginning.

Our benchmark set contains different classes of combinatorial optimization problems. All tests were run on a Linux system with kernel 2.6.15 and gcc 3.3.5 on a 64 bit AMD Opteron CPU with 2.4 GHz and 4 GB memory. Table 1 shows the comparisson of the runtimes in seconds. We set a time limit of 4 hours. An asterisk marks the exceedance of the time limit.

In fields like verification and real-time systems specification counting the solutions of SAT instances has many applications. From several SAT competitions [6,9] we took the instances aim, hole, ca004 and hfo6, converted them to linear constraint sets and counted their satisfying solutions. The aim instances are 3-SAT instances and the hole instances encode the pigeonhole principle. There are 20 satisfiable hfo6 instances for which the results are similiar. For convenience we only show the first 4 of them.

Counting the number of matchings in a graph is one of the most prominent counting problems with applications in physics in the field of statistical mechanics. We counted the number of matchings for the urquhart instance, which comes from a particular family of bipartite graphs [15], and for f2, which is a bipartite graph encoding a projective plane known as the Fano plane.

The two instance classes OA and TC were taken from a collection of $0/1$ polytopes that has been compiled in connection with [17]. Starting from the convex hull of these polytopes as input we counted their $0/1$ vertices.

For instances with a large number of constraints `azove` 2.0 clearly outperforms version 1.1. Due to the explosion in size during the sequential AND operation `azove` 1.1 is not able to solve some instances within the given time limit. The parallel AND operation in `azove` 2.0 successfully overcomes this problem.

Table 1. Comparisson of the tools `azove 1.1` and `azove 2.0`

Name	Dim	Inequalities	0/1 solutions	azove 1.1	azove 2.0
aim-50-3_4-yes1-2	50	270	1	77.26	50.23
aim-50-6_0-yes1-1	50	400	1	43.97	9.59
aim-50-6_0-yes1-2	50	400	1	179.05	1.62
aim-50-6_0-yes1-3	50	400	1	97.24	4.58
aim-50-6_0-yes1-4	50	400	1	164.88	13.08
hole6	42	217	0	0.15	0.09
hole7	56	316	0	4.16	1.57
hole8	72	441	0	5572.74	29.69
ca004.shuffled	60	288	0	53.07	20.38
hfo6.005.1	40	1825	1	*	1399.57
hfo6.006.1	40	1825	4	*	1441.56
hfo6.008.1	40	1825	2	*	1197.91
hfo6.012.1	40	1825	1	*	1391.39
f2	49	546	151200	*	49.50
urquhart2_25.shuffled	60	280	0	*	12052.10
OA:9-33	9	1870	33	0.05	0.03
OA:10-44	10	9708	44	0.51	0.34
TC:9-48	9	6875	48	0.16	0.15
TC:10-83	10	41591	83	1.96	1.24
TC:11-106	11	250279	106	26.41	11.67

References

1. Becker, B., Behle, M., Eisenbrand, F., Wimmer, R.: BDDs in a branch and cut framework. In: Nikoletseas, S. (ed.) WEA 2005. LNCS, vol. 3503, pp. 452–463. Springer, Heidelberg (2005)
2. Behle, M.: Another Zero One Vertex Enumeration tool Homepage (2007), http://www.mpi-inf.mpg.de/~behle/azove.html
3. Behle, M., Eisenbrand, F.: 0/1 vertex and facet enumeration with BDDs. In: Workshop on Algorithm Engineering and Experiments (ALENEX'07), New Orleans (to appear) (January 2007)
4. Bollig, B., Wegener, I.: Improving the variable ordering of OBDDs is NP-complete. IEEE Transactions on Computers 45(9), 993–1002 (1996)
5. Bryant, R.E.: Graph-based algorithms for Boolean function manipulation. IEEE Transactions on Computers C-35, 677–691 (1986)
6. Buro, M., Kleine Büning, H.: Report on a SAT competition. Bulletin of the European Association for Theoretical Computer Science 49, 143–151 (1993)
7. Ebendt, R., Günther, W., Drechsler, R.: An improved branch and bound algorithm for exact BDD minimization. IEEE Transactions on Computer-Aided Design of Integrated Circuits and Systems 22(12), 1657–1663 (2003)
8. Friedman, S., Supowit, K.: Finding the optimal variable ordering for binary decision diagrams. In: Proceedings of the 24th ACM/IEEE Design Automation Conference, pp. 348–356. IEEE Computer Society Press, Los Alamitos (1987)
9. Hoos, H.H., Stützle, T.: SATLIB: An online resource for research on SAT. In: Gent, I.P., Walsh, T. (eds.) Satisfiability in the year 2000, pp. 283–292. IOS Press, Amsterdam (2000)

10. Hosaka, K., Takenaga, Y., Kaneda, T., Yajima, S.: Size of ordered binary decision diagrams representing threshold functions. Theoretical Computer Science 180, 47–60 (1997)
11. Lee, C.Y.: Representation of switching circuits by binary-decision programs. The Bell Systems Technical Journal 38, 985–999 (1959)
12. Meinel, C., Theobald, T.: Algorithms and Data Structures in VLSI Design. Springer, Heidelberg (1998)
13. Schrijver, A.: Theory of Linear and Integer Programming. John Wiley, Chichester (1986)
14. Somenzi, F.: CU Decision Diagram Package Release 2.4.1 Homepage. Department of Electrical and Computer Engineering, University of Colorado at Boulder (May 2005) `http://vlsi.colorado.edu/~fabio/CUDD`
15. Urquhart, A.: Hard examples for resolution. Journal of the ACM 34(1), 209–219 (1987)
16. Wegener, I.: Branching Programs and Binary Decision Diagrams. SIAM Monographs on Discrete Mathematics and Applications. SIAM, Philadelphia, PA (2000)
17. Ziegler, G.M.: Lectures on Polytopes. LNCS. Springer, Heidelberg (1995)

Communication Leading to Nash Equilibrium
Through Robust Messages
– S5-Knowledge Model Case –

Takashi Matsuhisa*

Department of Liberal Arts and Sciences, Ibaraki National College of Technology
Nakane 866, Hitachinaka-shi, Ibaraki 312-8508, Japan
mathisa@ge.ibaraki-ct.ac.jp

Abstract. A communication process in the **S5**-knowledge model is presented which leads to a Nash equilibrium of a strategic form game through robust messages. In the communication process each player predicts the other players' actions under his/her private information. The players communicate privately their conjectures through message according to the communication graph, where each recipient of the message learns and revises his/her conjecture. The emphasis is on that each player sends not exact information about his/her individual conjecture but robust information about the conjectures to an accuracy ε.

Keywords: Communication, Robust message, Nash equilibrium, Protocol, Conjecture, Non-corporative game, **S5**-knowledge model.

AMS 2000 Mathematics Subject Classification: Primary 91A35, Secondary 03B45.

Journal of Economic Literature Classification: C62, C78.

1 Introduction

This article presents the communication system leading to a mixed strategy Nash equilibrium for a strategic form game as a learning process through robust messages in the **S5**-knowledge model associated with a partitional information structure. We show that

Main theorem. *Suppose that the players in a strategic form game have the knowledge structure associated a partitional information with a common prior distribution. In a communication process of the game according to a protocol with revisions of their beliefs about the other players' actions, the profile of their future predictions converges to a mixed strategy Nash equilibrium of the game in the long run.*

Recently, researchers in economics, AI, and computer science become entertained lively concerns about relationships between knowledge and actions. At what point does an economic agent sufficiently know to stop gathering information and make

---

* Partially supported by the Grant-in-Aid for Scientific Research(C)(No.18540153) in the Japan Society for the Promotion of Sciences.

A. Dress, Y. Xu, and B. Zhu (Eds.): COCOA 2007, LNCS 4616, pp. 136–145, 2007.

decisions? There are also concerns about the complexity of computing knowledge. The most interest to us is the emphasis on the considering the situation involving the knowledge of a group of agents rather than of just a single agent.

In game theoretical situations, the concept of mixed strategy Nash equilibrium (J.F. Nash [12]) has become central. Yet a little is known about the process by which players learn if they do. This article will give a communication protocol run by the mutual learning leading to a mixed strategy Nash equilibrium of a strategic form game from the point of distributed knowledge system.

Let us consider the following protocol: The players start with the same prior distribution on a state-space. In addition they have private information which is given by a partitional of the state space. Each player predicts the other players' actions as the posterior of the others' actions given his/her information. He/she communicates privately their beliefs about the other players' actions through robust messages, which message is approximate information about his/her individual conjecture about the others' actions to an accuracy ε. The recipients update their belief according to the messages. Precisely, at every stage each player communicates privately not only his/her belief about the others' actions but also his/her rationality as messages according to a protocol,[1] and then the recipient updates their private information and revises her/his prediction. The main theorem says that the players' predictions regarding the future beliefs converge in the long run, which lead to a mixed strategy Nash equilibrium of a game. The emphasis is on the three points: First that each player sends not exact information about his/her individual conjecture but robust information about the actions to an accuracy ε, secondly that each player's prediction is not required to be common-knowledge among all players, and finally that the communication graph is not assumed to be acyclic.

Many authors have studied the learning processes modeled by Bayesian updating. The papers by E. Kalai and E. Lehrer [5] and J. S. Jordan [4] (and references in therein) indicate increasing interest in the mutual learning processes in games that leads to equilibrium: Each player starts with initial erroneous belief regarding the actions of all the other players. They show the two strategies converges to an ε-mixed strategy Nash equilibrium of the repeated game.

As for as J.F. Nash's fundamental notion of strategic equilibrium is concerned, R.J. Aumann and A. Brandenburger [1] gives epistemic conditions for mixed strategy Nash equilibrium: They show that the common-knowledge of the predictions of the players having the partitional information (that is, equivalently, the S5-knowledge model) yields a Nash equilibrium of a game. However it is not clear just what learning process leads to the equilibrium.

To fill this gap from epistemic point of view, Matsuhisa ([6], [8], [9]) presents his communication system for a strategic game, which leads a mixed Nash equilibrium in several epistemic models. The articles [6], [8] [10] treats the communication system in the S4-knowledge model where each player communicates to other players by sending exact information about his/her conjecture on the

---

[1] When a player communicates with another, the other players are not informed about the contents of the message.

others' action. In Matsuhisa and Strokan [10], the communication model in the p-belief system is introduced:[2]Each player sends exact information that he/she believes that the others play their actions with probability at least his/her conjecture as messages. Matsuhisa [9] extended the communication model to the case that the sending messages are non-exact information that he/she believes that the others play their actions with probability at least his/her conjecture.

This article is in the line of [9]; each player sends his/her robust information about the actions to an accuracy ε in the **S5**-knowledge model.

This paper organizes as follows. Section 2 recalls the knowledge structure associated with a partition information structure, and we extend a game on knowledge structure. The communication process for the game is introduced where the players send robust messages about their conjectures about the other players' action. In Section 3 we give the formal statement of the main theorem (Theorem 1) and sketch the proof. In Section 4 we conclude with remarks. The illustrated example will be shown in the lecture presentation in COCOA 2007.

2   The Model

Let Ω be a non-empty *finite* set called a *state-space*, N a set of finitely many *players* $\{1, 2, \ldots n\}$ at least two ($n \geq 2$), and let 2^{Ω} be the family of all subsets of Ω. Each member of 2^{Ω} is called an *event* and each element of Ω called a *state*. Let μ be a probability measure on Ω which is common for all players. For simplicity it is assumed that (Ω, μ) is a *finite* probability space with μ *full support*.[3]

2.1   Information and Knowledge[4]

A *partitional information structure* $\langle \Omega, (\Pi_i)_{i \in N} \rangle$ consists of a state space Ω and a class of the mappings Π_i of Ω into 2^{Ω} such that

(i) $\{\Pi_i(\omega) | \omega \in \Omega\}$ is a partition of Ω;

(ii) $\omega \in \Pi_i(\omega)$ for every $\omega \in \Omega$.

Given our interpretation, an player i for whom $\Pi_i(\omega) \subseteq E$ knows, in the state ω, that some state in the event E has occurred. In this case we say that in the state ω the player i knows E.

Definition 1. The *knowledge structure* $\langle \Omega, (\Pi_i)_{i \in N}, (K_i)_{i \in N} \rangle$ consists of a partitional information structure $\langle \Omega, (\Pi_i)_{i \in N} \rangle$ and a class of i's *knowledge operator* K_i on 2^{Ω} such that $K_i E$ is the set of states of Ω in which i knows that E has occurred; that is,

$$K_i E = \{\omega \in \Omega \mid \Pi_i(\omega) \subseteq E\}.$$

The set $\Pi_i(\omega)$ will be interpreted as the set of all the states of nature that i knows to be possible at ω, and $K_i E$ will be interpreted as the set of states of

---

[2] C.f.: Monderer and Samet [11] for the p-belief system.

[3] That is; $\mu(\omega) \neq 0$ for every $\omega \in \Omega$.

[4] C.f.; Bacharach [2], Binmore [3] for the information structure and the knowledge operator.

nature for which i knows E to be possible. We will therefore call Π_i i's *possibility operator* on Ω and also will call $\Pi_i(\omega)$ i's *information set* at ω.

We record the properties of i's knowledge operator[5]: For every E, F of 2^Ω,

N $K_i\Omega = \Omega$ and $K_i\emptyset = \emptyset$;   **K** $K_i(E \cap F) = K_iE \cap K_iF$;

T $K_iF \subseteq F$;   **4** $K_iF \subseteq K_iK_iF$;

5 $\Omega \setminus K_i(E) \subseteq K_i(\Omega \setminus K_i(E))$.

Remark 1. i's possibility operator Π_i is uniquely determined by i's knowledge operator K_i satisfying the above five properties: For $\Pi_i(\omega) = \bigcap_{\omega \in K_iE} E$.

2.2   Game on Knowledge Structure[6]

By a *game* G we mean a *finite* strategic form game $\langle N, (A_i)_{i \in N}, (g_i)_{i \in N} \rangle$ with the following structure and interpretations: N is a finite set of players $\{1, 2, \ldots, i, \ldots n\}$ with $n \geq 2$, A_i is a finite set of i's *actions* (or i's pure strategies) and g_i is an i's *payoff function* of A into $\mathbb{R}$, where A denotes the product $A_1 \times A_2 \times \cdots \times A_n$, A_{-i} the product $A_1 \times A_2 \times \cdots \times A_{i-1} \times A_{i+1} \times \cdots \times A_n$. We denote by g the n-tuple $(g_1, g_2, \ldots g_n)$ and by a_{-i} the $(n-1)$-tuple $(a_1, \ldots, a_{i-1}, a_{i+1}, \ldots, a_n)$ for a of A. Furthermore we denote $a_{-I} = (a_i)_{i \in N \setminus I}$ for each $I \subset N$.

A probability distribution ϕ_i on A_{-i} is said to be i's *overall conjecture* (or simply i's *conjecture*). For each player j other than i, this induces the marginal distribution on j's actions; we call it i's *individual conjecture* about j (or simply i's conjecture *about j*.) Functions on Ω are viewed like random variables in the probability space (Ω, μ). If $\mathbf{x}$ is a such function and x is a value of it, we denote by $[\mathbf{x} = x]$ (or simply by $[x]$) the set $\{\omega \in \Omega | \mathbf{x}(\omega) = x\}$.

The information structure (Π_i) with a common prior μ yields the distribution on $A \times \Omega$ defined by $\mathbf{q}_i(a, \omega) = \mu([\mathbf{a} = a]|\Pi_i(\omega))$; and the i's overall conjecture defined by the marginal distribution $\mathbf{q}_i(a_{-i}, \omega) = \mu([\mathbf{a}_{-i} = a_{-i}]|\Pi_i(\omega))$ which is viewed as a random variable of ϕ_i. We denote by $[\mathbf{q}_i = \phi_i]$ the intersection $\bigcap_{a_{-i} \in A_{-i}}[\mathbf{q}_i(a_{-i}) = \phi_i(a_{-i})]$ and denote by $[\phi]$ the intersection $\bigcap_{i \in N}[\mathbf{q}_i = \phi_i]$. Let $\mathbf{g}_i$ be a random variable of i's payoff function g_i and $\mathbf{a}_i$ a random variable of an i's action a_i.

According to the Bayesian decision theoretical point of view we assume that each player i absolutely knows his/her own actions; i.e., letting $[a_i] := [\mathbf{a}_i = a_i]$, $[a_i] = K_i([a_i])$ (or equivalently, $\Pi_i(\omega) \subseteq [a_i]$ for all $\omega \in [a_i]$ and for every a_i of A_i.) i's action a_i is said to be *actual* at a state ω if $\omega \in [\mathbf{a}_i = a_i]$; and the profile a_I is said to be *actually played* at ω if $\omega \in [\mathbf{a}_I = a_I] := \bigcap_{i \in I}[\mathbf{a}_i = a_i]$ for $I \subset N$. The pay off functions $g = (g_1, g_2, \ldots, g_n)$ is said to be *actually played* at a state ω if $\omega \in [\mathbf{g} = g] := \bigcap_{i \in N}[\mathbf{g}_i = g_i]$. Let **Exp** denote the expectation defined by

$$\mathbf{Exp}(g_i(b_i, \mathbf{a}_{-i}); \omega) := \sum_{a_{-i} \in A_{-i}} g_i(b_i, a_{-i})\, \mathbf{q}_i(a_{-i}, \omega).$$

---

[5] According to these we can say the structure $\langle \Omega, (K_i)_{i \in N} \rangle$ is a model for the multi-modal logic **S5**.

[6] C.f., Aumann and Brandenburger [1].

A player i is said to be *rational* at ω if each i's actual action a_i maximizes the expectation of his actually played pay off function g_i at ω when the other players actions are distributed according to his conjecture $\mathbf{q}_i(\,\cdot\,;\omega)$.

Formally, letting $g_i = \mathbf{g}_i(\omega)$ and $a_i = \mathbf{a}_i(\omega)$, $\mathbf{Exp}(g_i(a_i,\mathbf{a}_{-i});\omega) \geq \mathbf{Exp}(g_i(b_i,\mathbf{a}_{-i});\omega)$ for every b_i in A_i. Let R_i denote the set of all of the states at which i is rational.

2.3   Protocol [7]

We assume that the players communicate by sending *messages*. Let T be the time horizontal line $\{0,1,2,\cdots t,\cdots\}$. A *protocol* is a mapping $\mathrm{Pr} : T \to N \times N, t \mapsto (s(t), r(t))$ such that $s(t) \neq r(t)$. Here t stands for *time* and $s(t)$ and $r(t)$ are, respectively, the *sender* and the *recipient* of the communication which takes place at time t. We consider the protocol as the directed graph whose vertices are the set of all players N and such that there is an edge (or an arc) from i to j if and only if there are infinitely many t such that $s(t) = i$ and $r(t) = j$.

A protocol is said to be *fair* if the graph is strongly-connected; in words, every player in this protocol communicates directly or indirectly with every other player infinitely often. It is said to contain a *cycle* if there are players $i_1, i_2, \ldots, i_k$ with $k \geq 3$ such that for all $m < k$, i_m communicates directly with i_{m+1}, and such that i_k communicates directly with i_1. The communications is assumed to proceed in *rounds*[8].

2.4   Communication on Knowledge Structure

Let ε be a real number with $0 \leq \varepsilon < 1$. An ε-*robust communication process* $\pi^\varepsilon(G)$ with revisions of players' conjectures $(\phi_i^t)_{(i,t)\in N\times T}$ according to a protocol for a game G is a tuple

$$\pi^\varepsilon(G) = \langle G, (\Omega, \mu)\,\mathrm{Pr}, (\Pi_i^t)_{i\in N}, (K_i^t)_{i\in N}, (\phi_i^t)_{(i,t)\in N\times T}\rangle$$

with the following structures: the players have a common prior μ on Ω, the protocol Pr among N, $\mathrm{Pr}(t) = (s(t), r(t))$, is fair and it satisfies the conditions that $r(t) = s(t+1)$ for every t and that the communications proceed in rounds. The revised information structure Π_i^t at time t is the mapping of Ω into 2^Ω for player i. If $i = s(t)$ is a sender at t, the *message* sent by i to $j = r(t)$ is M_i^t. An n-tuple $(\phi_i^t)_{i\in N}$ is a revision process of individual conjectures. These structures are inductively defined as follows:

- Set $\Pi_i^0(\omega) = \Pi_i(\omega)$.
- Assume that Π_i^t is defined. It yields the distribution $\mathbf{q}_i^t(a,\omega) = \mu([\mathbf{a} = a]|\Pi_i^t(\omega))$. Whence
  - R_i^t denotes the set of all the state ω at which i is *rational* according to his conjecture $\mathbf{q}_i^t(\,\cdot\,;\omega)$; that is, each i's actual action a_i maximizes the

---

[7] C.f.: Parikh and Krasucki [13]

[8] There exists a time m such that for all t, $\mathrm{Pr}(t) = \mathrm{Pr}(t+m)$. The *period* of the protocol is the minimal number of all m such that for every t, $\mathrm{Pr}(t+m) = \mathrm{Pr}(t)$.

expectation of his pay off function g_i being actually played at ω when the other players actions are distributed according to his conjecture $\mathbf{q}_i^t(\,\cdot\,;\omega)$ at time t. Formally, letting $g_i = \mathbf{g}_i(\omega)$, $a_i = \mathbf{a}_i(\omega)$, the expectation at time t, $\mathbf{Exp}^t$, is defined by

$$\mathbf{Exp}^t(g_i(a_i, \mathbf{a}_{-i}); \omega) := \sum_{a_{-i} \in A_{-i}} g_i(a_i, a_{-i})\, \mathbf{q}_i^t(a_{-i}, \omega).$$

An player i is said to be rational according to his conjecture $\mathbf{q}_i^t(\,\cdot\,, \omega)$ at ω if for all b_i in A_i, $\mathbf{Exp}^t(g_i(a_i, \mathbf{a}_{-i}); \omega) \geq \mathbf{Exp}^t(g_i(b_i, \mathbf{a}_{-i}); \omega)$.

- The message $M_i^t : \Omega \to 2^\Omega$ sent by the sender i at time t is defined as a robust information:

$$M_i^t(\omega) = \bigcap_{a_{-i} \in A_{-i}} \left\{ \xi \in \Omega \,\big|\, \left| \mathbf{q}_i^t(a_{-i}, \xi) - \mathbf{q}_i^t(a_{-i}, \omega) \right| < \varepsilon \right\}.$$

Then:
- The revised knowledge operator $K_i^t : 2^\Omega \to 2^\Omega$ is defined by $K_i^t(E) = \{\omega \in \Omega \mid \Pi_i^t(\omega) \subseteq E\}$.
- The revised partition Π_i^{t+1} at time $t+1$ is defined as follows:
  - $\Pi_i^{t+1}(\omega) = \Pi_i^t(\omega) \cap M_{s(t)}^t(\omega)$ if $i = r(t)$;
  - $\Pi_i^{t+1}(\omega) = \Pi_i^t(\omega)$ otherwise,
- The revision process $(\phi_i^t)_{(i,t) \in N \times T}$ of conjectures is inductively defined by the following way:
  - Let $\omega_0 \in \Omega$, and set $\phi_{s(0)}^0(a_{-s(0)}) := \mathbf{q}_{s(0)}^0(a_{-s(0)}, \omega_0)$
  - Take $\omega_1 \in M_{s(0)}^0(\omega_0) \cap K_{r(0)}([g_{s(0)}] \cap R_{s(0)}^0),$[9] and set $\phi_{s(1)}^1(a_{-s(1)}) := \mathbf{q}_{s(1)}^1(a_{-s(1)}, \omega_1)$
  - Take $\omega_{t+1} \in M_{s(t)}^t(\omega_t) \cap K_{r(t)}^t([g_{s(t)}] \cap R_{s(t)}^t)$, and set $\phi_{s(t+1)}^{t+1}(a_{-s(t+1)}) := \mathbf{q}_i^{t+1}(a_{-s(t+1)}, \omega_{t+1})$.

The specification is that a sender $s(t)$ at time t informs the recipient $r(t)$ his/her prediction about the other players' actions as approximate information of his/her individual conjecture to an accuracy ε. The recipient revises her/his information structure under the information. She/he predicts the other players action at the state where the player knows that the sender $s(t)$ is rational, and she/he informs her/his the predictions to the other player $r(t+1)$.

We denote by ∞ a sufficient large $\tau \in T$ such that for all $\omega \in \Omega$, $\mathbf{q}_i^\tau(\,\cdot\,;\omega) = \mathbf{q}_i^{\tau+1}(\,\cdot\,;\omega) = \mathbf{q}_i^{\tau+2}(\,\cdot\,;\omega) = \cdots$. Hence we can write $\mathbf{q}_i^\tau$ by $\mathbf{q}_i^\infty$ and ϕ_i^τ by ϕ_i^∞.

Remark 2. This communication model is a variation of the model introduced by Matsuhisa [6].

---

[9] We denote $[g_i] := [\mathbf{g}_i = g_i]$

3   The Result

We can now state the main theorem :

Theorem 1. *Suppose that the players in a strategic form game G have the knowledge structure with μ a common prior. In the ε-robust communication process $\pi^\varepsilon(G)$ according to a protocol Pr among all players in the game, the n-tuple of their conjectures $(\phi_i^t)_{(i,t)\in N\times T}$ converges to a mixed strategy Nash equilibrium of the game in finitely many steps.*

The proof is based on the below proposition:

Proposition 1. *Notation and assumptions are the same in Theorem 1. For any players $i, j \in N$, their conjectures $\mathbf{q}_i^\infty$ and $\mathbf{q}_j^\infty$ on $A \times \Omega$ must coincide; that is, $\mathbf{q}_i^\infty(a;\omega) = \mathbf{q}_j^\infty(a;\omega)$ for every $a \in A$ and $\omega \in \Omega$.*

Proof. On noting that Pr is fair, it suffices to verify that $\mathbf{q}_i^\infty(a;\omega) = \mathbf{q}_j^\infty(a;\omega)$ for $(i,j) = (s(\infty), r(\infty))$. Since $\Pi_i(\omega) \subseteq [a_i]$ for all $\omega \in [a_i]$, we can observe that $\mathbf{q}_i^\infty(a_{-i};\omega) = \mathbf{q}_i^\infty(a;\omega)$, and we let define the partitions of Ω, $\{W_i^\infty(\omega) \mid \omega \in \Omega\}$ and $\{Q_j^\infty(\omega) \mid \omega \in \Omega\}$, as follows:

$$W_i^\infty(\omega) := \bigcap_{a_{-i}\in A_{-i}} [\mathbf{q}_i^\infty(a_{-i},*) = \mathbf{q}_i^\infty(a_{-i},\omega)] = \bigcap_{a\in A}[\mathbf{q}_i^\infty(a,*) = \mathbf{q}_i^\infty(a,\omega)],$$

$$Q_j^\infty(\omega) := \Pi_j^\infty(\omega) \cap W_i^\infty(\omega).$$

It follows that

$$Q_j^\infty(\xi) \subseteq W_i^\infty(\omega) \quad \text{for all } \xi \in W_i^\infty(\omega),$$

and hence $W_i^\infty(\omega)$ can be decomposed into a disjoint union of components $Q_j^\infty(\xi)$ for $\xi \in W_i^\infty(\omega)$;

$$W_i^\infty(\omega) = \bigcup_{k=1,2,\ldots,m} Q_j^\infty(\xi_k) \quad \text{for } \xi_k \in W_i^\infty(\omega).$$

It can be observed that

$$\mu([\mathbf{a} = a]| \ W_i^\infty(\omega)) = \sum_{k=1}^m \lambda_k \mu([\mathbf{a} = a]| \ Q_j^\infty(\xi_k)) \tag{1}$$

for some $\lambda_k > 0$ with $\sum_{k=1}^m \lambda_k = 1$.[10]

On noting that $W_j^\infty(\omega)$ is decomposed into a disjoint union of components $\Pi_j^\infty(\xi)$ for $\xi \in W_j^\infty(\omega)$, it can be observed that

$$\mathbf{q}_j^\infty(a;\omega) = \mu([\mathbf{a} = a]| \ W_j^\infty(\omega)) = \mu([\mathbf{a} = a]| \ \Pi_j^\infty(\xi_k)) \tag{2}$$

---

[10] This property is called the *convexity* for the conditional probability $\mu(X|*)$ in Parikh and Krasucki [13].

for any $\xi_k \in W_i^\infty(\omega)$. Furthermore we can verify that for every $\omega \in \Omega$,

$$\mu([\mathbf{a} = a]|\ W_j^\infty(\omega)) = \mu([\mathbf{a} = a]|\ Q_j^\infty(\omega)). \tag{3}$$

In fact, we first note that $W_j^\infty(\omega)$ can also be decomposed into a disjoint union of components $Q_j^\infty(\xi)$ for $\xi \in W_j^\infty(\omega)$. We shall show that for every $\xi \in W_j^\infty(\omega)$, $\mu([\mathbf{a} = a]|\ W_j^\infty(\omega)) = \mu([\mathbf{a} = a]|\ Q_j^\infty(\xi))$. For: Suppose not, the disjoint union G of all the components $Q_j(\xi)$ such that $\mu([\mathbf{a} = a]|\ W_j^\infty(\omega)) = \mu([\mathbf{a} = a]|\ Q_j^\infty(\xi))$ is a proper subset of $W_j^\infty(\omega)$. It can be shown that for some $\omega_0 \in W_j^\infty(\omega) \setminus G$ such that $Q_j(\omega_0) = W_j^\infty(\omega) \setminus G$. On noting that $\mu([\mathbf{a} = a]|G) = \mu([\mathbf{a} = a]|\ W_j^\infty(\omega))$ it follows immediately that $\mu([\mathbf{a} = a]|\ Q_j^\infty(\omega_0)) = \mu([\mathbf{a} = a]|\ W_j^\infty(\omega))$, in contradiction. Now suppose that for every $\omega_0 \in W_j^\infty(\omega) \setminus G$, $Q_j(\omega_0) \neq W_j^\infty(\omega) \setminus G$. The we can take an infinite sequence of states $\{\omega_k \in W_j^\infty(\omega) \mid k = 0, 1, 2, 3, \ldots\}$ with $\omega_{k+1} \in W_j^\infty(\omega) \setminus (G \cup Q_j^\infty(\omega_0) \cup Q_j^\infty(\omega_1) \cup Q_j^\infty(\omega_2) \cup \cdots \cup Q_j^\infty(\omega_k))$ in contradiction also, because Ω is finite.

In viewing (1), (2) and (3) it follows that

$$\mathbf{q}_i^\infty(a; \omega) = \sum_{k=1}^{m} \lambda_k \mathbf{q}_j^\infty(a; \xi_k) \tag{4}$$

for some $\xi_k \in W_i^\infty(\omega)$. Let ξ_ω be the state in $\{\xi_k\}_{k=1}^m$ attains the maximal value of all $\mathbf{q}_j^\infty(a; \xi_k)$ for $k = 1, 2, 3, \cdots, m$, and let $\zeta_\omega \in \{\xi_k\}_{k=1}^m$ be the state that attains the minimal value. By (4) we obtain that $\mathbf{q}_j^\infty(a; \zeta_\omega) \leq \mathbf{q}_i^\infty(a; \omega) \leq \mathbf{q}_j^\infty(a; \xi_\omega)$ for $(i, j) = (s(\infty), t(\infty))$.

On continuing this process according to the *fair* protocol Pr, it can be plainly verified: For each $\omega \in \Omega$ and for any $t \geq 1$,

$$\mathbf{q}_i^\infty(a; \zeta_\omega') \leq \cdots \leq \mathbf{q}_j^\infty(a; \zeta_\omega) \leq \mathbf{q}_i^\infty(a; \omega) \leq \mathbf{q}_j^\infty(a; \xi_\omega) \leq \cdots \leq \mathbf{q}_i^\infty(a; \xi_\omega')$$

for some $\zeta_\omega', \cdots, \zeta_\omega, \xi_\omega, \cdots \xi_\omega' \in \Omega$, and thus $\mathbf{q}_i^\infty(a; \omega) = \mathbf{q}_j^\infty(a; \omega)$ because $\mathbf{q}_j^\infty(a; \zeta_\omega) \leq \mathbf{q}_i^\infty(a; \omega) \leq \mathbf{q}_j^\infty(a; \xi_\omega)$ and $\mathbf{q}_i^\infty(a; \zeta) = \mathbf{q}_j^\infty(a; \xi)$ for every $\zeta, \xi \in \Omega$. in completing the proof.

Proof of Theorem 1. We denote by $\Gamma(i)$ the set of all the players who directly receive the message from i on N; i.e., $\Gamma(i) = \{\ j \in N \mid (i, j) = \Pr(t)$ for some $t \in T\}$. Let F_i denote $[\phi_i^\infty] := \bigcap_{a_{-i} \in A_i} [\mathbf{q}_i^\infty(a_{-i}; *) = \phi_i^\infty(a_{-i})]$. It is noted that $F_i \cap F_j \neq \emptyset$ for each $i \in N$, $j \in \Gamma(i)$.

We observe the first point that for each $i \in N$, $j \in \Gamma(i)$ and for every $a \in A$, $\mu([\mathbf{a}_{-j} = a_{-j}]\mid F_i \cap F_j) = \phi_j^\infty(a_{-j})$. Then summing over a_{-i}, we can observe that $\mu([\mathbf{a}_i = a_i]\mid F_i \cap F_j) = \phi_j^\infty(a_i)$ for any $a \in A$. In view of Proposition 1 it can be observed that $\phi_j^\infty(a_i) = \phi_k^\infty(a_i)$ for each $j, k, \neq i$; i.e., $\phi_j^\infty(a_i)$ is independent of the choices of every $j \in N$ other than i. We set the probability distribution σ_i on A_i by $\sigma_i(a_i) := \phi_j^\infty(a_i)$, and set the profile $\sigma = (\sigma_i)$.

We observe the second point that for every $a \in \prod_{i \in N} \text{Supp}(\sigma_i)$,

$$\phi_i^\infty(a_{-i}) = \sigma_1(a_1) \cdots \sigma_{i-1}(a_{i-1})\sigma_{i+1}(a_{i+1}) \cdots \sigma_n(a_n):$$

In fact, viewing the definition of σ_i we shall show that

$$\phi_i^\infty(a_{-i}) = \prod_{k \in N \setminus \{i\}} \phi_i^\infty(a_k).$$

To verify this it suffices to show that for every $k = 1, 2, \cdots, n$,

$$\phi_i^\infty(a_{-i}) = \phi_i^\infty(a_{-I_k}) \prod_{k \in I_k \setminus \{i\}} \phi_i^\infty(a_k) :$$

We prove it by induction on k. For $k = 1$ the result is immediate. Suppose it is true for $k \geq 1$. On noting the protocol is fair, we can take the sequence of sets of players $\{I_k\}_{1 \leq k \leq n}$ with the following properties:

(a) $I_1 = \{i\} \subset I_2 \subset \cdots \subset I_k \subset I_{k+1} \subset \cdots \subset I_m = N$:

(b) For every $k \in N$ there is a player $i_{k+1} \in \bigcup_{j \in I_k} \Gamma(j)$ with $I_{k+1} \setminus I_k = \{i_{k+1}\}$.

We let take $j \in I_k$ such that $i_{k+1} \in \Gamma(j)$. Set $H_{i_{k+1}} := [\mathbf{a}_{i_{k+1}} = a_{i_{k+1}}] \cap F_j \cap F_{i_{k+1}}$. It can be verified that

$$\mu([\mathbf{a}_{-j-i_{k+1}} = a_{-j-i_{k+1}}] \mid H_{i_{k+1}}) = \phi_{-j-i_{k+1}}^\infty(a_{-j}).$$

Dividing $\mu(F_j \cap F_{i_{k+1}})$ yields that

$$\mu([\mathbf{a}_{-j} = a_{-j}] \mid F_j \cap F_{i_{k+1}}) = \phi_{i_{k+1}}^\infty(a_{-j})\mu([\mathbf{a}_{i_{k+1}} = a_{i_{k+1}}] \mid F_j \cap F_{i_{k+1}}).$$

Thus $\phi_j^\infty(a_{-j}) = \phi_{i_{k+1}}^\infty(a_{-j-i_{k+1}})\phi_j^t(a_{i_{k+1}})$; then summing over a_{I_k} we obtain $\phi_j^\infty(a_{-I_k}) = \phi_{i_{k+1}}^\infty(a_{-I_k-i_{k+1}})\phi_j^\infty(a_{i_{k+1}})$. It immediately follows from Proposition 1 that $\phi_i^\infty(a_{-I_k}) = \phi_i^\infty(a_{-I_k-i_{k+1}})\phi_i^\infty(a_{i_{k+1}})$, as required.

Furthermore we can observe that all the other players i than j agree on the same conjecture $\sigma_j(a_j) = \phi_i^\infty(a_j)$ about j. We conclude that each action a_i appearing with positive probability in σ_i maximizes g_i against the product of the distributions σ_l with $l \neq i$. This implies that the profile $\sigma = (\sigma_i)_{i \in N}$ is a mixed strategy Nash equilibrium of G, in completing the proof. $\square$

4   Concluding Remarks

We have observed that in a communication process with revisions of players' beliefs about the other actions, their predictions induces a mixed strategy Nash equilibrium of the game in the long run. Matsuhisa [6] and [8] established the same assertion in the S4-knowledge model. Furthermore Matsuhisa [7] showed a similar result for ε-mixed strategy Nash equilibrium of a strategic form game in the S4-knowledge model, which gives an epistemic aspect in Theorem of E. Kalai and E. Lehrer [5]. This article highlights a communication among the players in a game through sending rough information, and shows that the convergence to an exact Nash equilibrium is guaranteed even in such communication on approximate information after long run. It is well to end some remarks on the

ε-robust communication. The main theorem in this article can be extended into the **S4**-knowledge model and the p-belief system. The extended theorem in **S4**-knowledge model coincides with the theorems in Matsuhisa [6] and [8] when $\varepsilon = 0$. Can we unify all the communication models in the preceding papers ([6], [8], [10], [9])? The answer is yes; I will present the unified communication system leading to a Nash equilibrium in a near future paper.

References

1. Aumann, R.J., Brandenburger, A.: Epistemic conditions for mixed strategy Nash equilibrium. Econometrica 63, 1161–1180 (1995)
2. Bacharach, M.: Some extensions of a claim of Aumann in an axiomatic model of knowledge. Journal of Economic Theory 37, 167–190 (1985)
3. Binmore, K.: Fun and Games, p. xxx+642. D.C. Heath and Company, Lexington, Massachusetts (1992)
4. Jordan, J.S.: Bayesian learning in normal form games. Games and Economic Behavior 3, 60–81 (1991)
5. Kalai, E., Lehrer, E.: Rational learning to mixed strategy Nash equilibrium. Econometrica 61, 1019–1045 (1993)
6. Matsuhisa, T.: Communication leading to mixed strategy Nash equilibrium I. In: Maruyama, T. (ed.) Mathematical Economics, Suri-Kaiseki-Kenkyusyo Kokyuroku, vol. 1165, pp. 245–256 (2000)
7. Matsuhisa, T.: Communication leading to epsilon-mixed strategy Nash equilibrium, Working paper. IMGTA XIV. The extended abstract was presented in the XIV Italian Meeting of Game Theory and Applications (July 11-14, 2001)
8. Matsuhisa, T.: Communication leading to a Nash equilibrium without acyclic condition (S4-knowledge case). In: Bubak, M., van Albada, G.D., Sloot, P.M.A., Dongarra, J.J. (eds.) ICCS 2004. LNCS, vol. 3039, pp. 884–891. Springer, Heidelberg (2004)
9. Matsuhisa, T.: Bayesian communication under rough sets information. In: Butz, C.J., et al. (eds.) IEEE/WIC/ACM International Conference on Web Intelligence and Intelligent Agent Technology. WI-ITA 2006 Workshop Proceedings, pp. 378–381. IEEE Computer Society, Los Alamitos (2006)
10. Matsuhisa, T., Strokan, P.: Bayesian belief communication leading to a Nash equilibrium in belief. In: Deng, X., Ye, Y. (eds.) WINE 2005. LNCS, vol. 3828, pp. 299–306. Springer, Heidelberg (2005)
11. Monderer, D., Samet, D.: Approximating common knowledge with common beliefs. Games and Economic Behaviors 1, 170–190 (1989)
12. Nash, J.F.: Equilibrium points in n-person games. In: Proceedings of the National Academy of Sciences of the United States of America, vol. 36, pp. 48–49 (1950)
13. Parikh, R., Krasucki, P.: Communication, consensus, and knowledge. Journal of Economic Theory 52, 178–189 (1990)

Fundamental Domains for Integer Programs with Symmetries

Eric J. Friedman

Cornell University Ithaca, NY 14850
ejf27@cornell.edu
http://www.people.cornell.edu/pages/ejf27/

Abstract. We define a fundamental domain of a linear programming relaxation of a combinatorial integer program which is symmetric under a group action. We then provide a construction for the polytope of a fundamental domain defined by the maximization of a linear function. The computation of this fundamental domain is at worst polynomial in the size of the group. However, for the special case of the symmetric group, whose size is exponential in the size of the integer program, we show how to compute a separating hyperplane in polynomial time in the size of the integer program.

Fundamental domains may provide a straightforward way to reduce the computation difficulties that often arise in integer programs with symmetries. Our construction is closely related to the constructions of orbitopes by Kaibel and Pfetch, but are simpler and more general, at a cost of creating new non-integral extreme points.

1 Introduction

Combinatorial integer programs with symmetries arise in many standard problem formulations. Unfortunately, these symmetries often make the problems difficult to solve because integer programming algorithms can repeatedly examine solutions that are equivalent under the symmetry. For example, in a simple bin packing problem with multiple bins of the same size one often uses the variable x_{ij} to represent whether item i is in bin j. However, if bins j and k are the same size then any solution x is equivalent to the solution x' when x' is derived from x by exchanging the two columns, j and k. One way of resolving these problems is to restrict the search space to eliminate the additional equivalent copies of a solution. This can be done either by adding additional constraints to the integer program [2,1,3,8] or by modifying the branch and bound or branch and cut algorithms [6,7].

In this paper, we consider the problem of removing the multiple symmetrically equivalent solutions. We construct a polytope for a "minimal fundamental domain", which is a subset of the feasible region and contains only a single "representative" from each equivalence class of symmetrically equivalent extreme points.

A. Dress, Y. Xu, and B. Zhu (Eds.): COCOA 2007, LNCS 4616, pp. 146–153, 2007.

Our work is motivated by Kaibel and Pfetch's [5] recent study of orbitopes. In that paper they considered the fundamental domain generated by a lexicographic ordering. They provided a complete description for orbitopes for the cyclic and symmetric groups under packing and partitioning constraints.

In this paper, we consider a different approach to this problem: finding fundamental domains defined by maximizing a linear function. This leads to simple constructions and straightforward proofs. It also allows these techniques to be extended to more complex settings.

For example, consider a bin packing problem with multiple bins. Previous methods have considered the case when all bins are identical, in which the problem is invariant under the full symmetric group; however, our methods apply to arbitrary sets of identical bins, e.g., three bins of size 10, six bins of size 14 and one bin of size 22. In addition, our methods extend directly to covering problems without the combinatorial complexities that arise in the related orbitopes.

Our methods also apply to other groups, such as cyclic groups, which arise in transportation scheduling problems [10] or even problems for which several different group actions are combined. For example, consider a periodic bus scheduling problem with multiple bus sizes. This problem is invariant under the exchange of equal capacity buses (symmetric groups) and under time transformations (cyclic group).

In this paper, we present the general theory of fundamental domains. In the following section we provide the basic construction and then in Section 3 discuss the separation problem for the cyclic and symmetric groups. Section 4 compares fundamental domains to orbitopes, Section 5 discusses combinations of groups and Section 6 considers the linear optimization criterion used for generating the fundamental domains. We conclude in Section 7.

2   Group Actions and Fundamental Domains

Let G be a finite group and given a set $X \subset \Re^n$ consider a group action $\phi_g : X \to X$. A group action must satisfy $\phi_{g \circ g'} = \phi_g \circ \phi_{g'}$. Given $x \in X$, define the orbit of x, $orb(x)$, to be the set $\phi_g(x)$ for all $g \in G$. A (generalized) fundamental domain of X is a subset $F \subset X$ such that its orbit $orb(F) = X$, where

$$Orb(F) = \{x \in X \mid \exists y \in F, \quad \exists g \in G \quad s.t. \quad x = \phi_g y\}.$$

A fundamental domain is minimal if in addition, there is no closed subset of F that is also a fundamental domain.

To specialize to polytopes, assume that $X \subseteq [0,1]^n$ is a binary polytope. Let $Ext(X)$ be the extreme points of X, which are assumed to be integral. In addition, we will require that for all $g \in G$, the group action is an affine transformation of X. Thus, for each $g \in G$ we can assume that $\phi_{gx} = A_g x + b_g$ where A_g is an $n \times n$ matrix and b_g is an n-vector.

We first note to basic facts about affine group actions of finite groups.

Lemma 1. *Let G be a finite group and $\phi : G \times X \to X$ be an affine group action of G. Then $\forall g \in G$ the determinant of A_g has the absolute value of 1.*

Proof: Since $\phi_{g^{-1}} = (\phi_g)^{-1}$ the action ϕ_g is invertible so A_g must have nonzero determinant. In addition, since G is finite and for all $g \in G$, the composition of ϕ_g with itself k times, $(\phi_g)^k = \phi_{g^k} = \phi_{g'}$, for some $g' \in G$. Now the determinant of $\phi_{g'}$ must satisfy $det(\phi_{g'}) = det(\phi_g)^k$, so unless $|det(\phi_g)| = 1$, $(\phi_g)^k$ will be different for all k, contradicting the assumption that G is finite.   QED

Given an "ordering vector" $c \in \Re_n$, we define the fundamental domain of X, F_c, with respect to G by

$$F_c = \{x \in X \mid c^t x \geq c^t \phi_g x \ \ \forall g \in G\}$$

Lemma 2. *For any ordering vector c, the fundamental domain, F_c is a polytope.*

Proof: The fundamental domain is defined by a finite set of affine inequalities.
                QED

For example, consider the case with $X = [0,1]^2$ where G is the additive group Z_2 with elements $\{0,1\}$, and $0 + 0 = 0$, $0 + 1 = 1$, and $1 + 1 = 0$. Define the action of this group by setting ϕ_0 to be the identity and ϕ_1 being the exchange operator, $\phi_1(x_1, x_2) = (x_2, x_1)$. Let $c = (2,1)$. Then

$$F_c = \{x \in X \mid 2x_1 + x_2 \geq x_1 + 2x_2\},$$

which implies that

$$F_c = \{x \in X \mid x_1 \geq x_2\}.$$

Thus, F_c include the extreme points $(1,1)$ and $(1,0)$, while different choices of c can lead to different fundamental domains. For example if $c = (1,2)$ then the fundamental domain now includes $(0,1)$ instead of $(1,0)$.

First we note that a fundamental domain always contains a "representative" for each extreme point of X.

Theorem 1. *Let $x \in Ext(X)$. For any ordering vector c, the there exists a $g \in G$ such that $\phi_g x \in Ext(F_c)$.*

Proof: This follows immediately from the definition of F_c since for each $x \in Ext(X)$ must have at least one largest element in it's orbit, $c^t \phi_g x \ \forall g \in G$, since $|G|$ is finite.   QED

Note that, unlike orbitopes [5], there can exist extreme points of F_c which are not integral. For example, consider the case with $X = [0,1]^2$ and $G = Z_2$, where ϕ_1 inverts the first element of x, $\phi_1 = (x_1, x_2) = (1 - x_1, x_2)$. Let $c = (2,1)$. Then

$$F_c = \{x \in X \mid 2x_1 + x_2 \geq 2(1 - x_1) + x_2\},$$

which implies that

$$F_c = \{x \in X \mid x_1 \geq 1/2\}$$

which has $(1/2, 0)$ as an extreme point.

Note that a fundamental domain generated in this way need not be minimal. For example, when $c = 0$ we get $F_c = X$. However, even if c is nontrivial, the fundamental domain need not be minimal. In fact, unless the ordering vector c is chosen carefully, the fundamental domain will not be minimal.

First we show that there is a universal ordering vector $\hat{c}$ which generates minimal fundamental domains.

Theorem 2. *Let* $\hat{c} = (2^{n-1}, 2^{n-2}, \ldots, 2, 1)$ *be the "universal ordering vector". Then* $F_{\hat{c}}$ *will be minimal.*

Proof: First note that $F_{\hat{c}}$ contains a unique element of any orbit of an extreme point. This follows because $\hat{c}$ induces a lexicographic order on extreme points.

Next, we note that $F_{\hat{c}}$ must be full dimensional. i.e., the same dimension as X. This is because $Orb(F_{\hat{c}}) = X$ and each $Orb(x)$ contains a finite number of points.

Suppose that for some point $x \in F_{\hat{c}}$ there exists some $g \in G$ such that $\hat{c}^t \phi_g x = \hat{c}^t x$ and $\phi_g x \neq x$. However, this implies that the constraint from ϕ_g is tight, so unless the constraint is trivial ($0 \geq 0$) x will not be an interior point.

Since $x \in X$ and X is convex, we can write $x = \sum_{j=1}^{n+1} \alpha_j w^j$ where $\alpha \geq 0$, $\sum_{j=1}^{n+1} \alpha_j = 1$ and w^j are all extreme points of x. Since $\phi_g x \neq x$ and $\phi_g x = \sum_{j=1}^{n+1} \alpha_j \phi_g w^j$, there exists at least one j such that $w^j \neq \phi_g w^j$, and call this extreme point v.

Since $\hat{c}^t y \neq \hat{c}^t y'$ for any pair of extreme point $y \neq y'$ this implies that $\hat{c}^t v \neq \hat{c}^t \phi_g v$ which implies that the constraint is not trivial, since ϕ_g is affine.   QED

Note that the universal ordering vector $\hat{c}$ requires $O(n^2)$ bits. Although, in many cases, one can reduce the number of bits required, often c will require many bits, a topic we discuss in Section 6.

3   Separation for the Cyclic and Symmetric Groups

Two of the most common groups arising is practice are the cyclic and symmetric groups. The cyclic groups of order k are simply the additive group of integers modulo k and are denoted by Z_k. These are generated by a single element $1 \in Z_k$. The most natural action can be most easily described by viewing $x \in X$ as a matrix with r rows and $t \geq k$ columns, where $n = rt$. Then the action of ϕ_1 is given by cyclicly rotating the first k columns of this matrix, i.e, the first column becomes the second, the second becomes the third, column $k-1$ becomes column k and column k becomes column 1. Let $A = A_1$ be the matrix representation of ϕ_1 and note that $b_1 = 0$. Since $|G| = n$ the fundamental domain can be concisely represented by

$$F_c = \{x \in X \mid c^t x \geq c^t M^j x \quad j = 1..k - 1\}$$

and clearly given a point $x \in X$ but $x \notin F_c$ one can find a separating hyperplane by checking all $k - 1$ inequalities.

Theorem 3. *For the cyclic group (as described above), given a point $x \in X$ but $x \notin F_c$ one can find a separating hyperplane between x and F_c in $O(nks)$ time where s is the maximum number of bits in a component of c.*

The symmetric group is more complicated. As above, consider the vector x as a matrix, but in this case the group S_k is the set of all permutations of k elements and note that $|G| = k!$ which is exponential in k. Now, the group action consists of permutations of the first k columns of the matrix representation of x and the fundamental domain requires $k!$ additional inequalities. However, one can find a separating hyperplane efficiently as follows.

For simplicity, assume that $c > 0$. Construct a bipartite graph where the one set of vertices represent the current column ordering and the second set represents a permuted ordering. Let the value of an edge from i to j be the value represent the inner product of c and the i'th column of x if it were the j'th column. Then the maximum matching gives the optimal permutation. The separating hyperplane is simply given by the constraint related to this permutation. Since a maximum matching can be computed k^3 operations, we have proven the following theorem.

Theorem 4. *For the symmetric group action (as described above), given a point $x \in X$ but $x \notin F_c$ one can find a separating hyperplane between x and F_c in $O(nk^3s)$ time where s is the maximum number of bits in a component of c.*

Note that if we use the universal $\hat{c} = (2^{n-1}, \dots, 1)$ then the time to find a separating hyperplane for the cyclic group is $O(n^2k)$ while the time for the symmetric group is $O(n^2k^3)$. We note that this appears to be slower than the time required to compute a separating hyperplane for the related orbitopes [5].

4   Partitioning, Packing, Covering and Relations to Orbitopes

Now we discuss some important applications in which symmetry arises. Consider an optimization problem where there are r objects which must be put into k groups. Let x_{ij} be the variable that indicates that item i is in group j. Thus, the j'th column identifies the elements that are in group j. Problems can then be classified into three classes: partitioning (in which each item is in exactly one group), packing (in which each item is in at most one group), and covering (where each item is in at least one group).

When groups are identical, as in many combinatorial graph theory problems (such as coloring or partitioning), the IP is invariant under the full symmetry group of column permutations. Thus, our results from the previous section provide polynomial representations that remove much of the redundancy in the natural formulations.

However, in periodic scheduling problems, the same matrices arise but are only invariant under the cyclic group.

These problems are the subject of study by Kaibel and Pfetch [5] and the motivation behind orbitopes. Orbitopes are constructed by taking the convex hull of the set of $x \in Ext(X)$ which are maximal under the lexicographic ordering. While orbitopes are more refined than minimal fundamental domains, their analysis is significantly more complicated. In particular, the orbitopes for the covering problems appear to be quite complex and their explicit construction is not known. However, as can be seen from the analysis in the previous sections, the minimal fundamental domains can be easily characterized in all of these cases.

In fact, given the simplicity of their construction, our analysis easily extends to a wide variety of group actions.

5   Products of Groups

In this section we show that our analysis can be easily extended to special cases involving products of groups.

Given a group G and with a group action ϕ_g^G define the null space of the action to be the set of indices for which

$$(\phi_g^G x)_i = x_i \ \ \forall g \in F, x \in X.$$

Define the active space of the action to be the complement of the null space.

Now consider a second group action, H, ϕ^H such that the active space of H does not intersect the active space of G. The if we define the product action GH, ϕ^{GH} where GH is the direct product of the two groups, so an element of GH is the pair (g, h) with $g \in G$ and $h \in H$. The action is then given by $\phi_{(g,h)}^{GH} = \phi_g^G \phi_h^H$ and note that this is equal to $\phi_h^H \phi_g^G$ since the actions ϕ^G and ϕ^H must commute.

Then, the fundamental domain of the product action is simply the intersection of the fundamental domains and thus the required number of constraints in only $(|G| - 1) + (|H| - 1)$ instead of $(|G| - 1)(|H| - 1)$.

Theorem 5. *If active spaces of a set of group actions do not intersect then the fundamental domain of the product action is the intersection of the fundamental domains of the individual actions.*

For example, in the case where pairs of groups $\{(1, 2), (3, 4) \cdots, (n - 1, n)\}$ are interchangeable, the product action has $2^{n/2}$ constraints while the representation of the fundamental domain only requires $n/2$ constraints using the above result.

It appears that non-intersection of the group actions, although quite stringent, is necessary for these simplifications. One natural conjecture, that commutativity of the group actions is sufficient can be seen to be false from the following example.

Consider $X = [0, 1]^2$ and the action of G is interchanging the two components, $\phi_1^G(x_1, x_2) = (x_2, x_1)$ while H flips both bits, $\phi_1^H(x_1, x_2) = (1 - x_1, 1 - x_2)$. It is easy to see that the two group actions commute; However, the constraints for the two fundamental domains when taken separately with $c = (2, 1)$ are:

$$G: \ 2x_1 + x_2 \geq x_1 + 2x_2 \quad \rightarrow x_1 \geq x_2$$

$$H: \ 2x_1 + x_2 \geq 2(1 - x_1) + (1 - x_2) \quad \rightarrow 4x_1 + 2x_2 \geq 3$$

However, the constraint for the joint action, $\phi^{GH}_{(1,1)}$ is

$$2x_1 + x_2 \geq 2(1 - x_2) + (1 - x_1) \quad \rightarrow x_1 + x_2 \geq 1$$

which removes additional points from the intersection of the two separate group actions.

6   Choosing Ordering Vectors

The universal ordering vector $\hat{c}$ requires $O(n)$ bits which can be problematic for standard IP solvers. Consider the full symmetric group on $X \subseteq [0,1]^n$. Since a permutation preserves the number of zero elements, $c = (n - 1, n - 2, \ldots, 1, 0)$ is the smallest ordering vector such that F_c is minimal.

However, for the symmetric group operating by exchanging groups of elements, as discussed in Section 3, the ordering vector must be significantly larger in order to get a minimal fundamental domain.

Theorem 6. *Let $X \subseteq [0,1]^{rt}$ and assume that some element of the group G acts on X by exchanging the any two columns of the matrix $x_{ij} \in X$. If $c \geq 0$ is an integral vector and $max_i|c_i| < 2^{r+1}/r$ then F_c is not minimal.*

Proof: Assume that $g \in G$ acts on X by exchanging the first two columns. Then consider $x \in X$ such that the two exchangeable columns of x differ. In order for F_c to be minimal it must be true that $c^t x \neq c^t x'$ where x' is generated by exchanging the two exchangeable columns of x. Thus, all sums of subsets of the r elements corresponding to the first exchangeable column must differ from all sums from the second set of r elements. Since there are 2^r possible subsets, there must be at least $2 * 2^r$ possible values for the sums. Thus, there must be some sum which is greater than $2 * 2^r$ which implies that there is an element of size $2 * 2^r/r$. QED

A similar argument yields an exponential bound for the cyclic group; however, in many cases it might make more sense to choose c with smaller entries, and use non-minimal fundamental domains. For example, a simple random choice of c, while not yielding a minimal fundamental domain, still significantly reduces the number of symmetric copies of any extreme point, which could lead to computational savings.

7   Conclusions

We have provided a direct method for finding a minimal fundamental domain for a group action on a binary polytope. We note that our method can be easily extended to arbitrary polytopes and group actions. The only impediment to complete generality is the need to find separating hyperplanes, which might not be efficiently computable in some cases.

While this problem is of theoretical interest, it remains to be seen whether it is of practical value in solving real integer programs. However, recent results on the use symmetries in solving packing and partitioning problems [4,9], suggest fundamental domains might prove useful.

References

1. Barnhart, C., Johnson, E.L., Nemhauser, G.L., Savelsbergh, M.W.P., Vance, P.H.: Branch and price: Column generation for solving huge integer programs. Operations Research 46, 316–329 (1998)
2. Bazaraa, M.S., Kirca, O.: A branch-and-bound heuristic for solving the quadratic assignment problem. Naval Research Logistics Quarterly 30, 287–304 (1983)
3. Holm, S., Srensen, M.: The optimal graph partitioning problem: Solution method based on reducing symmetric nature and combinatorial cuts. OR Spectrum 15, 1–8 (1993)
4. Kaibel, V., Peinhardt, M., Pfetsch, M.: Orbitopal fixing. In: Proc. 12th Conference on Integer Programming and Combinatorial Optimization (IPCO). LNCS, Springer, Heidelberg (forthcoming, 2007)
5. Kaibel, V., Pfetsch, M.: Packing and partitioning orbitopes. Math. Program., Ser. A (in press) (2006), doi:10.1007/s10107-006-0081-5
6. Margot, F.: Pruning by isomorphism in branch-and-cut. Mathematical Programming 94(1), 71–90 (2002)
7. Margot, F.: Small covering designs by branch-and-cut. Mathematical Programming 94(2), 207–220 (2003)
8. Méndez-Díaz, I., Zabala, P.: A branch-and-cut algorithm for graph coloring. Discrete Applied Mathematics 154(5), 826–847 (2006)
9. Ostrowski, J., Linderoth, J., Rossi, F., Smriglio, S.: Orbital branching. In: Proc. 12th Conference on Integer Programming and Combinatorial Optimization (IPCO). LNCS, Springer, Heidelberg (forthcoming, 2007)
10. Serafini, P., Ukovich, W.: A mathematical model for periodic scheduling problems. SIAM J. Discrete Math 2(4), 550–581 (1989)

Exact Algorithms for Generalized Combinatorial Optimization Problems

Petrica C. Pop[1], Corina Pop Sitar[2], Ioana Zelina[1], and Ioana Taşcu[1]

[1] Department of Mathematics and Computer Science, Faculty of Sciences,
North University of Baia Mare, Romania
pop_petrica@yahoo.com, ioanazelina@yahoo.com, itascu@yahoo.com
[2] Department of Economics, Faculty of Sciences,
North University of Baia Mare, Romania
sitarcorina@yahoo.com

Abstract. We discuss fast exponential time exact algorithms for generalized combinatorial optimization problems. The list of discussed NP-complete generalized combinatorial optimization problems includes the generalized minimum spanning tree problem, the generalized subset assignment problem and the generalized travelling salesman problem.

Keywords: generalized combinatorial optimization problems, exact algorithms, dynamic programming.

1 Introduction

Classical combinatorial optimization problems can be generalized in a natural way by considering a related problem relative to a given partition of the nodes of the graph into node sets.

In the literature one finds generalized problems such as the generalized minimum spanning tree problem, the generalized travelling salesman problem, the generalized Steiner tree problem, the generalized (subset) assignment problem, etc. These generalized problems typically belong to the class of NP-complete problems, are harder than the classical ones and nowadays are intensively studied due to the interesting properties and applications in the real world.

Every NP-complete problem can be solved by exhaustive search. Unfortunately, when the size of the instances grows the running time for exhaustive search soon becomes very large, even for instances of rather small size. For some problems it is possible to design exact algorithms that are significantly faster than the exhaustive search, though still not polynomial time.

In this paper we present such fast exact algorithms that solve some NP-complete generalized combinatorial optimization problems.

Nowadays, there is a growing interest in the design and analysis of such fast exact algorithms because fast algorithms with exponential running time may lead to practical algorithms, at least for moderate size.

The techniques that we are using in order to provide exact algorithms for the generalized combinatorial optimization problems are dynamic programming

A. Dress, Y. Xu, and B. Zhu (Eds.): COCOA 2007, LNCS 4616, pp. 154–162, 2007.

combined with a local-global approach for the generalized minimum spanning tree problem, the local-global approach for the generalized subset assignment problem and the construction of a layered network for the generalized travelling salesman problem.

2   The Generalized Minimum Spanning Tree Problem

The minimum spanning tree (MST) problem can be generalized in a natural way by considering instead of nodes set of nodes (clusters) and asking for a minimum cost tree spanning *exactly* one node from each cluster. This problem is called the generalized minimum spanning tree problem (GMSTP) and it was introduced by Myung *et al.* [13]. The MST is a special case of the GMSTP where each cluster consists of exactly one node.

Meanwhile, the GMSTP have been studied by several authors w.r.t. heuristics and metaheuristics, LP-relaxations, polyhedral aspects and approximability, cf., e.g. Feremans, Labbe, and Laporte [4], Feremans [3], Pop, Kern and Still [17,18] and Pop [14,15].

Two variants of the generalized minimum spanning tree problem were considered in the literature: one in which in addition to the cost attached to the edges, we have costs attached also to the nodes, called the *prize collecting generalized minimum spanning tree problem* see [16] and the second one consists in finding a minimum cost tree spanning *at least* one node from each cluster, denoted by L-GMSTP and introduced by Dror *et al.* [2]. The same authors have proven that the L-GMSTP is *NP-hard*.

Let $G = (V, E)$ be an n-node undirected graph and $V_1, \ldots, V_m$ a partition of V into m subsets called *clusters* (i.e., $V = V_1 \cup V_2 \cup \ldots \cup V_m$ and $V_l \cap V_k = \emptyset$ for all $l, k \in \{1, \ldots, m\}$ with $l \neq k$). We assume that edges are defined between all nodes which belong to different clusters. We denote the cost of an edge $e = (i, j) \in E$ by c_{ij} or by $c(i, j)$ and the costs of the edges are chosen integers.

The *generalized minimum spanning tree* problem asks for finding a minimum-cost tree T spanning a subset of nodes which includes exactly one node from each cluster V_i, $i \in \{1, \ldots, m\}$. We will call such a tree a *generalized spanning tree*.

In [13], it is proved that the GMSTP is $\mathcal{NP}$-hard. A stronger result was presented in [15], namely the generalized minimum spanning tree problem even defined on trees is $\mathcal{NP}$-hard.

The proof of this result is based on a polynomial reduction of the set covering problem, which is known to be $\mathcal{NP}$-hard (see for example [8]), to the GMSTP defined on trees.

Let G' be the graph obtained from G after replacing all nodes of a cluster V_i with a supernode representing V_i. We will call the graph G' the global graph. For convenience, we identify V_i with the supernode representing it. Edges of the graph G' are defined between each pair of the graph vertices $\{V_1, ..., V_m\}$.

The *local-global approach* to the generalized minimum spanning tree problem aims at distinguishing between *global connections* (connections between clusters)

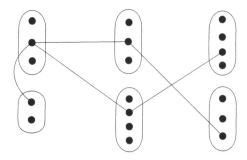

Fig. 1. Example showing a generalized spanning tree in the graph $G = (V, E)$

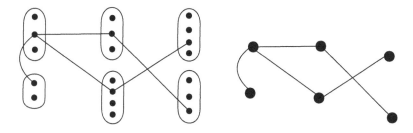

Fig. 2. Example showing a generalized spanning tree corresponding to a global spanning tree

and *local connections* (connections between nodes from different clusters). As we will see, having a global tree connection of the clusters it is rather easy to find the corresponding best (w.r.t. cost minimization) generalized spanning tree.

There are several generalized spanning trees corresponding to a global spanning tree. Between these generalized spanning trees there exists one called the best generalized spanning tree (w.r.t. cost minimization) that can be determined either by dynamic programming or by solving a linear integer program [18].

3   An Exact Algorithm for the Generalized Minimum Spanning Tree Problem

In this section, we present an algorithm that finds an exact solution to the GMST problem based on dynamic programming.

Given a spanning tree of the global graph G', which we shall refer to as a *global spanning tree*, we use dynamic programming in order to find the corresponding best (w.r.t. cost minimization) generalized spanning tree.

Fix an arbitrary cluster V_{root} as the root of the global spanning tree and orient all the edges away from vertices of V_{root} according to the global spanning tree. A directed edge $\langle V_k, V_l \rangle$ of G', resulting from the orientation of edges of the global spanning tree defines naturally an orientation $\langle i, j \rangle$ of an edge $(i, j) \in E$ where

$i \in V_k$ and $j \in V_l$. Let v be a vertex of cluster V_k for some $1 \leq k \leq m$. All such nodes v are potential candidates to be incident to an edge of the global spanning tree. On the graph G, we denote by $T(v)$ the subtree rooted at such a vertex v from G; $T(v)$ includes all vertices reachable from v under the above orientation of the edges of G based on the orientation of the edges of the global spanning tree. The *children* of $v \in V_k$, denoted by $C(v)$, are those vertices $u \in V_l$ which are heads of the directed edges $\langle v, u \rangle$ in the orientation. The leaves of the tree are those vertices that have no children.

Let $W(T(v))$ denote the minimum weight of a generalized subtree rooted at v. We want to compute

$$\min_{r \in V_{root}} W(T(r)).$$

We are now ready for giving the dynamic programming recursion to solve the subproblem $W(T(v))$. The initialization is:

$$W(T(v)) = 0, \ \text{ if } v \in V_k \text{ and } V_k \text{ is a leaf of the global spanning tree.}$$

To compute $W(T(v))$ for an interior to a cluster vertex $v \in V$, i.e., to find the optimal solution of the subproblem $W(T(v))$, we have to look at all vertices from the clusters V_l such that $C(v) \cap V_l \neq \emptyset$. If u denotes a child of the interior vertex v, then the recursion for v is as follows:

$$W(T(v)) = \sum_{l, C(v) \cap V_l \neq \emptyset} \min_{u \in V_l} [c(v, u) + W(T(u))].$$

Hence, for fixed v we have to check at most n vertices. Consequently, for the given global spanning tree, the overall complexity of this dynamic programming algorithm is $O(n^2)$. Since by Cayley's formula, the number of all distinct global spanning trees is m^{m-2}, we have established the following.

Theorem 1. *There exists a dynamic programming algorithm which provides an exact solution to the generalized minimum spanning tree problem in $O(m^{m-2}n^2)$ time, where n is the number of nodes and m is the number of clusters in the input graph.*

Clearly, the above is an exponential time algorithm unless the number of clusters m is fixed.

Remark 1. A similar dynamic programming algorithm can provide an exact solution to the prize collecting generalized minimum spanning tree problem with the difference that the dynamic programming recursion to solve the subproblem $W(T(v))$ for a node $v \in G$ should be considered as follows:

$$W(T(v)) = \sum_{l, C(v) \cap V_i \neq \emptyset} \min_{u \in V_l} [c(v, u) + d(u) + W(T(u))],$$

where by $d(u)$ we denoted the cost associated to the node u.

4   The Generalized Subset Assignment Problem

First, we present the classical Assignment Problem (AP). In this problem we are asking to minimize the total cost of assigning n workers to n tasks, where each worker has his own pay-scale associated with each task. No worker is allowed to perform more than one task and no task can be executed by more than one worker.

A polynomial time solution for the assignment problem of complexity $O(n^3)$ (the so-called Hungarian method) is credited to Kuhn [9].

Traditionally, in the literature (see for instance Fisher et al. [7]), the name *Generalized Assignment Problem* is associated with the problem of finding the minimum-cost assignment of n jobs to m agents such that each job is assign to one and only one agent subject to capacity constraints on the agents.

With respect to our definition of generalized combinatorial optimization problems, we assume instead of single workers a grouping of workers into subsets of skill categories. In addition, we assume that the tasks are also grouped into subsets of job categories and each worker from each group has his own pay-scale associated with each task in each of the task categories. More formally, let I_k, $1 \leq k \leq n$ be the grouping of skills (worker) categories and J_l, $1 \leq l \leq n$ be the grouping of task categories. Now the problem is no longer to match one worker to one job, but has to minimize the total cost of selecting exactly one worker from each group of workers to cover exactly one task from each of the task categories.

This problem version is called the *generalized subset assignment problem*. Clearly, when each worker is an individual category and each task is an individual, the problem reduces to the classical assignment problem.

In [1], it is proved that the problem is $\mathcal{NP}$-hard. The proof of this result is based on a polynomial reduction of the SATISFIABILITY problem, which is known to be $\mathcal{NP}$-hard (see for example [8]), to the generalized subset assignment problem.

In the case of the generalized subset assignment problem, based on the local-global approach defined in Section 2, an exact algorithm can be easily derived.

The idea is to replace all the workers corresponding to a skill category I_k with a supernode representing I_k, $1 \leq k \leq n$ and all the tasks corresponding to a job category J_l with a supernode representing J_l, $1 \leq l \leq n$. In this case the global graph G' will be a bipartite graph. The number of all distinct global assignments is $n!$. The best generalized subset assignment (w.r.t. cost minimization) corresponding to a global assignment is obtained by taking the worker from each group with smallest pay-scale associated with a task from the task categories, i.e. if there is a global assignment between supernodes I_k and J_l then we choose the worker from the group I_k with minimum cost associated to a task which belongs to the task category J_l.

5   The Generalized Travelling Salesman Problem

The generalized travelling salesman problem (GTSP), introduced by Laporte and Nobert [11] and by Noon and Bean [12] is defined on a complete undirected graph G whose nodes are partitioned into a number of subsets (clusters) and whose edges have a nonnegative cost. The GTSP asks for finding a minimum-cost Hamiltonian tour H in the subgraph of G induced by S, where $S \subseteq V$ such that S contains *at least* one node from each cluster.

A different version of the problem called E-GTSP arises when imposing the additional constraint that *exactly* one node from each cluster must be visited.

Both problems GTSP and E-GTSP are NP-hard, as they reduce to travelling salesman problem when each cluster consists of exactly one node.

The GTSP has several applications to location problems, telecommunication problems, railway optimization, etc. More information on these problems and their applications can be found in Fischetti, Salazar and Toth [5,6], Laporte, Asef-Vaziri and Sriskandarajah [10], Laporte and Nobert [11]. It is worth to mention that Fischetti, Salazar and Toth [6] solved the GTSP to optimality for graphs with up to 442 nodes using a branch-and-cut algorithm.

Let $G = (V, E)$ be an n-node undirected graph whose edges are associated with non-negative costs. We will assume w.l.o.g. that G is a complete graph (if there is no edge between two nodes, we can add it with an infinite cost). Let $V_1, ..., V_p$ be a partition of V into m subsets called *clusters* (i.e. $V = V_1 \cup V_2 \cup ... \cup V_p$ and $V_l \cap V_k = \emptyset$ for all $l, k \in \{1, ..., p\}$). We denote the cost of an edge $e = \{i, j\} \in E$ by c_{ij} or by $c(i, j)$. Let $e = \{i, j\}$ be an edge with $i \in V_l$ and $j \in V_k$. If $l \neq k$ the e is called an *inter-cluster* edge; otherwise e is called an *intra-cluster* edge.

The *generalized travelling salesman problem* (E-GTSP) asks for finding a minimum-cost tour H spanning a subset of nodes such that H contains exactly one node from each cluster V_i, $i \in \{1, ..., p\}$.

We will call such a cycle a *Hamiltonian tour*. An example of a Hamiltonian tour for a graph with the nodes partitioned into 6 clusters is presented in the next figure.

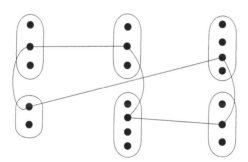

Fig. 3. Example of a Hamiltonian tour

The E-GTSP involves two related decisions:

- choosing a node subset $S \subseteq V$, such that $|S \cap V_k| = 1$, for all $k = 1, ..., p$.
- finding a minimum cost Hamiltonian cycle in the subgraph of G induced by S.

6 An Exact Algorithm for the Generalized Travelling Salesman Problem

In this section, we present an algorithm that finds an exact solution to the generalized travelling salesman problem.

Given a sequence $(V_{k_1}, ..., V_{k_p})$ in which the clusters are visited, we want to find the best feasible Hamiltonian tour H^* (w.r.t cost minimization), visiting the clusters according to the given sequence. This can be done in polynomial time, by solving $|V_{k_1}|$ shortest path problems as we will describe below.

We construct a layered network, denoted by LN, having $p + 1$ layers corresponding to the clusters $V_{k_1}, ..., V_{k_p}$ and in addition we duplicate the cluster V_{k_1}. The layered network contains all the nodes of G plus some extra nodes v' for each $v \in V_{k_1}$. There is an arc (i, j) for each $i \in V_{k_l}$ and $j \in V_{k_{l+1}}$ $(l = 1, ..., p-1)$, having the cost c_{ij} and an arc (i, h), $i, h \in V_{k_l}$, $(l = 2, ..., p)$ having cost c_{ih}. Moreover, there is an arc (i, j') for each $i \in V_{k_p}$ and $j' \in V_{k_1}$ having cost $c_{ij'}$.

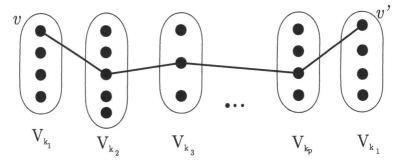

Fig. 4. Example showing a Hamiltonian tour in the constructed layered network LN

For any given $v \in V_{k_1}$, we consider paths from v to v', $w' \in V_{k_1}$, that visits exactly one node from each cluster $V_{k_2}, ..., V_{k_p}$, hence it gives a feasible Hamiltonian tour.

Conversely, every Hamiltonian tour visiting the clusters according to the sequence $(V_{k_1}, ..., V_{k_p})$ corresponds to a path in the layered network from a certain node $v \in V_{k_1}$ to $w' \in V_{k_1}$.

Therefore, it follows that the best (w.r.t cost minimization) Hamiltonian tour H^* visiting the clusters in a given sequence can be found by determining all the shortest paths from each $v \in V_{k_1}$ to the corresponding $v' \in V_{k_1}$ with the property that visits exactly one node from each of the clusters $(V_{k_2}, ..., V_{k_p})$.

The overall time complexity is then $|V_{k_1}|O(m + \log n)$, i.e. $O(nm + n\log n)$ in the worst case, where by m in this problem we denoted the number of edges. We can reduce the time by choosing $|V_{k_1}|$ as the cluster with minimum cardinality.

Notice that the above procedure leads to an $O((p-1)!(nm + nlogn))$ time exact algorithm for the GTSP, obtained by trying all the $(p-1)!$ possible cluster sequences. So, we have established the following result:

Theorem 2. *The above procedure provides an exact solution to the generalized travelling salesman problem in $O((p-1)!(nm + nlogn))$ time, where n is the number of nodes, m is the number of edges and p is the number of clusters in the input graph.*

Clearly, the algorithm presented, is an exponential time algorithm unless the number of clusters p is fixed.

7   Conclusions

In this paper we present fast exponential time exact algorithms for generalized combinatorial optimization problems. The list of discussed NP-complete generalized combinatorial optimization problems includes the generalized minimum spanning tree problem, the generalized subset assignment problem and the generalized travelling salesman problem. The techniques that we are using in order to provide exact algorithms are dynamic programming combined with a local-global approach to the generalized combinatorial optimization problems.

References

1. Dror, M., Haouari, M.: Generalized Steiner Problems and Other Variants. Journal of Combinatorial Optimization 4, 415–436 (2000)
2. Dror, M., Haouari, M., Chaouachi, J.: Generalized Spanning Trees. European Journal of Operational Research 120, 583–592 (2000)
3. Feremans, C.: Generalized Spanning Trees and Extensions, PhD thesis, Universite Libre de Bruxelles, Belgium (2001)
4. Feremans, C., Labbe, M., Laporte, G.: A Comparative Analysis of Several Formulations of the Generalized Minimum Spanning Tree Problem. Networks 39(1), 29–34 (2002)
5. Fischetti, M., Salazar, J.J., Toth, P.: The symmetric generalized traveling salesman polytope. Networks 26, 113–123 (1995)
6. Fischetti, M., Salazar, J.J., Toth, P.: A branch-and-cut algorithm for the symmetric generalized traveling salesman problem. Operations Research 45, 378–394 (1997)
7. Fisher, M.L., Jaikumar, R., van Wassenhove, L.N.: A multiplier adjustment method for generalized assignment problem. Management Science 32/9, 1095–1103 (1986)
8. Garey, M.R., Johnson, D.S.: Computers and Intractability: A guide to the theory of NP-Completeness. Freeman, San Francisco, California (1979)
9. Kuhn, H.W.: The hungarian method for the assignment problem. Naval Research Logistic Quarterly 2, 83–97 (1955)
10. Laporte, G., Asef-Vaziri, A., Sriskandarajah, C.: Some applications of the generalized traveling salesman problem. J. Oper. Res. Soc. 47, 1461–1467 (1996)
11. Laporte, G., Nobert, Y.: Generalized Traveling Salesman through n sets of nodes: an integer programming approach. INFOR 21, 61–75 (1983)

12. Noon, C.E., Bean, J.C.: A Lagrangian based approach for the asymmetric generalized traveling salesman problem. Operations Research 39, 623–632 (1991)
13. Myung, Y.S., Lee, C.H., Tcha, D.w.: On the Generalized Minimum Spanning Tree Problem. Networks 26, 231–241 (1995)
14. Pop, P.C.: The Generalized Minimum Spanning Tree Problem, PhD thesis, University of Twente, The Netherlands (2002)
15. Pop, P.C.: New Models of the Generalized Minimum Spanning Tree Problem. Journal of Mathematical Modelling and Algorithms 3(2), 153–166 (2004)
16. Pop, P.C.: On the Prize-Collecting Generalized Minimum Spanning Tree Problem. In Annals of Operations Research (to appear)
17. Pop, P.C., Kern, W., Still, G.: Approximation Theory in Combinatorial Optimization. Application to the Generalized Minimum Spanning Tree Problem, Revue d'Analyse Numerique et de Theorie de l'Approximation, Tome 34(1), 93–102 (2005)
18. Pop, P.C., Kern, W., Still, G.: A New Relaxation Method for the Generalized Minimum Spanning Tree Problem. European Journal of Operational Research 170, 900–908 (2006)
19. Yannakakis, M.: Expressing combinatorial optimization problems by linear programs. Journal of Computer and System Sciences 43, 441–466 (1991)

Approximation Algorithms for k-Duplicates Combinatorial Auctions with Subadditive Bidders

Wenbin Chen and Jiangtao Meng

Department of Computer Science, Nanjing University of Aeronautics and
Astronautics, Nanjing 210016, P.R. China
cwbiscas@yahoo.com, globangrilion@yahoo.com

Abstract. In this paper, we study the problem of maximizing welfare in
combinatorial auctions with k duplicates of each item, where bidders are
subadditive. We present two approximation algorithms for k-duplicates
combinatorial auctions with subadditive bidders. First, we give a factor-
$O(\sqrt{m})$ approximation algorithm for k-duplicates combinatorial auctions
with subadditive valuations using value queries. This algorithm is also
incentive compatible. Secondly, we give a factor-$O(\log m)$ approximation algorithm for k-duplicates combinatorial auctions with subadditive
valuations using demand queries.

1 Introduction

We consider the allocation problem in combinatorial auctions with k-duplicates
of of each item. In a combinatorial auction, a set M of m items is sold to n
bidders. Every bidder i has a valuation function $v_i : 2^M \to R^+$. We suppose that
the valuation is monotone, which means for every two bundles $S, T, S \subseteq T \subseteq M$
such that $v(S) \leq v(T)$, and normalized $v(\emptyset) = 0$. The goal is to find a partition
$(S_1, \ldots, S_n)$ of the m items that maximizes the total utility of *social welfare*,
i.e., $\Sigma_i v_i(S_i)$ is maximized. We call such an allocation an optimal allocation.

The k-duplicates combinatorial auction is the generalization of a combinatorial auction (where $k = 1$), which is the allocation problem in combinatorial
auctions by allowing k-duplicates of of each item. In the k-duplicates combinatorial auction, every bidder is still interested in at most one unit of each item
and every valuation is still defined on the subsets of M.

Notice that the size of the "input" is exponential in m (since each v_i is described by 2^m real numbers). Thus, we are interested in algorithms that are
polynomial time in m and n. Since the size of the input is exponential, we suppose that we have oracles for accessing it. Two common types of query methods
have been considered. One common type of queries is the "value queries" , which
answers $v(S)$ for a valuation v given a bundle of S. From a "computer science"
perspective, the kind of query is very natural. Another kind of query is the
"demand queries". Given a vector $p = (p_1, \ldots, p_m)$ of item prices, a demand
query replies a set that maximizes the profit, i.e. maximizes $v_i(S) - \Sigma_{j \in S} p_j$.

A. Dress, Y. Xu, and B. Zhu (Eds.): COCOA 2007, LNCS 4616, pp. 163–170, 2007.
© Springer-Verlag Berlin Heidelberg 2007

Demand queries are very natural from an economic point of view. It is known that demand queries can simulate values queries in polynomial time [3].

For general utility functions, the combinatorial auction problem is NP-hard. If value queries are used, it has been shown that there are no polynomial time algorithms with an approximation factor better than $O(\frac{\log m}{m})$ in [3]. It has also been shown that there are no polynomial time algorithms with an approximation factor better than $O(\frac{1}{m^{1/2-\epsilon}})$ even for single minded bidders in [11] and [14]. In [13], it is shown than achieving any approximation factors better than $O(\frac{1}{m^{1/2-\epsilon}})$ requires exponential communication if demand queries are used. In [6], more results on the combinatorial auction problems with general utilities can be found.

In this paper we study the important cases where all bidders are known to have subadditive valuations. A valuation is called subadditive (also called complement-free) if $v(S \cup T) \leq v(S) + v(T)$ for all $S, T \subseteq M$;

It is still NP-hard for the allocation problem with subadditive utility functions. In [7], a $O(\log m)$ approximation algorithm is given for combinatorial auctions with subadditive utility function if demand queries are used and an incentive compatible $O(\sqrt{m})$ approximation algorithm is also given if value queries are used. Recently, Feige improve the approximation ratio to 2 in [9]. As for negative results, it is shown that achieving an approximation ratio better than 2 require for an exponential amount of communication in [7]. In [9], it is shown that it is NP-hard to approximate the maximum welfare within a factor $2 - \epsilon$ unless $P = NP$, when bidders are subadditive.

For multi-unit combinatorial auctions, an incentive compatible mechanisms is given in [2]. In particular, this includes the case where each good has exactly k units, i.e. k-duplicates combinatorial auctions. For k-duplicates combinatorial auctions, Dobzinski and Schapira give a polynomial time approximation algorithm obtaining $\min\{\frac{n}{k}, O(m^{\frac{1}{k+1}})\}$ approximation ratio using demand queries in [8], and show that exponential communication is required for achieving an approximation ratio better than $\min\{\frac{n}{k}, O(m^{\frac{1}{k+1}-\epsilon})\}$, where $\epsilon > 0$. In the same paper, they also give an algorithm that achieves an approximation ratio of $O(\frac{m}{\sqrt{\log m}})$ using only a polynomial number of value queries and show that it is impossible to approximate a k-duplicates combinatorial auction to a factor of $O(\frac{m}{\log m})$ using a polynomial number of value queries. They studied the case where all valuations are general utility functions. The case about subadditive and submodular valuations are studied in [4]. In [4], it is shown that it is NP-hard to approximate the maximum welfare for k-duplicates combinatorial auctions with subadditive bidders within a factor of $2 - \epsilon$ where $\epsilon > 0$ unless $P = NP$. It is also shown that for any $\epsilon > 0$, any $(2 - \epsilon)$-approximation algorithm for a k-duplicates combinatorial auction with subadditive bidders requires an exponential amount of communication. A 2-approximation algorithm for k-duplicates combinatorial auctions with submodular bidders is also given in [4]. For k-duplicates combinatorial auctions with subadditive, they didn't study the approximation algorithm. In this paper, we give two approximation algorithms about k-duplicates combinatorial auctions with subadditive valuations.

Our Result

In this paper, we exhibit some upper approximation bounds for k-duplicates combinatorial auctions with subadditive valuations. First, we give a $O(\sqrt{m})$ approximation algorithm for k-duplicates combinatorial auctions with subadditive valuations using value queries. The approximation ratio improve on the upper bound $O(\frac{m}{\sqrt{\log m}})$ for general valuations [8]. This algorithm is also incentive compatible. Secondly, we give a $O(\log m)$ approximation algorithm for k-duplicates combinatorial auctions with subadditive valuations using demand queries. Thus, when $\min\{\frac{n}{k}, O(m^{\frac{1}{k+1}})\} \geq \log m$, our approximation algorithm is better than that for general valuations [8].

Structure of the Paper

In section 2, we propose a $O(\sqrt{m})$ approximation algorithm for k-duplicates combinatorial auctions with subadditive valuations using value queries. In section 3 we present a $O(\log m)$-approximation algorithm for k-duplicates combinatorial auctions with subadditive valuations using demand queries. Finally, in section 4 we present some conclusions and some open problems.

2 A $O(\sqrt{m})$ Approximation Algorithm for Subadditive Valuations with Value Queries

In this section we propose a $O(\sqrt{m})$ approximation algorithm that is incentive compatible for k-duplicates combinatorial auctions in which all valuations are submadditive, which extends the algorithm of [7]. The approximation ratio improve on the upper bound $O(\frac{m}{\sqrt{\log m}})$ for general valuations [8].

First we give the definition of b-matching problem. We adopt the definition in [15]. Let $G = (V, E)$ be a graph, where V is the set of nodes and E is the set of edges. Each $e \in E$ is assigned a real number cost w_e. Let $b = ((l_1, b_1), (l_2, b_2), \cdots, (l_{|V|}, b_{|V|}))$, where b_i's are integers and l_i equals b_i or 0 $(1 \leq i \leq |V|)$. A b-matching is a set $M \subseteq E$ such that the number of edges incident with i is no more than b_i and no less than l_i. The value of a b-matching is the sum of costs of its edges, i.e. $\Sigma_{e \in M} w_e$. The b-matching problem is to find a b-matching of maximum value.

In the following, we describe the approximation algorithm.

Input: The input is given as a set of n value oracles for the n valuations v_i.

Output: An allocation $S_1, \ldots, S_n$ which is an $O(\sqrt{m})$ approximation to the optimal allocation.

The Algorithm:

1. Query each bidder i for $v_i(M)$, for $v_i(\{j\})$, for each item j.
2. Construct a bipartite graph by defining a vertex a_j for each item j, and a vertex b_i for each bidder i. Let the set of edges be $E = \cup_{i \in N, j \in M}(a_j, b_i)$.

Define the cost of each edge (a_j, b_i) to be $v_i(\{j\})$. Compute the maximum b-matching $|P|$ in the graph. The pair of b values of a node a_j associated with the item j will be (k, k). The b value of a vertex b_i will be $(0, m)$.

3. Let $I \subseteq \{1, \ldots, n\}$ be the set of indices of the k bidders who offer the highest bids on M. If $\Sigma_{i \in I} v_i(M)$ is higher than the value of $|P|$, allocating all items to each bidder in I. Otherwise, for each edge $(a_j, b_i) \in P$ allocate the jth item to the ith bidder.

4. Let each bidder pay his VCG price.

In the following we give the analysis of the algorithm, which extends the analysis of theorem 2.2 of [7].

Lemma 1. If $x_1 \geq x_2 \cdots \geq x_r$, then $\frac{x_1 + \cdots x_k}{k} \geq \frac{x_1 + \cdots x_r}{r}$ for $1 \leq k \leq r$.

Proof: Since $x_1 \geq x_2 \cdots \geq x_k \geq x_{k+1}$, $x_1 + x_2 + \cdots + x_k \geq k x_{k+1}$. So $(k + 1)(x_1 + x_2 + \cdots + x_k) \geq k(x_1 + x_2 + \cdots + x_k + x_{k+1})$. Thus $\frac{x_1 + x_2 + \cdots + x_k}{k} \geq \frac{x_1 + x_2 + \cdots + x_k + x_{k+1}}{k+1}$. Since k is any integer, we get $\frac{x_1 + x_2 + \cdots + x_k}{k} \geq \frac{x_1 + x_2 + \cdots + x_k + x_{k+1}}{k+1} \geq \frac{x_1 + x_2 + \cdots + x_k + x_{k+1} + x_{k+2}}{k+2} \geq \cdots \geq \frac{x_1 + \cdots x_r}{r}$.

Theorem 5. *The algorithm provides a $O(\sqrt{m})$ approximation to the optimal allocation for k-duplicates combinatorial auctions in which all valuations are subadditive valuations and is incentive compatible.*

Proof: It is easy to know that the algorithm runs in polynomial time in n and m: firstly, finding the maximal b-matching is solved in polynomial time in m and n ([1] and [5]). Secondly, the calculation of the VCG prices requires solving only an additional auction for each of the bidders. Since these additional auctions are smaller in size (one bidder less), and thus can also be done in polynomial time.

Let us now prove that the algorithm obtains the desired approximation ratio.

Let $\{T_1, \ldots, T_r, Q_1, \ldots, Q_l\}$ denote the optimal allocation OPT, where for each $1 \leq t \leq r$, $|T_t| \geq \sqrt{m}$, and for each q, $1 \leq q \leq l$, $Q_q \leq \sqrt{m}$. Let $|OPT| = \Sigma_{t=1}^{r} v_t(T_t) + \Sigma_{q=1}^{l} v_q(Q_q)$.

Firstly, we consider the case where $\Sigma_{t=1}^{r} v_t(T_t) \geq \frac{|OPT|}{2}$. Let $I \subseteq [r]$ be the set of indices of the k bidders who offer the highest bids on M. Since $r\sqrt{m} \leq \Sigma_{t=1}^{r} |T_t| \leq mk$, $r \leq \frac{mk}{\sqrt{m}}$. Thus $\Sigma_{i \in I} v_i(M)/k \geq \Sigma_{i=1}^{r} v_i(M)/r \geq \Sigma_{i=1}^{r} v_i(T_i)/r \geq \frac{|OPT|/2}{mk/\sqrt{m}}$. So $\Sigma_{i \in I} v_i(M) \geq \frac{|OPT|}{2\sqrt{m}}$. Thus, by allocating all items to each bidder in I we get the desired approximation ratio.

The second case is when $\Sigma_{q=1}^{l} v_q(Q_q) > \frac{|OPT|}{2}$. For each i, $1 \leq i \leq l$, let $c_i = $arg $max_{j \in Q_i} v_i(\{j\})$. By the subadditive property: $|Q_i| v_i(\{c_i\}) \geq \Sigma_{j \in Q_i} v_i(\{j\}) \geq v_i(Q_i)$. Thus $v_i(\{c_i\}) \geq \frac{v_i(Q_i)}{|Q_i|}$. Since $|Q_i| < \sqrt{m}$ for all i, we have that $\Sigma_{i=1}^{l} v_i(c_i) > \frac{\Sigma_i v_i(Q_i)}{\sqrt{m}} \geq \frac{|OPT|}{2\sqrt{m}}$. By assigning c_i to bidder i we get an allocation with a social welfare of $\Sigma_{i=1}^{l} v_i(c_i) \geq \frac{|OPT|}{2\sqrt{m}}$. Since $\Sigma_{i=1}^{l} v_i(c_i)$ is less than the value of the maximum b-matching $|P|$ in the step 2 of above algorithm, the second allocation also obtain the desired approximation ratio.

Thus, the approximation ratio the algorithm produces is at least $O(\sqrt{m})$. Incentive compatibility is guaranteed because of the use of the VCG prices.

3   A $O(\log m)$ Approximation Algorithm for Subadditive Valuations with Demand Queries

In this section we propose a $O(\log m)$ approximation algorithm for k-duplicates combinatorial auctions in which all valuations are submadditive, which extends the algorithm of [7]. Thus, when $\min\{\frac{n}{k}, O(m^{\frac{1}{k+1}})\} \geq \log m$, our approximation algorithm is better than that for general valuations [8].

In the following, we describe the approximation algorithm.

Input: The input is given as a set of n demand oracles for the n valuations v_i.

Output: An allocation $T_1, \ldots, T_n$ which is an $O(\log m)$ approximation to the optimal allocation.

The Algorithm: We first describe the basic steps of the algorithm and then provide the details necessary for its implementation.

1. Solve the linear relaxation of the problem:
   $Maximize: \Sigma_{i,S} x_{i,S} v_i(S)$ subject to:
   (a) For each item j : $\Sigma_{i,S|j \in S} x_{i,S} \leq k$
   (b) For each bidder i : $\Sigma_S x_{i,S} \leq 1$
   (c) For each i, S : $x_{i,S} \geq 0$
2. Use randomized rounding to find a pre-allocation $S_1, \ldots, S_n$ with the following properties, where $d = (1 + (c-1)\log m)k$, and $c > 0$ is a constant to be chosen later:
   (a) Each item j appears at most d times in $\{S_i\}_i$, with $j \in S_i$.
   (b) $\Sigma_i v_i(S_i) \geq \frac{1}{3} \cdot (\Sigma_{i,S} x_{i,S} v_i(S))$.
3. For each bidder i, partition S_i into a disjoint union $S_i = S_i^1 \cup \ldots \cup S_i^k$ such that for each $1 \leq i_1 < i_2 \leq n$ and $1 \leq r \leq k$, it holds that $S_{i_1}^r \cap S_{i_2}^r = \emptyset$. This is done as follows: for each $i = 1, \ldots, n$ and each $r = 1, \ldots, k$, we let $S_i^r = \{j \in S_i | j$ appears in exactly $r - 1$ of the sets $S_1, \ldots, S_{i-1}\}$.
4. Find the r that maximizes $\Sigma_i v_i(S_i^r)$, and for each i allocate S_i^r to bidder $i, i + 1, \ldots, (i + (k-1)) \mod n$.

Even though the linear program has an exponential number of variable, it is shown in [13] and [3] that the linear program may be solved in polynomial time, which is done by solving the dual linear program using the ellipsoid method. The ellipsoid method requires a "separation" oracle, and this may be directly implemented using the demand oracles of the bidders.

The randomized rounding step is implemented as follows: For each bidder i we independently choose a set S_i by performing the following random experiment: every set S is chosen with probability $x_{i,S}$, and the empty set is chosen with probability $(1 - \Sigma_S x_{i,S})$. If any of the required constraint is violated, then this stage is repeated from scratch. Using the generator of [12] as explained in [14], this randomized step may be converted to be deterministic by derandomizing .

We show that $S_{i_1}^r \cap S_{i_2}^r = \emptyset$ for each $1 \leq i_1 < i_2 \leq n$ and $1 \leq r \leq k$ as follows. By the definition of $S_{i_1}^r$, j appears in exactly $r-1$ of the sets $S_1, \ldots, S_{i_1-1}$ and $j \in S_{i_1,}$. Thus j appears in at least r of the sets $S_1, \ldots, S_{i_2-1}$. So, if $j \in S_{i_1}^r$ then $j \notin S_{i_2}^r$. Hence, $S_{i_1}^r \cap S_{i_2}^r = \emptyset$.

Theorem 1. *The algorithm produces an allocation that is a $O(\log m)$ approximation to the optimal allocation.*

Proof: First, without loss of generality, we assume that $\max_i\{v_i(M)\} = 1$ (otherwise we can simply divide all valuation by $\max_i\{v_i(M)\}$).

The first step of the algorithm returns the optimal fractional solution $OPT^* = \Sigma_{i,S} x_{i,S} v_i(S)$, which is an upper bound on the value of the optimal allocation, OPT.

We prove that with constant probability the second step produces an pre-allocation $\{S_1, \ldots, S_n\}$ in which $\Sigma_i v_i(S_i) \geq \frac{1}{3} \cdot OPT^*$.

We will require the following version of the Chernoff bounds:

Lemma 2 (Chernoff Bound). *Let $X_1, \ldots, X_n$ be independent Bernoulli trials such that for $1 \leq i \leq n$, $Pr[X_i = 1] = p_i$. Then for $X = X_1 + \cdots + X_n, \mu \leq p_1 + \cdots + p_n$, and any $\delta \geq 2e - 1$ we have:$Pr[X > (1+\delta)\mu] < 2^{-\mu\delta}$.*

For each $j \in M$, let E_j denote the random variable that indicates whether j was allocated more than d times. Let B be the random variables that indicates whether $v_i(S_i) < \frac{1}{3} \cdot OPT^*$. We will prove that $Pr[\vee_j E_j \vee B] < 1$.

Firstly, we show that $Pr[\vee_j E_j] < \frac{1}{20}$. The proof is similar to that in [7].

Fix an item j, let $Z_{i,j}$ be the random variable that determines whether $j \in S_i$. Obviously, its value is in $\{0, 1\}$. Because of the randomized rounding method we used, we have that the variables $\{Z_{i,j}\}_i$ are independent. We define $Z_j = \Sigma_i Z_{i,j}$ (i.e., Z_j is the number of times item j appears in $\{S_i\}$). By the linearity of expectation and the first condition of the LP formulation, we have that $E[Z_j] = \Sigma_i E[Z_{i,j}] = \Sigma_{i,S|j \in S} x_{i,S} \leq k$. We can now use the Chernoff bound, and choose a c such that:

Pr[item j appears in more than d bundles in $\{S_i\}$]$=Pr[Z_j > (1 + (c - 1)\log m)k] < 2^{-(c-1)k\log m} = \frac{1}{m^{(c-1)k}} < \frac{1}{20m}$ (e.g. if $(c-1)k \geq 6$, then $m^{(c-1)k} \geq m^6 > 20m$).

By applying the union bound we get that the probability that any one of the items appears in more than d bundles in $\{S_i\}$ is smaller than $m \cdot \frac{1}{20m} = \frac{1}{20}$.

In [7], it is shown that $Pr[B] < \frac{3}{4}$. In our case, the conclusion holds also.

Therefore, using the union bound: $Pr[\vee_{j=1}^m E_j \vee B] \leq \Sigma_{j \in M} Pr[E_j] + Pr[B] \leq \frac{1}{20} + \frac{3}{4} = \frac{4}{5}$.

Since for each fixed i, $S_i = \cup_r S_i^r$ and v_i is subadditive, $\Sigma_r v_i(S_i^r) \geq v_i(S_i)$. By summing over all i we get that $\Sigma_r \Sigma_i v_i(S_i^r) = \Sigma_i \Sigma_r v_i(S_i^r) \geq \Sigma_i v_i(S_i) \geq \frac{1}{3} \cdot OPT^*$. Since r is chosen that maximizes $\Sigma_i v_i(S_i^r)$ we get that $\Sigma_i v_i(S_i^r) \geq \frac{OPT^*}{3k}$.

Since for every fixed r, the sets $\{S_i^r\}_i$ are pairwise disjoint, thus the sets $\{T_i\}$ are a valid allocation. Since for each i, $S_i^r \subseteq T_i$, $\Sigma_i(T_i) \geq \Sigma_i v_i(S_i^r) \geq \frac{OPT^*}{3k}$.

Thus the allocation $T_1, \ldots, T_n$ is an $O(\log m)$ approximation to the optimal allocation.

4   Conclusion and Open Problems

In this paper, we exhibit some upper approximation bounds for k-duplicates combinatorial auctions with subadditive valuations. We give an incentive compatible $O(\sqrt{m})$ approximation algorithm for k-duplicates combinatorial auctions with subadditive valuations using value queries. We also give a $O(\log m)$ approximation algorithm for k-duplicates combinatorial auctions with subadditive valuations using demand queries.

We conjecture that the $O(\log m)$ approximation ratio can be improved to 2. Obtaining a truthful mechanisms with better approximation ratio is also interesting open problem. The upper bound for k-duplicates combinatorial auctions with XOS valuations is also unknown. The problem should be further studied.

Acknowledgments

We would like to thank the anonymous referees for their careful readings of the manuscripts and many useful suggestions. The research was partially supported by young teachers research funds.

References

1. Anstee, R.: A polynomial algorithm for b-matchings: An alternative approach. Inform. Process. Lett. 24, 153–157 (1987)
2. Bartal, Y., Gonen, R., Nisan, N.: Incentive compatible multi unit combinatorial auctions. In: TARK 03 (2003)
3. Blumrosen, L., Nisan, N.: On the computational power of iterative auctions I: Demand queries. In: ACM Conference on Electornic Commerce, ACM Press, New York (2005)
4. Chen, W.B., Meng, J.T., Yin, D.P.: The upper and lower approximation bounds for k-duplicates combinatorial auctions with submodular and subadditive bidders (in submission)
5. Cook, W.J., Cunningham, W.H., Pulleyblank, W.R., Schrijver, A.: Combinatorial Optimization. Wiley, Chichester (1998)
6. Cramton, P., Shoham, Y., Steinberg, R. (eds.): Combinatorial Auctions. MIT Press, Cambridge (Forthcoming) (2005)
7. Dobzinski, S., Nisan, N., Schapira, M.: Approximation algorithm for combinatorial auctions with complement-free bedders. In: Proceedings of 37th STOC, pp. 610–618 (2005)
8. Dobzinski, S., Schapira, M.: Optimal upper and lower approximation bound for k-duplicates combinatorial auctions. Working paper (2005)

9. Feige, U.: On maximizing welfare when utility functions are subadditive. In: Proceedings of the 38th Annual ACM Symposium on Theory of Computing, pp. 41–50. ACM Press, New York (2006)

10. Lehmann, B., Lehmann, D., Nisan, N.: Combinatorial auctions with decreasing marginal utilities. In: ACM conference on electronic commerce, ACM Press, New York (2001)

11. Lehmanm, D., O'Callaghan, L., Shoham, Y.: Truth revelation in approximately efficient combinatorial auctions. In: ACM Conference On Electronic Commerce, ACM Press, New York (1999)

12. Nisan, N.: $RL \subseteq SC$. In: Proceedings of the twenty-fourth annual ACM symposium on theory of computing, pp. 619–623. ACM Press, New York (1992)

13. Nisan, N., Segal, I.: The communication requirements of efficient allocations and supporting prices. In Journal of Economic Theory (to appear)

14. Sandholm, T.: An algorithm for optimal winner determination in combinatorial auctions. In: IJCAI (1999)

15. Tennenholtz, M.: Tractable combinatorial auctions and b-matching. Artificial Intelligence 140(1/2), 231–243 (2002)

A Grid Resource Discovery Method Based on Adaptive k-Nearest Neighbors Clustering

Yan Zhang[1], Yan Jia[1], Xiaobin Huang[2], Bin Zhou[1], and Jian Gu[1]

[1] School of Computer, National University of Defense Technology,
410073 Changsha, China
jane325@tom.com
[2] Department of Information Engineering, Air Force Radar Academy,
430019 Wuhan, China
hxbtougao@gmail.com

Abstract. Several features of today's grid are based on centralized or hierarchical services. However, as the grid size increasing, some of their functions especially resource discovery should be decentralized to avoid performance bottlenecks and guarantee scalability. A novel grid resource discovery method based on adaptive k-Nearest Neighbors clustering is presented in this paper. A class is formed by a collection of nodes with some similarities in their characteristics, each class is managed by a leader and consists of members that serve as workers. Resource requests are ideally forwarded to an appropriate class leader that would then direct it to one of its workers. This method can handle resource requests by searching a small subset out of a large number of nodes by resource clustering which can improve the resource query efficiency; on the other hand, it also achieves well scalability by managing grid resources with adaptive mechanism. It is shown from a series of experiments that the method presented in this paper achieves more scalability and efficient lookup performance than other existing methods.

1 Introduction

The goal of the grid [1] is to pool resources. Fast discovery of available resources and efficient maintenance of resource states are key requirements for grids to achieve optimal utilization of the system and to balance load among the participating computers. Resources in grid are characterized as diverse, dynamic, heterogeneous, and geographically distributed. The autonomy of resource owners needs to be honored with their local management and usage policies. Each node in grid has one or more resources with such features mentioned above, and the task of grid resource discovery is to find such a node which satisfies all users' requests. Therefore, it is challenging to develop efficient methods to discover grid resource.

The desirable features of grid resource discovery method should be both scalability and efficiency. To achieve scalability, resource discovery should not rely on only a few centralized nodes, which could be potential performance and security

A. Dress, Y. Xu, and B. Zhu (Eds.): COCOA 2007, LNCS 4616, pp. 171–181, 2007.

bottleneck. To achieve efficiency, qualified resource providers should be located rapidly without too much network and processing overhead.

In this paper we demonstrate a novel grid resource discovery method which based on clustering algorithm to achieve the above features. In our method, a class is formed by a collection of nodes with some similarities in their characteristics, and each class is managed by a leader and consists of members that serve as workers. Resource requests are firstly forwarded to an appropriate class leader that would then direct them to one of its workers. By doing so, the efficiency of grid resource discovery can be guaranteed. Ideally, the resources which requested by users should be found with two hops.

In this paper, distributed mechanism is used when transferring messages through different classes, and centralized mechanism is used to process query in each class. Compared with centralized mechanism or decentralized respectively, the strategy taken in this paper achieves more scalability and efficiency. It is shown from a series of experiments that the method presented in this paper receives high lookup performance.

The rest of this paper is organized as follows: in section 2, we introduce the related work, in section 3 we introduce how to design the grid resource discovery method in detail which is mentioned above, in section 4, we present algorithms which are used in our method, and in section 5, the experimental result is analyzed, finally, we conclude this paper in section 6.

2   Related Work

A lot of research and experimental work has been done on grid resource discovery. And here, we will analyze and discuss some typical work.

Early grid resource discovery is based on centralized mechanism, Globus Toolkit [2], Condor [3] and Legion [4] are the excellent examples. The MDS-4 (Monitoring and Discovery Services) of Globus Toolkit provides a Web Service Resource Framework (WSRF) [5] compliant implementation of the Index Service, as well as novel mechanisms for delivering notifications in the presence of events that match a set of specified rules (Trigger Service). Matchmaker in Condor use a centre server to match the attributes in the user's specification and those in the service providers' declaration. Such approach has a single point of failure and scales poorly. In Legion, Collections, the information database, are populated with resource description. The scheduler queries the Collection and finds proper resource for applications. A few global Collections will prohibit the scalability of the system.

With resolving problems that exist in centralized mechanisms, some researchers prompt decentralized resource discovery methods. Iamnitchi [6] proposes resource discovery based on an unstructured network similar to Gnutella combined with more sophisticated query forwarding strategies taken from the Freenet network. Requests are forwarded to one neighbor only based on experiences obtained from previous requests, thus trying to reduce network traffic and the number of requests per peer compared to simple query flooding as used by Gnutella. The approach suffers from higher numbers of required hops to resolve a query compared to our approach and provides no lookup guarantees. A grid resource discovery model based on routing-forwarding is proposed and analyzed in [7]. The model may suffer from scalability problem because all the resource routers are equal peers and resource routing information need to be propagated across the whole network.

3   Overview of the Method

3.1   Architecture

In order to form and maintain the class which consists of nodes and process resource query, we design and implement a two-layer overlay network similar to the architecture mentioned in paper [8]. The first layer consists of many leader nodes, we call it upper layer, and the lower layer is a class, it consists of one leader node and many worker nodes which belong to this leader node. In this two-layer overlay network, each node has a link to all leader nodes, and each leader node has a link to all of its own worker nodes, Fig.1 shows a visual representation of the overlay network. There exist two extreme cases in the overlay network. One is that there exists only one class in the grid, which is similar to typical client/server structure; the other is each node forms a class, thus, all nodes in grid form a complete connection graph. In this paper, the management of nodes is based on the two extreme cases mentioned above. We design a similarity

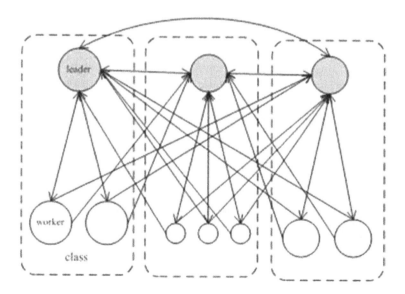

Fig. 1. Two layers overlay network. The upper layer consists of leader nodes, and the lower layer is a class.

threshold to control the class number formed in grid, and this design is based on two factors:

a) Communication cost load on leader node. If the class number is too small, then the leader nodes of these classes should process superfluous load, which may be performance bottleneck.
b) If there are too many classes in the grid, the cost for forming and maintaining these classes would be large.

3.2   Components of Grid Resource Discovery

In this paper we take the clustering method shown as follows: firstly, we use clustering algorithm to group the nodes already existed in grid into some classes. Secondly, as nodes joining and leaving, the algorithm can adjust the formation of class adaptively by class splitting and combination. The whole grid resource discovery process is made up of three parts:

a) Publish resource information.
b) Form and maintain classes.
c) Handle resource query message.

The first part publishes the resource information of one node to its own leader node periodically.

The second part creates and maintains the classes and updates resource information periodically to ensure that each node belongs to correct class. When a new node entering the grid, it calculating its statistical characteristic, and finding one correct class to join in. If the statistical characteristic is very similar to some leader, then it joins this class. Otherwise, this node should be a new class with itself being the leader.

The resource query message is processed by the third part. The querying mechanism taken in this paper limits the querying ideally to two hops. In the first step, resources query is transferred to the right class, and in the second step, it finds the requested resource in the class. This process is similar to looking up words in dictionary.

In order to get lower communication load when maintaining the formation of classes and processing resource request, we treat leader nodes and other nodes differently.

In summary, the grid resource discovery method presented in this paper has following characteristics:

a) Each class includes a collection of nodes with a single representative node called leader.
b) Nodes in the same class have some similarity on resource statistical characteristics.
c) The scale of each class is controlled by a threshold, and this threshold is chosen based on some factors such as grid resource number, resource density and load balancing of resource discovery request.

4   Algorithms

In this section, we will introduce algorithms in detail which are taken in our resource discovery method.

4.1   Data and Request Model

In this paper, each resource is described as a triple of (type, attribute, value), and each node is composed as a conjunction of a number of such resources.

Similarly, the resource request is also represented as a triple of (attribute, operator, value). We support operators such as $>, <, \leq, \geq, =, \neq$ currently.

4.2   Class Formation and Maintenance

We mainly discuss a few questions shown as follows from algorithm:

a) Formation of original overlay network.
b) Resource joining and leaving.
c) Update of resource information.
d) Node classes combination.

Formation of Original Overlay Network. We assume that there are N nodes in the grid. In this section, we will form the overlay network by using the algorithm introduced as follows.

Normalized Process of Data Set. We use $\hat{S} = \{\hat{s}_1, \hat{s}_2, \cdots, \hat{s}_h, \cdots, \hat{s}_N\}$ to represent the set with N nodes, where each node has M properties, and $\hat{s}_h = \{\hat{s}_{h1}, \hat{s}_{h2}, \cdots, \hat{s}_{hM}\}$ represents the hth node $(1 \leq h \leq N)$, $\hat{s}_{hi}$ denotes the ith property of the node h $(1 \leq i \leq M)$. Then we use formula (1) to process $\hat{S}$, and the data set after normalization is $S = \{s_1, s_2, \cdots, s_N\}$.

$$s_{hi} = \frac{\hat{s}_{hi}}{\max_{1 \leq i \leq M}[\max_{1 \leq h \leq N}(\hat{s}_{hi}) - \min_{1 \leq h \leq N}(\hat{s}_{hi})]} \tag{1}$$

s_{hi} is the ith component of the hth sample. From formula (1), we can see that each node set is located in a unit space.

Formation of Original Class. Actually, the original overlay network formation is the process of detecting the distribution characteristics of grid resources, and k-Nearest Neighbors algorithm [9] is one kind of effective local detection technology. In traditional k-Nearest Neighbors algorithm, the value of k is fixed, which is not suitable for detecting the grid with complex resources distribution. For example, when the resource density is not distributed evenly, if k is chosen higher than the distributed density of "small" pattern class, the original class will be wrong, else, if k is too small, the number of original classes would be large, which increase computational load of latter process, so how to choose the value of k adaptively based on pattern class density is important for resolving problem. The within-classes covariance distance matrix is an efficient method for characterizing cohesion of samples. Research [9] shows that the trace of within-classes covariance matrix more smaller, the cohesion of sample set more better. So, in this paper, we present an adaptive k-Nearest Neighbors clustering algorithm for grid resource discovery.

Before introduce the adaptive k-Nearest Neighbors clustering algorithm, we define:

a) $E_p : (l_p, k_p\text{-NN of } l_p, c_p, T_p)$ — the pth class.
b) l_p — the leader node of the class.
c) k_p-NN of l_p — k_p nearest neighbor nodes of l_p.
d) c_p — the centre of k_p+1 nodes, and its computing formula is $c_p = \frac{\sum_{s_k \in E_p} s_k}{k_p+1}$.
e) T_p — the trace of within-classes covariance matrix of k_p+1 nodes.

The original classes formation is finished through 5 steps shown as follows:

Step 1: Use formula (2) to get the distance matrix $\boldsymbol{D} = [d_{ij}]_{N \times N}$ of data set $\boldsymbol{S}$.

$$d_{ij} = \|\boldsymbol{s}_i - \boldsymbol{s}_j\|^2 = \sqrt{\sum_{k=1}^{M} (s_{ik} - s_{jk})^2} \tag{2}$$

Where d_{ij} denotes the distance between $\boldsymbol{s}_i$ and $\boldsymbol{s}_j$.
Step 2: If $p=1$, l_1 is obtained with the following restrictions:

a) $l_1 \in E_1$.
b) l_1 is the farthest sample from the global centre c_0.
c) $c_0 = \frac{\sum_{s_k \in S} s_k}{N}$.

If $p \geq 2$, l_p is obtained with the following restrictions:

a) $l_p \in E_p$ and $l_p \notin E_\alpha, 1 \leq \alpha < p$.
b) l_p is the farthest sample from the local centre c_{p-1}.

Step 3: Obviously $l_p \in \boldsymbol{S}$, so we might assume it as $\boldsymbol{s}_l$. Extract the lth row elements from the distance matrix $\boldsymbol{D}$ to form the distance vector $\{d_{l1}, d_{l2}, \cdots, d_{lN}\}$, then delete the elements that denotes the distance between $\boldsymbol{s}_l$ and $\boldsymbol{s}_j$ where $\boldsymbol{s}_j \in E_\alpha$. So we get the abridged distance vector $\Psi = \{d_{lj}|1 \leq j \leq N, s_j \notin E_\alpha\}$, sort Ψ increasingly as:

$$\Psi' = \{d_{lj_1}, d_{lj_2}, \cdots, d_{lj_\beta}|d_{lj_m} \leq d_{lj_n}, 1 \leq m \leq n \leq \beta\} \tag{3}$$

where β denotes that there are β samples which have not been put into the initial patterns.

Step 4: Use the following algorithm to get the k_p nearest neighbors of l_p and the trace of within-classes covariance matrix T_p.

$\Omega = l_p$;
set the *threshold* of the trace of within-classes covariance matrix;
for i=1 to β begin
$\quad \Omega = \{\Omega, s_{j_i} | s_{l_{j_i}} \in \Psi'\}$;
$\quad t_i = \text{trace}(\text{conv}(\Omega))$;
$\quad$ if $t_i \leq threshold$
$\quad\quad T_i = t_i$;
$\quad$ else begin
$\quad\quad \Omega = \Omega \backslash s_{j_i}$;
$\quad\quad$ break;
$\quad$ end.
end.

Step 5: If there are still samples that have not been put into the initial patterns, then go to step 2, otherwise the constitution of initial patterns is finished.

Resource Joining and Leaving. A node s prepares for joining the grid through node n, which it learns about offline. Because node n is connected to all leaders, it will get information from all leaders and return this information to s, then, by using the following formula, we know which class that s should join in:

$$s \in E_p, \quad \text{if} \quad \|c_p - s\|_2 = \min_{i=1,\cdots,K} \|c_i - s\|_2 \tag{4}$$

in the above formula, K denotes the number of the classes, $s \in E_p$ denotes that s should be put into the pth class, at the same time, s publishes its information to l_p.

When a node leaving from some class, it deletes its resource information directly.

Update of Resource Information. In this paper, class checking takes place periodically in order to ensure that each node belongs to an appropriate class as a selected resource characteristic dynamically changes. Class checking consists of worker node process and leader node process.

Each worker node periodically checks to see if it is placed in an appropriate class, based on its temporal usage/availability characteristics and the statistical similarity measure. If not, it tries to search for a class to which it should belong. If there is no existing class for the node to join, it creates a new class with itself as leader and its current characteristic as the class's characteristic.

If a node is a leader, it will first find a replacement leader before it leaves the class, so as to maintain class information.

Class Combination. As nodes joining and leaving in grid frequently, there may be too many small classes, then the cost of communication might be large and thus become performance bottlenecks. So, it is necessary to combine classes. In this paper, we use the following method to combine classes.

Step 1: Normalize leader nodes, and compute the distance matrix of these leader nodes denoted by $\boldsymbol{W} = [w_{ij}]_{P \times P}$, where $w_{ij} = \|\boldsymbol{c}_i - \boldsymbol{c}_j\|_2$ and P denotes the number the original classes.

Step 2: Use the following algorithm to get relevant matrix.

$\boldsymbol{G} = [g_{ij}]_{P \times P}$;
set the threshold d_h of the trace of within-classes covariance matrix;
for i=1 to P begin
    for j=1 to P begin
        if $\big((i \neq j)$ and $(w_{ij} < d_h)$ and $(w_{ij} = \min_{k=1,\cdots,P} w_{ik})$
        and $(w_{kj}|_{k=1,\cdots,P,k \neq i} \neq 1)\big)$ then begin
          $g_{ij} = 1$;
          break;
        end.
    end.
end.

Step 3: Combine classes by using information of matrix $\boldsymbol{G}$: notes that the row number and column number is the corresponding class number. After processed by step2, in matrix $\boldsymbol{G}$, there is at most one element be 1 in each line and each column, then combine these two line and column which has elements be 1 into one class, and the new leader can be any one leader of them, at the same time, we also combine the information of these two leader nodes.

4.3   Resource Query Processing and Message Dispatching Mechanism

When user querying a resource in the grid, the process of the query is completed by three steps shown as follows:

Assume $\boldsymbol{v}$ is the node which sending resource query, and this node always choose the nearest node denoted by $\boldsymbol{s}_k$ in grid to process the query.

Step 1: From the topology of the overlay network, we can see that node $\boldsymbol{s}_k$ connects to all leader nodes, so it can reach all clustering centre through $\boldsymbol{s}_k$. Using the following formula, we know which class the resource query should be sent to.

$$\boldsymbol{v} \to E_p, \quad \text{if} \quad \|\boldsymbol{c}_p - \boldsymbol{v}\|_2 = \min_{i=1,\cdots,K} \|\boldsymbol{c}_i - \boldsymbol{v}\|_2 \tag{5}$$

in the above formula, $\boldsymbol{v} \to E_p$ denotes dispatching resource query to the pth class.

Step 2: In E_p class, the node which satisfies the following conditions is the proper one that user required, and we call it $\boldsymbol{s}_n$:

$$\begin{cases} \|\boldsymbol{s}_n - \boldsymbol{v}\|_2 = \min_{\boldsymbol{s}_i \in E_p} \|\boldsymbol{s}_i - \boldsymbol{v}\|_2 \\ \|\boldsymbol{s}_n - \boldsymbol{v}\|_2 < d_g \end{cases} \tag{6}$$

actually, the first condition means to find node which confirms to user's request best in E_p, the second condition means that $\boldsymbol{s}_n$ must has some similarity to $\boldsymbol{v}$, where d_g means the threshold of the similarity. If these two conditions are both satisfied, then return the node s_n which has been found, and the search is end, else, go to step3.

Step 3: If it didn't find the node which satisfy user's request in class E_p, l_p will send the request to other leaders, and other leaders will search in their classes. If l_p receive search results from many leaders, then, it will choose the best one and return it to the requester; if no suitable node is found, then search failed.

5 Experimental Evaluation

5.1 Grid Environment Modeling

We use the experimental environment set in paper [8], which used the following parameters to specify our grid environment:

a) Resource density: this parameter reflects the abundance/scarcity of resources. It represents the percentage of nodes that have resources.
b) Resource type: we define continuous-valued resources, and this type resources are able to satisfy multiple requests at the same time.
c) User query: in this paper, we use average number of hops taken by a query before it is satisfied to evaluate resource query performance.
d) Statistical characteristic and similarity measure: statistical characteristic and similarity measure can not only control the number of classes for resource, but also the size of each class.
e) Resource distribution: this parameter describes how resources distribute in node, in our experiment, we assume that it is a random distribution.

5.2 Experiment Result

Our experiments were conducted in 10 computers with Linux operation system (namely 10 physical nodes). On these 10 computers, we simulated 1024 and 256 virtual nodes respectively and each virtual node has one resource at most. The number of nodes containing the requested resource was varied by changing resource density from 1/4 to 1/256. No limit was placed on the number of times a query gets forward. An average of 20000 queries was handled for each simulation run.

Fig.2 shows the average number of hops taken when querying resource at different resource density and system size. From Fig.2, we can see that at different grid size and resource density, the average number of hops taken by queries stay at around 2. This observation indicates that most queries are dispatched to an appropriate leader, which is able to find a suitable node to satisfy them. This also demonstrates that query handling performance in our mechanism is unaffected by resource density, resource type, or system size.

The method presented in this paper achieves better scalability and efficiency in query handling than [7] and [10]. Because in [7] there is an increasing trend in average number of hops as resource density decreases, while in [10] there is an increasing trend in average number of hops as system size increases.

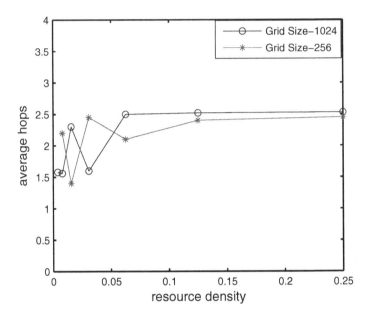

Fig. 2. The relationships between the average of number hops taken by each query and resource density

6   Conclusion

In this paper, we present a grid resource discovery method based on adaptive k-Nearest Neighbors clustering algorithm. This method can handle resource requests by searching a small subset out of a large number of nodes by resource clustering which can improve the resource query efficiency, on the other hand, it also achieves well scalability by managing grid resources with adaptive mechanism. It is shown from a series of computational experiments that the method presented in this paper achieves more scalability and efficient lookup performance than other existing methods.

Acknowledgements.   This work is supported by 973 project (No. 2005CB321800) of China, and 863 project (No. 2006AA01Z198) of China.

References

1. Globus project: http://www.globus.org
2. Raman, R., Livny, M., Solomon, M.: Matchmaking: Distributed Resource Management for High Throughput Computing. In: Proc. of the 7th IEEE HPDC, Washington DC, pp. 140–146. IEEE Computer Society Press, Los Alamitos (1998)
3. Iamnitchi, A., Foster, I.: On Fully Decentralized Resource Discovery in Grid Environments. In: Lee, C.A. (ed.) GRID 2001. LNCS, vol. 2242, pp. 51–62. Springer, Heidelberg (2001)

4. Chapin, S.J., Katramatos, D., Karpovich, J., Grimshaw, A.: Resource Management in Legion. Future Generation Computer System 5, 583–594 (1999)
5. Czajkowski, K., Ferguson, Foster, I., et al.: The WS-Resource Framework Version 1.0. http://www-106.ibm.com/developerworks/library/ws-resource/ws-wsrf.pdf
6. Hauswirth, M., Schmidt, R.: An Overlay Network for Resource Discovery in Grids. In: Andersen, K.V., Debenham, J., Wagner, R. (eds.) DEXA 2005. LNCS, vol. 3588, pp. 343–348. Springer, Heidelberg (2005)
7. Li, W., Xu, Z., Dong, F., Zhang, J.: Grid Resource Discovery Based on a Routing-transferring Model. In: Parashar, M. (ed.) GRID 2002. LNCS, vol. 2536, pp. 145–156. Springer, Heidelberg (2002)
8. Anand, P., Shaowen, W., Sukuma, G.: A Self-organized Grouping (SOG) Method for Efficient Grid Resource Discovery. In: Proceedings of the The 6th IEEE/ACM International Workshop on Grid Computing, Washington, pp. 312–317. ACM Press, New York (2005)
9. Bian, Z.Q., Zhang, X.G.: Pattern Recognition. The Publishing Company of Tsinghua University, Bei Jin (1999)
10. Iamnitchi, A., Foster, I., Numi, D.: A Peer-to-Peer Approach to Resource Location in Grid Environments. In: Proceedings of the 11th Symposium on High Performance Distributed Computing, Edinburgh, Scotland, pp. 413–429 (2002)

Algorithms for Minimum m-Connected k-Dominating Set Problem[*]

Weiping Shang[1,2], Frances Yao[2], Pengjun Wan[3], and Xiaodong Hu[1]

[1] Institute of Applied Mathematics, Chinese Academy of Sciences, Beijing, China
[2] Department of Computer Science, City University of Hong Kong
[3] Department of Computer Science, Illinois Institute of Technology, Chicago, USA

Abstract. In wireless sensor networks, virtual backbone has been proposed as the routing infrastructure to alleviate the broadcasting storm problem and perform some other tasks such as area monitoring. Previous work in this area has mainly focused on how to set up a small virtual backbone for high efficiency, which is modelled as the minimum Connected Dominating Set (CDS) problem. In this paper we consider how to establish a small virtual backbone to balance efficiency and fault tolerance. This problem can be formalized as the minimum m-connected k-dominating set problem, which is a general version of minimum CDS problem with $m = 1$ and $k = 1$. In this paper we will propose some approximation algorithms for this problem that beat the current best performance ratios.

Keywords: Connected dominating set, approximation algorithm, k-vertex connectivity, wireless sensor networks.

1 Introduction

A Wireless Sensor Network (WSN) consists of wireless nodes (transceivers) without any underlying physical infrastructure. In order to enable data transmission in such networks, all the wireless nodes need to frequently flooding control messages thus causing a lot of redundancy, contentions and collisions. To support various network functions such as multi-hop communication and area monitoring, some wireless nodes are selected to form a *virtual backbone*. Virtual backbone has been proposed as the routing infrastructure of WSNs. In many existing schemes (e.g., [1]) virtual backbone nodes form a Connected Dominating Set (CDS) of the WSN. With virtual backbones, routing messages are only exchanged between the backbone nodes, instead of being broadcasted to all the nodes. Prior work (e.g., [8]) has demonstrated that virtual backbones could dramatically reduce routing overhead.

In WSNs, a node may fail due to accidental damage or energy depletion and a wireless link may fade away during node movement. Thus it is desirable to

---

[*] This work was supported in part by the Research Grants Council of Hong Kong under Grant No. CityU 1165/04E, the National Natural Science Foundation of China under Grant No. 70221001 and 10531070.

A. Dress, Y. Xu, and B. Zhu (Eds.): COCOA 2007, LNCS 4616, pp. 182–190, 2007.

have several sensors monitor the same target, and let each sensor report via different routes to avoid losing an important event. Hence, how to construct a fault tolerant virtual backbone that continues to function when some nodes or links break down is an important research problem.

In this paper we assume as usual that all nodes have the same transmission range (scaled to 1). Under such an assumption, a WSN can be modelled as a Unit Disk Graph (UDG) that consists of all nodes in the WSN and there exists an edge between two nodes if the distance between them is at most 1. Fault tolerant virtual backbone problem can be formalized as a combinatorial optimization problem: Given a UDG $G = (V, E)$ and two nonnegative integers m and k, find a subset of nodes $S \subseteq V$ of minimum size that satisfies: i) each node u in $V \setminus S$ is *dominated* by at least k nodes in S, ii) S is *m-connected* (there are at least m disjoint paths between each pair of nodes in S). Every node in S is called a *backbone node* and every set S satisfying (i-ii) is called *m-connected k-dominating set* ((m, k)-CDS), and the problem is called *minimum m-connected k-dominating set* problem.

In this paper, we will first study the minimum m-connected k-dominating set problem for $m = 1, 2$, which is important for fault tolerant virtual backbone problem in WSNs. (When $m = 1$ and $k = 1$ the problem is reduced to well known minimum connected dominating set problem.) We propose three centralized approximation algorithms to construct k-dominating set and m-connected k-dominating sets for $m = 1, 2$. Our performance analysis show that the algorithms have small approximation ratio improving the current best result for small k. Then for $3 \leq m \leq k$, we discuss the relation between (m, k)-CDS and (m, m)-CDS. The remainder of this paper is organized as follows: In Section 2 and 3 we first give some definitions and then present some related works. In Section 4 we present our algorithms with theoretical analysis on guaranteed performances. In Section 5 we conclude the paper.

2 Preliminaries

Let G be a graph with vertex-set $V(G)$ and edge-set $E(G)$. For any vertex $v \in V$, the neighborhood of v is defined by $N(v) \equiv \{u \in V(G) : uv \in E(G)\}$ and the closed neighborhood of v is defined by $N[v] \equiv \{u \in V(G) : uv \in E(G)\} \cup \{v\}$. The minimum degree of vertices in $V(G)$ is denoted by $\delta(G)$.

A subset $U \subseteq V$ is called an *independent set* (IS) of G if all vertices in U are pairwise non-adjacent, and it is further called a *maximal independent set* (MIS) if each vertex $V \setminus U$ is adjacent to at least one vertex in U.

A *dominating set* (DS) of a graph $G = (V, E)$ is a subset $S \subseteq V$ such that each vertex in $V \setminus S$ is adjacent to at least one vertex in S. A DS is called a *connected dominating set* (CDS) if it also induces a connected subgraph. A *k-dominating set* (*k*-DS) $S \subseteq V$ of G is a set of vertices such that each vertex $u \in V$ is either in S or has at least k neighbors in S.

A *cut-vertex* of a connected graph G is a vertex v such that the graph $G \setminus \{v\}$ is disconnected. A *block* is a maximal connected subgraph having no cut-vertex (so a graph is a block if and only if it is either 2-connected or equal to K_1 or K_2). The block-cut-vertex graph of G is a graph H where $V(H)$ consists of all cut-vertices of G and all blocks of G, with a cut-vertex v adjacent to a block G_0 if v is a vertex of G_0. The block-cut-vertex graph is always a forest. A *2-connected graph* is a graph without cut-vertices. Clearly a block with more than three nodes is a 2-connected component. A *leaf block* of a connected graph is a subgraph of which is a block with only one cut-vertex.

3   Related Work

Lots of efforts have been made to design approximation algorithms for minimum connected dominating set problem. Wan et al. [10] proposed a two-phase distributed algorithm for the problem in UDGs that has a constant approximation performance ratio of 8. The algorithm first constructs a spanning tree, and then at the first phase, each node in a tree is examined to find a *Maximal Independent Set* (MIS) and all the nodes in the MIS are colored black. At the second phase, more nodes are added (color blue) to connect those black nodes. Recently, Li et al. [6] proposed another two-phase distributed algorithm with a better approximation ratio of $(4.8 + ln5)$. As in [10], at the first phase, an MIS is computed. At the second phase, a Steiner tree algorithm is used to connect nodes in the MIS. The Steiner tree algorithm is based on the property that any node in UDG is adjacent to at most 5 independent nodes.

In [3], Dai et al address the problem of constructing k-connected k-dominating virtual backbone which is k-connected and each node not in the backbone is dominated by at least k nodes in the backbone. They propose three localized algorithms. Two algorithms, k-gossip algorithm and color based (k, k)-CDS algorithm, are probabilistic. In k-Gossip algorithm, each node decides its own backbone status with a probability based on the network size, deploying area size, transmission range, and k. Color based (k, k)-CDS algorithm proposes that each node randomly selects one of the k colors such that the network is divided into k-disjoint subsets based on node colors. For each subset of nodes, a CDS is constructed and (k, k)-CDS is the union of k CDS's. The deterministic algorithm, k-Coverage condition, only works in very dense network and no upper bound on the size of resultant backbone is analyzed.

Recently, Wang et al. [11] proposed a 64-approximation algorithm for the minimum $(2, 1)$-CDS problem. The basic idea of this centralized algorithm is as follows: i) Construct a small-sized CDS as a starting point of the backbone; ii) iteratively augment the backbone by adding new nodes to connect a leaf block in the backbone to other block (or blocks); iii) the augmentation process stops when all backbone nodes are in the same block, i.e., the backbone nodes are 2-connected. The augmentation process stops in at most $|CDS| - 1$ steps and each step at most 8 nodes are added.

Most recently, in work [7] we proposed three centralized approximation algorithms to construct k-tuple dominating set and m-connected k-tuple dominating sets for $m = 1, 2$, respectively.

4   Approximation Algorithms

We first prove the following lemma, which will be used in our performance analysis of proposed algorithms.

Lemma 1. *Let $G = (V, E)$ be a unit disk graph and k a constant such that $\delta(G) \geq k - 1$. Let D_k^* be a minimum k-dominating set of G and S a maximal independent set of G. Then $|S| \leq \max\{\frac{5}{k}, 1\}|D_k^*|$.*

Proof Let $S_0 = S \bigcap D_k^*$, $X = S \setminus S_0$ and $Y = D_k^* \setminus S_0$. It is clearly that X and Y are two disjoint subsets. For all $u \in X$, let $c_u = |N(u) \bigcap Y|$. As D_k^* is a k-dominating set of G, $c_u \geq k$ for each $u \in X$ and we have: $\sum_{u \in X} c_u \geq k|X|$. For all $v \in Y$, let $d_v = |N(v) \bigcap X|$. As G is a unit disk graph, for all $v \in Y$ there are at most 5 independent vertices in its neighborhood and $d_v \leq 5$. We have: $5|Y| \geq \sum_{v \in Y} d_v$. For $\sum_{u \in X} c_u = |\{uv \in E : u \in X, v \in Y\}| = \sum_{v \in Y} d_v$, we have $|X| \leq \frac{5}{k}|Y|$. Hence, $|S| = |X| + |S_0| \leq \frac{5}{k}|D_k^* \setminus S_0| + |S_0| \leq \max\{\frac{5}{k}, 1\}|D_k^*|$, which proves the lemma. □

Corollary 1. *Let $G = (V, E)$ be a unit disk graph and k a constant such that $\delta(G) \geq k-1$. Let D_k^* be a minimum k-dominating set of G and S an independent set of G satisfying that $S \bigcap D_k^* = \emptyset$. Then $|S| \leq \frac{5}{k}|D_k^*|$.*

4.1   Algorithm for Computing $(1, k)$-CDS

The basic idea of our algorithm for the minimum $(1, k)$-CDS problem is as follows: First choosing a CDS and then sequentially choosing an MIS $k - 1$ times such that all vertices in $V \setminus D^c$ are k-dominated by set D^c. The algorithm is more formally presented as follows.

Algorithm A. for computing $(1, k)$-CDS

---

1. Choose an MIS I_1 of G and a set C such that $I_1 \cup C$ is a CDS (refer to [10])
2. **for** $i := 2$ **to** k
3.   Construct an MIS I_i in $G \setminus I_1 \cup \cdots \cup I_{i-1}$
4. **end for**
5. $D^c := I_1 \cup \cdots \cup I_k \cup C$
6. **return** D^c

Theorem 1. *Algorithm A returns a solution that is a $(5 + \frac{5}{k})$-approximate solution to the minimum connected k-dominating set problem for $k \leq 5$ and 7-approximate solution for $k > 5$.*

Proof: Suppose that Algorithm A, given graph $G = (V, E)$ and a natural number $k \geq 1$, returns $D^c = I_1 \cup \cdots \cup I_k \bigcup C$. Let D_k^* be a minimum k-dominating set of G. We will show that D is a connected k-dominating set of G. For all $u \in G \setminus D^c$, at the i-th iteration, u is not in I_i and thus it is dominated by one vertex of I_i. At the end, u is dominated by at least k different vertices of $I_1 \cup \cdots \cup I_k$. By the first step of Algorithm A, $C \cup I_1$ is a CDS and thus $I_1 \cup \cdots \cup I_k \bigcup C$ is connected. So, D is a connected k-dominating set of G.

Let $S_i = I_i \bigcap D_k^*$ for $i = 1, 2, \cdots, k$. By the rule of Algorithm A, we have each $I_i \setminus S_i$ is an independent set and $(I_i \setminus S_i) \bigcap D_k^* = \emptyset$. Thus it follows from Corollary 1 that $|I_i \setminus S_i| \leq \frac{5}{k}|D_k^* \setminus S_i|$. Let us prove now the approximation ratio.

$$|I_1 \cup \cdots \cup I_k| = \sum_{i=1}^{k} |S_i| + \sum_{i=1}^{k} |I_i \setminus S_i|$$

$$\leq \sum_{i=1}^{k} |S_i| + \sum_{i=1}^{k} \frac{5}{k}|D_k^* \setminus S_i|$$

$$= (1 - \frac{5}{k}) \sum_{i=1}^{k} |S_i| + 5|D_k^*|.$$

And $\sum_{i=1}^{k} |S_i| \leq |D_k^*|$. Hence we have $|I_1 \cup \cdots \cup I_k| \leq 5|D_k^*|$ for $k \leq 5$ and $|I_1 \cup \cdots \cup I_k| \leq 6|D_k^*|$ for $k > 5$.

In the end, let C be the set constructed from the first step of Algorithm A. By using the argument for the proof of Lemma 10 in [10], we can deduce $|C| \leq |I_1|$. Hence it follows from Lemma 1 that $|C| \leq \max\{\frac{5}{k}, 1\}|D_k^*|$, and the size of connected k-dominating set D is bounded by $(5 + \frac{5}{k})|D_k^*|$ for $k \leq 5$ and $7|D_k^*|$ for $k > 5$. The size of the optimal solution of connected k-dominating set is at least $|D_k^*|$. The proof is then finished. □

4.2 Algorithm for Computing $(2, k)$-CDS

The basic idea of our algorithm for the minimum $(2, k)$-CDS problem with $k \geq 2$ is similar to the method proposed in [11]. It essentially consists of following four steps:

Step 1. Apply Algorithm A to construct a connected k-dominating set D.

Step 2. Compute all the blocks in D by computing the 2-connected components through the depth first search.

Step 3. Produce the shortest path in the original graph such that it can connect a leaf block in D with other part of D but does not contain any vertices in D except the two endpoints. Then add all intermediate vertices in this path to D.

Step 4. Repeat Step 2 and Step 3 until D is 2-connected.

In Step 2, we can apply the standard algorithm proposed in [9] to compute all blocks in D, denote the number of blocks in D by ComputeBlock(D). The algorithm is more formally presented as follows:

Algorithm B. for computing a 2-connected k-dominating set $(k \geq 2)$

---

   1. Choose a connected k-dominating set D^c using Algorithm A
   2. $D := D^c$ and $B:=$ ComputeBlocks(D)
   3. **while** $B > 1$ **do**
   4.   Choose a leaf block L
   5.   **for** vertex $v \in L$ not a cut-vertex **do**
   6.   **for** vertex $u \in V \setminus L$ **do**
   7.   Construct G' from G by deleting all nodes in D except u and v
   8. $P_{uv} :=$shortestPath($G'; v, u$) and $P := P \cup P_{uv}$
   9.   **end-for**
  10.   **end-for**
  11. $P_{ij}:=$ the shortest path in P
  12. $D := D\cup$ the intermediate vertices in P_{ij}
  13.   ComputeBlocks(D)
  14. **end-while**
  15. **return** D

Lemma 2. *For $k \geq 2$, at most two new vertices are added into D at each augmenting step.*

Proof Suppose that L is a leaf block of D and w is the cut-vertex. Consider two vertices u and v in D with $u \in L \setminus \{w\}$ and $v \in V \setminus L$, let P_{uv} be the shortest path that connects u and v. We claim that P_{uv} has at most two intermediate vertices. Suppose, by contradiction, that P_{uv} contains $u, x_1, x_2, ..., x_l, v$, where $l \geq 3$. Since each vertex x_i has at least 2 neighbors in D and $N(x_i) \cap D \subseteq L$ or $N(x_i) \cap D \subseteq (V \setminus L) \cup \{w\}$, $N(x_1) \cap D \subseteq L$. If $N(x_2) \cap D \subseteq L$, x_2 must have a neighbor s in $L \setminus \{w\}$, then the path between sv has a shorter distance than P_{uv}. Otherwise $N(x_2) \bigcap D \subseteq (V \setminus L) \cup \{w\}$, x_2 must have a neighbor s in $V \setminus L$, then the path between us has a shorter distance than P_{uv}. Which contradicts that P_{uv} has the shortest distance. $\qquad\square$

Lemma 3. *The number of cut-vertices in the connected k-dominating set D^c by Algorithm A is no bigger than the number of connected dominating sets in $I_1 \cup C$ chosen in Step 1 of Algorithm A.*

Proof Let $S = I_1 \cup C$ be the connected domination set. We will show that no vertex in $D^c \setminus S$ is a cut-vertex. For any two vertices $u, v \in S$, there is a path P_{uv} between them that contains only vertices in S. Since any vertex in $D^c \setminus S$ is dominated by at least one vertex in S, Hence, for any two vertices $u, v \in D^c$, there is a path P_{uv} between them that contains only vertices in $S \bigcup \{u, v\}$. Hence, any vertex in $D^c \setminus S$ is not a cut-vertex. $\qquad\square$

Theorem 2. *Algorithm B returns a $(5 + \frac{25}{k})$-approximate solution to the minimum 2-connected k-dominating set problem for $2 \leq k \leq 5$ and 11-approximate solution for $k > 5$.*

Proof Let D_k^* and D_{opt} be the optimal k-dominating set and 2-connected k-dominating set, respectively. It is clearly that $|D_k^*| \leq |D_{opt}|$. After S is constructed, by Lemmas 2-3, the algorithm terminates in at most $|C| + |I_1|$ steps, and in each step at most two vertices are added. Since $|C| + |I_1| \leq 2|I_1| \leq 2 \max\{\frac{5}{k}, 1\}|D_k^*|$, we have $|D| \leq |D^c| + 4\max\{\frac{5}{k}, 1\}|D_k^*|$. It follows from Theorem 1 that $|D^c| \leq (5 + \frac{5}{k})|D_k^*|$ for $k \leq 5$ and $|D^c| \leq 7|D_k^*|$ for $k > 5$. Hence we obtain $|D| \leq (5 + \frac{25}{k})|D_{opt}|$ for $2 \leq k \leq 5$ and $|D| \leq 11|D_{opt}|$ for $k > 5$.   □

4.3   Algorithm for Computing (2, 1)-CDS

The main idea of our algorithm is as follows: First, construct a connected dominating set C using the algorithm in [6], and then construct a maximal independent set D in $G \setminus C$, in the end make $C \cup D$ to be 2-connected by adding some new vertices to it.

Algorithm C. for computing 2-connected dominating set

---

1. Produce a connected dominating set C of G using the algorithm in [6].
2. Construct a maximal independent set D in $G \setminus C$
3. $S := C \cup D$
4. Augment S using Steps 2-14 of Algorithm B

Theorem 3. *Algorithm C returns a 2-connected dominating set whose size is at most $(18.2 + 3\ln 5)|D_2^*| + 4.8$, where $|D_2^*|$ is the size of the optimal 2-connected dominating set.*

Proof Let D_1^* and D_2^* be the optimal (1, 1)-CDS and (2, 1)-CDS, respectively. It is clear that $|D_1^*| \leq |D_2^*|$. After C and D is constructed, which are a connected dominating set of G and a dominating set of $G \setminus C$, respectively, each vertex in $V \setminus S$ is dominated by at least two vertices in S. Thus, Lemmas 2-3 also hold true for Algorithm C. Thus it follows from Lemmas 2-3 that at most $|C|$ steps are needed before the algorithm terminates, and at each step at most two vertices are added. Hence, we obtain $|S| \leq 3|C| + |D|$. Using the same argument for Theorem 1 in [6,12], we could show $|C| \leq (4.8 + \ln 5)|D_1^*| + 1.2$ and $|D| \leq 3.8|D_1^*| + 1.2$ respectively. Thus we obtain $|S| \leq (18.2 + 3\ln 5)|D_2^*| + 4.8$.   □

Observe that $(18.2 + 3\ln 5) < 23.03$. So Algorithm C has a better guaranteed performance than the 64-approximation algorithm in [11] for the same problem (when the size of the optimal 2-connected dominating set is not very big).

4.4 (m, k)-CDS for $3 \leq m \leq k$

Let $A_{(m,m)}$ be an α-approximation algorithm for the (m, m)-CDS problem. The basic idea of algorithm $A_{(m,k)}$ for the minimum (m, k)-CDS problem is as follows: First choosing a (m, m)-CDS and then sequentially choosing an MIS $k-m$ times. The algorithm is more formally presented as follows.

Algorithm $A_{(m,k)}$**.** for computing (m, k)-CDS

---

1. Choose an (m, m)-CDS S of G using algorithm $A_{(m,m)}$
2. **for** $i := 1$ **to** $k - m$
3.   Construct an MIS I_i in $G \setminus S \cup I_1 \cup \cdots \cup I_{i-1}$
4. $D := I_1 \cup \cdots \cup I_{k-m} \cup S$
5. **return** D

Theorem 4. *If there exists an α-approximation algorithm for the (m, m)-CDS problem, then there exists a $(\alpha + 6)$-approximation algorithm for the (m, k)-CDS problem, where $k > m$.*

Proof: We first show that D is a (m, k)-CDS of G. For all $u \in G \setminus D$, u is not in S and thus it is dominated by at least m vertices of S. And at the i-th iteration, u is not in I_i and thus it is dominated by one vertex of I_i for $i = 1, ..., k - m$. At the end, u is dominated by at least k different vertices of D. Now we show that D is m-connected, suppose there exist $m - 1$ vertices in D such that the induced subgraph D is disconnected by removing the $m - 1$ vertices. Let X be the vertex set. For S is a (m, m)-CDS, $S \setminus X$ is a connected dominating set. So, $D \setminus X$ is connected, a contraction. Hence, D is a (m, k)-CDS of G.

Let D^* be the optimal solution of (m, k)-CDS. It is clearly that $|S| \leq \alpha|D^*|$, and $|I_1 \cup \cdots \cup I_{k-m}| \leq 6|D^*|$ by similar argument of Theorem 1. This gives a $(\alpha + 6)$-approximation algorithm for the (m, k)-CDS problem, where $k > m$. The proof is then finished. □

5 Conclusion

In this paper we have proposed centralized approximation algorithms for the minimum m-connected k-dominating set problem for $m = 1, 2$. Although the approximation performance ratios of Algorithms A and B are dependent on k, they are very small when k is not very big, that, in fact, is the case of virtual backbone construction in wireless sensor networks. For $3 \leq m \leq k$, we discuss the relation between (m, k)-CDS and (m, m)-CDS. Our future work is to extend our study to the more general case of $m \geq 3$, and design distributed and localized algorithms for minimum m-connected k-dominating set problem.

References

1. Alzoubi, K.M., Wan, P.-J., Frieder, O.: Distributed heuristics for connected dominating sets in wireless ad hoc networks. Journal of Communications and Networks 4(1), 22–29 (2002)
2. Bredin, J.L., Demaine, E.D., Hajiaghayi, M., Rus, D.: Deploying sensor networks with guaranteed capacity and fault tolerance. In: Proceedings of the 6th ACM International Symposium on Mobile Ad Hoc Networking and Computing (MobiHoc), pp. 309–319. ACM Press, New York (2005)
3. Dai, F., Wu, J.: On constructing k-connected k-dominating set in wireless networks. IEEE International Parallel and Distributed Processing Symposium. IEEE Computer Society Press, Los Alamitos (2005)
4. Kuhn, F., Moscibroda, T., Wattenhofer, R.: Fault-Tolerant Clustering in Ad Hoc and Sensor Networks. In: Proceedings 26th International Conference on Distributed Computing Systems (ICDCS) (2006)
5. Koskinen, H., Karvo, J., Apilo, O.: On improving connectivity of static ad-hoc networks by adding nodes. In: Proceedings of the 4th annual Mediterranean Workshop on Ad Hoc Networks (Med-Hoc-Net), pp. 169–178 (2005)
6. Li, Y.S., Thai, M.T., Wang, F., Yi, C.-W., Wan, P.-J., Du, D.-Z.: On greedy construction of connected dominating sets in wireless networks. Wiley Journal on Wireless Communications and Mobile Computing 5(8), 927–932 (2005)
7. Shang, W.-P., Yao, F., Wan, P.-J., Hu, X.-D.: Algorithms for minimum m-connected k-tuple dominating set problem. Theoretical Computer Science (to submitted)
8. Sinha, P., Sivakumar, R., Bharghavan, V.: Enhancing ad hoc routing with dynamic virtual infrastructures. In: Proceedings of the 20th Annual Joint Conference of the IEEE Computer and Communications Societies, vol. 3, pp. 1763–1772 (2001)
9. Tarjan, R.: Depth first search and linear graph algorithms. SIAM Journal on Computing 1(2), 146–160 (1972)
10. Wan, P.-J., Alzoubi, K.M., Frieder, O.: Distributed construction of connected dominating set in wireless ad hoc networks. Mobile Networks and Applications 9(2), 141–149 (2004)
11. Wang, F., Thai, T.: On the construction of 2-connected virtual backbone in wireless networks. IEEE Transactions on Wireless Communications (to appear)
12. Wu, W., Du, H., Jia, X., Li, Y., Huang, C.-H.: Minimum connected dominating sets and maximal independent sets in unit disk graphs. Theoretical Computer Science 352(1), 1–7 (2006)

Worst Case Analysis of a New Lower Bound for Flow Shop Weighted Completion Time Problem

Danyu Bai and Lixin Tang[*]

The Logistics Institute, Northeastern University, Shenyang,
110004, P.R. China
Tel/ Fax: 0086(24)83680169
qhjytlx@mail.neu.edu.cn

Abstract. For the m-machine Flow Shop Weighted Completion Time problem, a New Lower Bound (NLB) is derived to improve the original lower bound which was given by Kaminsky and Simchi-Levi. 1) For the case of arbitrary weight, the NLB is asymptotically equivalent to the optimality solution, as the total number of jobs goes to infinity. Specially, when the processing times of jobs are all equal, the NLB is just the optimal solution. 2) For the case of equal-weight, a tight worst case performance ratio of the optimal solution to the NLB is obtained. At the end of the paper, computational results show the effectiveness of NLB on a set of random test problems.

Keywords: Asymptotic analysis, Worst case analysis, Flow shop weighted completion time problem, WSPT rule.

1 Introduction

The scheduling problem of *m-machine Flow Shop Weighted Completion Time* is considered. In the problem a set of n jobs has to be sequentially processed through m machines without preemption. It is assumed that at any given time each machine can handle at most one job and a job can only be processed on one machine. Every machine processes the arriving jobs in a first come first served manner. The objective is to find a sequence of jobs to minimize the total weighted completion time on the final machine for the given processing time of each job on each machine and weight associated with each job. Garey et al. [2] pointed out that this problem is strongly NP-hard even in the two-machine case with all weights equal.

In the previous research, *Branch-and-bound* and *local search* strategies are combined to deal with the small size problems (Kohler and Steiglitz [5]). For larger problems, it is pointed out that using *dispatch rules* to find the reasonable sequences is typical (Bhakaran and Pinedo [1]). Shakhlevich et al. [7] pointed out that *WSPT with Minimum Cost Insertion* (WSPT-MCI) is optimal when the processing time of each operation of a job is equal. With the tools of *probabilistic analysis*, Kaminsky and Simchi-Levi [3] proved that the general flow shop weighted completion time problem is asymptotically equal to a certain single machine scheduling problem with an assumption that the processing times on all m machines for all jobs must be *independently and identically distributed* (i.i.d.) and extended Weighted Shortest

---

[*] Corresponding author.

A. Dress, Y. Xu, and B. Zhu (Eds.): COCOA 2007, LNCS 4616, pp. 191–199, 2007.

Processing Time (WSPT) first rule which is asymptotically optimal to the flow shop problem. And they also pointed out that the *Shortest Processing Time* (SPT) algorithm is asymptotic optimality for any continuous, independent, and identically distributed job processing times [4]. Xia et al [8] provides an alternative proof using *martingales* which extends the result of Kaminsky and Simchi-Levi and simplifies their argument.

Specially, Kaminsky and Simchi-Levi presented a *Lower Bound* (LB) to estimate the optimality solution in their paper [3]. In this paper, a *New Lower Bound* (NLB) is derived to improve the original LB. For the case of arbitrary weight, the NLB is asymptotically equivalent to the optimality solution, as the total number of jobs goes to infinity. Specially, when the processing times of jobs are all equal, the NLB is just the optimal solution. For the case of equal-weight, a tight worst case performance ratio of the optimal solution to the NLB is obtained. At the end of the paper, computational results show the effectiveness of NLB on a set of random test problems.

The remainder of the paper is organized as follows. The problem is formulated in section 2, and the NLB and its asymptotic optimality are provided in section 3. In section 4, a tight worst case performance ratio is given in equal-weight case. Some computational results are presented in section 5, and this paper is closed by the conclusions in section 6.

2   Problem Specification and the Main Results

For convenience, we quote the descriptions and notations of the problem given by Kaminsky and Simchi-Levi without any change. There is a set of n jobs which have to be processed on m machines. Job i, $i=1, 2,..., n$, has a processing time t_i^l on machine l, $l=1, 2,..., m$, and an associated weight w_i. The processing times are i.i.d. random variables, defined on the interval $(0, 1]$. Similarly, the weights are i.i.d. random variables, defined on the interval $(0, 1]$. Each job must be processed without preemption on each machine in the same order. At time 0, jobs are available, and the machine processes the jobs in a first come first served fashion, that is, a *permutation schedule*. Also, the intermediate storage between successive machines is unlimited. The objective is to find a schedule that minimizes the total weighed completion times of all the jobs on the final machine. This problem is called *Problem P*, in which the optimal objective function value is denoted as Z^*. By letting $t_i = \sum_{l=1}^{m} t_i^l$, $l=1, 2,..., m$,

$i=1, 2,..., n$, we can associate the Problem P with *Problem P_1*, the *Single Machine Weighted Completion Time* problem with n tasks, which has the optimal value Z_1^*. In the problem P_1, each task has a processing time t_i and weight w_i, $i=1, 2,..., n$.

These two problems, Problem P and Problem P_1, are related through the following theorem, whose proof can be found in Kaminsky and Simchi-Levi [3].

Theorem 2.1. *Let the processing times* t_i^1, t_i^2,..., t_i^m, $i=1, 2,..., n$, *be independent random variables having the same continuous distribution with bounded density* $\phi(\bullet)$ *defined on* $(0, 1]$. *Let the weights* w_i, $i=1, 2,..., n$, *be i.i.d. according to a cumulative distribution function* $\Phi(\bullet)$ *defined on* $(0, 1]$. *Then with probability one, we have*

$$\lim_{n \to \infty} Z^*/n^2 = \lim_{n \to \infty} Z^*/mn^2 = \theta \qquad (1)$$

for some constant θ.

The theorem means that the optimal solution of Problem P is asymptotically equivalent to the value of the optimal solution of Problem P_1 divided by the number of machines m. As the WSPT heuristic algorithm is optimal for Problem P_1, Kaminsky and Simchi-Levi extend the WSPT heuristic to Problem P.

The WSPT heuristics for Problem P can be described as follows:

WSPT algorithm for Problem P

Step 1, calculate $t_i = \sum_{l=1}^{m} t_i^l$, $i=1, 2,\ldots, n$, $l=1, 2,\ldots, m$;

Step2, calculate the ratio $r_i=t_i/w_i$, $i=1, 2,\ldots, n$, *and sequence them in non-decreasing order*;

Step3, reindex the jobs so that $r_1 \le r_2 \le \cdots \le r_n$;

Step4, process the jobs from 1 to n, and calculate the total weighted completion time $Z^{WSPT} = \sum_{i=1}^{n} w_i C_i$, *where C_i is the completion time of job i.*

To compare the Z^{WSPT} with Z^*, Kaminsky and Simchi-Levi [3] presented a LB value,

$$Z^{LB} = \frac{1}{m} Z_1^* + \frac{1}{m} \sum_{i=1}^{n} \sum_{k=2}^{m} w_i (k-1) t_i^k \qquad (2)$$

to estimate the optimality solution Z^*. But in some case, the LB does not work well. Consider an instance. For example, $m=4$, $n=2$, and the processing time of the job on each machine is 1, we have $Z^*=9$ and $Z^{LB}=6$, whereas $Z^*/Z^{LB}=150\%>1$.

3 The New Lower Bound and Its Asymptotic Analysis

To improve the original LB, a NLB is presented as the following theorem.

Theorem 3.1. *For any instance of Problem P and its associated Problem P_1, we have*

$$Z^* \ge Z^{NLB}$$

where

$$Z^{NLB} = \frac{1}{m} Z_1^* + \frac{1}{m} \sum_{i=1}^{n} w_i \sum_{k=1}^{m-1} (m-k) t_1^k + \frac{1}{m} \sum_{i=1}^{n} w_i \sum_{k=2}^{m} (k-1) t_i^k \qquad (3)$$

Proof. For an optimal schedule of Problem P, jobs are indexed according to their completion time, that is, the departure time from the last machine. Let C_i denote the completion time of ith job in the sequence and t_i^k denote the processing times of job i on machine k, $i=1, 2,\ldots, n$, $k=1, 2,\ldots, m$. From these definitions, we have

$$mC_i \ge \left(\sum_{j=1}^{i} t_j^1 + \sum_{k=2}^{m} t_i^k \right) + \left(t_1^1 + \sum_{j=1}^{i} t_j^2 + \sum_{k=3}^{m} t_i^k \right) + \cdots + \left(\sum_{k=1}^{m-1} t_1^k + \sum_{j=1}^{i} t_j^m \right)$$

$$= \sum_{k=1}^{m-1} (m-k) t_1^k + \sum_{k=1}^{m} \sum_{j=1}^{i} t_j^k + \sum_{k=2}^{m} (k-1) t_i^k \qquad (4)$$

We note that

$$\sum_{k=1}^{m} \sum_{j=1}^{i} t_j^k = \sum_{j=1}^{i} \sum_{k=1}^{m} t_j^k = \sum_{j=1}^{i} t_j \qquad (5)$$

where t_j is the processing time of job j on all m machines. Combining (4) and (5), rewriting inequality (4) and multiplying by the weight of job i, we get

$$w_i C_i \geq \frac{1}{m} w_i \sum_{j=1}^{i} t_j + \frac{1}{m} w_i \sum_{k=1}^{m-1} (m-k) t_1^k + \frac{1}{m} w_i \sum_{k=2}^{m} (k-1) t_i^k$$

Summing over all of the jobs, we have

$$\sum_{i=1}^{n} w_i C_i \geq \frac{1}{m} \sum_{i=1}^{n} w_i \sum_{j=1}^{i} t_j + \frac{1}{m} \sum_{i=1}^{n} w_i \sum_{k=1}^{m-1} (m-k) t_1^k + \frac{1}{m} \sum_{i=1}^{n} w_i \sum_{k=2}^{m} (k-1) t_i^k$$

With

$$Z^* = \sum_{i=1}^{n} w_i C_i \text{ and } Z_1^* \leq \sum_{i=1}^{n} w_i \sum_{j=1}^{i} t_j$$

we obtain the result of the theorem.   □

Computing the example presented in section 2 with the NLB, we can easily get that $Z^* = Z^{NLB} = 9$. Specially, when processing times of the jobs are all equal, the optimal solution is just the NLB. And how is the performance of the NLB in the infinite case? Does it still approach to the optimal solution? It is discussed in the following theorem.

Theorem 3.2. *Let the processing time t_i^k, $i=1, 2,..., n$, $k=1, 2,..., m$, be independent random variables having the same continuous and bounded distribution $\phi(\bullet)$ defined on $(0, 1]$. Let the weights w_i, $i=1, 2,..., n$, be i.i.d. according to a cumulative distribution function $\Phi(\bullet)$ defined on $(0, 1]$. Then with probability one, we have*

$$\lim_{n \to \infty} Z^* / Z^{NLB} = 1 \tag{6}$$

Proof. It is easy to see that

$$Z^* \geq Z^{NLB} \geq Z_1^* / m$$

Combining (1), we have

$$\lim_{n \to \infty} Z^* / n^2 = \lim_{n \to \infty} Z^{NLB} / n^2 = \lim_{n \to \infty} Z_1^* / mn^2 \tag{7}$$

Multiplying n^2 on the two sides of (7), we can get the result.   □

4 Worst Case Analysis of NLB

In section 3, we have proved that the NLB works as well as the optimal solution when the size of the problem goes to infinity. But does the NLB always perform effectively in any case? This question can be answered when all weights are equal.

Theorem 4.1. *Let all weights be equal. For the optimal solution Z^* of Problem P and its associated NLB value Z^{NLB}, we have*

$$Z^* / Z^{NLB} \leq m \tag{8}$$

and this bound is tight.

Proof. Without loss of generality, let $w_i = 1$, $i = 1, 2, \ldots, n$, and C_i^* denote the completion time of job i associated with Z^*; C_i' denote the completion time of job i associated with Z^{NLB}. We express the processing times appeared in the NLB by matrix A when job i departs from the last machine.

$$A = \begin{bmatrix} t_1^1 & \cdots & t_{i-1}^1 & t_i^1 & t_i^2 & \cdots & t_i^{m-2} & t_i^{m-1} & t_i^m \\ t_1^1 & t_1^2 & \cdots & t_{i-1}^2 & t_i^2 & t_i^3 & \cdots & t_i^{m-1} & t_i^m \\ & \cdots & & & \cdots & & & \cdots & \\ t_1^1 & t_1^2 & t_1^3 & \cdots & t_1^{m-1} & t_1^m & t_2^m & \cdots & t_i^m \end{bmatrix}$$

Since the difference of the processing time, idle times maybe exist in mC_i^*, where mC_i^* means the m times of C_i^*. Let I_i be the total idle times in mC_i^*. Therefore, we have

$$mC_i^* - mC_i' = I_i$$

Summing over all of the jobs, we have

$$Z^* - Z^{NLB} = \frac{1}{m} \sum_{i=1}^{n} I_i$$

Hence,

$$Z^* / Z^{NLB} \leq 1 + \left(\sum_{i=1}^{n} I_i \right) \Big/ m Z^{NLB} \tag{9}$$

So, if we obtain that

$$\left(\sum_{i=1}^{n} I_i \right) \Big/ m Z^{NLB} \leq m - 1$$

the proof of inequality (8) is completed.

We reindex the elements of matrix A as

$$A' = \begin{bmatrix} x_{1,1} & x_{1,2} & \cdots & x_{1,i+m-1} \\ x_{2,1} & x_{2,2} & \cdots & x_{2,i+m-1} \\ \cdots & & \cdots & \cdots \\ x_{m,1} & x_{m,2} & \cdots & x_{m,i+m-1} \end{bmatrix}$$

where the element in A and the element in A' are equal when they have the same location.

For flow shop scheduling problem, there must exist one critical path at least, from t_1^1 to t_i^m, on which the total sum of processing times is equal to the C_i^* (see Pinedo [6] or Xia et al. [8]). Obviously, as A' is the same as the A, the critical path crosses the first column to the last column of A', that is, the critical path is composed of the certain elements which come from each column of A'. Without loss of generality, denote the length of the critical path as

$$C_i^* = \sum_{h=1}^{i+m-1} x_{c_l, h}$$

where $x_{c_l,h}$ belongs to the critical path, and it is a certain element of hth column of A', and $c_l \in Q = \{1,2,...,m\}$, $l=1, 2,..., m$.

Now, subtracting each row of A' from mC_i^*, we get

$$I_i = \sum_{h=1}^{i+m-1} \left(x_{c_l,h} - x_{1,h} \right) + \sum_{h=1}^{i+m-1} \left(x_{c_l,h} - x_{2,h} \right) + \cdots + \sum_{h=1}^{i+m-1} \left(x_{c_l,h} - x_{m,h} \right)$$

$$= \sum_{h=1}^{i+m-1} \left(mx_{c_l,h} - \sum_{l=1}^{m} x_{l,h} \right) = \sum_{h=1}^{i+m-1} \left((m-1) x_{c_l,h} - \sum_{Q\backslash\{c_l\}} x_{l,h} \right)$$

$$\leq (m-1) \sum_{h=1}^{i+m-1} \left(x_{c_l,h} + \sum_{Q\backslash\{c_l\}} x_{l,h} \right) = (m-1) mC_i'$$

Therefore,

$$I_i / mC_i' \leq m - 1$$

Summing over all the jobs, we have

$$(\sum_{i=1}^{n} I_i) / mZ^{NLB} \leq (n-1)(m-1)/n \leq m-1 \tag{10}$$

The second term of (10) holds because there is no idle time in the case of $i=1$. Combining the inequality (9), we obtain the (8).

To see that the bound of m is tight, consider the following instance of the problem. There are 2 jobs, J_1 and J_2, and m machines. For J_1, the processing times $t_1^1 = t_2^1 = \cdots = t_m^1 = \varepsilon/m$, where ε is an arbitrarily small number on $(0, 1)$. The processing times of J_2 are $t_2^1 = 1$ and $t_2^2 = t_3^2 = \cdots = t_m^2 = \varepsilon/m$. We deduce that

$$Z^*/Z^{NLB} = (1+2\varepsilon)/((1+(2m-1)(m+1)\varepsilon/m)/m) \to m$$

as $\varepsilon \to 0$.   □

Obviously, the NLB is better than the LB since $Z^{NLB} > Z^{LB}$, and especially when the processing times of the jobs are all equal the NLB is just the optimal solution. But in some extreme cases, the optimal solution may be much larger than the NLB, which is caused by the idle times. However, as the size of the problem becomes large enough, the total idle times can be ignored (see Kaminsky and Simchi-Levi [4]).

5 Computational Results

To study the effectiveness of the NLB, a series of computational experiments are designed and conducted. The theme of the experiments mainly rests on revealing the better performances of the NLB. As WSPT rule is asymptotically optimal for the Flow Shop Weighted Completion Time problem (see Xia et al. [8]), in table 1, 2, 3 and 4, we compare the objective values obtained from WSPT rule with their associated NLB values and LB values for various numbers of machines respectively. The percentages given in the four tables are the ratios of the objective values of WSPT to their associated NLB values and LB values, respectively. For the data which were presented in the four tables, the processing times and weights were generated from a uniform (0, 1] distribution.   And we performed three different random trials for each of the combination, and the averages are shown in the four tables. In table 1 and 2, from 500 to 5000 jobs, 3, 6 and 12 machines were tested with general weights and equal weights. The numerical results in the tables clearly evince that the NLB works as well as the LB when the instances get larger enough. For example, for 5000 jobs and 3 machines with general weights, the percentages of Z^{WSPT}/Z^{NLB} and Z^{WSPT}/Z^{LB} in table 1 and 2 are 101.87, 102.50, 102.55 and 101.87, 102.51, 102.55; and the averages of them are 102.31 and 102.31 respectively.

The second part of the experiments was conducted to evaluate the NLB when the number of jobs is less than the number of machines, that is, $n<m$. In table 3 and 4, for 6, 12 and 18 machines, the jobs which satisfy $n<m$ were tested with general weights and equal weights. And the percentages presented in table 3 and 4 are also the ratios of the objective values of WSPT to their NLB values and LB values, respectively. The numerical results provide the information that the NLB works better than the LB as $n<m$. For example, for 11 jobs and 18 machines with general weights, the percentages of Z^{WSPT}/Z^{NLB} and Z^{WSPT}/Z^{LB} in table 3 and 4 are 130.12, 126.42, 126.56 and 210.79, 202.01, 197.62; and the averages of them are 127.70 and 203.47 respectively.

6 Summary and Conclusions

In this paper, we improve the original LB which was presented by Kaminsky and Simchi-Levi [3] and give a better NLB. When the size of the problem, Flow shop weighted Completion Time problem, goes to infinity, we prove that the NLB is

Table 1. The percentages of Z^{WSPT}/Z^{NLB} for $n>m$ (%)

Weights		Uniform			Equal		
Machines		3	6	12	3	6	12
500 Jobs	Trial 1	107.43	112.21	119.15	106.54	111.89	117.51
	Trial 2	106.54	111.52	118.04	108.41	113.88	121.49
	Trial 3	107.35	112.83	118.71	106.40	113.27	120.64
	Average	107.11	112.19	118.63	107.12	113.01	119.88
1000 Jobs	Trial 1	103.99	107.84	115.04	105.55	107.59	114.26
	Trial 2	106.09	109.93	114.83	105.48	110.21	113.87
	Trial 3	104.42	109.14	114.11	103.57	109.52	115.93
	Average	104.83	108.97	114.66	104.87	109.11	114.69
2500 Jobs	Trial 1	103.43	105.89	108.36	103.24	106.43	109.15
	Trial 2	103.87	106.29	109.23	103.08	105.04	109.05
	Trial 3	103.78	105.24	109.16	102.62	105.44	109.64
	Average	103.69	105.81	108.92	102.98	105.64	109.28
5000 jobs	Trial 1	101.87	104.93	105.27	101.92	104.27	106.81
	Trial 2	102.50	104.70	106.68	101.55	103.17	107.09
	Trial 3	102.55	104.25	106.04	102.05	104.00	106.83
	Average	102.31	104.63	106.00	101.84	103.81	106.91

Table 2. The percentages of Z^{WSPT}/Z^{LB} for $n>m$ (%)

Weights		Uniform			Equal		
Machines		3	6	12	3	6	12
500	Trial 1	107.48	112.60	120.73	106.63	112.29	118.89
Jobs	Trial 2	106.66	111.88	119.97	108.50	114.42	122.77
	Trial 3	107.40	113.54	120.23	106.40	113.67	121.96
	Average	107.18	112.67	120.31	107.18	113.46	121.21
1000	Trial 1	104.00	108.09	115.95	105.57	107.82	115.04
Jobs	Trial 2	106.11	110.09	115.72	105.51	110.43	114.38
	Trial 3	104.45	109.37	114.62	103.58	109.68	116.56
	Average	104.85	109.18	115.43	104.89	109.31	115.33
2500	Trial 1	103.44	105.94	108.56	103.25	106.49	109.29
Jobs	Trial 2	103.89	106.35	109.58	103.09	105.09	109.29
	Trial 3	103.79	105.29	109.42	102.62	105.54	109.83
	Average	103.71	105.86	109.19	102.99	105.71	109.47
5000	Trial 1	101.87	104.95	105.38	101.92	104.30	106.94
jobs	Trial 2	102.51	104.72	106.80	101.55	103.20	107.20
	Trial 3	102.55	104.28	106.14	102.05	104.05	106.91
	Average	102.31	104.65	106.11	101.84	103.85	107.02

Table 3. The percentages of Z^{WSPT}/Z^{NLB} for $n<m$ (%)

Weights		Uniform			Equal		
Jobs		3	4	5	3	4	5
6	Trial 1	115.10	114.74	123.71	111.82	146.79	116.23
Machines	Trial 2	104.06	111.08	123.33	110.41	116.60	121.23
	Trial 3	109.34	132.06	118.90	117.00	131.46	109.62
	Average	109.50	119.29	121.98	113.08	131.62	115.69
Jobs		7	9	11	7	9	11
12	Trial 1	118.67	129.95	126.38	121.20	131.20	130.17
Machines	Trial 2	123.96	137.14	128.12	121.48	120.25	128.14
	Trial 3	116.66	119.41	133.44	120.96	122.53	136.80
	Average	119.76	128.83	129.31	121.21	124.66	131.70
Jobs		11	14	17	11	14	17
18	Trial 1	130.12	128.20	135.54	124.63	126.38	133.92
Machines	Trial 2	126.42	131.36	145.08	120.47	133.83	135.77
	Trial 3	126.56	132.67	136.87	124.66	128.40	131.67
	Average	127.70	130.74	139.16	123.25	129.54	133.79

Table 4. The percentages of Z^{WSPT}/Z^{LB} for $n<m$ (%)

Weights		Uniform			Equal		
Jobs		3	4	5	3	4	5
	Trial 1	156.08	153.07	147.63	161.23	170.78	156.47
6	Trial 2	146.78	166.75	172.04	153.12	164.72	162.36
Machines	Trial 3	191.26	188.32	184.20	161.44	193.69	146.06
	Average	164.71	169.38	167.96	158.60	176.40	154.96
Jobs		7	9	11	7	9	11
	Trial 1	191.97	189.52	191.82	198.83	178.60	173.10
12	Trial 2	203.49	215.84	176.55	165.58	169.85	172.21
Machines	Trial 3	196.37	182.17	186.38	179.55	171.10	179.99
	Average	197.28	195.84	184.92	181.32	173.18	175.10
Jobs		11	14	17	11	14	17
	Trial 1	210.79	210.34	196.48	184.95	183.61	178.70
18	Trial 2	202.01	198.76	194.46	183.14	186.41	183.86
Machines	Trial 3	197.62	205.12	187.71	168.56	191.83	194.31
	Average	203.47	201.74	192.88	178.88	187.28	185.62

asymptotically optimal to the optimal solution. For the case of equal-weight, a worst case performance ratio, m, of the optimal solution to the NLB is obtained and this bound is tight. The experiments and numerical results show that the NLB works as well as the LB when the number of jobs is large enough, and as $n<m$ the NLB is better than the LB.

Acknowledgements. We are grateful to the graduate student Liangjun Wang for testing the data. This research is partly supported by National Natural Science Foundation for Distinguished Young Scholars of China (Grant No. 70425003), National 863 High-Tech Research and Development Program of China (Grant No. 2006AA04Z174) and National Natural Science Foundation of China (Grant No. 60674084).

References

[1] Bhaskaran, K., Pinedo, M.: Dispatching. In: Salvendy, G. (ed.) Handbook of industrial engineering, Wiley, New York (1992)
[2] Garey, M.R., Johnson, D.S., Sethi, R.: The complexity of flow shop and job shop scheduling. Mathematics of Operations Research 1, 117–129 (1976)
[3] Kaminsky, P., Simchi-Levi, D.: Probabilistic analysis and practical algorithms for the flow shop weighted completion time problem. Operations Research 46, 872–882 (1998)
[4] Kaminsky, P., Simchi-Levi, D.: The asymptotic optimality of the SPT rule for the flow shop mean completion time problem. Operations Research 49, 293–304 (2001)
[5] Kohler, W., Steiglitz, K.: Exact, approximate, and guaranteed accuracy algorithms for the flow shop problem $n/2/F/\bar{F}$. Journal of the Association of Computer Machinery 22, 106–114 (1975)
[6] Pinedo, M.: Scheduling: Theory, Algorithms and Systems. Prentice-Hall, Englewood Cliffs, New Jersey (1995)
[7] Shakhlevich, N., Hoogeveen, H., Pinedo, M.: Minimizing total weighted completion time in a proportionate flow shop. Journal of Scheduling 1, 157–168 (1998)
[8] Xia, C.H., Shanthikumar, J.G., Glynn, P.W.: On the asymptotic optimality of the SPT rule for the flow shop average completion time problem. Operations Research 48, 615–622 (2000)

Scaling, Renormalization, and Universality in Combinatorial Games: The Geometry of Chomp

Eric J. Friedman[1] and Adam Scott Landsberg[2]

[1] School of ORIE, Cornell University, Ithaca, NY 14853, USA
[2] Joint Science Department, Claremont McKenna, Pitzer, and Scripps Colleges, Claremont, California 91711, USA

Abstract. Combinatorial games pose an extreme challenge to combinatorial optimization. Several combinatorial games have been shown to be PSPACE-hard and many more are believed to be so. In this paper, we present a new approach to analyzing combinatorial games, which differs dramatically from current approaches. Using the combinatorial game Chomp as a model system, we employ ideas from physics and dynamical systems theory to unveil deep connections between such games and nonlinear phenomena commonly seen in nature.

Combinatorial games, which include Chess, Go, Checkers, Chomp, and Nim, have both captivated and challenged mathematicians, computer scientists, and players alike [1-10]. Analysis of these two-player games has generally relied upon a few beautiful analytical results [1,11-14] or on numerical algorithms that combine heuristics with look-ahead approaches (α–β pruning) [15,16]. Using Chomp as a prototype, we report on a new geometrical approach which unveils unexpected parallels between combinatorial games and key ideas from physics and dynamical systems, most notably notions of scaling, renormalization, universality, and chaotic attractors. Our central finding is that underlying the game of Chomp is a probabilistic geometric structure (Fig. 2) that encodes essential information about the game, and that this structure exhibits a type of scale invariance: Loosely speaking, the geometry of "small" winning positions and "large" winning positions are the same after rescaling (cf., Fig. 2a,b). This geometric insight not only provides (probabilistic) answers to some open questions about Chomp, but it suggests a natural pathway toward a new class of algorithms for general combinatorial games, and hints at deeper links between such games and nonlinear science.

The game of Chomp, introduced by Gale [17] and Schuh [18] over 30 years ago, makes an ideal candidate for study, since it is among the simplest in the class of "unsolved" combinatorial games. It has thus far defied a complete analysis (and is conjectured to potentially be PSPACE-hard), yet at the same time it is not entirely intractable, as evidenced by some significant theoretical advances [19-22]. The rules of Chomp are easily explained. Play begins with an N x M array of counters (Fig. 1a). On each turn a player selects a counter and removes it along with all counters to the north and east of it (Fig. 1b). Play alternates between the two players until one player takes the last ("poison") counter, thereby losing the game. An intriguing feature of Chomp, as shown by Gale, is that although it is very easy to prove that the

A. Dress, Y. Xu, and B. Zhu (Eds.): COCOA 2007, LNCS 4616, pp. 200–207, 2007.
© Springer-Verlag Berlin Heidelberg 2007

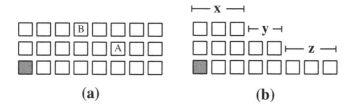

Fig. 1. The game of Chomp.   (a) play begins with an MxN rectangular array of counters (three-row Chomp is illustrated). At each turn, a player selects a counter and removes it along with all counters lying in the northeast quadrant extending from the selected counter. Play alternates between the two players until one player is forced to take the 'poison' counter (shown in black) in the southwest corner, thereby losing the game. (b) a sample game configuration after player 1 selects counter A, followed by player 2 selecting counter B. More generally, an arbitrary game configuration can be specified by coordinates [x,y,z], as shown.

player who moves first can always win (under optimal play), what this opening move should be has been an open question! Our methodology will in fact provide a probabilistic answer to this question.

For simplicity, we will focus here on the case of three-row (M=3) Chomp, a subject of recent study by Zeilberger [19-20] and Sun [21]. (Generalizations to four-row and higher Chomp are analogous.) To start, we note that the configuration of the counters at any stage of the game can be described (using Zeilberger's coordinates) by the position p=[x,y,z], where x specifies the number of columns of height three, y specifies the number of columns of height two, and z the number with height one (Fig. 1b). Each position p may be classified as either a *winner*, if a player starting from that position can always force a win (under optimal play), or as a *loser* otherwise. (This classification is well defined by Zermelo's theorem.) We may group the losing positions according to their x values by defining a "loser sheet" L_x to be an infinite two-dimensional matrix whose $(y,z)^{th}$ component is a 1 if position [x,y,z] is a loser, and a 0 otherwise. (As noted by Zeilberger, it is formally possible to express L_x in terms of all preceding loser sheets L_{x-1}, L_{x-2}, ..., L_0.) The set of all L_x's contains the information for solving the game.

Studies by Zeilberger [19,20] and others [21-23] have detected several patterns and analytical features about losing positions, and their interesting but non-obvious properties have even led to a conjecture that Chomp may be "chaotic in a yet-to-be-made-precise sense" [20]. To provide broader insight into the general structure of the game, we depart from the usual analytic/algebraic/algorithmic approaches. Our approach will be distinguished by its decidedly geometric flavor, but equally importantly, it also introduces *probabilistic* elements into the analysis, despite the fact that the combinatorial games we consider are all games of *no chance*, which lack any inherent probabilistic components to them whatsoever!

To proceed, we must consider "instant-winner sheets", defined as follows: A position p=[x,y,z] is called an *instant* winner (in Zeilberger's terminology) if from that position a player can legally move to a losing position with a smaller x-value.

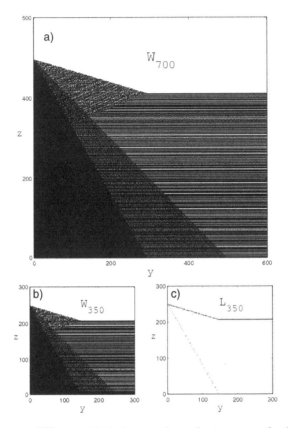

Fig. 2. The geometry of Chomp. (a) the instant-winner sheet geometry for three-row Chomp, shown for x=700. Instant winner locations in the y-z plane are shown in black. The horizontal bands in the figure extend to infinity. (b) the instant-winner sheet for x=350. Comparing W_{350} to W_{700} illustrates the crucial scaling property of the instant winner sets: Their overall geometry is identical up to a scaling factor; i.e., all values of boundary-line slopes and densities of the various internal regions are preserved (although the actual point-by-point locations of the instant winners are different). (c) the loser-sheet geometry L_x, shown for x=350. Note that losers are confined to lie near one of three boundary lines (compare L_{350} to W_{350}): a lower line of slope $m_L = -1 - 1/\sqrt{2}$, density $\lambda_L = 1 - 1/\sqrt{2}$; an upper (tilted) line of slope $m_U = -1 + 1/\sqrt{2}$, density $\lambda_U = 1/\sqrt{2}$; and an upper flat line (of density one) which only exists for some x-values. The probability that a flat line in L_x exists for a randomly chosen x is $\gamma = \sqrt{2} - 1$. The lower and upper tilted lines both emanate from a point near $(y,z)=(0, \alpha x)$, where $\alpha = 1/\sqrt{2}$. The geometrical structure of the L_x's, like that of the W_x's, remains invariant (up to a scale factor) as one goes to progressively larger x values. As described in the text, the analysis of this invariance allows for a complete geometrical/probabilistic characterization of the structures shown in these figures.

We then define an instant-winner sheet W_x to be the infinite, two-dimensional matrix consisting of all instant winners with the specified x-value, i.e., the $(y,z)^{th}$ component of matrix W_x is a 1 if position [x,y,z] is an instant winner, and a 0 otherwise. The crucial insight is seen in Fig. 2, which reveals the geometric structure of these instant-winner sheets. Each sheet exhibits a nontrivial internal structure characterized by several distinct regions, and most importantly, the sheets as a group possess a striking scaling property – upon rescaling, the overall geometry of the sheets become identical (in a probabilistic sense).

We can show that the instant-winner sheets obey an analytical recursion relation $W_{x+1} = \mathbf{R}\, W_x$, where $\mathbf{R}$ denotes the recursion operator. (The operator $\mathbf{R}$ can in fact be decomposed as $\mathbf{R}=\mathbf{L}(\mathbf{I}+\mathbf{DM})$, where $\mathbf{L}$ is a left-shift operator, $\mathbf{I}$ is the identity operator, $\mathbf{D}$ is a diagonal element-adding operator, and $\mathbf{M}$ is a "sheet-valued" version of the standard mex operator which is often used for combinatorial games. However, for our purposes here it is enough to note simply that a well-defined recursion operator $\mathbf{R}$ relating the instant-winner sheets exists.) The loser sheets can be readily found via $L_x = \mathbf{M}\, W_x$. The characteristic geometry of these loser sheets is revealed in Fig. 2c. It consists of three (diffuse) lines: a lower line of slope m_L and density of points λ_L, an upper line of slope m_U and density λ_U, and a flat line extending to infinity. The upper and lower lines originate from a point whose height (i.e., z-value) is αx. The flat line (with density one) is only present with probability γ in randomly selected loser sheets.

Stepping back for a moment, what we now have is a renormalization problem akin to those so often encountered in physics and the nonlinear sciences, such as the famous period-doubling cascade described by May [24] in a biological mapping and analyzed by Feigenbaum using renormalization techniques [25]. In particular, we have objects (instant winner matrices) that exhibit similar structure at different size scales (cf., Figs. 2a,b), and a recursion operator relating them. Our task therefore is to show that (for large x's) if we act with the recursion operator followed by an appropriately-defined rescaling operator $\mathbf{S}$, we get W_x back again: $W_x = \mathbf{S}\,\mathbf{R}\, W_x$ (i.e., we seek a fixed point of the renormalization-group operator $\mathbf{S}\,\mathbf{R}$.) This can be done, but before doing so we point out a critical feature of the analysis. Even though the recursion operator $\mathbf{R}$ is exact and the game itself has absolutely no stochastic aspects to it, it is necessary to adopt a probabilistic framework in order to solve this recursion relation. Namely, our renormalization procedure will show that the slopes of all boundary lines and densities of all regions in the W_x's (and L_x's) are preserved – not that there exists a point-by-point equivalence. In essence, we bypass consideration of the random-looking 'scatter' of points surrounding the various lines and regions of W_x and L_x by effectively averaging over these 'fluctuations'.

The key to the renormalization analysis is to observe from Figs. 2b,c that the losers in L_x are constrained to lie along certain boundary lines of the W_x plot; the various interior regions of W_x are "forbidden" to the losers (by the recursion operator). As will be described in more detail elsewhere, each such forbidden region imposes a constraint on the structural form that the W_x, L_x's can take, and can be formulated as an algebraic equation relating the hitherto unknown parameters m_L, λ_L, m_U, λ_U, γ, α that define the loser sheets. We find

$$\lambda_U - m_U = 1, \quad \lambda_U + \lambda_L = 1, \quad \frac{1}{\alpha+1} - \frac{\lambda_L}{m_L+1} = 1, \quad (\gamma-1)\frac{m_L}{\alpha - m_L} + \frac{1}{\alpha+1} = 1,$$

$$\frac{\alpha\lambda_L}{\alpha - m_L}\left(\frac{m_U - m_L}{m_U\alpha - m_L\alpha + m_L\gamma}\right) + \frac{1}{\alpha+1} = 1, \quad \frac{\lambda_L}{\alpha - m_L} - \frac{\alpha}{\alpha+1}\left(1 - \frac{\lambda_U}{\alpha - m_U}\right) = 0. \tag{1}$$

Stated differently, these are the necessary conditions for the instant-winner sheets to be fixed points of the renormalization operator **S R**. Solving, we find

$$\alpha = \tfrac{1}{\sqrt{2}}, \quad \lambda_L = 1 - \tfrac{1}{\sqrt{2}}, \quad \lambda_U = \tfrac{1}{\sqrt{2}}, \quad m_L = -1 - \tfrac{1}{\sqrt{2}}, \quad m_U = -1 + \tfrac{1}{\sqrt{2}}, \quad \gamma = \sqrt{2} - 1 \tag{2}$$

The densities associated with the various regions of W_x can all be readily calculated from these six key parameters. We thus have a fundamental (probabilistic) description of the overall geometry of the game. Our results also confirm several numerical conjectures on loser properties by Brouwer [23]. We mention that only a single assumption was needed to construct the six preceding parameter relations; namely, that fluctuations associated with the diagonal operator **D** were uncorrelated with the fluctuations surrounding the upper line in L_x.

Several interesting results immediately follow. First, having identified the geometric structure of the loser sheets, we can now easily show that the best opening move in Chomp from the initial position $[x_0, 0, 0]$ must lie in the vicinity of the positions $[x_0\sqrt{2}/(\sqrt{2}+1), 0, x_0/(\sqrt{2}+1)]$ or $[x_0(\sqrt{2}+1)/(\sqrt{2}+2), x_0/(\sqrt{2}+2), 0]$.

Second, for most winning positions (except those near a boundary), knowing their location within W_x allows us to compute the expected number of winning moves based on which lines in the loser sheets are accessible. Third, knowledge of the geometrical structure of the loser sheets suggests a natural pathway to more efficient algorithms by simply designing the search algorithm to aim directly for the known loser lines in L_x. This is in fact a general feature of our methodology (not limited to just Chomp): once the geometry of a combinatorial game has been identified by the renormalization procedure, efficient geometrically-based search algorithms can be constructed. Lastly, as seen in Fig. 2c, the co-existence of order (i.e., analytically well-defined loser lines) and disorder (i.e., the scatter of points around these lines) signifies that combinatorial games such as Chomp may be unsolvable yet still informationally compressible, in the language of Chaitin [26].

The probabilistic renormalization approach we have employed naturally gives rise to a whole new set of interesting questions about combinatorial games. For instance, we can construct variants of standard games simply by perturbing an instant-winner sheet by the addition of a finite number of new points. (Such additions effectively modify the game by declaring new positions to be automatic winners.) We can then examine whether the previously observed geometry of the W_x's is preserved for these variants (i.e., are they attractors?). Simulations show that for a sizeable class of variants of Chomp the geometric structure of Fig. 2 re-emerges. Hence it appears stable (in a probabilistic sense). In the language of renormalization we would say that such game variants fall into the same universality class as the original game. A related issue concerns sensitivity to initial conditions, a hallmark of chaos in dynamical

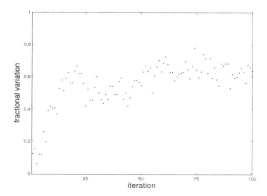

Fig. 3. Dependence on initial conditions. The figure illustrates how perturbing an instant winner matrix by a single point subsequently spreads and "infects" the loser sheets at higher x values (i.e., altering the precise locations of the losing positions). The graph shows the fraction of losers along the upper tilted and lower lines (e.g., Fig. 2c) that are affected when one adds a single new point to W_{80} and then iterates. Note that the effects can be pronounced – e.g., after only about 15 iterations of the recursion operator, the locations of nearly half of all losing positions have shifted.

systems theory. Using our recursion operator $W_{x+1} = \mathbf{R}\, W_x$, we can examine how small perturbations to W_x propagate. Although the overall instant-winner geometry of the perturbed and unperturbed systems will be the same if they lie in the same universality class, they will differ on a point-by-point basis. We find (see Fig. 3) that small initial perturbations can in fact significantly alter the actual loser locations quite dramatically, highly reminiscent of chaotic systems.

We can also apply our methodology to other combinatorial games. Consider the game of Nim [1,27]. It is straightforward to construct the recursion and renormalization operators for this game, and to analyze its properties analogously. Fig. 4a shows the geometry of an instant-winner sheet W_x for three-heap Nim. As in Chomp, this structure exhibits a geometric scaling property (although the W_x's do depend on their x-values). Unlike Chomp however, ordinary Nim is a completely solvable game, and we find that the geometry of its W_x's is unstable. Indeed, if we add just a few random perturbations to one of the sheets, then a very different-looking instant winner structure of the form shown in Fig. 4b emerges. This striking new structure, just as for Chomp, is remarkably stable, generic (i.e., it seems to naturally emerge for most perturbations), and scale invariant. In fact, we speculate that the ordinary game of Nim has an unstable, nongeneric geometry precisely because of its solvable nature, and that the robust geometry of Fig. 4b for variants of Nim is much more typical. It is not unreasonable to conjecture that generic combinatorial games will have robust underlying geometric structures, while those of solvable games will be structurally unstable to perturbations.

Lastly, we remark that the "growth" (with increasing x) of the geometric structures W_x (Figs. 4b and 2a) for games such as Nim and Chomp is reminiscent of certain crystal growth and aggregation processes in physics [28] and activation- inhibition cellular automata models in biology [29]. Such a semblance likely arises because the

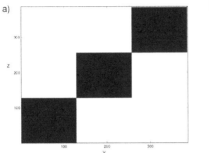
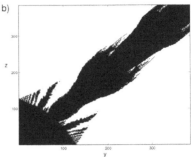

Fig. 4. The geometries of ordinary and variant Nim. In ordinary (3-heap) Nim, play begins with counters stacked into three piles (heaps). Coordinates (x,y,z) denote here the number of counters in each heap. At each turn, a player removes one or more counters from one heap. Play alternates until one of the two players removes all remaining counters, thereby winning the game. Ordinary Nim is completely solvable. **(a)** the instant winner structure W_x at x=128 for ordinary Nim. As in Chomp, this geometrical structure is preserved (up to an overall scale factor) with increasing x values; i.e., W_x and W_{2x} look identical (not shown). However, unlike Chomp, the geometry is highly unstable to perturbations, and also exhibits an internal periodicity such that W_x and W_{x+1} are similar but not wholly identical in structure. **(b)** the instant winner structure W_x at x=128 for a generic Nim variant. Recall that Nim variants are similar to ordinary Nim, except that various heap configurations are arbitrarily declared to be automatic winners. The striking geometrical structure shown in the figure is both stable and reproducible, i.e., it typically emerges whenever one or more random heap configurations are declared automatic winners. As in Chomp, this attracting structure is preserved (up to scale factors) as one goes to increasingly large x-values. (We note, however, that the scaling behavior appears more pronounced for $W_x \rightarrow W_{2x}$ than it is for $W_x \rightarrow W_{x+1}$, a remnant, we believe, of the underlying solvable structure of ordinary Nim upon which these Nim variants are based.)

recursion operators governing the game evolution typically act by attaching new points to the boundaries of the existing structures, thereby transforming the study of a combinatorial game into that of a growth process.

We hope that this novel (renormalization-based) approach to combinatorial games and the tantalizing connections it raises to key ideas from the nonlinear sciences will stimulate further research along these lines. In addition, we expect that these approaches might prove useful for ordinary combinatorial optimization problems too. In particular, the recursive formulations currently provide the data structure for the most efficient (in both time and memory) known rigorous algorithm, while the renormalization solution leads to the most efficient non-rigorous one.

References

1. Berlekamp, E.R., Conway, J.H., Guy, R.K.: Winning Ways for Your Mathematical Plays. Academic Press, London (1982)
2. Nowakowski, R.J.: Games of No Chance. Cambridge University Press, Cambridge (1996)
3. Nowakowski, R.J.: More Games of No Chance. Cambridge University Press, Cambridge (2002)

4. Fleischer, R.H., Nowakowski, R.J.: Algorithmic combinatorial game theory. Elsevier, Amsterdam (2004)
5. Berlekamp, E.: The Game of Dots and Boxes - sophisticated child's play. A K Peters Ltd, Natick, MA (2000)
6. Berlekamp, E., Wolfe, D.: Mathematical Go: Chilling Gets the Last Point. A K Peters Ltd, Natick, MA (1994)
7. Demaine, E.D.: Playing games with algorithms: Algorithmic combinatorial game theory. In: Sgall, J., Pultr, A., Kolman, P. (eds.) MFCS 2001. LNCS, vol. 2136, Springer, Heidelberg (2001)
8. Fraenkel, A.S.: Complexity, appeal and challenges of combinatorial games. Theoretical Computer Science 313, 393–415 (2004)
9. Demaine, E.D., Fleischer, R., Fraenkel, A.S., Nowakowski, R.J.: Open problems at the 2002 Dagstuhl Seminar on Algorithmic combinatorial game theory. Theoretical Computer Science 313, 539–543 (2004)
10. Guy, R., Nowakowski, R.: Unsolved Problems in Combinatorial Games. In: Nowakowski, R.J. (ed.) More Games of No Chance, Cambridge University Press, Cambridge (2002)
11. Sprague, R.: Uber mathematische Kampfspiele. Tohoku Mathematical Journal 41, 438–444 (1936)
12. Grundy, P.M.: Mathematics and games. Eureka 2, 6–8 (1939)
13. Smith, C.: Graphs and composite games. Journal of Combinatorial Theory 1, 51–81 (1966)
14. Conway, J.H.: On Numbers and Games. AK Peters Ltd., Natick, Mass (2000)
15. Newborn, M.: Kasparov versus Deep Blue: Computer Chess Comes of Age. Springer, New York (1997)
16. Schaeffer, J.: One Jump Ahead: Challenging Human Supremacy in Checkers. Springer, New York (1997)
17. Gale, D.: A Curious Nim-type game. Amer. Math. Monthly 81, 876–879 (1974)
18. Schuh, F.: Spel van delers. Nieuw Tijdschrift voor Wiskunde 39, 299–304 (1952)
19. Zeilberger, D.: Three-rowed Chomp. Adv. in Appl. Math. 26, 168–179 (2001)
20. Zeilberger, D.: Chomp, Recurrences, and Chaos. J. Difference Equations and its Applications 10, 1281–1293 (2004)
21. Sun, X.: Improvements on Chomp. Integers 2 G1, 8 (2002), http://www.integers-ejcnt.org/
22. Byrnes, S.: Poset Games Periodicity. Integers 3 G3, 8 (2003), http://www.integers-ejcnt.org/
23. Brouwer, A.E.: The game of Chomp (2004), On-line document http://www.win.tue.nl/~aeb/games/chomp.html

Mechanism Design by Creditability*

Raphael Eidenbenz, Yvonne Anne Oswald, Stefan Schmid, and Roger Wattenhofer

Computer Engineering and Networks Laboratory
ETH Zurich, Switzerland

Abstract. This paper attends to the problem of a mechanism designer seeking to influence the outcome of a strategic game based on her creditability. The mechanism designer offers additional payments to the players depending on their mutual choice of strategies in order to steer them to certain decisions. Of course, the mechanism designer aims at spending as little as possible and yet implementing her desired outcome. We present several algorithms for this optimization problem both for singleton target strategy profiles and target strategy profile regions. Furthermore, the paper shows how a bankrupt mechanism designer can decide efficiently whether strategy profiles can be implemented at no cost at all. Finally, risk-averse players and dynamic games are examined.

1 Introduction

Game theory is a powerful tool for analyzing decision making in systems with autonomous and rational (or selfish) participants. It is used in a wide variety of fields such as economics, politics, biology, or computer science. A major achievement of game theory is the insight that networks of self-interested agents often suffer from inefficiency due to effects of selfishness. Popular problems in computer science studied from a game theoretic point of view include *virus propagation* [1], *congestion* [2], or *network creation* [6], among many others.

If a game theoretic analysis reveals that a system suffers from the presence of selfish participants, mechanisms to encourage cooperation have to be devised. The field of *mechanism design* [5,9] is also subject to active research; for example, Cole et al. [3,4] have studied how incentive mechanisms can influence selfish behavior in a routing system.

In many distributed systems, a mechanism designer cannot change the rules of interactions. However, she may be able to influence the agents' behavior by offering payments for certain outcomes. On this account, Monderer and Tennenholtz [10] have initiated the study of a mechanism designer whose power is to some extent based on her monetary assets, primarily, though, on her *creditability*, i.e., the players trust her to pay the promised payments. Thus, a certain subset of outcomes is *implemented* in a given game if, by expecting additional non-negative payments, rational players will necessarily choose one of the desired outcomes. The designer faces the following optimization problem: How can a desired outcome be implemented at minimal cost? Surprisingly, it

---

* Supported in part by the Swiss National Science Foundation (SNF). A full version including all proofs, more simulation results and an appendix is available as TIK Report 270 at http://www.tik.ee.ethz.ch/.

A. Dress, Y. Xu, and B. Zhu (Eds.): COCOA 2007, LNCS 4616, pp. 208–219, 2007.

is sometimes possible to improve the performance of a given system merely by creditability, i.e., without any payments at all.

This paper extends [10] in various respects. First, an algorithm for finding an exact, incentive compatible implementation of a desired set of outcomes is given. We also show how a bankrupt mechanism designer can decide in polynomial time if a set of outcomes can be implemented at no costs at all, and an interesting connection to best response graphs is established. We propose and analyze efficient heuristic algorithms and demonstrate their performance. Furthermore, we extend our analysis for risk-averse behavior and study dynamic games where the mechanism designer offers payments in each round.

2 Model

Game Theory. A *strategic game* can be described by a tuple $G = (N, X, U)$, where $N = \{1, 2, \ldots, n\}$ is the set of *players* and each Player $i \in N$ can choose a *strategy* (action) from the set X_i. The product of all the individual players' strategies is denoted by $X := X_1 \times X_2 \times \ldots \times X_n$. In the following, a particular outcome $x \in X$ is called *strategy profile* and we refer to the set of all other players' strategies of a given Player i by $X_{-i} = X_1 \times \ldots \times X_{i-1} \times X_{i+1} \times \ldots \times X_n$. An element of X_i is denoted by x_i, and similarly, $x_{-i} \in X_{-i}$; hence x_{-i} is a vector consisting of the strategy profiles of x_i. Finally, $U = (U_1, U_2, \ldots, U_n)$ is an n-tuple of *payoff functions*, where $U_i : X \to \mathbb{R}$ determines Player i's payoff arising from the game's outcome. Let $x_i, x_i' \in X_i$ be two strategies available to Player i. We say that x_i *dominates* x_i' iff $U_i(x_i, x_{-i}) \geq U_i(x_i', x_{-i})$ for every $x_{-i} \in X_{-i}$ and there exists at least one x_{-i} for which a strict inequality holds. x_i is the *dominant* strategy for Player i if it dominates every other strategy $x_i' \in X_i \backslash \{x_i\}$. x_i is a *non-dominated* strategy if no other strategy dominates it. By $X^* = X_1^* \times \ldots \times X_n^*$ we will denote the set of non-dominated strategy profiles, where X_i^* is the set of non-dominated strategies available to the individual Player i. The set of *best responses* $B_i(x_{-i})$ for Player i given the other players' actions is defined as $B_i(x_{-i}) := \{x_i | \arg\max_{x_i \in X_i} U_i(x_i, x_{-i})\}$. A *Nash equilibrium* is a strategy profile $x \in X$ such that for all $i \in N$, $x_i \in B_i(x_{-i})$.

Mechanism Design by Creditability. This paper acts on the classic assumption that players are rational and always choose a non-dominated strategy. Additionally, it is assumed that players do not cooperate. We examine the impact of payments to players offered by a *mechanism designer* (an interested third party) who seeks to influence the outcome of a game. These payments are described by a tuple of non-negative payoff functions $V = (V_1, V_2, \ldots, V_n)$, where $V_i : X \to \mathbb{R}^+$, i.e. the payments depend on the strategy Player i selects as well as on the choices of all other players. Thereby, we assume that the players trust the mechanism designer to finally pay the promised amount of money, i.e., consider her trustworthy (*mechanism design by creditability*). The original game $G = (N, X, U)$ is modified to $G(V) := (N, X, [U + V])$ by these payments, where $[U + V]_i(x) = U_i(x) + V_i(x)$, that is, each Player i obtains the payoff of V_i in addition to the payoffs of U_i. The players' choice of strategies changes accordingly: Each player now selects a non-dominated strategy in $G(V)$. Henceforth, the set of non-dominated strategy profiles of $G(V)$ is denoted by $X^*(V)$. A *strategy profile set* – also called *strategy profile region* – $O \subseteq X$ of G is a subset of all strategy profiles X, i.e., a region in the payoff matrix consisting of one or multiple strategy profiles. Similarly to X_i and X_{-i}, we define $O_i := \{x_i | \exists x_{-i} \in X_{-i} \text{ s.t. } (x_i, x_{-i}) \in O\}$ and $O_{-i} := \{x_{-i} | \exists x_i \in X_i \text{ s.t. } (x_i, x_{-i}) \in O\}$.

The mechanism designer's main objective is to force the players to choose a certain strategy profile or a set of strategy profiles. For a desired strategy profile region O, we say that payments V *implement* O if $\emptyset \subset X^*(V) \subseteq O$. V is called a *k-implementation* if, in addition $\sum_{i=1}^n V_i(x) \leq k, \forall x \in X^*(V)$. That is, the players' non-dominated strategies are within the desired strategy profile, and the payments do not exceed k for any possible outcome. Moreover, V is an *exact k-implementation* of O if $X^*(V) = O$ and $\sum_{i=1}^n V_i(x) \leq k \ \forall x \in X^*(V)$. The *cost* $k(O)$ of implementing O is the lowest of all non-negative numbers q for which there exists a q-implementation. If an implementation meets this lower bound, it is optimal, i.e., V is an *optimal implementation* of O if V implements O and $\max_{x \in X^*(V)} \sum_{i=1}^n V_i(x) = k(O)$. The cost $k^*(O)$ of implementing O exactly is the smallest non-negative number q for which there exists an exact q-implementation of O. V is an *optimal exact implementation* of O if it implements O exactly and requires cost $k^*(O)$. The set of all implementations of O will be denoted by $\mathcal{V}(O)$, and the set of all exact implementations of O by $\mathcal{V}^*(O)$. Finally, a strategy profile region $O = \{z\}$ of cardinality one – consisting of only one strategy profile – is called a *singleton*. Clearly, for singletons it holds that non-exact and exact k-implementations are equivalent. For simplicity's sake we often write z instead of $\{z\}$ and $V(z)$ instead of $\sum_{i \in N} V_i(z)$. Observe that only subsets of X which are in $2^{X_1} \times 2^{X_2} \times \ldots \times 2^{X_n} \subset 2^{X_1 \times X_2 \times \ldots \times X_n}$ can be implemented exactly. We call such a subset of X a *convex strategy profile region*.[1]

3   Algorithms and Analysis

3.1   Exact Implementation

Algorithm and Complexity. Recall that in our model each player classifies the strategies available to her as either dominated or non-dominated. Thereby, each dominated strategy $x_i \in X_i \backslash X_i^*$ is dominated by at least one non-dominated strategy $x_i^* \in X_i^*$. In other words, a game determines for each Player i a relation M_i^G from dominated to non-dominated strategies $M_i^G : X_i \backslash X_i^* \to X_i^*$, where $M_i^G(x_i) = x_i^*$ states that $x_i \in X_i \backslash X_i^*$ is dominated by $x_i^* \in X_i^*$. See Fig. 1 for an example.

When implementing a strategy profile region O exactly, the mechanism designer creates a modified game $G(V)$ with a new relation $M_i^V : X_i \setminus O_i \to O_i$ such that all strategies outside O_i map to at least one strategy in O_i. Therewith, the set of all newly non-dominated strategies of Player i must constitute O_i. As every $V \in \mathcal{V}^*(O)$ determines a set of relations $M^V := \{M_i^V : i \in N\}$, there must be a set M^V for every V implementing O optimally as well. If we are given such an optimal relation set M^V without the corresponding optimal exact implementation, we can compute a

---

[1] These regions define a convex area in the n-dimensional hyper-cuboid, provided that the strategies are depicted such that all o_i are next to each other.

V with minimal payments and the same relation M^V, i.e., given an optimal relation we can find an optimal exact implementation. As an illustrating example, assume an optimal relation set for G with $M_i^G(x_{i1}^*) = o_i$ and $M_i^G(x_{i2}^*) = o_i$. Thus, we can compute V such that o_i must dominate x_{i1}^* and x_{i2}^* in $G(V)$, namely, the condition $U_i(o_i, o_{-i}) + V_i(o_i, o_{-i}) \geq \max_{s \in (x_{i1}^*, x_{i2}^*)}(U_i(s, o_{-i}) + V_i(s, o_{-i}))$ must hold $\forall o_{-i} \in O_{-i}$. In an optimal implementation, Player i is not offered payments for strategy profiles of the form $(\bar{o}_i, x_{-i})$ where $\bar{o}_i \in X_i \backslash O_i$, $x_{-i} \in X_{-i}$. Hence, the condition above can be simplified to

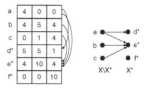

Fig. 1. A single player's game's view and its domination relation M^G

$V_i(o_i, o_{-i}) = \max(0, \max_{s \in \{x_{i1}^*, x_{i2}^*\}}(U_i(s, o_{-i}))) - U_i(o_i, o_{-i})$. Let $S_i(o_i) := \{s \in X_i \backslash O_i | M_i^V(s) = o_i\}$ be the set of strategies where M^V corresponds to an optimal exact implementation of O. Then, an implementation V with $V_i(\bar{o}_i, x_{-i}) = 0$, $V_i(o_i, \bar{o}_{-i}) = \infty$ for any Player i, and $V_i(o_i, o_{-i}) = \max\{0, \max_{s \in S_i(o_i)}(U_i(s, o_{-i}))\} - U_i(o_i, o_{-i})$ is an optimal exact implementation of O as well. Therefore, the problem of finding an optimal exact implementation V of O corresponds to the problem of finding an optimal set of relations $M_i^V : X_i \backslash O_i \to O_i$.

Our algorithm $\mathcal{ALG}_{exact}$ (cf. Algorithm 1) exploits this fact and constructs an implementation V for all possible relation sets, checks the cost that V would entail and returns the lowest cost found.

Algorithm 1 Exact k-Implementation ($\mathcal{ALG}_{exact}$)

Input: Game G, convex region O with $O_{-i} \subset X_{-i} \forall i$
Output: $k^*(O)$
1: $V_i(x) := 0$, $W_i(x) := 0 \ \forall x \in X$, $i \in N$;
2: $V_i(o_i, \bar{o}_{-i}) := \infty \ \forall i \in N$, $o_i \in O_i$, $\bar{o}_{-i} \in X_{-i} \backslash O_{-i}$;
3: compute X^*;
4: **return** ExactK(V, 1);

ExactK(V, i):
Input: payments V, current Player i
Output: minimal r s.t. $\exists$ exact r-implementation $W \in \{W | W(x) \geq V(x) \ \forall x \in X\}$
1: **if** $|X_i^*(V) \backslash O_i| > 0$ **then**
2:     $s :=$ any strategy in $X_i^*(V) \backslash O_i$; $k_{best} := \infty$;
3:     **for all** $o_i \in O_i$ **do**
4:         **for all** $o_{-i} \in O_{-i}$ **do**
5:             $W(o_i, o_{-i}) := \max(0, U_i(s, o_{-i}) - (U_i(o_i, o_{-i}) + V(o_i, o_{-i})))$;
6:         $k := $ ExactK($V + W$, i);
7:         **if** $k < k_{best}$ **then**
8:             $k_{best} := k$;
9:         **for all** $o_{-i} \in O_{-i}$ **do**
10:            $W(o_i, o_{-i}) := 0$;
11:     **return** k_{best};
12: **else if** $i < n$ **then**
13:     **return** $ExactK(V, i+1)$;
14: **else**
15:     **return** $\max_{o \in O} \sum_i V_i(o)$;

Algorithm 2 Exact 0-Implementation ($\mathcal{ALG}_{bankrupt}$)

Input: Game G, convex region O with $O_{-i} \subset X_{-i} \forall i$
Output: $\top$ if $k^*(O) = 0$, $\bot$ otherwise
1: compute X^*;
2: **for all** $i \in N$ **do**
3:     **for all** $s \in X_i^* \backslash O_i$ **do**
4:         dZero $:= \bot$;
5:         **for all** $o_i \in O_i$ **do**
6:             b $:= \top$;
7:             **for all** $o_{-i} \in O_{-i}$ **do**
8:                 b $:=$ b $\wedge (U_i(s, o_{-i}) \leq U_i(o_i, o_{-i}))$;
9:             dZero $:=$ dZero $\vee$ b;
10:         **if** $\neg$ dZero **then**
11:             **return** $\bot$;
12: **return** true;

Theorem 1. $\mathcal{ALG}_{exact}$ *computes a strategy profile region's optimal exact implementation cost in time* $O\left(n|X|^2 + n|O|(\max_{i \in N}|O_i|^{n \max_{i \in N}|X_i^*|})\right)$.

Note that $\mathcal{ALG}_{exact}$ has a large time complexity. In fact, a faster algorithm for this problem, called *Optimal Perturbation Algorithm* has been presented in [10]. In a nutshell, this algorithm proceeds as follows: After initializing V similarly to our algorithm, the values of the region O in the matrix V are increased slowly for every Player i, i.e., by all possible differences between an agent's payoffs in the original game. The algorithm terminates as soon as all strategies in $X_i^* \setminus O_i$ are dominated. Unfortunately, this algorithm does not always return an optimal implementation. Sometimes, as we show in Appendix A of the full version, the optimal perturbation algorithm increases the values unnecessarily. In fact, we even conjecture that deciding whether an k-exact implementation exists is **NP**-hard.

Conjecture 1. Finding an optimal exact implementation of a strategy region is **NP**-hard.

Bankrupt Mechanism Designers. Imagine a mechanism designer who is broke. At first sight, it seems that without any money, she will hardly be able to influence the outcome of a game. However, this intuition ignores the power of creditability: a game can have 0-implementable regions.

Let V be an exact implementation of O with exact costs $k^*(O)$. It holds that if $k^*(O) = 0$, V cannot contain any payments larger than 0 in O. Consequently, for an region O to be 0-implementable exactly, any strategy s outside O_i must be dominated within the range of O_{-i} by a o_i, or there must be one o_i for which no payoff $U_i(s, o_{-i})$ is larger than $U_i(o_i, o_{-i})$. In the latter case, the strategy o_i can still dominate s by using a payment $V(o_i, x_{-i})$ with $x_{-i} \in X_{-i} \setminus O_{-i}$ outside O. Note that this is only possible under the assumption that $O_{-i} \subset X_{-i} \ \forall i \in N$.

$\mathcal{ALG}_{bankrupt}$ (cf. Algorithm 2) describes how a bankrupt designer can decide in polynomial time whether a certain region is 0-implementable. It proceeds by checking for each Player i if the strategies in $X_i^* \setminus O_i$ are dominated or "almost" dominated within the range of O_{-i} by at least one strategy inside O_i. If there is one strategy without such a dominating strategy, O is not 0-implementable exactly. On the other hand, if for every strategy $s \in X_i^* \setminus O_i$ such a dominating strategy is found, O can be implemented exactly without expenses.

Theorem 2. *Given a convex strategy profile region O where $O_{-i} \subset X_{-i} \ \forall i$, Algorithm $\mathcal{ALG}_{bankrupt}$ decides whether O has an exact 0-implementation in time $O\left(n |X|^2\right)$.*

Best Response Graphs. Best response strategies maximize the payoff for a player given the other players' decisions. For now, let us restrict our analysis to games where the sets of best response strategies consist of only one strategy for each $x_{-i} \ \forall i \in N$. Given a game G, we construct a directed *best response graph* $\mathcal{G}_G$ with vertices v_x for strategy profiles $x \in X$ iff x is a best response for at least one player, i.e., if $\exists i \in N$ such that $x_i \in B_i(x_{-i})$. There is a directed edge $e = (v_x, v_y)$ iff $\exists i \in N$ such that $x_{-i} = y_{-i}$ and $\{y_i\} = B_i(y_{-i})$. In other words, an edge from v_x to v_y, indicates that it is better to play y_i instead of x_i for a player if for the other players' strategies $x_{-i} = y_{-i}$. A strategy profile region $O \subset X$ has a *corresponding subgraph* $\mathcal{G}_{G,O}$ containing the vertices $\{v_x | x \in O\}$ and the edges which both start and end in a vertex of the subgraph. We say $\mathcal{G}_{G,O}$ has an *outgoing edge* $e = (v_x, v_y)$ if $x \in O$ and $y \notin O$. Note that outgoing edges are not in the edge set of $\mathcal{G}_{G,O}$. Clearly, it holds that if a singleton x's corresponding subgraph $\mathcal{G}_{G,\{x\}}$ has no outgoing edges then x is a *Nash equilibrium*. More generally, we make the following observation.

Theorem 3. *Let G be a game and $|B_i(x_{-i})| = 1 \ \forall i \in N, x_{-i} \in X_{-i}$. If a convex region O has an exact 0-implementation, then the corresponding subgraph $\mathcal{G}_{G,O}$ in the game's best response graph has no outgoing edges.*

In order to extend best response graphs to games with multiple best responses, we modify the edge construction as follows: In the general best response graph $\mathcal{G}_G$ of a game G there is a directed edge $e = (v_x, v_y)$ iff $\exists i \in N$ s.t. $x_{-i} = y_{-i}$, $y_i \in B_i(y_{-i})$ and $|B_i(y_{-i})| = 1$.

Corollary 1. *Theorem 3 holds for arbitrary games.*

Note that Theorem 3 is a generalization of Monderer and Tennenholtz' Corollary 1 in [10]. They discovered that for a singleton x, it holds that x has a 0-implementation if and only if x is a Nash equilibrium. While their observation covers the special case of singleton-regions, our theorem holds for any strategy profile region. Unfortunately, for general regions, one direction of the equivalence holding for singletons does not hold anymore due to the fact that 0-implementable regions O must contain a player's best response to any o_{-i} but they need not contain best responses exclusively.

Fig. 2. Sample game G with best response graph $\mathcal{G}_G$. The Nash equilibrium in the bottom left corner has no outgoing edges. The dotted arrows do not belong to the edge set of $\mathcal{G}_G$ as the row has multiple best responses.

3.2 Non-exact Implementation

In contrast to exact implementations, where the complete set of strategy profiles O must be non-dominated, the additional payments in non-exact implementations only have to ensure that a *subset* of O is the newly non-dominated region. Obviously, it matters which subset this is. Knowing that a subset $O' \subseteq O$ bears optimal costs, we could find $k(O)$ by computing $k^*(O')$. Apart from the fact that finding an optimal implementation includes solving the – believed to be **NP**-hard – optimal exact implementation cost problem for at least one subregion of O, finding this subregion might also be **NP**-hard since there are exponentially many possible subregions. In fact, a reduction from the SAT problem is presented in [10]. The authors show how to construct a 2-person game in polynomial time given a CNF formula such that the game has a 2-implementation if and only if the formula has a satisfying assignment. However, their proof is not correct: While there indeed exists a 2-implementation for every satisfiable formula, it can be shown that 2-implementations also exist for non-satisfiable formulas. E.g., strategy profiles $(x_i, x_i) \in O$ are always 1-implementable. Unfortunately, we were not able to correct their proof. However, we conjecture the problem to be **NP**-hard, i.e., we assume that no algorithm can do much better than performing a brute force computation of the exact implementation costs (cf. Algorithm 3.1) of all possible subsets, unless **NP = P**.

Conjecture 2. Finding an optimal implementation of a strategy region is **NP**-hard.

For the special case of zero cost regions, Theorem 3 implies the following result.

Corollary 2. *If a strategy profile region O has zero implementation cost then the corresponding subgraph $\mathcal{G}_{G,O}$ in the game's best response graph contains a subgraph $\mathcal{G}_{G,O'}, O' \subseteq O$, with no outgoing edges.*

Corollary 2 is useful to a bankrupt mechanism designer since searching the game's best response graph for subgraphs without outgoing edges helps her spot candidates for regions which can be implemented by mere creditability. In general though, the fact that finding optimal implementations seems computationally hard raises the question whether there are polynomial time algorithms achieving good approximations. As mentioned in Section 3.1, each V implementing a region O defines a domination relation $M_i^V : X_i \setminus O_i \to O_i$. This observation leads to the idea of designing heuristic algorithms that find a correct implementation by establishing a corresponding relation set $\{M_1, M_2, \ldots, M_n\}, M_i : X_i^* \setminus O_i \mapsto O_i$ where each $x_i^* \in X_i^* \setminus O_i$ maps to at least one $o_i \in O_i$. These algorithms are guaranteed to find a correct implementation of O, however, the corresponding implementations may not be cost-optimal.

Our greedy algorithm $\mathcal{ALG}_{greedy}$ (cf. Algorithm 3) associates each strategy x_i^* yet to be dominated with the o_i with minimal distance Δ_G to x_i^*, i.e., the maximum value that has to be added to $U_i(x_i', x_{-i})$ such that x_i' dominates x_i: $\Delta_G(x_i, x_i') := \max_{x_{-i} \in X_{-i}} \max(0, U_i(x_i, x_{-i}) - U_i(x_i', x_{-i}))$. Similarly to the greedy approximation algorithm for the *set cover problem* [7,8] which chooses in each step the subset covering the most elements not covered already, $\mathcal{ALG}_{greedy}$ selects a pair of (x_i^*, o_i) such that by dominating x_i^* with o_i, the number of strategies in $X_i^* \setminus O_i$ that will be dominated therewith is maximal. Thus, in each step there will be an o_i assigned to dominate x_i^* which has minimal dominating cost. Additionally, $\mathcal{ALG}_{greedy}$ takes any opportunity to dominate multiple strategies. $\mathcal{ALG}_{greedy}$ is described in detail in Algorithm 3.2. It returns an implementation V of O; to determine V's cost, one needs to compute $\max_{x^* \in X^*(V)} \sum_{i \in N} V_i(x^*)$.

Theorem 4. $\mathcal{ALG}_{greedy}$ *returns an implementation of a convex strategy profile region* $O \in X$ *in time* $O\left(n |X|^2 + |O| \sum_{i \in N} |X_i^* \setminus O_i|^2 |O_{-i}|\right)$.

$\mathcal{ALG}_{red}$ (cf. Algorithm 4) is a more sophisticated algorithm applying $\mathcal{ALG}_{greedy}$. Instead of terminating when the payment matrix V implements O, this algorithm continues to search for a payment matrix inducing even less cost. It uses $\mathcal{ALG}_{greedy}$ to approximate the cost repeatedly, varying the region to be implemented. As $\mathcal{ALG}_{greedy}$ leaves the while loop if $X_i^*(V) \subseteq O_i$, it might miss out on cheap implementations where $X_i^*(V) \subseteq Q_i, Q_i \subset O_i$. $\mathcal{ALG}_{red}$ examines some of these subsets as well by calling $\mathcal{ALG}_{greedy}$ for some Q_i. If we manage to reduce the cost, we continue with

$O_i := Q_i$ until neither the cost can be reduced anymore nor any strategies can be deleted from any O_i.

Theorem 5. $\mathcal{ALG}_{red}$ *returns an implementation of O in time* $O(nT_g|O| \max_{i \in N} |O_i|)$, *where T_g denotes the runtime of $\mathcal{ALG}_{greedy}$.*

Algorithm 3 Greedy Algorithm $\mathcal{ALG}_{greedy}$

Input: Game G, convex target region O
Output: Implementation V of O
1: $V_i(x) := 0; W_i(x) := 0 \ \forall x \in X$, $i \in N$;
2: compute X^*;
3: **for all** $i \in N$ **do**
4: $\quad V_i(o_i, \bar{o}_{-i}) := \infty \ \forall o_i \in O_i$, $\bar{o}_{-i} \in X_{-i} \backslash O_{-i}$;
5: $\quad$ **while** $X_i^*(V) \not\subseteq O_i$ **do**
6: $\quad\quad c_{best} := 0; m_{best} :=$null; $s_{best} :=$null;
7: $\quad\quad$ **for all** $s \in X_i^*(V) \backslash O_i$ **do**
8: $\quad\quad\quad m := \arg\min_{o_i \in O_i}(\Delta_{G(V)}(s, o_i))$;
9: $\quad\quad\quad$ **for all** $o_{-i} \in O_{-i}$ **do**
10: $\quad\quad\quad\quad W(m, o_{-i}) := \max(0, U_i(s, o_{-i}) - (U_i(m, o_{-i}) + V(m, o_{-i})))$;
11: $\quad\quad\quad c := 0$;
12: $\quad\quad\quad$ **for all** $x \in X_i^* \backslash O_i$ **do**
13: $\quad\quad\quad\quad$ **if** m dominates x in $G(V + W)$ **then**
14: $\quad\quad\quad\quad\quad c + +$;
15: $\quad\quad\quad$ **if** $c > c_{best}$ **then**
16: $\quad\quad\quad\quad c_{best} := c$; $m_{best} := m$; $s_{best} := s$;
17: $\quad\quad$ **for all** $o_{-i} \in O_{-i}$ **do**
18: $\quad\quad\quad V(m_{best}, o_{-i}) += \max(0, U_i(s_{best}, o_{-i}) - (U_i(m_{best}, o_{-i}) + V(m_{best}, o_{-i})))$;
19: **return** V;

Algorithm 4 Reduction Algorithm $\mathcal{ALG}_{red}$

Input: Game G, convex target region O
Output: Implementation V of O
1: $[k, V] := greedy(G, O)$;
2: $k_{temp} := -1; c_i := \bot \ \forall i; T_i := \{\}$;
3: **while** $(k > 0) \wedge (\exists i : |O_i| > 1) \wedge (\exists i : O_i \not\subseteq T_i)$ **do**
4: $\quad$ **for all** $i \in N$ **do**
5: $\quad\quad x_i := \arg\min_{o_i \in O_i} (\max_{o_{-i} \in O_{-i}} U_i(o_i, o_{-i}))$;
6: $\quad\quad$ **if** $(O_i \not\subseteq T_i) \wedge \neg(\forall j : |T_j| = 0 \vee c_j) \wedge (x_i \in T_i)$ **then**
7: $\quad\quad\quad x_i := \arg\min_{o_i \in O_i \backslash \{x_i\}} (\max_{o_{-i} \in O_{-i}} (U_i(o_i, o_{-i})))$;
8: $\quad\quad$ **if** $|O_i| > 1$ **then**
9: $\quad\quad\quad O_i := O_i \backslash \{x_i\}$;
10: $\quad\quad\quad [k_{temp}, V] := greedy(G, O)$;
11: $\quad\quad\quad$ **if** $k_{temp} \geq k$ **then**
12: $\quad\quad\quad\quad O_i := O_i \cup \{x_i\}; T_i := T_i \cup \{x_i\}; c_i := \bot$;
13: $\quad\quad\quad$ **else**
14: $\quad\quad\quad\quad k := k_{temp}; T_i := \{\} \ \forall i; c_i := \top$;
15: **return** V;

An alternative heuristic algorithm for computing a region O's implementation cost retrieves the region's cheapest singleton, i.e., $\min_{o \in O} k(o)$, where a singleton's implementation cost is $k(o) = \min_{o \in O} \sum_{i \in N} \max_{x_i \in X_i} (U_i(x_i, o_{-i}) - U_i(o_i, o_{-i}))$ [10]. The best singleton heuristic algorithm performs quite well for randomly generated games as our simulations reveal (cf. Section 4), but it can result in an arbitrarily large k in the worst case: Fig. 3 depicts a game where each singleton o in the region O consisting of the four bottom left profiles has cost $k(o) = 11$ whereas V implements O at cost 2.

$G = $ 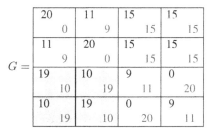 $V = $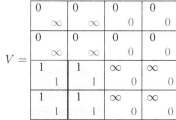

Fig. 3. 2-player game where O's optimal implementation V yields a region $|X^*(V)| > 1$

4 Simulation

All our algorithms return correct implementations of the desired strategy profile sets and – apart from the recursive algorithm $\mathcal{ALG}_{exact}$ for the optimal exact implementation

– run in polynomial time. In order to study the quality of the resulting implementations, we performed several simulations comparing the implementation costs computed by the different algorithms. We have focused on two-person games using random game tables where both players have payoffs chosen uniformly at random from the interval $[0, max]$, for some constant max.

We can modify an implementation V of O, which yields a subset of O, without changing any entry $V_i(o), o \in O$, such that the resulting V implements O *exactly*.

Theorem 6. *If $O_{-i} \subset X_{-i} \; \forall i \in N$, it holds that $k^*(O) \leq \max_{o \in O} V(o)$ for an implementation V of O.*

Theorem 6 enables us to use $\mathcal{ALG}_{greedy}$ for an exact cost approximation by simply computing $\max_{o \in O} V(o)$ instead of $\max_{x \in X^*(V)} V(x)$.

Non-Exact Implementation. We observe that implementing the best singleton often yields low costs. In other words, especially when large sets have to be implemented, our greedy algorithms tend to implement too many strategy profiles and consequently incur unnecessarily high costs. On average, the singleton algorithm performed much better than the other two, with $\mathcal{ALG}_{greedy}$ being the worst of the candidates. We presume that the $\mathcal{ALG}_{red}$ might improve relatively to the best singleton heuristic algorithm for larger player sets.

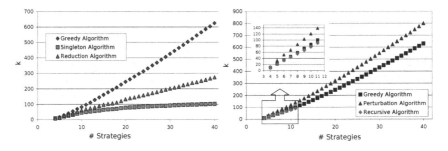

Fig. 4. The average implementation cost k of sets O over 100 random games where $|O_i| = \lfloor n/3 \rfloor$ (left: Non-exact, right: exact). The utility values are chosen uniformly at random from $[0, 20]$. For different intervals we obtain approximately the same result when normalizing k with the maximal possible value.

Exact Implementation. Due to the large runtime of $\mathcal{ALG}_{exact}$, we were only able to compute k for a small number of strategies. However, for these cases, our simulations reveals that $\mathcal{ALG}_{greedy}$ often finds implementations which are close to optimal and is better than the perturbation algorithm. For different payoff value intervals $[0, max]$, we observe a faster increase in k than in the non-exact implementation case. This suggests that implementing a smaller region entails lower costs for random games on average.

Finally, we tested different options to choose the next strategy in Line 8 of $\mathcal{ALG}_{red}$ and $\mathcal{ALG}_{greedy}$. However, none of the alternatives we tested performed better than the ones described in Section 3.

In conclusion, our simulations have shown that for the case of non-exact implementations, there are interesting differences between the algorithms proposed in Section 3. In particular, the additional reductions by $\mathcal{ALG}_{red}$ are beneficial. For the case of exact implementations, our modified greedy algorithm yields good results. As a final remark we want to mention that, although $\mathcal{ALG}_{greedy}$ and $\mathcal{ALG}_{red}$ may find cheap implementations in the average case, there are examples where the approximation ratio of these algorithms is large.

5   Variations

Mechanism design by creditability offers many interesting extensions. In this section, two alternative models of rationality are introduced. If we assume that players do not just select *any* non-dominated strategy, but have other parameters influencing their decision process, our model has to be adjusted. In many (real world) games, players typically do not know which strategies the other players will choose. In this case, a player cannot do better than assume the other players to select a strategy *at random*. If a player wants to maximize her gain, she will take the *average payoff* of strategies into account. This kind of decision making is analyzed in the subsequent section. Afterwards, risk-averse players are examined. Finally, we take a brief look at the dynamics of repeated games with an interested third party offering payments *in each round*.

5.1   Average Payoff Model

As a player may choose any non-dominated strategy, it is reasonable to compute the payoff which each of her strategy will yield *on average*. Thus, assuming no knowledge on the payoffs of the other players, each strategy x_i has an average payoff of $p_i(x_i) := \frac{1}{|X_{-i}|} \sum_{x_{-i} \in X_{-i}} U_i(x_i, x_{-i})$ for Player i. Player i will then select the strategy $s \in X_i$ with the largest $p_i(s)$, i.e., $s = \arg\max_{s \in X_i} p_i(s)$. If multiple strategies have the same average payoff, she plays one of them uniformly at random. For such average strategy games, we say that x_i *dominates* x_i' iff $p_i(x_i) > p_i(x_i')$. Note that with this modified meaning of domination, the region of non-dominated strategies, X^*, differs as well.

The average payoff model has interesting properties, e.g., singleton profiles can be implemented for free.

Theorem 7. *If players maximize their average payoff, singleton strategy profiles are always 0-implementable if there are at least two players with at least two strategies.*

Theorem 7 implies that entire strategy profile regions O are 0-implementable as well: we just have to implement any singleton inside O.

Corollary 3. *In average strategy games where every player has at least two strategies, every strategy profile region can be implemented for free.*

Exact implementations can be implemented at no costs as well.

Theorem 8. *In average strategy games where $O_{-i} \subset X_{-i}$ $\forall i \in N$, each strategy profile region has an exact 0-implementation.*

5.2   Risk-Averse Players

Instead of striving for a high payoff on average, the players might be cautious or *risk-averse*. To account for such behavior, we adapt our model by assuming that the players seek to minimize the risk on missing out on benefits. In order to achieve this objective, they select strategies where the minimum gain is not less than any other strategy's minimum gain. If there is more than one strategy with this property, the risk-averse player can choose a strategy among these, where the average of the benefits is maximal. More formally, let $min_i := \max_{x_i \in X_i}(\min_{x_{-i} \in X_{-i}}(U_i(x_i, x_{-i})))$ and $\oslash_X f(x) := \frac{1}{|X|} \cdot \sum_{x \in X} f(x)$. Then Player i selects a strategy m satisfying $m = \arg\max_{m \in M}(\oslash_{X_{-i}} U_i(m, x_{-i}))$, where $M = \{x_i | \forall x_{-i} \ U_i(x_i, x_{-i}) = min_i\}$.

Theorem 9. *For risk-averse players the implementation cost of a singleton $z \in X$ is* $k(z) = \sum_{i=1}^{N} \max(0, min_i - U_i(z))$

For strategy profile regions, the situation with risk-averse players differs from the standard model considerably.

Theorem 10. *For risk-averse players the implementation cost for a strategy profile region $O \subset X$ is $k(O) = \min_{o \in O} \sum_{i=1}^{n} \max(0, min_i - U_i(o))$.*

In Section 3, we conjectured the problem of computing $k(O)$ to be **NP**-complete for both general and exact implementations. This is not the case for risk-averse players, as the following theorem states.

Theorem 11. $\mathcal{ALG}_{risk}$ *computes* $k(O)$ *in time* $\mathrm{O}(n|X|^2)$, *thus the problem of computing k for risk-averse agents is in* **P**.

Algorithm 5 Risk-averse Players: Exact Implementation

Input: Game G, target region O, $O_i \cap X_i^* = \emptyset \ \forall i \in N$

Output: V

1:   compute X^*;
2:   $V_i(z) = 0$ for all $i \in N, z \in X$;
3:   **for all** $i \in N$ **do**
4:     $V_i(x_i, x_{-i}) := \infty \quad \forall x_i \in O_i, \ x_{-i} \in X_{-i} \setminus O_{-i}$;
5:     $V_i(x_i, x_{-i}) := \max(0, min_i - U_i(x_i, x_{-i})) \quad \forall x_i \in O_i$,
       $x_{-i} \in X_{-i}$;
6:     **if** $O_{-i} = X_{-i}$ **then**
7:       **if** $\tau(O_i) > \tau(X_i^*)$ **then**
8:         **if** $|X_i| + \epsilon|O_i| > |X_i| + \sum_{o_i} \delta(o_i)$ **then**
9:           $V_i(o_i, x_{-i}) := V_i(o_i, x_{-i}) + \delta(o_i) \quad \forall o_i, x_{-i}$;
10:         **else**
11:           $V_i(o_i, x_{-i}) := V_i(o_i, x_{-i}) + \epsilon \quad \forall o_i, x_{-i}$;
12:       **else**
13:         **if** $\epsilon|O_i| > \sum_{o_i}[\epsilon + \delta(o_i)]$ **then**
14:           $V_i(o_i, x_{-i}) := V_i(o_i, x_{-i}) + \epsilon + \delta(o_i) \quad \forall o_i, x_{-i}$;
15:         **else**
16:           $V_i(o_i, x_{-i}) := V_i(o_i, x_{-i}) + \epsilon \quad \forall o_i, x_{-i}$;
17:   **return** V;

5.3   Round-Based Mechanisms

The previous sections dealt with static models only. Now, we extend our analysis to dynamic, round-based games, where the designer offers payments to the players after each round in order to make them change strategies. This opens many questions: For example, imagine a concrete game such as a *network creation game* [6] where all players are stuck in a costly Nash equilibrium. The goal of a mechanism designer could then be to guide the players into another, better Nash equilibrium. Many such extensions are reasonable; due to space constraints, only one model is presented in more detail.

In a dynamic game, we regard a strategy profile as a state in which the participants find themselves. In a network context, each $x \in X$ could represent one particular network topology. We presume to find the game in an initial starting state $s^{T=0} \in X$ and that, in state $s^{T=t}$, each Player i only sees the states she can reach by changing her strategy given the other players remain with their chosen strategies. Thus Player i sees only strategy profiles in $X_{visible,i}^{T=t} = X_i \times \{s_{-i}^{T=t}\}$ in round t. In every round t, the mechanism designer offers the players a payment matrix $V^{T=t}$ (in addition to the game's static payoff matrix U). Then all players switch to their best visible strategy (which is any best response $B_i(s_{-i}^{T=t})$), and the game's state changes to $s^{T=t+1}$. Before the next round starts, the mechanism designer disburses the payments $V^{T=t}(s^{T=t+1})$ offered for the newly reached state. The same procedure is repeated until the mechanism designer decides to stop the game. We prove that a mechanism designer can guide the players to any strategy profile at zero costs in two rounds.

Theorem 12. *Starting in an arbitrary strategy profile, a dynamic mechanism can be designed to lead the players to any strategy profile without any expenses in at most two rounds if $|X_i| \geq 3\ \forall i \in N$.*

6 Conclusions

It is widely believed that live streaming is difficult in heterogeneous networks where some peers have poor Internet connections with upload rates smaller than the streaming rate, or where peers do not upload on purpose because they are selfish. We demonstrated that several fairness mechanisms cause an intolerable number of underflows even if the network consists entirely of honest peers. This paper has proposed a mechanism which provides good streaming quality to those peers which are sufficiently strong by minimizing the influence of weak peers. Moreover, it is not worthwhile for freeloading peers to remain in the network as they are never able to obtain the needed data blocks in time.

References

1. Aspnes, J., Chang, K., Yampolskiy, A.: Inoculation Strategies for Victims of Viruses and the Sum-of-Squares Partition Problem. In: Proc. 16th Annual ACM-SIAM Symposium on Discrete Algorithms (SODA), pp. 43–52. ACM Press, New York (2005)
2. Christodoulou, G., Koutsoupias, E.: The Price of Anarchy of Finite Congestion Games. In: Proc. 37th Annual ACM Symposium on Theory of Computing (STOC), pp. 67–73. ACM Press, New York (2005)
3. Cole, R., Dodis, Y., Roughgarden, T.: How Much Can Taxes Help Selfish Routing? In: Proc. 4th ACM Conference on Electronic Commerce (EC), pp. 98–107. ACM Press, New York (2003)
4. Cole, R., Dodis, Y., Roughgarden, T.: Pricing Network Edges for Heterogeneous Selfish Users. In: Proc. 35th Annual ACM Symposium on Theory of Computing (STOC), pp. 521–530. ACM Press, New York (2003)
5. Dash, R., Parkes, D., Jennings, N.: Computational Mechanism Design: A Call to Arms. In: IEEE Intelligent Systems, IEEE Computer Society Press, Los Alamitos (2003)
6. Fabrikant, A., Luthra, A., Maneva, E., Papadimitriou, C.H., Shenker, S.: On a Network Creation Game. In: Proc. 22nd Annual Symposium on Principles of Distributed Computing (PODC), pp. 347–351 (2003)
7. Johnson, D.S.: Approximation Algorithms for Combinatorial Problems. Journal of Computer and System Sciences 9, 256–278 (1974)
8. Lovász, L.: On the Ratio of Optimal Integral and Fractional Covers. Discrete Mathematics 13, 391–398 (1975)
9. Maskin, E., Sjöström, T.: Handbook of Social Choice and Welfare (Implementation Theory), vol. 1. North-Holland, Amsterdam (2002)
10. Monderer, D., Tennenholtz, M.: k-Implementation. In: Proc. 4th ACM Conference on Electronic Commerce (EC), pp. 19–28. ACM Press, New York (2003)

Infinite Families of Optimal Double-Loop Networks

Xiaoping Dai, Jianqin Zhou, and Xiaolin Wang*

Department of Computer Science, Anhui University of Technology
Ma'anshan, 243002, Anhui, China
xpdai@ahut.edu.cn, zhou9@yahoo.com, wxl@ahut.edu.cn

Abstract. A double-loop network(DLN) $G(N; r, s)$ is a digraph with
the vertex set $V = \{0, 1, \ldots, N - 1\}$ and the edge set $E = \{v \to v + r($
mod $N)$ and $v \to v + s(\mod N)|v \in V \}$. Let $D(N; r, s)$ be the diam-
eter of G, $D(N) = \min\{D(N; r, s)|1 \leq r < s < N$ and $\gcd(N; r, s) =
1\}$ and $D_1(N) = \min\{D(N; 1, s)|1 < s < N\}$. Xu and Aguiló et al.
gave some infinite families of 0-tight non-unit step(nus) integers with
$D_1(N) - D(N) \geq 1$. In this paper, an approach is proposed for finding infinite families of k-tight($k \geq 0$) optimal double-loop networks
$G(N; r, s)$, and two infinite families of k-tight optimal double-loop networks $G(N; r, s)$ are presented. We also derive one infinite family of 1-
tight nus integers with $D_1(N) - D(N) \geq 1$ and one infinite family of
1-tight nus integers with $D_1(N) - D(N) \geq 2$. As a consequence of these
works, some results by Xu are improved.

Keywords: Double-loop network, tight optimal, L-shaped tile, non-unit
step integer.

1 Introduction

Double-loop digraphs $G = G(N; r, s)$, with $1 \leq r < s < N$ and $\gcd(N; r, s) = 1\}$,
have the vertex set $V = \{0, 1, \ldots, N - 1\}$ and the adjacencies are defined by
$v \to v + r(\mod N)$ and $v \to v + s(\mod N)$ for $v \in V$. These kinds of digraphs
have been widely studied as architectures for local area networks, known as
double-loop networks (DLN). For surveys about these networks, refer to [3,7].

From the metric point of view, the minimization of the diameter of G corresponds to a faster transmission of messages in the network. The diameter of G is
denoted by $D(N; r, s)$. As G is vertex symmetric, its diameter can be computed
from the expression $\max\{d(0; i)|i \in V\}$, where $d(u; v)$ is the distance from u to
v in G. For a fixed integer $N > 0$, the optimal value of the diameter is denoted
by

$$D(N) = \min\{D(N; r, s)|1 \leq r < s < N \text{ and } \gcd(N; r, s) = 1\}$$

* This research is supported by Chinese Natural Science Foundation (No. 60473142)
and Natural Science Foundation of Anhui Education Bureau of China (No.
2006KJ238B).

A. Dress, Y. Xu, and B. Zhu (Eds.): COCOA 2007, LNCS 4616, pp. 220–229, 2007.

Several works studied the minimization of the diameter (for a fixed N) with $r = 1$. Let us denote

$$D_1(N) = \min\{D(N; 1, s)|1 < s < N\}$$

Since the work of Wong and Coppersmith [10], a sharp lower bound is known for $D_1(N)$:

$$D_1(N) \geq \lceil \sqrt{3N} \rceil - 2 = lb(N)$$

Fiol et al. in [8] showed that $lb(N)$ is also a sharp lower bound for $D(N)$. A given DLN $G(N; r, s)$ is called k-tight if $D(N; r, s) = lb(N) + k(k \geq 0)$. A k-tight DLN is called optimal if $D(N) = lb(N) + k(k \geq 0)$, hence integer N is called k-tight optimal. The 0-tight DLN are known as tight ones and they are also optimal. A given DLN $G(N; 1, s)$ is called k-tight if $D(N; 1, s) = lb(N) + k(k \geq 0)$. A k-tight DLN is called optimal if $D_1(N) = lb(N) + k(k \geq 0)$.

The metrical properties of $G(N; r, s)$ are fully contained in its related L-shaped tile $L(N; l, h, x, y)$, where $N = lh - xy, l > y$ and $h \geq x$. In Figure 1, we illustrate generic dimensions of an L-shaped tile.

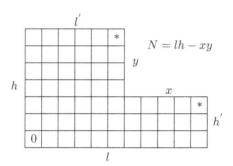

Fig. 1. Generic dimensions of an L-shaped tile

Let $D(L) = D(L(N; l, h, x, y)) = \max\{l + h - x - 2, l + h - y - 2\}$. For obvious reasons, the value $D(L)$ is called the diameter of the tile L. It is known that an L-shaped tile $L(N; l, h, x, y)$ can be assigned to a $G(N; r, s)$ without any confusion. However, we can not find double-loop network $G(N; r, s)$ from some L-shaped tiles. When an L-shaped tile $L(N; l, h, x, y)$ has diameter $lb(N) + k$, we say it is k-tight.

It is known that finding infinite families of k-tight optimal double-loop networks $G(N; r, s)$ is a difficult task as the value k increases. In this paper, an approach is proposed for finding infinite families of k-tight($k \geq 0$) optimal double-loop networks $G(N; r, s)$, and two infinite families of k-tight optimal double-loop networks $G(N; r, s)$ are presented.

Although the identity $D(N) = D_1(N)$ holds for infinite values of N, there are also another infinite set of integers with $D(N) < D_1(N)$. These other integral values of N are called non-unit step integers or nus integers.

Xu [11] presented three infinite families of 0-tight nus integers with $D_1(N) - D(N) \geq 1$. Aguiló et al. [2] derived a method for finding infinite families of nus integers and then presented some infinite families of 0-tight nus integers with $D_1(N) - D(N) \geq 1$. It is known that finding infinite families of nus integers with $D_1(N) - D(N) \geq k$ is a extremely difficult task as the value k increases. In this paper, we derive one infinite family of 1-tight nus integers with $D_1(N) - D(N) \geq 1$ and one infinite family of 1-tight nus integers with $D_1(N) - D(N) \geq 2$. As a consequence of these works, some results in [11] are improved.

2   Preliminary

The following Lemma 1, 2, 3 and 4 can be found in [6 or 8 or 9].

Lemma 1[6,9]. Let t be a nonnegative integer. We define $I_1(t) = [3t^2+1, 3t^2+2t]$, $I_2(t) = [3t^2 + 2t + 1, 3t^2 + 4t + 1]$ and $I_3(t) = [3t^2 + 4t + 2, 3(t + 1)^2]$. Then we have $[4, 3T^2 + 6T + 3] = \bigcup_{t=1}^{T} \bigcup_{i=1}^{3} I_i(t)$, where $T > 1$, and $lb(N) = 3t + i - 2$ if $N \in I_i(t)$ for $i = 1, 2, 3$.

Lemma 2[8,11]. Let $L(N; l, h, x, y)$ be an L-shaped tile, $N = lh - xy$. Then
(a) There exists $G(N; 1, s)$ realizing the L-shaped tile iff $l > y$, $h \geq x$ and $\gcd(h, y) = 1$, where $s \equiv \alpha l - \beta(l - x)(\mod N)$ for some integral values α and β satisfying $\alpha y + \beta(h - y) = 1$.
(b) There exists $G(N; s_1, s_2)$ realizing the L-shaped tile iff $l > y$, $h \geq x$ and $\gcd(l, h, x, y) = 1$, where $s_1 \equiv \alpha h + \beta y(\mod N)$, $s_2 \equiv \alpha x + \beta l(\mod N)$ for some integral values α and β satisfying $\gcd(N, s_1, s_2) = 1$.

Lemma 3 [9]. Let $L(N; l, h, x, y)$ be an L-shaped tile, $N = lh - xy$. Then
(a) If $L(N; l, h, x, y)$ is realizable, then $|y - x| < \sqrt{N}$;
(b) If $x > 0$ and $|y - x| < \sqrt{N}$, then
$$D(L(N; l, h, x, y)) \geq \sqrt{3N - \tfrac{3}{4}(y - x)^2} + \tfrac{1}{2}|y - x| - 2 ;$$
(c) Let $f(z) = \sqrt{3N - \tfrac{3}{4}z^2} + \tfrac{1}{2}z$. Then $f(z)$ is strictly increasing when $0 \leq z \leq \sqrt{N}$.

Lemma 4 [9]. Let $N(t) = 3t^2 + At + B \in I_i(t)$ and L be the L-shaped tile $L(N(t); l, h, x, y)$, where A and B are integral values; $l = 2t + a$, $h = 2t + b$, $z = |y-x|$, a, b, x, y are all integral polynomials of variable t, and $j = i+k(k \geq 0)$. Then L is k-tight iff the following identity holds,
$$(a + b - j)(a + b - j + z) - ab + (A + z - 2j)t + B = 0. \tag{1}$$

The following Lemma 5 is the generalization of Theorem 2 in [12], and can be found in [13].

Lemma 5 [13]. Let $H(z, j) = (2j - z)^2 - 3[j(j - z) + (A + z - 2j)t + B]$, and the identity (1) be an equation of a and b. A necessary condition for the equation (1) to have integral solution is that $4H(z, j) = s^2 + 3m^2$, where s and m are integers.

It is easy to show that the following Lemma 6 is equivalent to Theorem 1 in [12]. Lemma 6 can be found in [13].

Lemma 6 [13]. Let n, s and m be integers, $n = s^2 + 3m^2$. If n has a prime factor p, where $p \equiv 2(\mod 3)$, then there exists an even integer q, such that n is divisible by p^q, but not divisible by p^{q+1}.

3 Infinite Families of k-Tight Optimal Double-Loop Networks

We first describe our approach to generate infinite families of optimal double-loop networks.

Step 1. Find an integer N_0 , such that $G(N_0; s_1, s_2)$ is k-tight optimal $(k \geq 0)$;

Step 2. Find a polynomial $N(t) = 3t^2 + At + B$, such that $N(t_0) = N_0$ and $N(t) \in I_i(t), 1 \leq i \leq 3$;

Step 3. If $G(N_0; s_1, s_2)$ is 0-tight optimal and $A = 2i$, then $A + z - 2j = 0$ if $z = 0$. Find all integral solutions of equation (1), let $S = \{(a,b)|(a + b - j)(a + b - j + z) - ab + (A + z - 2j)t + B = 0$, where $z = 0, j = i\}$. Find all pairs (s_1, s_2) satisfying $\gcd(l, h, x, y) = 1$ and $\gcd(N, s_1, s_2) = 1$, then all $G(N(t); s_1, s_2)$ are infinite families of 0-tight optimal DLN.

Step 4. If $G(N_0; s_1, s_2)$ is 0-tight optimal and $A = 2i - 1$, find all integral solutions of equation (1), let $S = \{(a,b)|(a + b - j)(a + b - j + z) - ab + (A + z - 2j)t + B = 0$, where $t = t_0, z = 0, j = i\}$. Note that $A + z - 2j = -1$, let $(a_0, b_0) \in S, a = a_0, b = f + b_0$, then $t = f^2 + cf + t_0$. Let $l = 2t + a, h = 2t + b, x = t + a + b - i, y = x, \alpha = -1, \beta = 2, s_1 \equiv \alpha h + \beta y(\mod N), s_2 \equiv \alpha x + \beta l(\mod N)$, find integer p, such that when $f = pe, \gcd(l, h, x, y) = 1$ and $\gcd(N, s_1, s_2) = 1$. We have that $G(N(t); s_1, s_2)$ is an infinite family of 0-tight optimal DLN.

Step 5. If $G(N_0; s_1, s_2)$ is k-tight optimal$(k > 0)$. First we ensure that there is no i-tight optimal DLN for $0 \leq i < k$. By Lemma 4,5 and 6, to ensure that there is no i-tight L-shaped tile t must be in the form $t = pe + t_0$. By Lemma 2, to ensure that the i-tight L-shaped tile be not realizable t must be in the form $t = qe + t_0$. Therefore, let $lcm(p, q)$ be the lease common multiple of p and q, then t must be in the form $t = lcm(p, q)e + t_0$. If $A + z - 2j = 0$, continue proceeding in the same way like step 3. If $A + z - 2j \neq 0$, continue proceeding in the same way like step 4, here we need let $a = a_0$, $b = -(A + z - 2j)f + b_0$. Then we will get some infinite families of k-tight optimal DLN.

We now present some application examples to illustrate the above method.

Let $N_0 = 450, N(t) = 3t^2 + 2t - 6$. Then $N(12) = 450$. For $D(N; 3, 35) = lb(450) = 35$, then $G(N; 3, 35)$ is 0-tight optimal. For $i = 1, k = 0, z = 0, A = 2, B = -6$, the equation (1) becomes

$$(a + b - 1)(a + b - 1) - ab - 6 = 0,$$

which has integral solutions:

$$S = \{(-2, 1), (-2, 3), (1, -2), (1, 3), (3, -2), (3, 1)\}$$

Let $l = 2t + a, h = 2t + b, x = t + a + b - 1, y = x, \alpha = -1, \beta = 2, s_1 \equiv \alpha h + \beta y ($ mod $N), s_2 \equiv \alpha x + \beta l ($ mod $N)$. Then all $(a, b) \in S$ satisfy(or partly satisfy) $\gcd(l, h, x, y) = 1$ and $\gcd(N, s_1, s_2) = 1$.

All pairs of (s_1, s_2) corresponding to $(a, b) \in S$ are

$$\begin{aligned}
\{ \ &(-5, 3t - 2)|t \neq 4(\mod 5) \text{ and } t > 4, (-3, 3t - 4)|t > 3, \\
&(-2, 3t + 4)|t \neq 0(\mod 2) \text{ and } t > 4, (3, 3t - 1)|t > 3, \\
&(2, 3t + 6)|t \neq 0(\mod 2) \text{ and } t > 4, \\
&(5, 3t + 3)|t \neq 4(\mod 5) \text{ and } t > 4\}.
\end{aligned} \tag{2}$$

This leads to the following theorem.

Theorem 1. Let $N(t) = 3t^2 + 2t - 6$, (s_1, s_2) belong to the set (2). Then $G(N(t); s_1, s_2)$ is an infinite family of 0-tight optimal DLN.

Proof. Note that $3N(t) - 3/4 = (3t + 1/2)^2 + 3t - 19$. By Lemma 3, if $y - x \geq 1$ and $t > 6$, we have

$$D(N(t)) \geq \sqrt{3N(t) - \frac{3}{4}} > (3t + 1/2) + 1/2 - 2 = 3t - 1 = lb(N(t))$$

Therefore, all the 0-tight L-shaped tile $L(N(t); l, h, x, y)$ must satisfy $y - x = 0$. By Lemma 4, with $i = 1, k = 0, z = 0, A = 2, B = -6$, the equation (1) becomes

$$(a + b - 1)(a + b - 1) - ab - 6 = 0$$

which has integral solutions:

$$S = \{(-2, 1), (-2, 3), (1, -2), (1, 3), (3, -2), (3, 1)\}$$

For any $(a, b) \in S$, let $l = 2t + a, h = 2t + b, x = t + a + b - 1, y = x, L(N(t); l, h, x, y)$ is an L-shaped tile with $N(t) = lh - xy$. By Lemma 2, for any (s_1, s_2) belongs to the set (2), $G(N(t); s_1, s_2)$ is an infinite family of 0-tight optimal DLN.

We have this theorem. □

From Theorem 1, $\{G(3t^2 + 2t - 6; 3, 3t - 1) : t > 3\}$ is an infinite family of 0-tight optimal DLN, which is better than that of Theorem 1 in [11].

Observe that $G(N; 3, 35)$ is 0-tight optimal ensures that the equation (1) has integral solutions.

Note that for some other pairs of (α, β), we can also get many infinite families of 0-tight optimal DLN. For instance, for $(\alpha, \beta) = (1, 1), (a, b) = (-2, 1)$, we have that $G(N(t); 3t - 4, 3t - 1)$ is an infinite family of 0-tight optimal DLN, where $t > 3$.

Let $N_0 = 417289, N(t) = 3t^2 + 6t - 95$. Then $N(372) = 417289$. For

$$D(N; -33, 1165) = D(N) = lb(417289) + 5 = 1122.$$

Thus $G(N; -33, 1165)$ is 5-tight optimal.

For $A = 6, B = -95$, compute $H(z, j) = (2j-z)^2 - 3[j(j-z) + (A+z-2j)t + B]$ respectively:

For $j = 3, z = 0, H(0, 3) = 294 = 2 \times 147$, where 2 has power 1.
For $j = 4, z = 0, t = 372, H(0, 4) = 2533 = 17 \times 149$, where 17 has power 1.
For $j = 4, z = 1, t = 372, H(1, 4) = 1414 = 2 \times 707$, where 2 has power 1.
For $j = 4, z = 2, H(2, 4) = 297 = 11 \times 27$, where 11 has power 1.
For $j = 5, z = 0, t = 372, H(0, 5) = 4474 = 2 \times 2237$, where 2 has power 1.
For $j = 5, z = 1, t = 372, H(1, 5) = 3654 = 2 \times 1827$, where 2 has power 1.
For $j = 5, z = 2, t = 372, H(2, 5) = 2536 = 8 \times 317$, where 2 has power 3.
For $j = 5, z = 3, t = 372, H(3, 5) = 1420 = 5 \times 284$, where 5 has power 1.
For $j = 5, z = 4, H(4, 5) = 306 = 2 \times 153$, where 2 has power 1.
For $j = 6, z = 0, H(0, 6) = 3(6t + 107)$. Suppose that $3(6t + 107) = s^2 + 3m^2$, then $3|s^2$, let $s = 3q$, we have $6t + 107 = 3q^2 + m^2$, so $m^2 \equiv 107(\mod 3) \equiv 2$ ($\mod 3$), which is impossible. Thus $4H(0, 6)$ has no the form of $s^2 + 3m^2$.
For $j = 6, z = 1, t = 372, H(1, 6) = 5896 = 8 \times 737$, where 2 has power 3.
For $j = 6, z = 2, t = 372, H(2, 6) = 4777 = 17 \times 281$, where 17 has power 1.
For $j = 6, z = 3, t = 372, H(3, 6) = 3660 = 5 \times 732$, where 5 has power 1.
For $j = 6, z = 4, t = 372, H(4, 6) = 2545 = 5 \times 509$, where 5 has power 1.
For $j = 6, z = 5, t = 372, H(5, 6) = 1432 = 8 \times 179$, where 2 has power 3.
For $j = 6, z = 6, H(6, 6) = 321 = 107 \times 3$, where 107 has power 1.
For $j = 7, z = 0, t = 372, H(0, 7) = 9262 = 2 \times 4631$, where 2 has power 1.
For $j = 7, z = 1, t = 372, H(1, 7) = 8140 = 5 \times 1628$, where 5 has power 1.
For $j = 7, z = 2, t = 372, H(2, 7) = 7020 = 5 \times 1404$, where 5 has power 1.
For $j = 7, z = 3, t = 372, H(3, 7) = 5902 = 2 \times 2951$, where 2 has power 1.
For $j = 7, z = 4, t = 372, H(4, 7) = 4786 = 2 \times 2393$, where 2 has power 1.
For $j = 7, z = 5, t = 372, H(5, 7) = 3672 = 8 \times 459$, where 2 has power 3.
For $j = 7, z = 6, t = 372, H(6, 7) = 2560 = 5 \times 512$, where 5 5(mod6).
For $j = 7, z = 7, t = 372, H(7, 7) = 1450 = 2 \times 725$, where 2 has power 1.
For $j = 7, z = 8, H(8, 7) = 342 = 2 \times 171$, where 2 has power 1.
Let $t = 16 \times 5^2 \times 17^2 \times e + 372(e \geq 0)$.

For $0 \leq k \leq 4(3 \leq j \leq 7), 0 \leq z = y - x \leq 2k$, by Lemma 6, $4H(z, j)$ has no the form of $s^2 + 3m^2$. By Lemma 5, the equation (1) has no integral solutions of a and b. By Lemma 4, there is no k-tight L-shaped tile $L(N(t); l, h, x, y)$.

For $A = 6, B = -95, j = 8, z = 6$, the equation (1) becomes

$$(a + b - 8)(a + b - 2) - ab - 4t - 95 = 0$$

which has integral solutions for $t = 372$:

$$S = \{(-37, 4), (-37, 43), (4, -37), (4, 43), (43, -37), (43, 4)\}$$

Let $a = 4, b = 4f - 37$, then $t = 4f^2 - 80f + 372$.

Let $\alpha = -1, \beta = 2$, then $s_1 \equiv 4f - 33(\mod N)$, $s_2 = 3t - 4f + 49 = 12f^2 - 244f + 1165$.

Let $l = 2t + a, h = 2t + b, x = t + a + b - 8, y = x + 6$, then $\gcd(l, h, x, y) = 1$. Note that $12f^2 - 244f + 1165 = (3f - 36)(4f - 33) - f - 23$, and $4f - 33 = -4(-f - 23) - 125$. Thus,

$$\begin{aligned} \gcd(N, s_1, s_2) &= \gcd(N, 4f - 33, 24f^2 + 196f + 383) \\ &= \gcd(N, 4f - 33, -f - 23) \\ &= \gcd(N, -125, -f - 23) = 1 \end{aligned}$$

if $f = 5e(e$ is integral and $e \geq 0)$.

This leads to the following theorem.

Theorem 2. Let $N(t) = 3t^2 + 6t - 95$, $s_1 \equiv 4f - 33(\mod N)$, $s_2 = 3t - 4f + 49$, where $t = 4f^2 - 80f + 372$, $f = 5g, g = 2 \times 5 \times 17^2 \times e(e \geq 0)$. Then $G(N(t); s_1, s_2)$ is an infinite family of 5-tight optimal DLN.

The proof is omitted.

4   Infinite Families of Nus Integers

Aguiló et al. [2] gave the definition and the characterization of nus(non unit step) integers, and presented the 1-tight nus integer 2814, where $lb(2814) = 90, D(2814) = 91$, and $D_1(2814) = 92$.

Let $N(t) = 3t^2 + 4t - 6$. Then $N(30) = 2814$. For $A = 4, B = -6, j = 3, z = 2$, the equation (1) becomes $(a + b - 3)(a + b - 1) - ab - 6 = 0$, which has integral solutions: $S = \{(-2, 3), (3, -2), (3, 3)\}$.

Let $l = 2t + a, h = 2t + b, x = t + a + b - 3, y = x + 2, \alpha = -1, \beta = 2, s_1 \equiv \alpha h + \beta y(\mod N)$, $s_2 \equiv \alpha x + \beta l(\mod N)$. Then all $(a, b) \in S$ satisfy(or partly satisfy) $\gcd(, h, x, y) = 1$ and $\gcd(N, s_1, s_2) = 1$.

All pairs of (s_1, s_2) corresponding to $(a, b) \in S$ are

$$\begin{aligned} \{ \ &(-3, 3t - 2)|t > 3, (2, 3t + 8)|t \neq 0(\mod 2) \text{ and } t > 4, \\ &(7, 3t + 3)|t \neq 6(\mod 7) \text{ and } t > 3\} \end{aligned} \tag{3}$$

For $A = 4, B = -6, j = 2, z = 0, H(0, 2) = 22$. By Lemma 6, $4H(0, 2) = 88$ has no the form of $s^2 + 3m^2$. By Lemma 5, the equation (1) has no integral solutions of a and b. By Lemma 4, there is no 0-tight L-shaped tile $L(N(t); l, h, x, y)$.

This leads to the following theorem.

Theorem 3. Let $N(t) = 3t^2 + 4t - 6, (s_1, s_2)$ belong to the set (3). Then $G(N(t); s_1, s_2)$ is an infinite family of 1-tight optimal DLN.

Proof. Note that $3N(t) - 3/4 = (3t + 3/2)^2 + 3t - 21$. By Lemma 3, if $y - x \geq 1$ and $t > 7$, we have

$$D(N(t)) \geq \sqrt{3N(t) - \frac{3}{4}} > (3t + 3/2) + 1/2 - 2 = 3t = lb(N(t))$$

Therefore, all the 0-tight L-shaped tile L(N(t); , h, x, y) must satisfy $y - x = 0$. From $H(0,2) = 22$, there is no 0-tight L-shaped tile $L(N(t); l, h, x, y)$.

Note that $3N(t) - (3/4)3^2 = (3t + 3/2)^2 + 3t - 27$. By Lemma 3, if $y - x \geq 3$ and $t > 9$, we have

$$D(N(t)) \geq \sqrt{3N(t) - \frac{3}{4}3^2} > (3t + 3/2) + 3/2 - 2 = 3t + 1 = lb(N(t)) + 1$$

Therefore, all the 1-tight L-shaped tile $L(N(t); l, h, x, y)$ must satisfy $0 \leq y - x \leq 2$. By Lemma 4, with $A = 4, B = -6, j = 3, z = 2$, the equation (1) becomes $(a + b - 3)(a + b - 1) - ab - 6 = 0$, which has integral solutions: $S = \{(-2, 3), (3, -2), (3, 3)\}$.

By Lemma 1, for any (s_1, s_2) belongs to the set (3), $G(N(t); s_1, s_2)$ is an infinite family of 1-tight optimal DLN.

We have this theorem.   □

Continue the above discussion.

For (-2,3), $\gcd(h, y) = \gcd(2t + b, t + a + b - 1) = \gcd(2t + 3, t) = 3$ if $t \equiv 0$ (mod 3).

For (3,-2), $\gcd(h, y) = \gcd(2t + b, t + a + b - 1) = \gcd(2t - 2, t) = 2$ if $t \equiv 0$ (mod 2).

For (3,3), $\gcd(h, y) = \gcd(2t + b, t + a + b - 1) = \gcd(2t + 3, t + 5) = 7$ if $t \equiv 2$ (mod 7).

By Lemma 2(a) and consider the symmetry of L-shaped tile, for $t = 2 \times 3 \times 7 \times e + 30(e \geq 0)$, there is no $G(N; 1, s)$ realizing the 1-tight L-shaped tile $L(N(t); l, h, x, y)$ where $|y - x| = 2$.

For $A = 4, B = -6, j = 2 + 1, z = 0, t = 30, H(0,3) = 207 = 23 \times 9$, where $23 \equiv 2(mod3)$.

For $A = 4, B = -6, j = 2 + 1, z = 1, t = 30, H(1,3) = 115 = 23 \times 5$.

By Lemma 4,5,6 and consider the symmetry of L-shaped tile, for $t = 23^2 \times e + 30(e \geq 0)$, there is no 1-tight L-shaped tile $L(N(t); l, h, x, y)$ where $|y - x| = 0$ or $|y - x| = 1$.

Combining these arguments with theorem 3 will lead to the following theorem.

Theorem 4. The nodes $N(t) = 3t^2 + 4t - 6, t = 2 \times 3 \times 7 \times 23^2 \times e + 30(e \geq 0)$, of an infinite family of 1-tight optimal DLN correspond to 1-tight nus integers with $D_1(N) - D(N) \geq 1$.

Let $N_0 = 267360, N(t) = 3t^2 + 4t - 244$. Then $N(298) = 267360$. For $D(N; 9, 874) = D(N) = lb(267360) + 1 = 895$, then $G(N; 9, 874)$ is 1-tight optimal. For $A = 4, B = -244, j = 3, z = 2$, the equation (1) becomes $(a + b - 3)(a + b - 1) - ab - 244 = 0$, which has integral solutions:

$$S = \{(-11, -4), (-11, 19), (-4, -11), (-4, 19), (19, -11), (19, -4)\}$$

Let $l = 2t + a, h = 2t + b, x = t + a + b - 3, y = x + 2, \alpha = -1, \beta = 2, s_1 \equiv \alpha h + \beta y$ (mod N), $s_2 \equiv \alpha x + \beta l$(mod N), then all $(a, b) \in S$ satisfy(or partly satisfy) $\gcd(l, h, x, y) = 1$ and $\gcd(N, s_1, s_2) = 1$. All pairs of (s_1, s_2) corresponding to $(a, b) \in S$ are

$$\{ \; (-28, 3t-4)|t \neq 0(\mod 2) \text{ and } t \neq 6(\mod 7) \text{ and } t > 24,$$
$$(-5, 3t-27)|t \neq 4(\mod 5) \text{ and } t > 24,$$
$$(-21, 3t+10)|t \neq 6(\mod 7) \text{ and } t > 24,$$
$$(9, 3t-20)|t > 24, (25, 3t+33)|t \neq 4(\mod 5) \text{ and } t > 24,$$
$$(32, 3t+26)|t \neq 0(\mod 2) \text{ and } t > 24\} \tag{4}$$

For $A = 4, B = -244, j = 2, z = 0, H(0,2) = 736 = 23 \times 32$. Note that $23 \equiv 2(\mod 3)$, there is no 0-tight L-shaped tile $L(N(t); l, h, x, y)$. With a similar argument as Theorem 3, we have the following theorem.

Theorem 5. Let $N(t) = 3t^2 + 4t - 244, (s_1, s_2)$ belong to the set (4). Then $G(N(t); s_1, s_2)$ is an infinite family of 1-tight optimal DLN.

For (-11,-4), $\gcd(h, y) = \gcd(2t+b, t+a+b-1) = \gcd(2t-4, t-16) = 2$ if $t \equiv 0(\mod 2)$

For (-11,19), $\gcd(h, y) = \gcd(2t+b, t+a+b-1) = \gcd(2t+19, t+7) = 5$ if $t \equiv 3(\mod 5)$

For (-4,-11), $\gcd(h, y) = \gcd(2t+b, t+a+b-1) = \gcd(2t-11, t-16) = 3$ if $t \equiv 1(\mod 3)$

For (-4,19), $\gcd(h, y) = \gcd(2t+b, t+a+b-1) = \gcd(2t+19, t+14) = 3$ if $t \equiv 1(\mod 3)$

For (19,-11), $\gcd(h, y) = \gcd(2t+b, t+a+b-1) = \gcd(2t-11, t+7) = 5$ if $t \equiv 3(\mod 5)$

For (19,-4), $\gcd(h, y) = \gcd(2t+b, t+a+b-1) = \gcd(2t-4, t+14) = 2$ if $t \equiv 0(\mod 2)$

By Lemma 2(a) and consider the symmetry of L-shaped tile, for $t = 2 \times 3 \times 5 \times e + 298(e \geq 0)$, there is no $G(N; 1, s)$ realizing the 1-tight L-shaped tile $L(N(t); l, h, x, y)$ where $|y - x| = 2$.

For $A = 4, B = -244, j = 4, z = 2$, the equation (1) becomes $(a+b-4)(a+b-2) - ab - 2t - 244 = 0$.

It is easy to show that both a and b are even. Let $l = 2t+a, h = 2t+b, x = t+a+b-4, y = x+2$. Then $\gcd(l, h, x, y) \geq 2$ if $t \equiv 0(\mod 2)$, so $L(N(t); l, h, x, y)$ is not realizable.

For $j = 3, z = 0, t = 298, H(0,3) = 2529 = 281 \times 9$, where $281 \equiv 2(\mod 3)$.

For $j = 3, z = 1, t = 298, H(1,3) = 1633 = 23 \times 71$, where $23 \equiv 2(\mod 3)$.

For $j = 4, z = 0, t = 298, H(0,4) = 4324 = 23 \times 188$, where $23 \equiv 2(\mod 3)$.

For $j = 4, z = 1, t = 298, H(1,4) = 3427 = 23 \times 149$, where $23 \equiv 2(\mod 3)$.

For $j = 4, z = 3, t = 298, H(3,4) = 1639 = 11 \times 149$, where $11 \equiv 2(\mod 3)$.

For $j = 4, z = 4, H(4,4) = 748 = 11 \times 68$, where $11 \equiv 2(\mod 3)$.

Let $t = 11^2 \times 23^2 \times 281^2 \times e + 298(e \geq 0)$. For $(z, j) \in \{(0,2), (0,3), (1,3), (0,4), (1,4), (3,4), (4,4)\}$, by Lemma 6, $4H(z,j)$ has no the form of $s^2 + 3m^2$. By Lemma 5, the equation (1) has no integral solutions of a and b. By Lemma 4, there is no k-tight L-shaped tile $L(N(t); l, h, x, y)$ for (z, j).

Note that $t = 2 \times 3 \times 5 \times e + 298(e \geq 0)$, for $0 \leq k \leq 2(2 \leq j \leq 4)$, $0 \leq z = y - x \leq 2k$, there is no k-tight L-shaped tile or the k-tight L-shaped tile is not realizable by $G(N; 1, s)$.

Combining these arguments with Theorem 5 will lead to the following theorem.

Theorem 6. The nodes $N(t) = 3t^2 + 4t - 244, t = 2 \times 3 \times 5 \times 11^2 \times 23^2 \times 281^2 \times e + 298(e \geq 0)$, of an infinite family of 1-tight optimal DLN correspond to 1-tight nus integers with $D_1(N) - D(N) \geq 2$.

5 Remarks

In a way similar to those of Theorem 1, let $(a, b) = (4, -1), \alpha = -1, \beta = 2$. Then $\{G(3t^2 + 4t - 5; 3, 3t + 7) : t > 2\}$ is an infinite family of 0-tight optimal DLN, which is better than that of Theorem 2 in [11].

As $t \equiv 3(\mod 4)$ (in Theorem 3 [11]) can be changed to $t \equiv 1(\mod 2)$, similar to the proof of Theorem 3, we can have that the nodes $N(t) = 3t^2 + 4t - 11, t = 42 \times e + 17(e \geq 0)$, of an infinite family of 0-tight optimal DLN correspond to 0-tight nus integers with $D_1(N) - D(N) \geq 1$, which contains the second minimum nus integer 924 as its first element. This result is much better than that of Theorem 3 in [11].

References

1. Aguiló, F., Fiol, M.A.: An efficient algorithm to find optimal double loop networks. Discrete Mathematics 138, 15–29 (1995)
2. Aguiló, F., Simó, E., Zaragozá, M.: Optimal double-loop networks with non-unit steps. The Electronic Journal of Combinatorics 10, #R2 (2003)
3. Bermond, J.-C., Comellas, F., Hsu, D.F.: Distributed loop computer networks: a survey. J. Parallel Distribut. Comput. 24, 2–10 (1995)
4. Chan, C.F., Chen, C., Hong, Z.X.: A simple algorithm to find the steps of double-loop networks. Discrete Applied Mathematics 121, 61–72 (2002)
5. Erdös, P., Hsu, D.F.: Distributed loop networks with minimum transmission delay. Theoret. Comput. Sci. 100, 223–241 (1992)
6. Esqué, P., Aguiló, F., Fiol, M.A.: Double commutative-step digraphs with minimum diameters. Discrete Mathematics 114, 147–157 (1993)
7. Hwang, F.K.: A complementary survey on double-loop networks. Theoret. Comput. Sci. 263, 211–229 (2001)
8. Fiol, M.A., Yebra, J.L.A., Alegre, I., Valero, M.: A discrete optimization problem in local networks and data alignment. IEEE Trans. Comput. C-36, 702–713 (1987)
9. Li, Q., Xu, J., Zhang, Z.: The infinite families of optimal double loop networks. Discrete Applied Mathematics 46, 179–183 (1993)
10. Wong, C.K., Coppersmith, D.: A combinatorial problem related to multimode memory organizations. J. Ass. Comput. Mach. 21, 392–402 (1974)
11. Xu, J.: Designing of optimal double loop networks. Science in China, Series E E-42(5), 462–469 (1999)
12. Xu, J., Liu, Q.: An infinite family of 4-tight optimal double loop networks. Science in China, Series A A-46(1), 139–143 (2003)
13. Zhou, J., Xu, X.: On infinite families of optimal double-loop networks with non-unit steps, Ars Combinatoria (accepted)

Point Sets in the Unit Square and Large Areas of Convex Hulls of Subsets of Points

Hanno Lefmann

Fakultät für Informatik, TU Chemnitz, D-09107 Chemnitz, Germany
lefmann@informatik.tu-chemnitz.de

Abstract. In this paper generalizations of Heilbronn's triangle problem are considered. By using results on the independence number of linear hypergraphs, for fixed integers $k \geq 3$ and any integers $n \geq k$ a $o(n^{6k-4})$ time deterministic algorithm is given, which finds distributions of n points in the unit square $[0,1]^2$ such that, simultaneously for $j = 3, \ldots, k$, the areas of the convex hulls determined by any j of these n points are $\Omega((\log n)^{1/(j-2)}/n^{(j-1)/(j-2)})$.

1 Introduction

Distributions of n points in the unit square $[0,1]^2$ such that the minimum area of a triangle determined by three of these n points is large have been investigated by Heilbronn. Let $\Delta_3(n)$ denote the supremum over all distributions of n points in $[0,1]^2$ of the minimum area of a triangle among n points. Since no three of the points $1/n \cdot (i \bmod n, i^2 \bmod n)$, $i = 0, \ldots, n-1$, are collinear, we infer $\Delta_3(n) = \Omega(1/n^2)$, provided n is prime, as has been observed by Erdös. For a while this lower bound was believed to be also the upper bound. However, Komlós, Pintz and Szemerédi [14] proved that $\Delta_3(n) = \Omega(\log n/n^2)$, see [7] for a deterministic polynomial time algorithm achieving this lower bound. Upper bounds on $\Delta_3(n)$ were given by Roth [19]–[22] and Schmidt [23] and, improving these earlier results, the currently best upper bound $\Delta_3(n) = O(2^{c\sqrt{\log n}}/n^{8/7})$ for a constant $c > 0$, is due to Komlós, Pintz and Szemerédi [13]. We remark that the expected value of the minimum area of a triangle formed by three of n uniformly at random and independently of each other distributed points in $[0,1]^2$ has been shown in [12] to be equal to $\Theta(1/n^3)$.

Variants of Heilbronn's triangle problem in higher dimensions were investigated by Barequet [3,4], who considered the minimum volumes of simplices among n points in the d-dimensional unit cube $[0,1]^d$, see also [15] and Brass [8]. Recently, Barequet and Shaikhet [5] considered the on-line situation, where the points have to be positioned one after the other and suddenly this process stops. For this situation they obtained for the supremum of the minimum volume of $(d+1)$-point simplices among n points in $[0,1]^d$ the lower bound $\Omega(1/n^{(d+1)\ln(d-2)+2})$.

A generalization of Heilbronn's triangle problem to k-gons, see Schmidt [23], asks, given an integer $k \geq 3$, to maximize the minimum area of the convex hull of any k distinct points in a distribution of n points in $[0,1]^2$. In particular, let $\Delta_k(n)$ be the supremum over all distributions of n points in $[0,1]^2$

A. Dress, Y. Xu, and B. Zhu (Eds.): COCOA 2007, LNCS 4616, pp. 230–241, 2007.

of the minimum area of the convex hull determined by some k of n points. For $k = 4$, Schmidt [23] proved the lower bound $\Delta_4(n) = \Omega(1/n^{3/2})$, and in [7] the lower bound $\Delta_k(n) = \Omega(1/n^{(k-1)/(k-2)})$ has been shown for fixed integers $k \geq 3$. Also in [7] a deterministic polynomial time algorithm was given which achieves this lower bound. This has been improved in [16] to $\Delta_k(n) = \Omega((\log n)^{1/(k-2)}/n^{(k-1)/(k-2)})$ for any fixed integers $k \geq 3$.

We remark that for k a function of n, Chazelle proved in [9] in connection with some range searching problems $\Delta_k(n) = \Theta(k/n)$ for $\log n \leq k \leq n$.

In [17] a deterministic polynomial time algorithm has been given, which finds for fixed integers $k \geq 2$ and any integers $n \geq k$ a distribution of n points in the unit square $[0,1]^2$ such that, simultaneously for $j = 2, \ldots, k$, the areas of the convex hulls of any j among the n points are $\Omega((\log n)^{1/(j-2)}/n^{(j-1)/(j-2)})$. Recently, in [18] these (simultaneously achievable) lower bounds on the minimum areas of the convex hull of any j among n points in $[0,1]^2$ have been improved by using (non-discrete) probabilistic arguments by a polylogarithmic factor to $\Omega((\log n)^{1/(j-2)}/n^{(j-1)/(j-2)})$ for $j = 3, \ldots, k$. (Note that $\Delta_2(n) = \Theta(1/n^{1/2})$.) While this was an existence argument, here we give a deterministic polynomial time algorithm, which provides such a configuration of n points in $[0,1]^2$.

Theorem 1. *Let $k \geq 3$ be a fixed integer. For each integer $n \geq k$ one can find deterministically in time $o(n^{6k-4})$ some n points in the unit square $[0,1]^2$ such that, simultaneously for $j = 3, \ldots, k$, the minimum area of the convex hull determined by some j of these n points is $\Omega((\log n)^{1/(j-2)}/n^{(j-1)/(j-2)})$.*

Concerning upper bounds, we remark that for fixed $j \geq 4$ only the simple bounds $\Delta_j(n) = O(1/n)$ are known, compare [23].

2   The Independence Number of Linear Hypergraphs

In our considerations we transform the geometric problem into a problem on hypergraphs.

Definition 1. *A hypergraph is a pair $\mathcal{G} = (V, \mathcal{E})$ with vertex-set V and edge-set $\mathcal{E}$, where $E \subseteq V$ for each edge $E \in \mathcal{E}$. For a hypergraph $\mathcal{G}$ the notation $\mathcal{G} = (V, \mathcal{E}_2 \cup \cdots \cup \mathcal{E}_k)$ means that $\mathcal{E}_i$ is the set of all i-element edges in $\mathcal{G}$, $i = 2, \ldots, k$. A hypergraph $\mathcal{G} = (V, \mathcal{E})$ is called k-uniform if $|E| = k$ for each edge $E \in \mathcal{E}$. The independence number $\alpha(\mathcal{G})$ of $\mathcal{G} = (V, \mathcal{E})$ is the largest size of a subset $I \subseteq V$ which contains no edges from $\mathcal{E}$.*

For hypergraphs $\mathcal{G}$ a lower bound on the independence number $\alpha(\mathcal{G})$ is given by Turán's theorem for hypergraphs, see [24].

Theorem 2. *Let $\mathcal{G} = (V, \mathcal{E}_2 \cup \cdots \cup \mathcal{E}_k)$ be a hypergraph on $|V| = N$ vertices with average degree $t_i^{i-1} := i \cdot |\mathcal{E}_i|/|V|$ for the i-element edges, $i = 2, \ldots, k$. Let $t_{i_0} := \max \{t_i \mid 2 \leq i \leq k\} \geq 1/2$.*

Then, the independence number $\alpha(\mathcal{G})$ of $\mathcal{G}$ satisfies

$$\alpha(\mathcal{G}) \geq N/(4 \cdot t_{i_0}). \tag{1}$$

An independent set $I \subseteq V$ in $\mathcal{G}$ with $|I| \geq N/(4 \cdot t_{i_0})$ can be found deterministically in time $O(|V| + |\mathcal{E}_2| + \cdots + |\mathcal{E}_k|)$.

For fixed positive integers $k \geq 2$ one can show by Theorem 2 and Lemmas 2 and 4 (see below), that one can find deterministically in polynomial time n points in $[0,1]^2$ such that the areas of the convex hulls of any j of these n points are $\Omega(1/n^{(j-1)/(j-2)})$ simultaneously for $j = 2, \ldots, k$, compare [17]. However, we want to obtain better lower bounds. To achieve this, we consider the independence number of hypergraphs, which do not contain cycles of small lenghts.

Definition 2. *A j-cycle in a hypergraph $\mathcal{G} = (V, \mathcal{E})$ is a sequence $E_1, \ldots, E_j$ of distinct edges $E_1, \ldots, E_j \in \mathcal{E}$, such that $E_i \cap E_{i+1} \neq \emptyset$ for $i = 1, \ldots, j-1$, and $E_j \cap E_1 \neq \emptyset$, and a sequence $v_1, \ldots, v_j$ of distinct vertices with $v_{i+1} \in \mathcal{E}_i \cap \mathcal{E}_{i+1}$ for $i = 1, \ldots, j-1$, and $v_1 \in \mathcal{E}_1 \cap \mathcal{E}_j$. An unordered pair $\{E, E'\}$ of distinct edges $E, E' \in \mathcal{E}$ with $|E \cap E'| \geq 2$ is called a 2-cycle. For a hypergraph $\mathcal{G} = (V, \mathcal{E}_3 \cup \cdots \cup \mathcal{E}_k)$ a 2-cycle $\{E, E'\}$ in $\mathcal{G}$ is called $(2; (g, i, j))$-cycle if and only if $|E \cap E'| = g$, and $E \in \mathcal{E}_i$ and $E' \in \mathcal{E}_j$ for $2 \leq g \leq i \leq j$ but $g < j$. A hypergraph $\mathcal{G} = (V, \mathcal{E})$ is called linear if it does not contain any 2-cycles, and it is called uncrowded if it does not contain any 2-, 3- or 4-cycles.*

For k-uniform uncrowded hypergraphs the next lower bound on the independence number, which has been proved by Ajtai, Komlós, Pintz, Spencer and Szemerédi [1], is better than the one in (1), see also [2] and [10], and compare [6] and [11] for a deterministic polynomial time algorithm.

Theorem 3. *Let $k \geq 3$ be a fixed integer. Let $\mathcal{G} = (V, \mathcal{E}_k)$ be an uncrowded k-uniform hypergraph with $|V| = N$ vertices and average degree $t^{k-1} := k \cdot |\mathcal{E}_k|/N$. Then, for some constant $C_k > 0$ the independence number $\alpha(\mathcal{G})$ of $\mathcal{G}$ satisfies*

$$\alpha(\mathcal{G}) \geq C_k \cdot (N/t) \cdot (\log t)^{\frac{1}{k-1}} . \tag{2}$$

Hence, for fixed integers $k \geq 3$ and uncrowded k-uniform hypergraphs with average degree t^{k-1} the lower bound (2) improves (1) by a factor of $\Theta((\log t)^{1/(k-1)})$.

We use the following extension of Theorem 3 – instead of an uncrowded hypergraph we require only a linear one –, see [17].

Theorem 4. *Let $k \geq 3$ be a fixed integer. Let $\mathcal{G} = (V, \mathcal{E}_3 \cup \cdots \cup \mathcal{E}_k)$ be a linear hypergraph with $|V| = N$ such that the average degrees $t_i^{i-1} := i \cdot |\mathcal{E}_i|/|V|$ for the i-element edges satisfy $t_i^{i-1} \leq c_i \cdot S^{i-1} \cdot (\log S)^{(k-i)/(k-1)}$, where $c_i > 0$ are constants with $c_i < 1/32 \cdot \binom{k-1}{i-1}/(10^{(3(k-i))/(k-1)} \cdot k^6)$, $i = 3, \ldots, k$. Then for some constant $C_k > 0$, the independence number $\alpha(\mathcal{G})$ of $\mathcal{G}$ satisfies*

$$\alpha(\mathcal{G}) \geq C_k \cdot \frac{N}{S} \cdot (\log S)^{\frac{1}{k-1}}. \tag{3}$$

An independent set of size $\Omega((N/S) \cdot (\log S)^{1/(k-1)})$ can be found deterministically in time $O(N \cdot S^{4k-2})$.

Both Theorems 3 and 4 are best possible for a certain range of the parameters $k < T < N$ as can be seen by a random hypergraph argument.

Theorem 4 is helpful in our situation, since one has to take care only of the 2-cycles and not of 3- and 4-cycles anymore.

3   A Deterministic Algorithm

Here we prove Theorem 1. To give a polynomial time algorithm, which for fixed integers $k \geq 3$ finds for any integers $n \geq k$ deterministically n points in the unit square $[0,1]^2$ such that simultaneously for $j = 3, \ldots, k$, the areas of the convex hulls of any j of these n points are $\Omega((\log n)^{1/(j-2)}/n^{(j-1)/(j-2)})$, we discretize the unit square $[0,1]^2$ by considering the standard $T \times T$-grid, i.e., the set $\{(i,j) \in \mathbb{Z}^2 \mid 0 \leq i,j \leq T-1\}$, where $T = n^{1+\beta}$ for some constant $\beta > 0$, which will be specified later.

For distinct grid-points P, Q in the $T \times T$-grid let PQ denote the *line* through P and Q and let $[P, Q]$ denote the *segment* between P and Q. Let dist $(P,Q) := ((p_x - q_x)^2 + (p_y - q_y)^2)^{1/2}$ denote the *Euclidean distance* between the grid-points $P = (p_x, p_y)$ and $Q = (q_x, q_y)$. For grid-points $P_1, \ldots, P_l$ in the $T \times T$-grid let area $(P_1, \ldots, P_l)$ be the area of the convex hull of the points $P_1, \ldots, P_l$. A *strip* centered at the line PQ of width w is the set of all points in $\mathbb{R}^2$, which are at Euclidean distance at most $w/2$ from the line PQ. Let $\leq_l$ be a total order on the $T \times T$-grid, which is defined as follows: for grid-points $P = (p_x, p_y)$ and $Q = (q_x, q_y)$ in the $T \times T$-grid let $P \leq_l Q :\Longleftrightarrow (p_x < q_x)$ or $(p_x = q_x$ and $p_y < q_y)$. First notice the following simple fact.

Lemma 1. *Let* $P_1, \ldots, P_l$ *be grid-points in the* $T \times T$-*grid,* $l \geq 3$.

(i) *Then, it is area* $(P_1, \ldots, P_l) \geq$ *area* $(P_1, \ldots, P_{l-1})$.
(ii) *If area* $(P_1, \ldots, P_l) \leq A$, *then for any distinct grid-points* P_i, P_j *every grid-point* P_k, $k = 1, \ldots, l$, *is contained in a strip centered at the line* $P_i P_j$ *of width* $(4 \cdot A)/\text{dist }(P_i, P_j)$.

For suitable constants $c_j^* > 0$, $j = 3, \ldots, k$, we set

$$A_j := \frac{c_j^* \cdot T^2 \cdot (\log n)^{1/(j-2)}}{n^{(j-1)/(j-2)}} > 1. \tag{4}$$

Then, it is $0 < A_3 \leq \cdots \leq A_k$ for $n \geq n_0$. We form a hypergraph $\mathcal{G} = \mathcal{G}(A_3, \ldots, A_k) = (V, \mathcal{E}_3^0 \cup \mathcal{E}_3 \cup \mathcal{E}_4 \cup \cdots \cup \mathcal{E}_k)$, which contains two types of 3-element edges, and (one type of) j-element edges, $j = 4, \ldots, k$. The vertex-set V of $\mathcal{G}$ consists of the T^2 grid-points in the $T \times T$-grid. The edge-sets are defined as follows. For distinct grid-points $P, Q, R \in V$ in the $T \times T$-grid let $\{P, Q, R\} \in \mathcal{E}_3^0$ if and only if P, Q, R are collinear. Moreover, for $j = 3, \ldots, k$, and distinct grid-points $P_1, \ldots, P_j \in V$ in the $T \times T$-grid let $\{P_1, \ldots, P_j\} \in \mathcal{E}_j$ if and only if area $(P_1, \ldots, P_j) \leq A_j$ and no three of the grid-points $P_1, \ldots, P_j$ are collinear.

We want to find a large independent set in this hypergraph $\mathcal{G} = (V, \mathcal{E}_3^0 \cup \mathcal{E}_3 \cup \mathcal{E}_4 \cup \cdots \cup \mathcal{E}_k)$, as an independent set $I \subseteq V$ in $\mathcal{G}$ corresponds to $|I|$ many grid-points in the $T \times T$-grid, such that the areas of the convex hulls of any j distinct grid-points from these $|I|$ points are bigger than A_j, $j = 3, \ldots, k$. To find a suitable induced subhypergraph of $\mathcal{G}$ to which Theorem 4 may be applied, in a first step we estimate the numbers $|\mathcal{E}_3^0|$ and $|\mathcal{E}_j|$, $j = 3, \ldots, k$, of 3- and j-element edges, respectively, and the numbers of 2-cycles in $\mathcal{G}$. Then in a certain induced subhypergraph $\mathcal{G}^*$ of $\mathcal{G}$ we omit one vertex from each 3-element edge in $\mathcal{E}_3^0$ and from each 2-cycle. The resulting induced subhypergraph $\mathcal{G}^{**}$ contains no 2-cycles anymore, hence is linear, and then we may apply Theorem 4 to $\mathcal{G}^{**}$.

3.1   The Numbers of Edges in $\mathcal{G}$

The next estimate is quite crude but it suffices for our purposes.

Lemma 2. *The number $|\mathcal{E}_3^0|$ of 3-element edges in the hypergraph $\mathcal{G} = (V, \mathcal{E}_3^0 \cup \mathcal{E}_3 \cup \mathcal{E}_4 \cup \cdots \cup \mathcal{E}_k)$ satisfies*

$$|\mathcal{E}_3^0| \leq T^5. \tag{5}$$

Proof. For grid-points $P, Q, R \in V$ we have $\{P, Q, R\} \in \mathcal{E}_3^0$ if and only if P, Q, R are collinear. Each line is determined by two grid-points in the $T \times T$-grid, for which there are at most T^2 choices each, and each line contains at most T grid-points, and the upper bound T^5 on the number of collinear triples follows.   □

To estimate $|\mathcal{E}_j|$, $j = 3, \ldots, k$, we use the following result from [7].

Lemma 3. *For distinct grid-points $P = (p_x, p_y)$ and $R = (r_x, r_y)$ with $P \leq_l R$ from the $T \times T$-grid, where $s := r_x - p_x \geq 0$ and $h := r_y - p_y$, it holds:*

(a) *There are at most $4 \cdot A$ grid-points Q in the $T \times T$-grid such that*
  (i) $P \leq_l Q \leq_l R$, *and*
  (ii) P, Q, R *are not collinear, and area* $(P, Q, R) \leq A$.
(b) *The number of grid-points Q in the $T \times T$-grid which fulfill only (ii) from (a) is at most $(12 \cdot A \cdot T)/s$ for $s > 0$, and at most $(12 \cdot A \cdot T)/|h|$ for $|h| > s$.*

Lemma 4. *For $j = 3, \ldots, k$, the numbers $|\mathcal{E}_j|$ of unordered j-tuples $P_1, \ldots, P_j$ of distinct grid-points in the $T \times T$-grid with area $(P_1, \ldots, P_j) \leq A_j$, where no three of the grid-points $P_1, \ldots, P_j$ are collinear, satisfy for some constants $c_j > 0$:*

$$|\mathcal{E}_j| \leq c_j \cdot A_j^{j-2} \cdot T^4. \tag{6}$$

Proof. Let $P_1, \ldots, P_j$ be grid-points, no three on a line, in the $T \times T$-grid with area $(P_1, \ldots, P_j) \leq A_j$. We may assume that $P_1 \leq_l \cdots \leq_l P_j$. For $P_1 = (p_{1,x}, p_{1,y})$ and $P_k = (p_{j,x}, p_{j,y})$ let $s := p_{j,x} - p_{1,x} \geq 0$ and $h := p_{j,y} - p_{1,y}$. Then $s > 0$, as otherwise $P_1, \ldots, P_j$ are collinear.

There are T^2 choices for the grid-point P_1. Given P_1, any grid-point P_j with $P_1 \leq_l P_j$ is determined by a pair $(s, h) \neq (0, 0)$ of integers with $1 \leq s \leq T$ and $-T \leq h \leq T$. By Lemma 1 (i) we have area $(P_1, P_i, P_j) \leq A_j$ for $i = 2, \ldots, j-1$. Given the grid-points P_1 and P_j, since $P_1 \leq_l P_i \leq_l P_j$ for $i = 2, \ldots, j-1$, by

Lemma 3 (a) there are at most $4 \cdot A_j$ choices for each grid-point P_i, hence for a constant $c_j > 0$:

$$|\mathcal{E}_j| \leq T^2 \cdot \sum_{s=1}^{T} \sum_{h=-T}^{T} (4 \cdot A_j)^{j-2} \leq c_j \cdot A_j^{j-2} \cdot T^4. \qquad \square$$

By (6) the average degrees t_j^{j-1} for the j-element edges $E \in \mathcal{E}_j$, $j = 3, \ldots, k$, of $\mathcal{G}$ satisfy

$$t_j^{j-1} = j \cdot |\mathcal{E}_j|/|V| \leq j \cdot c_j \cdot A_j^{j-2} \cdot T^2 =: (t_j(0))^{j-1}. \qquad (7)$$

3.2   The Numbers of 2-Cycles in $\mathcal{G}$

Let $s_{2;(g,i,j)}(\mathcal{G})$ denote the number of $(2; (g, i, j))$-cycles, $2 \leq g \leq i \leq j \leq k$ with $g < j$ in the hypergraph $\mathcal{G}$, i.e., the number of unordered pairs $\{E, E'\}$ of edges with $E \in \mathcal{E}_i$ and $E' \in \mathcal{E}_j$ and $|E \cap E'| = g$.

Lemma 5. *For $2 \leq g \leq i \leq j \leq k$ with $g < j$, there exist constants $c_{2;(g,i,j)} > 0$ such that the numbers $s_{2;(g,i,j)}(\mathcal{G})$ of $(2; (g, i, j))$-cycles in the hypergraph $\mathcal{G} = (V, \mathcal{E}_3^0 \cup \mathcal{E}_3 \cup \mathcal{E}_4 \cup \cdots \cup \mathcal{E}_k)$ fulfill*

$$s_{2;(g,i,j)}(\mathcal{G}) \leq c_{2;(g,i,j)} \cdot A_i^{i-2} \cdot A_j^{j-g} \cdot T^4 \cdot (\log T)^3. \qquad (8)$$

Proof. Let the grid-points, which correspond to the vertices of an i-element edge $E \in \mathcal{E}_i$ and a j-element edge $E' \in \mathcal{E}_j$ and also yield a $(2; (g, i, j))$-cycle in $\mathcal{G}$, $2 \leq g \leq i \leq j \leq k$ with $g < j$, be $P_1, \ldots, P_i$ and $P_1, \ldots, P_g, Q_{g+1}, \ldots, Q_j$, where after renumbering $P_1 \leq_l \cdots \leq_l P_g$ and no three of the grid-points $P_1, \ldots, P_i$ and of $P_1, \ldots, P_g, Q_{g+1}, \ldots, Q_j$ are collinear, thus area $(P_1, \ldots, P_i) \leq A_i$ and area $(P_1, \ldots, P_g, Q_{g+1}, \ldots, Q_j) \leq A_j$.

There are T^2 choices for the grid-point $P_1 = (p_{1,x}, p_{1,y})$, any pair $(s, h) \neq (0, 0)$ of integers determines at most one grid-point $P_g = (p_{1,x} + s, p_{1,y} + h)$ in the $T \times T$-grid. By symmetry we may assume that $s > 0$ and $0 \leq h \leq s \leq T$, which is taken into account by an additional constant factor $c' > 1$. Given the grid-points P_1 and P_g, since area $(P_1, P_f, P_g) \leq A_i$ for $f = 2, \ldots, g - 1$ by Lemma 1, and $P_1 \leq_l P_f \leq_l P_g$, by Lemma 3 (a) there are at most $4 \cdot A_i$ choices for each grid-point P_f in the $T \times T$-grid, hence the number of choices for the grid-points $P_1, \ldots, P_{g-1}$ is at most

$$T^2 \cdot (4 \cdot A_i)^{g-2}. \qquad (9)$$

For the convex hulls of the grid-points $P_1, \ldots, P_i$ and $P_1, \ldots, P_g, Q_{g+1}, \ldots, Q_j$ let their (w.r.t $\leq_l$) *extremal* points be $P', P'' \in \{P_1, \ldots, P_i\}$ and $Q', Q'' \in \{P_1, \ldots, P_g, Q_{g+1}, \ldots, Q_j\}$, respectively, i.e., for $P' \leq_l P''$ and $Q' \leq_l Q''$ we have $P' \leq_l P_1, \ldots, P_i \leq_l P''$ and $Q' \leq_l P_1, \ldots, P_g, Q_{g+1}, \ldots, Q_j \leq_l Q''$.

Given the grid-points $P_1 \leq_l \cdots \leq_l P_g$, there are three possibilities for the convex hulls of the grid-points $P_1, \ldots, P_i$ and $P_1, \ldots, P_j, Q_{j+1}, \ldots, Q_k$ each:

(i) P_1 and P_g are extremal, or

(ii) exactly one grid-point, P_1 or P_g, is extremal, or

(iii) neither P_1 nor P_g is extremal.

We restrict our calculations to the convex hull of $P_1, \ldots, P_i$ as the considerations for the convex hull of $P_1, \ldots, P_g, Q_{g+1}, \ldots, Q_j$ are essentially the same.

In case (i) the grid-points P_1 and P_g are extremal for the convex hull of $P_1, \ldots, P_i$, hence $P_1 \leq_l P_{g+1}, \ldots, P_i \leq_l P_g$. By Lemma 3 (a), since area $(P_1, P_l, P_g) \leq A_i$, $l = g+1, \ldots, i$, and no three of the grid-points $P_1, \ldots, P_i$ are collinear, there are at most $4 \cdot A_i$ choices for each grid-point P_l, hence the number of choices for the grid-points $P_{g+1}, \ldots, P_i$ is at most

$$\text{case (i):} \qquad (4 \cdot A_i)^{i-g}. \qquad (10)$$

In case (ii) exactly one of the grid-points P_1 or P_g is extremal for the convex hull of $P_1, \ldots, P_i$. By Lemma 3 (b) there are at most $(12 \cdot A_i \cdot T)/s$ choices for the second extremal grid-point P' or P''. Having fixed this second extremal grid-point, for each grid-point $P_{g+1}, \ldots, P_i \neq P', P''$ there are by Lemma 3 (a) at most $4 \cdot A_i$ choices, hence the number of choices for the grid-points $P_{g+1}, \ldots, P_i$ is at most

$$\text{case (ii):} \qquad ((4 \cdot A_i)^{i-g-1} \cdot 12 \cdot A_i \cdot T)/s = ((4 \cdot A_i)^{i-g} \cdot 3 \cdot T)/s. \quad (11)$$

In case (iii) none of the grid-points P_1, P_g is extremal for the convex hull of $P_1, \ldots, P_i$. By Lemma 1 (ii) all grid-points $P_{g+1}, \ldots, P_i$ are contained in a strip S_i, which is centered at the line $P_1 P_g$, of width $(4 \cdot A_i)/\sqrt{h^2 + s^2}$. Consider the parallelogram $\mathcal{P}_0 = \{(p_x, p_y) \in S_i \mid p_{1,x} \leq p_x \leq p_{g,x}\}$ within the strip S_i, where $P_1 = (p_{1,x}, p_{1,y})$ and $P_g = (p_{g,x}, p_{g,y})$ and $s = p_{g,x} - p_{1,x}$.

We divide the strip S_i within the $T \times T$-grid into pairwise congruent parallelograms $\mathcal{P}_0, \mathcal{P}_i^+, \mathcal{P}_i^-$, $i = 1, \ldots, l \leq \lfloor T/s \rfloor + 2$, each of side-lengths $(4 \cdot A_i)/s$ and $\sqrt{h^2 + s^2}$ and of area $4 \cdot A_i$, where for $i \geq 1$ all parallelograms $\mathcal{P}_i^-$ are on the left of the parallelogram $\mathcal{P}_0$, and all parallelograms $\mathcal{P}_i^+$ are on the right of $\mathcal{P}_0$, in particular $\mathcal{P}_i^+ := \{(p_x, p_y) \in S_i \mid p_{g,x} + (i-1) \cdot s \leq p_x \leq p_{g,x} + i \cdot s\}$ and $\mathcal{P}_i^- := \{(p_x, p_y) \in S_i \mid p_{1,x} - i \cdot s \leq p_x \leq p_{1,x} - (i-1) \cdot s\}$. By Lemma 3 (a) each parallelogram $\mathcal{P}_i^+$ or $\mathcal{P}_i^-$ contains at most $4 \cdot A_i$ grid-points P, where P_1, P_j, P are not collinear. Each extremal grid-point, P' or P'', is contained in some parallelogram $\mathcal{P}_i^+$ or $\mathcal{P}_i^-$ for some $i \geq 1$, since by our assumption neither $P_1 \in \mathcal{P}_0$ nor $P_g \in \mathcal{P}_0$ are extremal. Each grid-point $P = (p_x, p_y) \in \mathcal{P}_i^+ \cup \mathcal{P}_i^-$, $i \geq 1$, satisfies $|p_x - p_{1,x}| \geq i \cdot s$ or $|p_x - p_{j,x}| \geq i \cdot s$. Thus, if $P' \in \mathcal{P}_i^+ \cup \mathcal{P}_i^-$ or $P'' \in \mathcal{P}_i^+ \cup \mathcal{P}_i^-$, by Lemma 3 (b) there are at most $(12 \cdot A_i \cdot T)/(i \cdot s)$ choices for the second extremal grid-point. Having chosen both extremal grid-points P' and P'' in at most $(4 \cdot A_i) \cdot ((12 \cdot A_i \cdot T)/(i \cdot s)) = (48 \cdot A_i^2 \cdot T)/(i \cdot s)$ ways, for the grid-points $P_{g+1}, \ldots, P_i \neq P', P''$ there are by Lemma 3 (a) at most $(4 \cdot A_i)^{i-g-2}$ choices. Hence, in case (iii) the number of choices for the grid-points $P_{g+1}, \ldots, P_i$ is at most case (iii): $(4 \cdot A_i)^{i-g-2} \cdot \sum_{i=1}^{\lfloor T/s \rfloor + 2} \dfrac{48 \cdot A_i^2 \cdot T}{i \cdot s} =$

$$= (4 \cdot A_i)^{i-g} \cdot \frac{3 \cdot T}{s} \cdot \sum_{i=1}^{\lfloor T/s \rfloor + 2} \frac{1}{i} \leq (4 \cdot A_i)^{i-g} \cdot \frac{5 \cdot T \cdot \log T}{s}. \quad (12)$$

By (10)–(12) and using $T \geq s$, in cases (i)–(iii) altogether the number of choices for the grid-points $P_{g+1}, \ldots, P_i$ is at most

$$(4 \cdot A_i)^{i-g} \cdot \left(1 + \frac{3 \cdot T}{s} + \frac{5 \cdot T \cdot \log T}{s} \right) \leq (4 \cdot A_i)^{i-g} \cdot \frac{9 \cdot T \cdot \log T}{s}. \quad (13)$$

Similar to (13), for the number of choices of the grid-points $Q_{g+1}, \ldots, Q_j$ the following upper bound holds:

$$((4 \cdot A_j)^{j-g} \cdot 9 \cdot T \cdot \log T)/s. \quad (14)$$

Hence with (9), (13) and (14) for $2 \leq g \leq i \leq j \leq k$ and $g < j$ we obtain for constants $c', c_{2;(g,i,j)} > 0$:

$$s_{2;(g,i,j)}(\mathcal{G}) \leq c' \cdot T^2 \cdot (4 \cdot A_i)^{g-2} \cdot \sum_{s=1}^{T} \sum_{h=0}^{s} \left(\frac{(4 \cdot A_i)^{i-g} \cdot 9 \cdot T \cdot \log T}{s} \right) \cdot$$

$$\cdot \left(\frac{(4 \cdot A_j)^{j-g} \cdot 9 \cdot T \cdot \log T}{s} \right) \leq$$

$$< 81 \cdot c' \cdot 4^{i+j-g-2} \cdot A_i^{i-2} \cdot A_j^{j-g} \cdot T^4 \cdot (\log T)^2 \cdot \sum_{s=1}^{T} \sum_{h=0}^{s} \frac{1}{s^2}$$

$$\leq c_{2;(g,i,j)} \cdot A_i^{i-2} \cdot A_j^{j-g} \cdot T^4 \cdot (\log T)^3. \qquad \square$$

3.3 Choosing a Subhypergraph in $\mathcal{G}$

With probability $p := T^\varepsilon / t_k(0) \leq 1$, hence $p = \Theta(T^\varepsilon / (A_k^{(k-2)/(k-1)} \cdot T^{2/(k-1)})$ by (7), where $\varepsilon > 0$ is a small constant, we pick uniformly at random and independently of each other vertices from V. Let $V^* \subseteq V$ be the random set of the picked vertices and let $\mathcal{G}^* = (V^*, \mathcal{E}_3^{0*} \cup \mathcal{E}_3^* \cup \mathcal{E}_4^* \cup \cdots \cup \mathcal{E}_k^*)$ with $\mathcal{E}_3^{0*} := \mathcal{E}_3^0 \cap [V^*]^3$ and $\mathcal{E}_j^* := \mathcal{E}_j \cap [V^*]^j$, $j = 3, \ldots, k$, be the on V^* induced random subhypergraph of $\mathcal{G}$. Let $E[|V^*|]$, $E[|\mathcal{E}_3^{0*}|]$, $E[|\mathcal{E}_j^*|]$, $j = 3, \ldots, k$, and $E[s_{2;(g,i,j)}(\mathcal{G}^*)]$, $2 \leq g \leq i \leq j \leq k$ but $g < j$, be the expected numbers of vertices, collinear triples of grid-points, j-element edges and $(2;(g,i,j))$-cycles, respectively, in $\mathcal{G}^* = (V^*, \mathcal{E}_3^{0*} \cup \mathcal{E}_3^* \cup \mathcal{E}_4^* \cup \cdots \cup \mathcal{E}_k^*)$. By (5), (6), and (8) we infer for constants $c_1', c_3^{0'} c_j', c_{2;(g,i,j)}' > 0$:

$$E[|V^*|] = p \cdot T^2 \geq (c_1' \cdot T^{\frac{2k-4}{k-1}+\varepsilon})/A_k^{\frac{k-2}{k-1}} \quad (15)$$

$$E[|\mathcal{E}_3^{0*}|] = p^3 \cdot |\mathcal{E}_3^0| \leq (c_3^{0'} \cdot T^{\frac{5k-11}{k-1}+3\varepsilon})/A_k^{\frac{3k-6}{k-1}} \quad (16)$$

$$E[|\mathcal{E}_j^*|] = p^j \cdot |\mathcal{E}_j| \leq (c_j' \cdot T^{\frac{4k-2j-4}{k-1}+j\varepsilon} \cdot A_j^{j-2})/A_k^{\frac{j(k-2)}{k-1}} \tag{17}$$

$$E[s_{2;(g,i,j)}(\mathcal{G}^*)] = p^{i+j-g} \cdot s_{2;(g,i,j)}(\mathcal{G}) \leq$$

$$leq \frac{c_{2;(g,i,j)}' \cdot T^{\frac{4k-4-2(i+j-g)}{k-1}+\varepsilon(i+j-g)} \cdot (\log T)^3 \cdot A_i^{i-2} \cdot A_j^{j-g}}{A_k^{\frac{(k-2)(i+j-g)}{k-1}}} \tag{18}$$

By (15)–(18) and by Chernoff's and Markov's inequality we obtain a subhypergraph $\mathcal{G}^* = (V^*, \mathcal{E}_3^{0*} \cup \mathcal{E}_3^* \cup \mathcal{E}_4^* \cup \cdots \cup \mathcal{E}_k^*)$ of $\mathcal{G}$ such that

$$|V^*| \geq ((c_1'/2) \cdot T^{\frac{2k-4}{k-1}+\varepsilon})/A_k^{\frac{k-2}{k-1}} \tag{19}$$

$$|\mathcal{E}_3^{0*}| \leq (k^3 \cdot c_3^{0'} \cdot T^{\frac{5k-11}{k-1}+3\varepsilon})/A_k^{\frac{3k-6}{k-1}} \tag{20}$$

$$|\mathcal{E}_j^*| \leq (k^3 \cdot c_j' \cdot T^{\frac{4k-2j-4}{k-1}+j\varepsilon} \cdot A_j^{j-2})/A_k^{\frac{j(k-2)}{k-1}} \tag{21}$$

$$s_{2;(g,i,j)}(\mathcal{G}^*) \leq \frac{k^3 \cdot c_{2;(g,i,j)}' \cdot T^{\frac{4k-4-2(i+j-g)}{k-1}+\varepsilon(i+j-g)} \cdot (\log T)^3 \cdot A_i^{i-2} \cdot A_j^{j-g}}{A_k^{\frac{(k-2)(i+j-g)}{k-1}}} \tag{22}$$

This probabilistic argument can be turned into a deterministic polynomial time algorithm by using the method of conditional probabilities. For $2 \leq g \leq i \leq j \leq k$ but $g < j$, let $\mathcal{C}_{2;(g,i,j)}$ be the set of all $(i+j-g)$-element subsets $E \cup E'$ of V such that $E \in \mathcal{E}_i$ and $E' \in \mathcal{E}_j$ and $|E \cap E'| = g$. Let the grid-points in the $T \times T$-grid be $P_1, \ldots, P_{T^2}$. To each grid-point P_i associate a variable $p_i \in [0,1]$, $i = 1, \ldots, T^2$, and let $F(p_1, \ldots, p_{T^2})$ be a function defined by

$$F(p_1, \ldots, p_{T^2}) := 2^{p \cdot T^2/2} \cdot \prod_{i=1}^{T^2}\left(1 - \frac{p_i}{2}\right) +$$

$$+ \frac{\sum_{\{i,j,k\}\in\mathcal{E}_3^0} p_i \cdot p_j \cdot p_k}{(k^3 \cdot c_3' \cdot T^{\frac{5k-11}{k-1}+3\varepsilon})/A^{\frac{3k-6}{k-1}}} + \sum_{j=3}^{k} \frac{\sum_{\{i_1,\ldots,i_j\}\in\mathcal{E}_j} \prod_{l=1}^{j} p_{i_l}}{(k^3 \cdot c_j' \cdot T^{\frac{4k-2j-4}{k-1}+j\varepsilon} \cdot A_j^{j-2})/A^{\frac{j(k-2)}{k-1}}} +$$

$$+ \sum_{2\leq g\leq i\leq j\leq k; g<j} \frac{A_k^{\frac{(k-2)(i+j-g)}{k-1}} \cdot \sum_{\{i_1,\ldots,i_{i+j-g}\}\in\mathcal{C}_j} \prod_{l=1}^{i+j-g} p_{i_l}}{k^3 \cdot c_{2;(g,i,j)}' \cdot T^{\frac{4k-4-2(i+j-g)}{k-1}+(i+j-g)\varepsilon} \cdot (\log T)^3 \cdot A_i^{i-2} \cdot A_j^{j-g}}.$$

With the initialisation $p_1 := \cdots := p_{T^2} := p = T^\varepsilon/t_0$, we infer by (15)–(18) that $F(p, \ldots, p) < (2/e)^{pT^2/2} + 1/3$, hence $F(p, \ldots, p) < 1$ for $p \cdot T^2 \geq 10$. By using the linearity of $F(p_1, \ldots, p_{T^2})$ in each p_i, we minimize $F(p_1, \ldots, p_{T^2})$ by choosing step by step $p_i := 0$ or $p_i := 1$, $i = 1, \ldots, T^2$, and finally we achieve $F(p_1, \ldots, p_{T^2}) < 1$. The set $V^* = \{P_i \in V \mid p_i = 1\}$ yields an induced subhypergraph $\mathcal{G}^* = (V^*, \mathcal{E}_3^{0*} \cup \mathcal{E}_3^* \cup \cdots \cup \mathcal{E}_k^*)$ of $\mathcal{G}$ with $\mathcal{E}_i^* := \mathcal{E}_i \cap [V^*]^i$ for $i = 3, \ldots, k$, and $\mathcal{E}_3^{0*} := \mathcal{E}_3^0 \cap [V^*]^3$ which satisfies (19)–(22), as otherwise $F(p_1, \ldots, p_{T^2}) > 1$ gives a contradiction. By (4)–(6) and (8) and using $T = n^{1+\beta}$ for fixed $\beta > 0$, the running time of this derandomization is given by

$$O(|V| + |\mathcal{E}_3^0| + \sum_{j=3}^{k} |\mathcal{E}_j| + \sum_{2 \le g \le i \le j \le k; g < j} |\mathcal{C}_{2;(g,i,j)}|) = O(|\mathcal{C}_{2;(2,k,k)}|) =$$

$$= O(A_k^{2k-4} \cdot T^4 \cdot (\log T)^3) = O\left((T^{4k-4} \cdot (\log n)^5)/n^{2k-2}\right). \tag{23}$$

Lemma 6. *For each fixed* $0 < \varepsilon < (\beta - 1)/(2 \cdot (1 + \beta))$ *and* $\beta > 1$ *it is*

$$|\mathcal{E}_3^{0*}| = o(|V^*|). \tag{24}$$

Proof. By (19), (20) and using $T = n^{1+\beta}$ with constants $\varepsilon > 0$ and $\beta > 1$ we have

$$|\mathcal{E}_3^{0*}| = o(|V^*|)$$
$$\Longleftarrow T^{\frac{5k-11}{k-1} + 3\varepsilon} \cdot \log T / A_k^{\frac{3k-6}{k-1}} = o(T^{\frac{2k-4}{k-1} + \varepsilon} / A_k^{\frac{k-2}{k-1}})$$
$$\Longleftrightarrow n^{2-(1+\beta)(1-2\varepsilon)} \cdot (\log n)^{1-\frac{2}{k-1}} = o(1)$$
$$\Longleftrightarrow (1 + \beta) \cdot (1 - 2 \cdot \varepsilon) > 2,$$

which holds for $\varepsilon < (\beta - 1)/(2 \cdot (1 + \beta))$. $\qquad\square$

Lemma 7. *For* $2 \le g \le i \le j \le k$ *but* $g < j$ *and each fixed* ε *with* $0 < \varepsilon < \frac{j-g}{(i+j-g-1)(j-2)(1+\beta)}$ *it is*

$$s_{2;(g,i,j)}(\mathcal{G}^*) = o(|V^*|). \tag{25}$$

Proof. For $2 \le g \le i \le j \le k$ but $g < j$ by (4), (19), (22) and using $T = n^{1+\beta}$ with fixed $\beta, \varepsilon > 0$ we infer

$$s_{2;(g,i,j)}(\mathcal{G}^*) = o(|V^*|)$$
$$\Longleftarrow \frac{T^{\frac{4k-4-2(i+j-g)}{k-1} + (i+j-g)\varepsilon} \cdot (\log T)^3 \cdot A_i^{i-2} \cdot A_j^{j-g}}{A_k^{\frac{(k-2)(i+j-g)}{k-1}}} = o\left(\frac{T^{\frac{2k-4}{k-1} + \varepsilon}}{A_k^{\frac{k-2}{k-1}}}\right)$$
$$\Longleftrightarrow n^{\varepsilon(1+\beta)(i+j-g-1) - \frac{j-g}{j-2}} \cdot (\log n)^{4 + \frac{j-g}{j-2} - \frac{i+j-g-1}{k-1}} = o(1)$$
$$\Longleftrightarrow \varepsilon < \frac{j-g}{(j-2)(i+j-g-1)(1+\beta)}. \qquad\square$$

By setting $\varepsilon := 1/(2 \cdot k^2 \cdot (1 + \beta))$ and $\beta > 1 + 1/k^2$ all assumptions in Lemmas 6 and 7 and also $p = T^\varepsilon / t_k(0) \le 1$ are fulfilled. We delete one vertex from each edge $E \in \mathcal{E}_3^{0*}$, and from each 2-cycle in $\mathcal{G}^*$. Let $V^{**} \subseteq V^*$ be the set of remaining vertices. By Lemmas 6 and 7 the induced subhypergraph $\mathcal{G}^{**} = (V^{**}, \mathcal{E}_3^{**} \cup \cdots \cup \mathcal{E}_k^{**})$ with $\mathcal{E}_j^{**} := \mathcal{E}_j^* \cap [V^{**}]^j$, $j = 3, \ldots, k$, where $|V^{**}| = (1 - o(1)) \cdot |V^*| \ge |V^*|/2$, contains no edges from $\mathcal{E}_3^0$ and no 2-cycles anymore, i.e., $\mathcal{G}^{**}$ is a linear hypergraph. Since $|\mathcal{E}_j^{**}| \le |\mathcal{E}_j^*|$ with (19) and (21), the average degrees $t_j^{j-1}(1)$ for the j-element edges of $\mathcal{G}^{**}$, $j = 3, \ldots, k$, fulfill by (4):

$$t_j^{j-1}(1) = \frac{j \cdot |\mathcal{E}_j^{**}|}{|V^{**}|} \leq \frac{(j \cdot k^3 \cdot c_j' \cdot T^{\frac{4k-2j-4}{k-1}+j\varepsilon} \cdot A_j^{j-2})/A_k^{\frac{j(k-2)}{k-1}}}{((c_1'/4) \cdot T^{\frac{2k-4}{k-1}+\varepsilon})/A_k^{\frac{k-2}{k-1}}} \leq$$

$$\leq \frac{4 \cdot k^4 \cdot c_j' \cdot (c_j^*)^{j-2}}{c_1' \cdot (c_k^*)^{\frac{(j-1)(k-2)}{k-1}}} \cdot T^{(j-1)\varepsilon} \cdot (\log n)^{\frac{k-j}{k-1}}. \tag{26}$$

As observed above, this subhypergraph $\mathcal{G}^{**}$ is linear. By choosing $S := c \cdot T^\varepsilon$ for a large enough constant $c > 0$, with $T = n^{1+\beta}$ with $T = n^{1+\beta}$ by (26) the assumptions in Theorem 4 are fulfilled, and we apply it, and, using (4) we find in time

$$O((T^{\frac{2k-4}{k-1}+\varepsilon}/A_k^{\frac{k-2}{k-1}}) \cdot S^{4k-2}) = O(n \cdot T^{(4k-1)\varepsilon}) = o(T^2) \tag{27}$$

an independent set I of size

$$|I| = \Omega((|V^{**}|/S) \cdot (\log S)^{\frac{1}{k-1}}) = \Omega((T^{\frac{2k-4}{k-1}+\varepsilon}/(A_k^{\frac{k-2}{k-1}} \cdot T^\varepsilon)) \cdot (\log T^\varepsilon)^{\frac{1}{k-1}}) =$$

$$= \Omega((n/(\log n)^{\frac{1}{k-1}}) \cdot (\log T)^{\frac{1}{k-1}}) = \Omega(n),$$

since $T = n^{1+\beta}$ and $\beta, \varepsilon > 0$ are constants. By choosing the constants $c_j^* > 0$, $j = 3, \ldots, k$, in (4) sufficiently small, we obtain an independent set of size n, which yields, after rescaling the areas A_j by the factor T^2, a desired set of n points in $[0,1]^2$ such that, simultaneously for $j = 3, \ldots, k$, the areas of the convex hulls of every j distinct of these n points are $\Omega((\log n)^{1/(j-2)}/n^{(j-1)/(j-2)})$. Adding the times in (23) and (27) we get the time bound $O(T^{4k-4} \cdot (\log n)^5/n^{2k-2} + T^2) = (n^{(2k-2)(1+2\beta)+1)}) = o(n^{6k-4})$ for $\beta > 1 + 1/k^2$ small enough. $\qquad\square$

We remark that the bound $o(n^{6k-4})$ on the running time might be improved a little, for example by using a better estimate on the number of collinear triples of grid-points in the $T \times T$-grid or by a random preselection of grid-points. However, we cannot do better than $O(n^{ck})$ for some constant $c > 0$.

References

1. Ajtai, M., Komlós, J., Pintz, J., Spencer, J., Szemerédi, E.: Extremal Uncrowded Hypergraphs. Journal of Combinatorial Theory Ser. A 32, 321–335 (1982)
2. Alon, N., Lefmann, H., Rödl, V.: On an Anti-Ramsey Type Result, Colloquia Mathematica Societatis János Bolyai. Sets, Graphs and Numbers 60, 9–22 (1991)
3. Barequet, G.: A Lower Bound for Heilbronn's Triangle Problem in d Dimensions, SIAM Journal on Discrete Mathematics 14, 230–236 (2001)
4. Barequet, G.: The On-Line Heilbronn's Triangle Problem. Discrete Mathematics 283, 7–14 (2004)
5. Barequet, G., Shaikhet, A.: The On-Line Heilbronn's Triangle Problem in d Dimensions. In: Chen, D.Z., Lee, D.T. (eds.) COCOON 2006. LNCS, vol. 4112, pp. 408–417. Springer, Heidelberg (2006)
6. Bertram–Kretzberg, C., Lefmann, H.: The Algorithmic Aspects of Uncrowded Hypergraphs. SIAM Journal on Computing 29, 201–230 (1999)

7. Bertram-Kretzberg, C., Hofmeister, T., Lefmann, H.: An Algorithm for Heilbronn's Problem. SIAM Journal on Computing 30, 383–390 (2000)
8. Brass, P.: An Upper Bound for the d-Dimensional Analogue of Heilbronn's Triangle Problem. SIAM Journal on Discrete Mathematics 19, 192–195 (2005)
9. Chazelle, B.: Lower Bounds on The Complexity of Polytope Range Searching. Journal of the American Mathematical Society 2, 637–666 (1989)
10. Duke, R.A., Lefmann, H., Rödl, V.: On Uncrowded Hypergraphs. Random Structures & Algorithms 6, 209–212 (1995)
11. Fundia, A.: Derandomizing Chebychev's Inequality to find Independent Sets in Uncrowded Hypergraphs. Random Structures & Algorithms 8, 131–147 (1996)
12. Jiang, T., Li, M., Vitany, P.: The Average Case Area of Heilbronn-type Triangles. Random Structures & Algorithms 20, 206–219 (2002)
13. Komlós, J., Pintz, J., Szemerédi, E.: On Heilbronn's Triangle Problem. Journal of the London Mathematical Society 24, 385–396 (1981)
14. Komlós, J., Pintz, J., Szemerédi, E.: A Lower Bound for Heilbronn's Problem. Journal of the London Mathematical Society 25, 13–24 (1982)
15. Lefmann, H.: On Heilbronn's Problem in Higher Dimension. Combinatorica 23, 669–680 (2003)
16. Lefmann, H.: Distributions of Points in the Unit-Square and Large k-Gons. In: Proceedings '16th ACM-SIAM Symposium on Discrete Algorithms SODA'2005, pp. 241–250. ACM and SIAM Press, New York
17. Lefmann, H.: Distributions of Points and Large Convex Hulls of k Points, In: Cheng, S.-W., Poon, C.K. (eds.) AAIM 2006. LNCS, vol. 4041, pp. 173–184. Springer, Heidelberg (2006)
18. Lefmann, H.: Convex Hulls of Point Sets and Non-Uniform Uncrowded Hypergraphs. In: Proceedings Third Int. Conference 'Algorithmic Aspects in Information and Management AAIM'2007 (to appear)
19. Roth, K.F.: On a Problem of Heilbronn. Journal of the London Mathematical Society 26, 198–204 (1951)
20. Roth, K.F.: On a Problem of Heilbronn, II, and III. In: Proc. of the London Mathematical Society, vol. 25, pp. 193–212, 543–549 (1972)
21. Roth, K.F.: Estimation of the Area of the Smallest Triangle Obtained by Selecting Three out of n Points in a Disc of Unit Area. In: Proc. of Symposia in Pure Mathematics, AMS, Providence, vol. 24, pp. 251–262 (1973)
22. Roth, K.F.: Developments in Heilbronn's Triangle Problem. Advances in Mathematics 22, 364–385 (1976)
23. Schmidt, W.M.: On a Problem of Heilbronn. Journal of the London Mathematical Society 4(2), 545–550 (1972)
24. Spencer, J.: Turán's Theorem for k-Graphs. Discrete Mathematics 2, 183–186 (1972)

An Experimental Study of Compressed Indexing and Local Alignments of DNA

Tak-Wah Lam[1,*], Wing-Kin Sung[2], Siu-Lung Tam[1],
Chi-Kwong Wong[1], and Siu-Ming Yiu[1]

[1] Department of Computer Science, University of Hong Kong, Hong Kong
{twlam,sltam,ckwong3,smyiu}@cs.hku.hk
[2] Department of Computer Science, National University of Singapore, Singapore
ksung@comp.nus.edu.sg

Abstract. Recent experimental studies on compressed indexes (BWT, CSA, FM-index) have confirmed their practicality for indexing long DNA sequences such as the human genome (about 3 billion characters) in the main memory [5,13,16]. However, these indexes are designed for exact pattern matching, which is too stringent for most biological applications. The demand is often on finding local alignments (pairs of similar substrings with gaps allowed). In this paper, we show how to build a software called BWT-SW that exploits a BWT index of a text T to speed up the dynamic programming for finding all local alignments with any pattern P. Experiments reveal that BWT-SW is very efficient (e.g., aligning a pattern of length 3,000 with the human genome takes less than a minute). We have also analyzed BWT-SW mathematically, using a simpler model (with gaps disallowed) and random strings. We find that the expected running time is $O(|T|^{0.628}|P|)$. As far as we know, BWT-SW is the first practical tool that can find all local alignments.

1 Introduction

The decoding of different genomes, in particular the human genome, has triggered a lot of bioinformatics research. In many cases, it is required to search the human genome (called the text below) for different patterns (say, a gene of another species). Exact matching is usually unlikely and may not make sense. Instead, one wants to find local alignments, which are pairs of similar substrings in the text and pattern, possibly with gaps (see, e.g., [7]). Typical biological applications require a minimum similarity of 75% (match 1 point; mismatch -3 points) and a minimum length of 18 to 30 characters (see Section 2 for details).

To find all local alignments, one can use the dynamic programming algorithm by Smith and Waterman [18], which uses $O(nm)$ time, where n and m are the length of the text and pattern, respectively. This algorithm is, however, too slow for a large text like the human genome. Our experiment shows that it takes more than 15 hours to align a pattern of 1000 characters against the human genome.

---

[*] This work was supported by the Hong Kong RGC Grant HKU7140/064.

A. Dress, Y. Xu, and B. Zhu (Eds.): COCOA 2007, LNCS 4616, pp. 242–254, 2007.

In real applications, patterns can be genes or even chromosomes, ranging from a few thousand to a few hundred million characters, and the SW algorithm would require days to weeks. As far as we know, there does not exist any practical solution for finding all local alignments in this scale. At present, a heuristic-based software called BLAST [1,17,2] is widely used by the biological community for finding local alignments. BLAST is very efficient (e.g., it takes only 10 to 20 seconds to align a pattern of 1000 characters against the human genome). Yet BLAST does not guarantee to find all local alignments. The past few years have witnessed a lot of work attempting to improve the heuristic used by BLAST (e.g. [12]). This paper, however, revisits the problem of finding all local alignments. We attempt to speed up the dynamic programming approach by exploiting the recent breakthrough on text indexing.

The indexing idea. Let T be a text of n characters and let P be a pattern of m characters. A naive approach to finding all of their local alignments is to examine all substrings of length cm of the text T, where c is a constant depending on the similarity model, and to align them one by one with P. Obviously, we want to avoid aligning P with the same substring at different positions of the text. A natural way is to build a suffix tree [14] (or a suffix trie) of the text. Then distinct substrings of T are represented by different paths from the root of the suffix tree. We align P against each path from the root up to cm characters using dynamic programming. The common prefix structure of the paths also gives a way to share the common parts of the dynamic programming on different paths. Specifically, we perform a pre-order traversal of the suffix tree; at each node, we maintain a dynamic programming table (DP table) for aligning the pattern and the path up to the node. We add more rows to the table as we go down the suffix tree, and delete the corresponding rows when going up the tree. Note that filling a row of the table costs $O(m)$ time. For very short patterns, the above approach performs dynamic programming only on a few layers of nodes and could be very efficient; but there are a few issues to be resolved for this approach to be successful in general.

- **Index size:** The best known implementation of a suffix tree requires $17.25n$ bytes for a text of length n [11]. For human genome, this is translated to 50G bytes of memory, which far exceeds the 4G capacity of a standard PC nowadays. An alternative is a hard-disk-based suffix tree, which would increase the access time several order of magnitude. In fact, even the construction of a suffix tree on a hard disk is already very time-consuming; for the human genome, it would need a week [10].
- **Running time and pruning effectiveness:** The above approach requires traversing each path of the suffix tree starting from the root up to $O(m)$ characters, and computing possible local alignments starting from the first character of the path. For long patterns, this may mean visiting many nodes of the tree, using $O(nm)$ or more time. Nevertheless, we can show that at an intermediate node u, if the DP table indicates that no substring of the pattern has a positive similarity score when aligned with the path to u, then it is useless to further extend the path and we can prune the subtree

rooted at u. It is interesting to study how effective, in practice, such a simple pruning strategy could be.

- **Dynamic programming table:** Recall that we have to maintain a dynamic programming table at each node, which is of size $m \times d$ where d is the length of the substring represented by the node. The worst-case memory requirement is $O(m^2)$. For long patterns, say, even a gene with tens of thousands characters, the table would demand several gigabytes or more and cannot be fit into the main memory.

Meek et al. [15] have attempted to use a suffix tree in the hard disk to speed up the dynamic programming for finding all local alignments. As expected, the success is limited to a small scale; their experiments are based on a text of length 40M and relatively short patterns with at most 65 characters. To alleviate the memory requirement of suffix trees, we exploit the recent breakthrough on compressed indexing, which reduces the space complexity from $O(n)$ bytes to $O(n)$ bits, while preserving similar searching time. FM-index [4,5], CSA (Compressed Suffix Array)[6,16], BWT (Burrow-Wheeler Text) [3] are among the best known examples. In fact, empirical studies have confirmed their practicality for indexing long biological sequences to support very efficient exact matching on a PC (e.g. [9,13]). For DNA sequences, BWT was found to be the most efficient and the memory requirement can be as small as $0.25n$ bytes. For the human genome, this requires only 1G memory and the whole index can reside in the main memory of a PC. Moreover, the construction time of a BWT index is shorter, our experiment shows that it takes only one hour for the human genome.

Our new tool. Based on a BWT index in the main memory, we have built a software called BWT-SW to find all local alignments using dynamic programming. This paper is devoted to the details of BWT-SW. Among others, we will present how to use a BWT index to emulate a suffix trie of the text (i.e., the tree structure of all the suffices of the text), how to modify the dynamic programming to allow pruning but without jeopardizing the completeness, and how to manage the DP tables.

BWT-SW performs very well in practice, even for long patterns. The pruning strategy is effective and terminates most of the paths at a very early stage. We have tested BWT-SW extensively with the human genome and random patterns of length from 500 to a hundred million. On average, a pattern of 500 characters [resp. 5,000 and 1M characters] requires at most 10 seconds [resp. 1 minute and 2.5 hours] (see Section 4.1 for more results). When compared with the Smith-Waterman algorithm, BWT-SW is at least a thousand times faster. As far as we know, BWT-SW is the first software that can find all local alignments efficiently in such a scale. We have also tested BWT-SW using different texts and patterns. In a rough sense, the timing figures of our experiments suggest that the time complexity could be in the order of $n^{0.628}m$. In Section 4, we will also present the experimental findings on the memory utilization due to the DP tables.

To better understand the efficiency of the dynamic programming and pruning, we have also analyzed mathematically the pure match/mismatch model (no gaps are allowed), and we found that for DNA alignment, the probability of a length-d path with positive score decreases exponentially with d, and the total number of entries filled in all the DP tables can be upper bounded by $O(n^{0.628}m)$. This also implies that the DP table is very sparse; in particular, when we extend a path, the number of positive entries also decreases exponentially and is eventually bounded by a constant. Thus, we can save a lot of space by storing only the entries with positive scores.

Note that BWT-ST is not meant to be a replacement of BLAST; BLAST is still several times faster than BWT-SW for long patterns and BLAST is accurate enough in most cases. Using BWT-SW, we found that BLAST may miss some significant alignments (with high similarity) that could be critical for biological research, but this occurs only rarely[1].

2 Preliminaries

2.1 Local Alignments with Affine Gap Penalty

Let x and y be two strings. A space is a special character not found in these strings.

- An alignment A of x and y maps x and y respectively to another two strings x' and y' that may contain spaces such that (i) $|x'| = |y'|$; and (ii) removing spaces from x' and y' should get back x and y, respectively; and (iii) for any i, $x'[i]$ and $y'[i]$ cannot be both spaces.
- A gap is a maximal substring of contiguous spaces in either x' or y'.
- An alignment A is composed of three kinds of regions. (i) Matched pair: $x'[i] = y'[i]$; (ii) Mismatched pair: $x'[i] \neq y'[i]$ and both are not spaces; (iii) Gap: either $x'[i..j]$ or $y'[i..j]$ is a gap. Only a matched pair has a positive score a, a mismatched pair has a negative score b and a gap of length r also has a negative score $g + rs$ where $g, s < 0$. For DNA, the most common scoring scheme (e.g. used by BLAST) makes $a = 1$, $b = -3$, $g = -5$, and $s = -2$.
- The score of the alignment A is the sum of the scores for all matched pairs, mismatched pairs, and gaps. The alignment score of x and y is defined as the maximum score among all possible alignments of x and y.

Let T be a text of n characters and let P be a pattern of m characters. The local alignment problem can be defined as follows. For any $1 \leq i \leq n$ and $1 \leq j \leq m$, compute the largest possible alignment score of $T[h..i]$ and $P[k..j]$ where $h \leq i$ and $k \leq j$ (i.e., the best alignment score of any substring of T ending at position i and any substring of P ending at position j). Furthermore, for biological applications, we are only interested in those $T[h..i]$ and $P[k..j]$ if their alignment score attains a score threshold H.

---

[1] We have conducted an experiment using 10,000 queries randomly selected from the human genome of length ranging from 100 to 1,000 and masked by DUST. If we ignore all practically insignificant alignments (with E-value less than 1×10^{-10}), BLAST only missed 0.03% of the alignments found by BWT-SW.

2.2 Suffix Trie and BWT

Suffix trie: Given a text T, a suffix trie for T is a tree comprising all suffices of T such that each edge is uniquely labeled with a character, and the concatenation of the edge labels on a path from the root to a leaf corresponds to a unique suffix of T. Each leaf stores the starting location of the corresponding suffix. Note that an pre-order traversal of a suffix trie can enumerate all suffices of T. Furthermore, if we compress every maximal path of degree-one nodes of the suffix trie, then we obtain the suffix tree of T.

BWT: The Burrows-Wheeler transform (BWT) [3] was invented as a compression technique. It was later extended to support pattern matching by Ferragina and Manzini [4]. Let T be a string of length n over an alphabet Σ. We assume that the last character of T is a special character $\$$ which is unique in T and is smaller than any character in Σ. The suffix array $SA[0, n-1]$ of T is an array of indexes such that $SA[i]$ stores the starting position of the ith-lexicographically smallest suffix, i.e., $T_{SA[i]} < T_{SA[i+1]}$ for all $i = 0, 1, \ldots, n-1$. And BWT of T is a permutation of T such that $BWT[i] = T[SA[i] - 1]$.

Given a string X, let $SA[i]$ and $SA[j]$ be the smallest and largest suffices of T that have X as the prefix. The range $[i, j]$ is referred to as the *SA range of* X. Given the SA range $[i, j]$ of X, finding the SA range $[p, q]$ of aX, for any character a, can be done using the backward search technique [4].

Lemma 1. *Let X be a string and a be a character. Suppose that the SA range of X and aX is $[i..j]$ and $[p..q]$, respectively. Then $p = C(a) + Occ(a, i-1) + 1$, and $q = C(a) + Occ(a, j)$, where $C(a)$ is the total number of characters in T that are lexicographically smaller than a and $Occ(a, i)$ is the total number of a's in $BWT[0..i]$.*

We can precompute $C(a)$ for all characters a and retrieve any entry in constant time. Using the auxiliary data structure introduced by [4], computing $Occ(x, i)$ also takes constant time. Then $[p, q]$ can be calculated from $[i, j]$ in constant time. As a remark, BWT can be constructed in a more efficient way than other indexes like suffix trees. We have implemented the construction algorithm of BWT described in [8]; it takes 50 minutes to construct the BWT of the human genome using a Pentinum D 3.6GHz PC.

3 DP Formulation and Pruning

We can solve the local alignment problem as follows. For any $1 \le i \le n$ and $1 \le j \le m$, we compute the answer in two phases.

- **Phase I.** For each $x \le i$, compute $A[x, i, j]$ which equals the largest alignment score of $T[x..i]$ and any substring of P ending at position j.
- **Phase II.** Return the largest score among all alignment scores $A[x, i, j]$ computed in Phase I for different x.

If we consider the computation involved in Phase I for all i, j together, it is equivalent to the following: for any substring X of T and for any $1 \leq j \leq m$, find the best alignment score of X and any substring of P ending at position j. In the following, we will show how to use a suffix trie of T to speed up this step. With a suffix trie, we can avoid aligning substrings of T that are identical. That is, we exploit the common prefix structure of a trie to avoid identical substrings to be aligned more than once. We use a pre-order traversal of the suffix trie to generate all distinct substrings of X. Also, we only need to consider substrings of T of length at most cm where c is usually a constant (for example, if the score for a match is 1 and the penalty for an insertion is a constant ≥ 1, $c \leq 2$).

For each node u of depth d ($d \leq cm$) in the suffix trie of T, let $X[1..d]$ be the substring represented by this node. There may be multiple occurrences of X in T and the starting positions of these occurrences, say $p_1, p_2, \ldots, p_w$, can be found by traversing the leaves of the subtree rooted at u. For each $1 \leq j \leq m$, we compute the best possible alignment score of X and any substring of P ending at position j. Effectively, we have computed $A[p_1 + d - 1, p_1, j], A[p_2 + d - 1, p_2, j], \ldots, A[p_w + d - 1, p_w, j]$ for all $1 \leq j \leq m$.

The rest of this section is divided into three parts: Section 3.1 shows how to make use of a BWT index to simulate a pre-order traversal of a suffix trie. Section 3.2 gives a simple dynamic programming to compute, for each node u on a path of the suffix trie and for all $1 \leq j \leq m$, the best alignment score of the substring represented by u and any substring of P ending at j. Section 3.3 shows that the dynamic programming on a path can be terminated as soon as we realize that no "meaningful" alignment can be produced.

3.1   Simulating Suffix Trie Traversal Using BWT

We can make use of the backward search technique on BWT to simulate the pre-order traversal of a suffix trie to enumerate the substrings. Based on Lemma 1, we have the following corollary.

Corollary 1. *Given the SA range $[i, j]$ of X in T, if the SA range $[p, q]$ of aX for a character a computed by Lemma 1 is invalid, that is, $p > q$, then aX does not exist in T.*

Since we use backward search, instead of constructing the BWT for T, we construct the BWT for the reversal of T. In other words, to check if the edge with label a exists from a node u representing the substring X in the suffix trie for T is equivalent to check if aX^{-1} exists in T^{-1}.

We can simulate the traversal of a suffix trie and enumerate the substrings represented by the nodes in the trie as follows. Assume that we are at node u that represents the substring X in the suffix trie for T and we have already found the SA range for X^{-1} in T^{-1} using the BWT. We can check the existence of an edge with label a from u based on the above corollary in $O(1)$ time by computing the SA range for aX^{-1} using the BWT of T^{-1}. Then, we enumerate the corresponding substring if the edge does exist and repeat the same procedure to traverse the tree.

3.2 Dynamic Programming

Consider a path from the root of the suffix trie. Below we present a dynamic programming to compute, for each node u on this path and for all $1 \leq j \leq m$, the best possible alignment score of the substring $X[1..d]$ represented by u and any substring of P ending at j.

For any $i \leq d$ and $j \leq m$, let $M^u(i,j)$ be the best alignment score of $X[1..i]$ and any substring of P ending at position j. Let $M_1^u(i,j)$ be the best possible alignment score of $X[1..i]$ and a substring of P ending at position j with $X[i]$ aligned with $P[j]$. Let $M_2^u(i,j)$ be the best possible alignment score of $X[1..i]$ and a substring of P ending at position j with $X[i]$ aligned with a space. Let $M_3^u(i,j)$ be the best possible alignment score of $X[1..i]$ and a substring of P ending at position j with $P[j]$ aligning with a space. The values of $M^u(d,j)$ shows the best alignment score of $X[1..d]$ and a substring of P ending at position j.

Initial conditions:

$$\begin{aligned}
M^u(0,j) &= 0 && \text{for } 0 \leq j \leq m. \\
M^u(i,0) &= -(g + is) && \text{for } 1 \leq i \leq d. \\
M_2^u(0,j) &= -\infty && \text{for } 0 \leq j \leq m. \\
M_3^u(i,0) &= -\infty && \text{for } 1 \leq i \leq d.
\end{aligned}$$

Recurrences (for $i > 1$, $j > 1$):

$$\begin{aligned}
M_1^u(i,j) &= M^u(i-1,j-1) + \delta(X[i], P[j]). \\
M_2^u(i,j) &= \max\{M_2^u(i-1,j) - s, M^u(i-1,j) - (g+s)\}. \\
M_3^u(i,j) &= \max\{M_3^u(i,j-1) - s, M^u(i,j-1) - (g+s)\}. \\
M^u(i,j) &= \max\{M_1^u(i,j), M_2^u(i,j), M_3^u(i,j)\}
\end{aligned}$$

where $\delta(X[i], P[j]) = a$ if $X[i] = P[j]$, otherwise $\delta(X[i], P[j]) = b$. (See Section 2 for definitions of a and b.)

Consider a child v of u. Denote the substring represented by v as $X[1..d]c$. Note that when we extend the dynamic programming from node u v, we only need to compute a new row at each dynamic programming table of u (e.g., $M^u(d+1,j)M_1^u(d+1,j)$, $M_2^u(d+1,j)$, $M_3^u(d+1,j)$ for all $1 \leq j \leq m$. If a traversal of the suffix trie would move from node u to its parent, we erase the last row of every dynamic programming table computed at u.

3.3 Modified Dynamic Programming and Pruning

In this subsection, we show how to modify the dynamic programming to enable an effective pruning. We first define what a meaningless alignment is.

Meaningless Alignment: Let A be an alignment of a substring $X = T[h..i]$ of T and a substring $Y = P[k..j]$ of P. If A aligns a prefix $X' = T[h..h']$ of X with a prefix $Y' = P[k..k']$ of Y such that the alignment score of X' and Y' is less than or equal to zero, A is said to be a *meaningless alignment*. Otherwise, A is said to be *meaningful*.

Lemma 2. *Suppose that A is a meaningless alignment of a substring $X = T[h..i]$ and a substring $Y = P[k..j]$ with a positive score C. Then there exists a meaningful alignment for some proper suffix $X' = T[s..i]$ of X and some proper suffix $Y' = P[t..j]$ of Y with score at least C, where $h < s \leq i$ and $k < t \leq j$.*

The proof of Lemma 2 will be given in the full paper. Below we show how to modify the dynamic programming to only compute the best possible score of meaningful alignments (*meaningful alignment score*). It is important to note that for any two strings, the best meaningful alignment score may not be the best alignment score. Nevertheless, we will show that the meaningful alignment scores are already sufficient for Phase II to report the correct answers.

DP for Meaningful Alignment Score: In the dynamic programming tables, entries with values less than or equal to zero will never be used.

Let u be a node in the suffix trie for T and $X[1..d]$ be the string represented by u. Let $N^u(i, j)$ be the best possible score of a *meaningful* alignment between $X[1..i]$ and a suffix of $P[1..j]$. Furthermore, $N_1^u(i, j)$, $N_2^u(i, j)$, and $N_w^u(i, j)$ are defined in a similar way as $M_1^u(i, j)$, $M_2^u(i, j)$, and $M_w^u(i, j)$. The recurrence equations are modified as follows. For any $i, j > 1$,

$$N_1^u(i, j) = \begin{cases} N^u(i-1, j-1) + \delta(X[i], P[j]) & \text{if } N^u(i-1, j-1) > 0 \text{ or } i = 1 \\ -\infty & \text{otherwise} \end{cases}$$

$$N_2^u(i, j) = \begin{cases} \max\{N_2^u(i-1, j) - s, \\ \quad\quad N^u(i-1, j) - (g+s)\} & \text{if } N_2^u(i-1, j) > 0, N^u(i-1, j) > 0 \\ N_2^u(i-1, j) - s & \text{if only } N_2^u(i-1, j) > 0 \\ N^u(i-1, j) - (g+s) & \text{if only } N^u(i-1, j) > 0 \\ -\infty & \text{otherwise} \end{cases}$$

$$N_3^u(i, j) = \begin{cases} \max\{N_3^u(i, j-1) - s, \\ \quad\quad N^u(i, j-1) - (g+s)\} & \text{if } N_3^u(i, j-1) > 0, N^u(i, j-1) > 0 \\ N_3^u(i, j-1) - s & \text{if only } N_3^u(i, j-1) > 0 \\ N^u(i, j-1) - (g+s) & \text{if only } N^u(i, j-1) > 0 \\ -\infty & \text{otherwise} \end{cases}$$

$$N^u(i, j) = \max\{N_1^u(i, j), N_2^u(i, j), N_3^u(i, j)\}$$

Next, we show that the scores computed by the modified dynamic programming are sufficient for Phase II to compute the correct answers, thus solving the local alignment problem.

Lemma 3. *Let u be a node in the suffix trie for T and let $X[1..d]$ be the string represented by u. If $M^u(d, j) = C \geq H$ where H is the score threshold, then, there exists h in $[1, d]$ such that $N^v(d - h + 1, j) = C$ where v is the node in the suffix trie representing the string $X[h..d]$.*

Proof. If there exists a meaningful alignment for $X[1..d]$ and $P[k..j]$ with score $= C$, $h = 1$ and $v = u$. Otherwise, based on Lemma 2, there exists h with $1 < h \leq d$ such that there is a meaningful alignment for $X[h..d]$ and $P[k..j]$ with score at least C. Since $M^u(d, j)$ is the best possible score for $X[1..d]$ and any substring of P ending at j, so $N^v(d - h + 1, j) = C$ where v is the node representing $X[h..d]$.

Corollary 2. *For any i, j, let C be the largest possible score between a substring of T ending at i and a substring of P ending at j (i.e., C is the answer for Phase II). Then there exists a node v representing a substring $X = T[s..i]$ $(s \leq i)$ of T such that $N^v(i - s + 1, j) = C$.*

Pruning Strategy: Since we only consider meaningful alignments, for each node in the suffix trie, when filling the dynamic programming tables, we ignore all entries with values less than or equal to zero. For a node u, if there is a row with all entries in all dynamic programming tables with values less than or equal to zero, we can stop filling the tables since all the rows below will only contain entries with values less than or equal to zero. Moreover, based on the same argument, we can prune the whole subtree rooted at u.

4   Experiments and Mathematical Analysis

Note that DNA sequence is composed of double strands (i.e., two complimentary sequences of the same length bind together). Instead of aligning a pattern with both strands, we first search the pattern P and then its reverse with one strand. The searching time reported in all experiments is the total time for searching both P and its reverse. To avoid meaningless alignment, some regions of the query (e.g. a long sequence of "a") that are expected to contain very little information (called *low complexity regions*) are masked by a standard software tool, DUST, before it is used for testing. This is also the default setting of existing software, such as BLAST. All experiments are performed on a Pentinum D 3.0GHz PC with 4G memory.

4.1   Performance of BWT-SW

We have constructed the BWT index for human genome and used BWT-SW to align patterns of length from 500 to 100M with the human genome. The query patterns are randomly selected from the mouse genome except for the query of length 100M which is the whole mouse chromosome 15. For queries of lengths of 10K or shorter, we have repeated the same experiment at least hundred times to get the average time. For longer patterns, we have repeated the experiments at least a few dozen times. The following table shows the results.

Query Length	100	200	500	1K	2K	5K	10K	100K	1M	10M	100M
Time (seconds)	1.91	4.02	9.89	18.86	35.93	81.60	161.04	1.4K	8.9K	34.4K	218.2K

For patterns with thousands of characters (which is common in biological research), BWT-SW takes about one to two minutes, which is very reasonable.

Even for extremely long patterns, say, a chromosome of 100M, it takes about 2.5 days. In the past, finding a complete set of local alignments for such long patterns is not feasible.

To investigate how the searching time depends on the text size, we fix the pattern length and conduct experiments using different texts (chromosomes) of length ranging from 100M to 3G. We have repeated the study for 4 different query lengths. The following table shows the results.

Pattern Length	Text Size				
	114M	307M	1.04G	2.04G	3.08G
500	1.33	2.41	5.21	7.89	9.89
1K	2.55	4.59	10.05	15.14	18.86
5K	10.74	19.53	42.20	65.67	81.60
10K	21.01	38.20	83.96	128.97	161.04

Using the above figures, we roughly estimate that the time complexity of BWT-SW is in the order of $n^{0.628}m$.[2] However, our experiments are limited, and such estimation is not conclusive. It only provides a rough explanation why BWT-SW is a thousand times faster than the Smith-Waterman algorithm when aligning the human genome.

4.2   Mathematical Analysis

To better understand the performance of BWT-SW, we have studied and analyzed a simplified model in which an alignment cannot insert spaces or gaps, and the scoring function is simply a weighted sum of the number of matched and mismatched pairs. We found that under this model, the expected total number of DP cells with positive values is upper bounded by $69n^{0.628}m$. The time required by BWT-SW is proportional to the number of DP cells to be filled, or equivalently, the number of cells with positive values. Thus, our analysis suggests that BWT-SW takes $O(n^{0.628}m)$ time under this model.[3]

We assume that strings are over an alphabet of σ characters where σ is a constant. Let x, y be strings with $d \geq 1$ characters. Suppose that x and y match in $a \leq d$ positions. Define $Score(x, y) = a - 3(d - a)$ (i.e., match = 1, mismatch = -3) and define $f(x)$ to be the number of length-d strings y such that $Score(x, y) > 0$. Note that $f(x) = f(x')$ for any length-d string x'. Thus, we define $f(d) = f(x)$.

Lemma 4. $f(d) \leq k_1 k_2^d$, where $k_1 = \frac{\sigma-1}{\sigma-2} \frac{4e}{\sqrt{2\pi}}$, and $k_2 = (4e(\sigma-1))^{1/4}$.

---

[2] For each pattern length, we fit the above data to the function $f(n) = cn^{0.628}$ where c is a fixed constant. The root mean square errors of the data in all four cases are within 1.21% to 1.63%.

[3] It is perhaps a coincidence that the dependency on n is found to match the experimental result in Section 4.1. Note that Sections 4.1 and 4.2 are based on different alignment models, one with gaps and one without. It has yet to be verified whether the analysis can be extended to the gapped model with a similar upper bound.

Proof. Let x be a string of length d. $f(d)$ is the number of strings of length-d that has at most $\lfloor d/4 \rfloor$ differences with x. Using the fact that $\binom{a}{b} \leq \frac{1}{\sqrt{2b\pi}}\left(\frac{ae}{b}\right)^b$ (derived from Stirling's approximation), we have $\binom{d}{\lfloor d/4 \rfloor} \leq \binom{d}{\lceil d/4 \rceil} \leq \frac{1}{\sqrt{2d\pi}}\left(\frac{de}{\lceil d/4 \rceil}\right)^{\lceil d/4 \rceil}$ $\leq \frac{1}{\sqrt{2\pi}}\left(\frac{de}{d/4}\right)^{d/4+1} = \frac{4e}{\sqrt{2\pi}}(4e)^{d/4}$. Then, $f(d) \leq \sum_{i=0}^{\lfloor d/4 \rfloor}(\sigma-1)^i\binom{d}{i} \leq \left(\frac{\sigma-1}{\sigma-2}\right)(\sigma-1)^{d/4}\binom{d}{\lfloor d/4 \rfloor} = k_1 k_2^d$, where $k_1 = \frac{\sigma-1}{\sigma-2}\frac{4e}{\sqrt{2\pi}}$ and $k_2 = (4e(\sigma-1))^{1/4}$.

There are σ^d strings of length d, and the following fact follows.

Fact 1. *Let x be a string of length d, for any randomly chosen length-d string y, the probability that $Score(x, y) > 0$ is $f(d)/\sigma^d$.*

Let T be a text of n characters and let R be the suffix trie of T. Let P be a pattern of m characters. For any node u in R, let $X[1..d]$ be the string represented by u. Let N^u be the dynamic programming table for u such that $N^u(d, j)$ denote $Score(X, P')$ where P' is a length-d substring of P ending at position j. In the following, we try to bound the expected total number of positive entries $N^u(d, j)$ for all u, d, j. Let $c = \lfloor \log_\sigma n \rfloor$.

Lemma 5. *The expected total number of positive entries $N^u(d, j)$ for all nodes u at depth d is at most $mf(d)$, if $d \leq c$, and $m\lceil n\frac{f(d)}{\sigma^d}\rceil$, if $d > c$.*

Lemma 6. *For any d in $[1, c-1]$, the expected total number of positive entries $N^u(d, j)$ for all nodes u of depth d is at most $\sum_{d=1}^{c-1} mf(d) \leq c_1 mn^{c_2}$ where c_1, c_2 are constants and $c_2 < 1$.*

Lemma 7. *For all d, j and node u of depth d in $[c, m]$, the expected total number of positive entries $N^u(d, j)$ for all nodes at depth d is at most $\sum_{d=c}^{m} nm\frac{f(d)}{\sigma^d}$ $\leq c_1' mn^{c_2'}$ where c_1', c_2' are constants and $c_2' < 1$.*

Based on Lemmas 6 and 7, we have the following corollary.

Corollary 3. *The expected total number of positive entries $N^u(d, j)$ for all u, d, j is $69mn^{0.628}$.*

4.3   Memory for DP Tables

We only need to store entries of DP tables with positive scores, we store them in a compact manner as follows. We use a big single array $\mathbf{B}$ to store entries of all rows of a DP table N. Let n_i be the number of entries, say $N(i, j_{i_1})$, $N(i, j_{i_2})$, ..., $N(i, v_{j_{n_1}})$, in the i-th row. Let $k_i = \sum_{r=1}^{i} n_r$, the entry $N(i, j_{i_\ell})$ is stored at $\mathbf{B}[k_{i-1} + \ell]$. For each entry, we store the coordinates (i, j_{i_ℓ}) and the score. We also store the starting index of $\mathbf{B}$ for each row. Moving from node u at depth y of the trie to its child v, we add a new row for v starting at $\mathbf{B}[k_y + 1]$. If we go up from node v to u, we reuse the entries in $\mathbf{B}$ from $\mathbf{B}[k_y + 1]$.

Regarding the memory required by the DP tables, it is related to the maximum number of table entries to be maintained throughout the whole searching process. This number is very small in all test cases. For example, for human genome as the text, the maximum number of entries in the DP tables are about 2600 and 22,000 for patterns of length 100K and 1M, respectively. The actual memory required are about 20K and 10M, respectively. In fact, based on the simplified model of Section 4.2, we can show that the expected number of DP table entries along any path of the suffix trie is bounded by cm where c is a constant and m is the pattern length. The memory required for DP tables is neglectable when compared to the memory for the BWT data structure: For a DNA sequence of n characters, $2n$ bits are needed for BWT entries; $2n$ bits for storing the original sequence; $1n$ bits for the auxiliary data structures. So, altogether it translates to about 2G memory for the human genome.

References

1. Altschul, S.F., Gish, W., Miller, W., Myers, E.W., Lipman, D.J.: Basic local alignment search tool. Journal of Molecular Biology 215(3), 403–410 (1990)
2. Altschul, S.F., Schaffer, T.L., Zhang, A.A., Miller, Z., Lipman, D.J.: Gapped BLAST and PSI-BLAST: A new generation of protein database search programs. Nucleic Acids Research 25, 3389–3402 (1997)
3. Burrow, M., Wheeler, D.J.: A block-sorting lossless data compression algorithm. Technical Report 124, Digital Equipment Corporation, California (1994)
4. Ferragina, P., Manzini, G.: Opportunistic data structures with applications. In: FOCS, pp. 390–398 (2000)
5. Ferragina, P., Manzini, G.: An experimental study of an opportunistic index. In: SODA, pp. 269–278 (2001)
6. Grossi, R., Vitter, J.S.: Compressed suffix arrays and suffix trees with applications to text indexing and string matching. In: STOC, pp. 397–406 (2000)
7. Gusfield, D.: Algorithms on Strings, Trees, and Sequences. Cambridge University Press, Cambridge (1997)
8. Hon, W.K., Lam, T.W., Sadakane, K., Sung, W.K.: Constructing compressed suffix arrays with large alphabets. In: ISAAC, pp. 240–249, 2003 (to appear in Algorithmica)
9. Hon, W.K., Lam, T.W., Sung, W.K., Tse, W.L., Wong, C.K., Yiu, S.M.: Practical aspects of compressed suffix arrays and FM-Index in searching DNA sequences. In: ALENEX/ANALC, pp. 31–38 (2004)
10. Hunt, E., Atkinson, M.P., Irving, R.W.: Database indexing for large DNA and protein sequence collections. The VLDB Journal 11, 256–271 (2002)
11. Kurtz, S.: Reducing the space requirement of suffix trees. Software - Practice and Experience 29(13), 1149–1171 (1999)
12. Li, M., Ma, B., Kisman, D., Tromp, J.: PatterHunter II: Highly sensitive and fast homology search. Journal of Bioinformatics and Computational Biology 2(3), 417–440 (2004)
13. Lippert, R.: Space-efficient whole genome comparisons with Burrows-Wheeler transforms. Journal of Computational Biology 12(4), 407–415 (2005)

254   T.-W. Lam et al.

14. McCreight, E.M.: A space-economical suffix tree construction algorithm. J. ACM 23(2), 262–272 (1976)
15. Meek, C., Patel, J.M., Kasetty, S.: OASIS: An online and accurate technique for local-alignment searches on biological sequences. In: VLDB, pp. 910–921 (2003)
16. Sadakane, K.: New text indexing functionalities of the compressed suffix arrays. Journal of Algorithms 48(2), 294–313 (2003)
17. The BLAST Web Site: http://130.14.29.110/BLAST/
18. Smith, T.F., Waterman, M.S.: Identification of common molecular subsequences. Journal of Molecular Biology 147, 195–197 (1981)

Secure Multiparty Computations
Using the 15 Puzzle
(Extended Abstract)

Takaaki Mizuki[1], Yoshinori Kugimoto[2], and Hideaki Sone[1]

[1] Information Synergy Center, Tohoku University,
Aramaki-Aza-Aoba 6-3, Aoba-ku, Sendai 980-8578, Japan
`tm-paper@rd.isc.tohoku.ac.jp`
[2] Sone Lab., Graduate School of Information Sciences, Tohoku University,
Aramaki-Aza-Aoba 6-3, Aoba-ku, Sendai 980-8578, Japan

Abstract. This paper first considers the use of the "15 puzzle," which is one of the most famous sliding-block puzzles, to provide secure multiparty computations. That is, we design a class of 15-puzzle-based protocols for securely computing Boolean functions. Specifically, we show that any function of 4 variables (or less) and any symmetric function of 14 variables (or less) can be securely computed by a 15-puzzle-based protocol; furthermore, we present a 5-variable function and a 15-variable symmetric function, both of which cannot be securely computed by any protocol in the class.

1 Introduction

This paper first considers the use of the "15 puzzle" to provide secure multiparty computations. The 15 puzzle (illustrated in Fig. 1) is one of the most famous sliding-block puzzles, and we feel that most people might either own the 15 puzzles in their homes or have played them at least once. Therefore, this paper will give an application of a "ubiquitous" physical device to designing a cryptographic protocol.

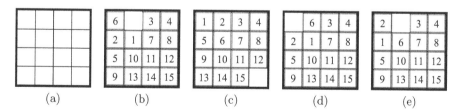

Fig. 1. Illustrations of the 15 puzzle

1.1 The 15 Puzzle

The 15 puzzle consists of a 4×4 *board* like in Fig. 1(a) and fifteen square *tiles* $\boxed{1}, \boxed{2}, \ldots, \boxed{15}$ numbered from 1 to 15. The goal of the 15 puzzle is as follows:

A. Dress, Y. Xu, and B. Zhu (Eds.): COCOA 2007, LNCS 4616, pp. 255–266, 2007.
© Springer-Verlag Berlin Heidelberg 2007

put these 15 tiles on the 4×4 board in an arbitrary order, say as in Fig. 1(b), then repeat to slide some tile into the empty space ☐ (which we will call the "blank") until the arrangement becomes *regularly ordered* as shown in Fig. 1(c). A comprehensive survey of the history of the 15 puzzle has been provided recently by Slocum and Sonneveld [9].

Throughout the paper, we regard 'sliding a tile' as 'moving the *blank* ☐,' and we denote such a *move* by an arrow $\uparrow$, $\downarrow$, $\rightarrow$ or $\leftarrow$. For example, each of three moves $\downarrow$, $\rightarrow$ and $\leftarrow$ is "applicable" to the arrangement (b) in Fig. 1; if one applies a move $\leftarrow$ to the arrangement (b), then the resulting arrangement is (d) in Fig. 1. For further example, applying a sequence $\leftarrow\downarrow\rightarrow\uparrow$ of four moves to the arrangement (b) in Fig. 1 results in "rotating" the three tiles 6, 2, 1 clockwise so that the resulting arrangement becomes (e) in Fig. 1. Note that a sequence

$$\leftarrow\downarrow\rightarrow\uparrow\leftarrow\downarrow\downarrow\rightarrow\rightarrow\rightarrow$$

is a solution to the 15 puzzle whose initial arrangement is (b) in Fig. 1: one can easily observe that, given the arrangement (b), this sequence rearranges the 15 tiles in the regular order (like in Fig. 1(c)).

1.2 An Example of a 15-Puzzle-Based Secure Computation

As mentioned in the beginning of this paper, our goal is to apply the physical property of the 15 puzzle to designing cryptographic protocols for secure multiparty computations. We now present a simple example of our protocols for a secure computation using the 15 puzzle. Assume that two honest-but-curious players P_1 and P_2, who hold one-bit private inputs $x_1, x_2 \in \{T, F\}$ respectively, wish to *securely compute* the Boolean AND function $\mathsf{AND}^2(x_1, x_2) = x_1 \wedge x_2$, i.e., they with to learn the value of $x_1 \wedge x_2$ without revealing more information about their inputs than necessary.

Before describing the protocol, we make an important assumption on a 15 puzzle used in this paper: the back sides of all the 15 tiles 1, 2, ..., 15 are assumed to be identical; we denote such an identical back side by ?. As will be seen later, one of the main ideas behind our protocols is to put a tile on the board with its face down. (Of course, we understand that not every 15 puzzle on the market satisfies such an assumption; however, we would believe that many toy stores sell a 15 puzzle satisfying the assumption above at a low price.)

The following protocol achieves a secure computation of AND^2, assuming that players P_1 and P_2 have a 15 puzzle whose arrangement is (b) in Fig. 1.

1. Player P_1 or P_2 draws the tile 1 from the board, turns over the tile, and puts it back to the board (with its face down). In addition, player P_1 or P_2 draws the two tiles 2 and 6 from the board, turns over the two tiles, shuffles them, and puts them back to the board (with keeping their faces down). Thus, the upper left four cells in the current arrangement satisfy either

or.

(Note that neither P_1 nor P_2 knows which is the current arrangement.)

2. Player P_1 moves (or does not move) the blank $\square$ depending on her private input $x_1 \in \{T, F\}$ without being seen by player P_2:

  (a) if $x_1 = T$, then P_1 applies a sequence $\leftarrow\downarrow\rightarrow\uparrow$ of four moves (remember that this sequence rotates the three tiles $\boxed{?}, \boxed{?}, \boxed{?}$ clockwise);

  (b) if $x_1 = F$, then P_1 never moves the blank.

3. Similarly, P_2 moves (or does not move) the blank $\square$ depending on her private input $x_2 \in \{T, F\}$ without being seen by P_1:

  (a) if $x_2 = T$, then P_2 applies a sequence $\leftarrow\downarrow\rightarrow\uparrow$;

  (b) if $x_2 = F$, then P_2 never moves the blank.

4. Notice that, only when $x_1 = x_2 = T$, the face-down tile $\boxed{?}$ of $\boxed{1}$ comes up to the uppermost leftmost cell:

Player P_1 or P_2 turns over the uppermost leftmost tile. If the face-up tile is $\boxed{1}$, then it implies $x_1 \wedge x_2 = T$; if the face-up tile is either $\boxed{2}$ or $\boxed{6}$, then it implies $x_1 \wedge x_2 = F$.

Note that, when the face-up tile is either $\boxed{2}$ or $\boxed{6}$ in step 4, this information gives only the fact that (x_1, x_2) is (F, F) or (T, F) or (F, T) because of shuffling the two face-down tiles $\boxed{?}, \boxed{?}$ of $\boxed{2}, \boxed{6}$ in step 1. Thus, the protocol above securely computes the function $\mathsf{AND}^2(x_1, x_2) = x_1 \wedge x_2$. We name this secure AND protocol the protocol $\mathcal{P}_{\mathsf{AND}^2}$.

It should be noted that, after the protocol $\mathcal{P}_{\mathsf{AND}^2}$ terminates, to keep the secrecy, one has to shuffle the remaining two face-down tiles $\boxed{?}, \boxed{?}$ (before turning over them) when $x_1 \wedge x_2 = F$.

1.3 Our Results and Related Work

Assume that n honest-but-curious players $P_1, P_2, \ldots, P_n$, who hold one-bit private inputs $x_1, x_2, \ldots, x_n \in \{T, F\}$ respectively, wish to securely compute a (Boolean) function $f(x_1, x_2, \ldots, x_n)$. (Hereafter, we use simply the term 'function' to refer to a Boolean function.) In this paper, to achieve the end, we first design a class of protocols using the 15 puzzle. Then, we show that any function $f(x_1, x_2, \ldots, x_n)$ with $n \leq 4$ can be securely computed by the 15 puzzle, i.e., any function of 4 variables or less is "15puz-computable" (as in Theorem 2). We also show that any symmetric function $f(x_1, x_2, \ldots, x_n)$ with $n \leq 14$ can be securely computed by the 15 puzzle, i.e., any symmetric function of 14 variables or less is 15puz-computable (as in Theorem 1). On the other hand, we prove that there exist a 5-variable function and a 15-variable symmetric function, both of which are not 15puz-computable (as in Theorems 3 and 4).

Our work falls into the area of *recreational cryptography* [1] or *human-centric cryptography* [8]. There exist other handy physical devices which can implement some cryptographic tasks (without any use of computers) in the literature: some examples are envelopes [4], cups [4], a deck of cards [2,3,5,6], a PEZ dispenser [1] and scratch-off cards [7].

The remainder of the paper is organized as follows. In Section 2, we formalize a class of 15-puzzle-based protocols. Then, we construct our protocols for symmetric functions in Section 3, and construct those for general functions in Section 4. (Since the protocols for symmetric functions are simpler than those for general functions, the former will be presented before the latter.) In Section 5, we present a few functions which are not 15puz-computable. This paper concludes in Section 6 with some discussions.

2   Formalizing 15-Puzzle-Based Protocols

In this section, we formalize secure multiparty computations using the 15 puzzle, that is, we straightforwardly extend the secure AND protocol $\mathcal{P}_{\mathsf{AND}^2}$ illustrated in Section 1.2 to a class of 15-puzzle-based protocols for securely computing n-variable (general) functions.

We first construct our class of 15-puzzle-based protocols in Section 2.1, and then abstract away the concrete 15 puzzles in Section 2.2.

2.1   Our Class of Protocols

First, remember the steps 2 and 3 of the protocol $\mathcal{P}_{\mathsf{AND}^2}$ described in Section 1.2: player P_1 (resp. P_2) applies a sequence $\leftarrow\downarrow\rightarrow\uparrow$ of four moves to the 15 puzzle when $x_1 = T$ (resp. $x_2 = T$); we call such a sequence of moves an "action" of a player.

Definition 1. *A finite string* $\alpha \in \{\uparrow, \downarrow, \rightarrow, \leftarrow\}^*$ *is called an* action.

For example, for every $i \in \{1, 2\}$ and $b \in \{T, F\}$, let α_i^b be the action of player P_i when $x_i = b$ in the protocol $\mathcal{P}_{\mathsf{AND}^2}$, then we have $\alpha_1^T = \alpha_2^T = \leftarrow\downarrow\rightarrow\uparrow$ and $\alpha_1^F = \alpha_2^F = \varepsilon$, where ε denotes the empty string. Notice that the location of the blank $\square$ does not change before and after the action $\leftarrow\downarrow\rightarrow\uparrow$; we make the following definition for changes of the location.

Definition 2. *For an action* α, *we define* $\|\alpha\|$ *as*

$$\|\alpha\| = (N_{\mathrm{up}} - N_{\mathrm{down}}, N_{\mathrm{right}} - N_{\mathrm{left}})$$

where N_{up}, N_{down}, N_{right} *and* N_{left} *are the numbers of arrows* $\uparrow$'s, $\downarrow$'s, $\rightarrow$'s *and* $\leftarrow$'s *in the string* α, *respectively.*

Note that $\|\alpha_1^T\| = \|\alpha_1^F\| (= (0,0))$ and $\|\alpha_2^T\| = \|\alpha_2^F\| (= (0,0))$ for the actions in $\mathcal{P}_{\mathsf{AND}^2}$ above. Furthermore, it should be noted that, if we designed P_1's actions with $\|\alpha_1^T\| \neq \|\alpha_1^F\|$, then player P_2 could get to know the value of P_1's input x_1 just by looking at the location of the blank $\square$ in her turn.

Next, remember the steps 1 and 4 of the protocol $\mathcal{P}_{\mathsf{AND}^2}$: the tile $\boxed{1}$ can be regarded virtually as a *true-tile* like "$\boxed{\mathrm{T}}$," and each of the tiles $\boxed{2}, \boxed{6}$ can be regarded virtually as a *false-tile* like "$\boxed{F}$." Furthermore, recall that the two face-down tiles $\boxed{?},\boxed{?}$ of $\boxed{2},\boxed{6}$ are shuffled in the step 1.

We are now ready to straightforwardly construct a general protocol as follows, assuming that n players $P_1, P_2, \ldots, P_n$ (holding private inputs $x_1, x_2, \ldots, x_n \in \{T, F\}$, respectively) have a 15 puzzle.

1. Divide all the 15 tiles into true-tiles and false-tiles arbitrarily. Turn over all the true-tiles, shuffle all the (face-down) true-tiles, and put these tiles on the board in arbitrary positions (with keeping their faces down). Similarly, turn over all the false-tiles, shuffle them, and put them in arbitrary remaining positions.
2. Set $i := 1$.
3. Player P_i moves the blank $\square$ depending on her private input $x_i \in \{T, F\}$ without being seen by any other player:
   (a) if $x_i = T$, then P_i applies an action α_i^T;
   (b) if $x_i = F$, then P_i applies an action α_i^F,
   where α_i^T and α_i^F must satisfy $\|\alpha_i^T\| = \|\alpha_i^F\|$.
4. Set $i := i + 1$. If $i \leq n$, then return to step 3.
5. Turn over the uppermost leftmost tile. If the face-up tile is a true-tile, then all the players recognize that the output is T; if the face-up tile is a false-tile, then they recognize that the output is F.

Thus, a protocol is specified by determining both an initial arrangement (of true-tiles and false-tiles) in step 1 and players' actions

$$(\alpha_1^T, \alpha_1^F), (\alpha_2^T, \alpha_2^F), \ldots, (\alpha_n^T, \alpha_n^F)$$

in step 3.

For simplicity, to represent such an arrangement of true-tiles and false-tiles, we use a 4×4 matrix as in the following Definition 3 (as if we considered a "binary 15 puzzle," whose tiles were $\boxed{T}$'s or $\boxed{F}$'s).

Definition 3. *A 4×4 matrix C is called a* configuration *if C contains exactly one element $\flat$ (denoting the blank $\square$) and the remaining 15 elements are T or F.*

For example,

$$C_0 = \begin{pmatrix} F & \flat & F & F \\ F & T & F & F \\ F & F & F & F \\ F & F & F & F \end{pmatrix} \tag{1}$$

can be an (appropriate) initial configuration for the protocol $\mathcal{P}_{\mathsf{AND}^2}$.

As mentioned above, to specify a protocol, it suffices to determine an initial configuration C_0 and players' actions $(\alpha_1^T, \alpha_1^F), (\alpha_2^T, \alpha_2^F), \ldots, (\alpha_n^T, \alpha_n^F)$; a formal treatment will be given in the next subsection.

2.2   Abstraction

In this subsection, we give a mathematical definition of a protocol, which abstracts away the concrete 15 puzzles.

We first define the "applicability" of moves and actions as in the following Definitions 4 and 5.

Definition 4. *Let C be a configuration such that the blank $\flat$ lies on an (i,j)-entry. If $2 \leq i \leq 4$, then we say that a move $\uparrow$ is applicable to C and we denote by $C \triangleleft \uparrow$ the configuration resulting from exchanging the (i,j)-entry and the $(i+1,j)$-entry. We also define $C \triangleleft \downarrow$, $C \triangleleft \rightarrow$ and $C \triangleleft \leftarrow$ in a similar way.*

Definition 5. *Let C be a configuration, and let α be an action. We define 'α is applicable to C' in a similar way as done in Definition 4, and we denote by $C \triangleleft \alpha$ the resulting configuration.*

We are now ready to give a formal definition of a protocol. Hereafter, we denote by $\circ$ a concatenation of two strings.

Definition 6. *A* protocol $\mathcal{P} = ((\alpha_1^T, \alpha_1^F), (\alpha_2^T, \alpha_2^F), \dots, (\alpha_n^T, \alpha_n^F); C_0)$ *consists of players' actions* $(\alpha_1^T, \alpha_1^F), (\alpha_2^T, \alpha_2^F), \dots, (\alpha_n^T, \alpha_n^F)$ *and an initial configuration C_0 such that*

- *$\|\alpha_i^T\| = \|\alpha_i^F\|$ for every i, $1 \leq i \leq n$; and*
- *$\alpha_1^{x_1} \circ \alpha_2^{x_2} \circ \cdots \circ \alpha_n^{x_n}$ is applicable to C_0 for every $(x_1, x_2, \dots, x_n) \in \{T, F\}^n$.*

For example, one can formally verify that the protocol

$$\mathcal{P}_{\mathsf{AND}^2} = ((\leftarrow\downarrow\rightarrow\uparrow, \varepsilon), (\leftarrow\downarrow\rightarrow\uparrow, \varepsilon); C_0) \tag{2}$$

satisfies the condition in Definition 6, where C_0 is given in Eq. (1).

The following Definitions 7 and 8 concern "secure computability," where upperleft(C) denotes the value of the $(1,1)$-entry in a configuration C.

Definition 7. *We say that a protocol*

$$\mathcal{P} = ((\alpha_1^T, \alpha_1^F), (\alpha_2^T, \alpha_2^F), \dots, (\alpha_n^T, \alpha_n^F); C_0)$$

securely computes an n-variable function f if

$$f(x_1, x_2, \dots, x_n) = \mathsf{upperleft}(C_0 \triangleleft \alpha_1^{x_1} \circ \alpha_2^{x_2} \circ \cdots \circ \alpha_n^{x_n})$$

for every $(x_1, x_2, \dots, x_n) \in \{T, F\}^n$.

Definition 8. *A function f is said to be* 15puz-computable *if there exists a protocol which securely computes f.*

For example, one can formally verify that the protocol $\mathcal{P}_{\mathsf{AND}^2}$ (given in Eq. (2)) securely computes the function $\mathsf{AND}^2(x_1, x_2) = x_1 \wedge x_2$, and hence AND^2 is 15puz-computable.

In the next three sections, we focus on 15puz-computability (and non-15puz-computability) of a function; specifically, we will show that

- any function of 4 variables is 15puz-computable (§4);
- any symmetric function of 14 variables is 15puz-computable (§3);
- there is a non-15puz-computable function of 5 variables (§5); and
- there is a non-15puz-computable symmetric function of 15 variables (§5).

3   Protocols for 14-Variable Symmetric Functions

In this section, we propose protocols for arbitrary symmetric functions of 14 variables or less. Note that, when a demand for a secure multiparty computation arises, it is a natural setting that all players are "symmetric," i.e., all the players have the same circumstances.

We first review symmetric functions in Section 3.1, and then specify our protocols in Section 3.2.

3.1   Symmetric Functions

In this subsection, we present a notation for symmetric functions.

An n-variable function f is said to be *symmetric* if it is unchanged by any permutation of its variables, that is,

$$f(x_1, \ldots, x_i, \ldots, x_j, \ldots, x_n) = f(x_1, \ldots, x_j, \ldots, x_i, \ldots, x_n)$$

for any variables x_i and x_j.

It is well-known that any n-variable symmetric function $f(x_1, x_2, \ldots, x_n)$ can be characterized by a unique set $A \subseteq \{0, 1, \ldots, n\}$: for any n-variable symmetric function f, there exists a set

$$A \subseteq \{0, 1, \ldots, n\}$$

such that

$$f(x_1, x_2, \ldots, x_n) = \begin{cases} T & \text{if } |\{i \mid x_i = T, 1 \le i \le n\}| \in A; \\ F & \text{otherwise.} \end{cases}$$

For such a set $A \subseteq \{0, 1, \ldots, n\}$, we denote the corresponding n-variable symmetric function by Sym_A^n.

3.2   Our Protocols

In this subsection, we describe our protocols for symmetric functions of 14 variables. Consider an arbitrary 14-variable symmetric function

$$\mathsf{Sym}_A^{14}(x_1, x_2, \ldots, x_{14})$$

(where a set $A \subseteq \{0, 1, \ldots, 14\}$ is arbitrary). We are going to construct a protocol $\mathcal{P}_{\mathsf{Sym}_A^{14}} = ((\alpha_1^T, \alpha_1^F), (\alpha_2^T, \alpha_2^F), \ldots, (\alpha_{14}^T, \alpha_{14}^F); C_0)$ which securely computes the symmetric function Sym_A^{14}.

To determine its initial configuration C_0, we first consider 15 binary values $b_0, b_1, \ldots, b_{14}$ such that

$$b_i = \begin{cases} T & i \in A; \\ F & i \notin A \end{cases}$$

Fig. 2. A "Hamilton cycle" on the board

for every i, $0 \le i \le 14$. (Note that, for an input vector $(x_1, x_2, \ldots, x_{14})$ having a number ℓ of the true values T's, $\mathsf{Sym}_A^{14}(x_1, x_2, \ldots, x_{14}) = b_\ell$.) Then, imagining a "Hamilton cycle" depicted in Figure 2, we place the 15 binary values $b_0, b_1, \ldots, b_{14}$ along the cycle, as follows:

$$
C_0 = \begin{pmatrix} b_0 & \flat & b_{14} & b_{13} \\ b_1 & b_2 & b_{11} & b_{12} \\ b_4 & b_3 & b_{10} & b_9 \\ b_5 & b_6 & b_7 & b_8 \end{pmatrix},
$$

which we adopt as the initial configuration of $\mathcal{P}_{\mathsf{Sym}_A^{14}}$.

Next, remember that the $(1,1)$-entry in the final configuration becomes the output. Then, given an input vector $(x_1, x_2, \ldots, x_{14})$ having a number ℓ of T's, in order for the "correct output" b_ℓ to come up to the $(1,1)$-entry, one may intuitively notice that each player P_i with $x_i = T$ should "rotate" the 15 binary values $b_0, b_1, \ldots, b_{14}$ clockwise along the "Hamilton cycle" (in Figure 2); hence, we determine the players' actions as follows:

$$
\alpha_1^T = \alpha_2^T = \cdots = \alpha_{14}^T = \ \leftarrow\downarrow\rightarrow\downarrow\leftarrow\downarrow\rightarrow\rightarrow\uparrow\leftarrow\uparrow\rightarrow\uparrow\leftarrow\leftarrow
$$

and

$$
\alpha_1^F = \alpha_2^F = \cdots = \alpha_{14}^F = \varepsilon.
$$

Thus, for each turn of player P_i, only when $x_i = T$, each of the 15 binary values "proceeds clockwise" along the cycle. Therefore, we have

$$
\mathsf{Sym}_A^{14}(x_1, x_2, \ldots, x_{14}) = \mathsf{upperleft}(C_0 \triangleleft \alpha_1^{x_1} \circ \alpha_2^{x_2} \circ \cdots \circ \alpha_{14}^{x_{14}})
$$

for every $(x_1, x_2, \ldots, x_{14}) \in \{T, F\}^{14}$, i.e., the protocol $\mathcal{P}_{\mathsf{Sym}_A^{14}}$ securely computes Sym_A^{14}.

For the case of a symmetric function Sym_A^n with $n < 14$, say Sym_A^{13}, it suffices to construct a protocol for $\mathsf{Sym}_A^{14}(x_1, x_2, \ldots, x_{13}, 0)$ by using the protocol $\mathcal{P}_{\mathsf{Sym}_A^{14}}$. Consequently, we obtain the following Theorem 1.

Theorem 1. *Let $n \le 14$. Then, any n-variable symmetric function is 15puz-computable.*

4   Protocols for 4-Variable General Functions

In this section, by proposing protocols for arbitrary (general) functions of 4 variables, we will prove the following Theorem 2.

Theorem 2. *Let $n \leq 4$. Then, any n-variable function is 15puz-computable.*

To prove Theorem 2, it suffices to construct protocols for any 4-variable function (because it immediately implies that any function of 3 variables or less is also 15puz-computable).

Let $f(x_1, x_2, x_3, x_4)$ be an arbitrary 4-variable function. If $f = T$ or $f = F$ or $f = x_1$ or $f = \overline{x}_1$, then f is obviously 15puz-computable. Therefore, one may assume that

$$f(b_1, b_2, b_3, b_4) \neq f(b_1, b_2', b_3', b_4')$$

for some $b_1, b_2, b_3, b_4, b_2', b_3', b_4' \in \{T, F\}$. Without loss of generality, we assume that $b_1 = F$, i.e.,

$$f(F, b_2, b_3, b_4) \neq f(F, b_2', b_3', b_4'). \tag{3}$$

We are going to construct a protocol

$$\mathcal{P}_f = ((\alpha_1^T, \alpha_1^F), (\alpha_2^T, \alpha_2^F), (\alpha_3^T, \alpha_3^F), (\alpha_4^T, \alpha_4^F); C_0)$$

which securely computes f.

We first determine its initial configuration C_0 as follows:

$$C_0 = \begin{pmatrix} f(F,F,F,F) & f(F,F,F,T) & f(T,T,T,F) & f(T,T,T,T) \\ f(F,F,T,T) & f(F,F,T,F) & f(T,T,F,T) & f(T,T,F,F) \\ f(F,T,F,F) & f(F,T,F,T) & f(T,F,T,F) & f(T,F,T,T) \\ f(F,T,T,T) & f(F,T,T,F) & \flat & f(T,F,F,F) \end{pmatrix}.$$

Note that the 8 values $f(F, *, *, *)$ lie on the left half,

$$f(T, F, F, T)$$

is "missing," and the remaining 7 values $f(T, *, *, *)$ lie on the right half.

We next determine player P_1's actions. Intuitively, our approach is to gather the "desired" values in the left side: we want to construct actions α_1^T and α_1^F so that

$$C_0 \triangleleft \alpha_1^{x_1} = \begin{pmatrix} f(x_1,F,F,F) & f(x_1,F,F,T) & - & - \\ f(x_1,F,T,T) & f(x_1,F,T,F) & - & - \\ f(x_1,T,F,F) & f(x_1,T,F,T) & - & - \\ f(x_1,T,T,T) & f(x_1,T,T,F) & \flat & - \end{pmatrix} \tag{4}$$

(where $-$ means some binary value). To this end, when $x_1 = F$, it suffices only to set $\alpha_1^F = \varepsilon$; hence, consider the case of $x_1 = T$. Since $f(T, F, F, T)$ is "missing," we first search the same value as $f(T, F, F, T)$ from the 8 values $f(F, *, *, *)$ in the left half (note that such a "desired" value must exist by Eq. (3)), and reposition it to the $(4, 2)$-entry: let $b = f(T, F, F, T)$, and define an action δ as

if $f(F,T,T,F) = b$ **then** $\delta := \varepsilon$;
else if $f(F,T,T,T) = b$ **then** $\delta := \uparrow\leftarrow\downarrow\leftarrow\uparrow\rightarrow\rightarrow\downarrow$;
else if $f(F,T,F,F) = b$ **then** $\delta := \uparrow\leftarrow\leftarrow\downarrow\rightarrow\uparrow\rightarrow\downarrow$;
else if $f(F,F,T,F) = b$ **then** $\delta := \uparrow\leftarrow\uparrow\leftarrow\downarrow\downarrow\rightarrow\uparrow\rightarrow\downarrow$;
else if $f(F,F,T,T) = b$ **then** $\delta := \uparrow\leftarrow\leftarrow\uparrow\rightarrow\downarrow\leftarrow\downarrow\rightarrow\uparrow\rightarrow\downarrow$;
else if $f(F,T,F,T) = b$ **then** $\delta := \leftarrow\uparrow\leftarrow\downarrow\rightarrow\rightarrow\uparrow\leftarrow\downarrow\leftarrow\uparrow\rightarrow\rightarrow\downarrow$;
else if $f(F,F,F,T) = b$ **then** $\delta := \uparrow\leftarrow\uparrow\uparrow\leftarrow\downarrow\downarrow\rightarrow\uparrow\leftarrow\downarrow\downarrow\rightarrow\uparrow\rightarrow\downarrow$;
else if $f(F,F,F,F) = b$ **then** $\delta := \uparrow\leftarrow\leftarrow\uparrow\uparrow\rightarrow\downarrow\leftarrow\downarrow\rightarrow\uparrow\leftarrow\downarrow\downarrow\rightarrow\uparrow\rightarrow\downarrow$.

Then, one can easily verify that

$$
C_0 \triangleleft \delta =
\begin{pmatrix}
- & - & f(T,T,T,F) & f(T,T,T,T) \\
- & - & f(T,T,F,T) & f(T,T,F,F) \\
- & - & f(T,F,T,F) & f(T,F,T,T) \\
- & f(T,F,F,T) & \flat & f(T,F,F,F)
\end{pmatrix}.
$$

In order for the configuration $C_0 \triangleleft \delta$ to become like Eq. (4), one may intuitively notice that it suffices to "rotate" the 15 binary values clockwise 7 times along the "Hamilton cycle" shown in Figure 3. Therefore, let

$$
\gamma = \rightarrow\uparrow\leftarrow\uparrow\rightarrow\uparrow\leftarrow\leftarrow\leftarrow\downarrow\rightarrow\downarrow\leftarrow\downarrow\rightarrow\rightarrow,
$$

and we set $\alpha_1^T = \delta \circ \gamma^7$. Thus, we have obtained player P_1's actions $\alpha_1^T = \delta \circ \gamma^7$ and $\alpha_1^F = \varepsilon$ which satisfy Eq. (4).

Fig. 3. Another "Hamilton cycle" on the board

We next determine player P_2's actions. Similarly as P_1's actions, we want to gather the "desired" values in the upper side: we want to construct actions α_2^T and α_2^F so that

$$
C_0 \triangleleft \alpha_1^{x_1} \circ \alpha_2^{x_2} =
\begin{pmatrix}
f(x_1,x_2,F,F) & f(x_1,x_2,F,T) & - & - \\
f(x_1,x_2,T,T) & f(x_1,x_2,T,F) & - & - \\
- & - & & - \\
- & - & \flat & -
\end{pmatrix}.
$$

To this end, it suffices to set $\alpha_2^T = \gamma^4$ and $\alpha_2^F = \varepsilon$.
Similarly, we set $\alpha_3^T = \gamma^2$ and $\alpha_3^F = \varepsilon$ so that

$$C_0 \triangleleft \alpha_1^{x_1} \circ \alpha_2^{x_2} \circ \alpha_3^{x_3} = \begin{pmatrix} f(x_1, x_2, x_3, F) & f(x_1, x_2, x_3, T) & - & - \\ - & - & - & - \\ - & - & - & - \\ - & - & \flat & - \end{pmatrix}.$$

Finally, we set

$$\alpha_4^T = \leftarrow\leftarrow\uparrow\uparrow\uparrow\rightarrow\downarrow\downarrow\downarrow\rightarrow$$

and $\alpha_4^F = \varepsilon$ so that

$$f(x_1, x_2, x_3, x_4) = \mathsf{upperleft}(C_0 \triangleleft \alpha_1^{x_1} \circ \alpha_2^{x_2} \circ \alpha_3^{x_3} \circ \alpha_4^{x_4})$$

for every $(x_1, x_2, x_3, x_4) \in \{T, F\}^4$.

Thus, we have given the description of our protocol $\mathcal{P}_f$ which securely computes f, and hence we have proved Theorem 2.

5   Non-15puz-computable Functions

In this section, we show that, unfortunately, there exist functions which are not 15puz-computable. Specifically, we prove the following Theorems 3 and 4.

Theorem 3. *There exists a 15-variable symmetric function which is not 15puz-computable.*

Theorem 4. *There exists a 5-variable function which is not 15puz-computable.*

Remember that any 4-variable function and any 14-variable symmetric function are 15puz-computable.

Due to the page limitation, we omit the proofs of Theorems 3 and 4 in this extended abstract.

6   Conclusions

In this paper, we first proposed a class of 15-puzzle-based cryptographic protocols for secure multiparty computations. Then, we showed that any function of 4 variables (or less) and any symmetric function of 14 variables (or less) can be securely computed by the 15 puzzle, i.e., they are 15puz-computable. Furthermore, we presented a 5-variable function and a 15-variable symmetric function, both of which are not 15puz-computable.

Although it is not so difficult to extend our results to dealing with an arbitrary m-puzzle (for some appropriate integer m), this paper addressed only the use of the 15 puzzle. The reason is that we are quite interested in "real" secure computing: the 15 puzzle has been fairly well-known around the world, and one can get a 15 puzzle at a low cost, perhaps, anywhere in the world.

Acknowledgement

This research was partially supported by the Grant-in-Aid for Exploratory Research No. 17650002 from the Ministry of Education, Culture, Sports, Science and Technology, Japan.

References

1. Balogh, J., Csirik, J.A., Ishai, Y., Kushilevitz, E.: Private computation using a PEZ dispenser. Theoretical Computer Science 306, 69–84 (2003)
2. den Boer, B.: More efficient match-making and satisfiability: the five card trick. In: Quisquater, J.-J., Vandewalle, J. (eds.) EUROCRYPT 1989. LNCS, vol. 434, pp. 208–217. Springer, Heidelberg (1990)
3. Crépeau, C., Kilian, J.: Discreet solitary games. In: Stinson, D.R. (ed.) CRYPTO 1993. LNCS, vol. 773, pp. 319–330. Springer, Heidelberg (1994)
4. Fagin, R., Naor, M., Winkler, P.: Comparing information without leaking it. Communications of the ACM 39(5), 77–85 (1996)
5. Fischer, M.J., Wright, R.N.: Bounds on secret key exchange using a random deal of cards. Journal of Cryptology 9, 71–99 (1996)
6. Mizuki, T., Shizuya, H., Nishizeki, T.: Dealing necessary and sufficient numbers of cards for sharing a one-bit secret key. In: Stern, J. (ed.) EUROCRYPT 1999. LNCS, vol. 1592, pp. 389–401. Springer, Heidelberg (1999)
7. Moran, T., Naor, M.: Basing cryptographic protocols on tamper-evident seals. In: Caires, L., Italiano, G.F., Monteiro, L., Palamidessi, C., Yung, M. (eds.) ICALP 2005. LNCS, vol. 3580, pp. 285–297. Springer, Heidelberg (2005)
8. Moran, T., Naor, M.: Polling with physical envelopes: a rigorous analysis of a human-centric protocol. In: Vaudenay, S. (ed.) EUROCRYPT 2006. LNCS, vol. 4004, pp. 88–108. Springer, Heidelberg (2006)
9. Slocum, J., Sonneveld, D.: The 15 Puzzle. Slocum Puzzle Foundation, Beverly Hills, CA (2006)

A Lagrangian Relaxation Approach for the Multiple Sequence Alignment Problem[*]

Ernst Althaus[1] and Stefan Canzar[2]

[1] Max-Planck Institut für Informatik, Stuhlsatzenhausweg 85,
D-66123 Saarbrücken, Germany
[2] Université Henri Poincaré, LORIA, B.P. 239, 54506 Vandœvre-lès-Nancy, France

Abstract. We present a branch-and-bound (bb) algorithm for the multiple sequence alignment problem (MSA), one of the most important problems in computational biology. The upper bound at each bb node is based on a Lagrangian relaxation of an integer linear programming formulation for MSA. Dualizing certain inequalities, the Lagrangian subproblem becomes a pairwise alignment problem, which can be solved efficiently by a dynamic programming approach. Due to a reformulation w.r.t. additionally introduced variables prior to relaxation we improve the convergence rate dramatically while at the same time being able to solve the Lagrangian problem efficiently. Our experiments show that our implementation, although preliminary, outperforms all exact algorithms for the multiple sequence alignment problem.

1 Introduction

Aligning DNA or protein sequences is one of the most important and predominant problems in computational molecular biology. Before we motivate this we introduce the following notation for the multiple sequence alignment problem.

Let $\mathcal{S} = \{s^1, s^2, \ldots, s^k\}$ be a set of k strings over an alphabet Σ and let $\bar{\Sigma} = \Sigma \cup \{-\}$, where "$-$" (dash) is a symbol to represent "gaps" in strings. Given a string s, we let $\mid s \mid$ denote the number of characters in the string and s_l the lth character of s, for $l = 1, \ldots, \mid s \mid$. We will assume that $\mid s^i \mid \geq 4$ for all strings s^i and let $n := \sum_{i=1}^{k} \mid s^i \mid$.

An *alignment* $\mathcal{A}$ of $\mathcal{S}$ is a set $\bar{\mathcal{S}} = \{\bar{s}^1, \bar{s}^2, \cdots, \bar{s}^k\}$ of strings over the alphabet $\bar{\Sigma}$ where each string can be interpreted as a row of a two dimensional *alignment matrix*. The set $\bar{\mathcal{S}}$ of strings has to satisfy the following properties: (1) the strings in $\bar{\mathcal{S}}$ all have the same length, (2) ignoring dashes, string $\bar{s}^i$ is identical to string s^i, and (3) none of the columns of the alignment matrix is allowed to contain only dashes.

If $\bar{s}_l^i$ and $\bar{s}_l^j$ are both different from "$-$", the corresponding characters in s^i and s^j are *aligned* and thus contribute a weight $w(\bar{s}_l^i, \bar{s}_l^j)$ to the value of $\mathcal{A}$. The pairwise scoring matrix w over the alphabet Σ models either costs or benefits, depending on whether we minimize distance or maximize similarity.

---

[*] Supported by the German Academic Exchange Service (DAAD).

A. Dress, Y. Xu, and B. Zhu (Eds.): COCOA 2007, LNCS 4616, pp. 267–278, 2007.
© Springer-Verlag Berlin Heidelberg 2007

In the following, we assume that we maximize the weight of the alignment. Moreover, a *gap* in s^i with respect to s^j is a maximal sequence $s^i_l \, s^i_{l+1} \, \ldots \, s^i_m$ of characters in s^i that are aligned with dashes "−" in row j. Associated with each of these gaps is a cost. In the *affine gap cost* model the cost of a single gap of length q is given by the affine function $c_{\text{open}} + q c_{\text{ext}}$, i.e. such a gap contributes a weight of $-c_{\text{open}} - q c_{\text{ext}} = w_{\text{open}} + q w_{\text{ext}}$ to the total weight of the alignment. The problem calls for an alignment $\mathcal{A}$ whose overall weight is maximized.

Alignment programs still belong to the class of the most important Bioinformatics tools with a large number of applications. Pairwise alignments, for example, are mostly used to find strings in a database that share certain commonalities with a query sequence but which might not be known to be biologically related. Multiple alignments serve a different purpose. Indeed, they can be viewed as solving problems that are *inverse* to the ones addressed by pairwise string comparisons [12]. The inverse problem is to infer certain shared patterns from known biological relationships.

The question remains how a multiple alignment should be scored. The model that is used most consistently by far is the so called *sum of pairs* (SP) score. The SP score of a multiple alignment $\mathcal{A}$ is simply the sum of the scores of the pairwise alignments induced by $\mathcal{A}$ [6].

If the number k of sequences is fixed the multiple alignment problem for sequences of length n can be solved in time and space $\mathcal{O}\left(n^k\right)$ with (quasi)-affine gap costs [11,15,19,20]. More complex gap cost functions add a polylog factor to this complexity [8,14]. However, if the number k of sequences is not fixed, Wang and Jiang [22] proved that multiple alignment with SP score is $\mathcal{NP}$-complete by a reduction from *shortest common supersequence* [10]. Hence it is unlikely that polynomial time algorithms exist and, depending on the problem size, various heuristics are applied to solve the problem approximately (see, e.g., [4,7]).

In [3,2] Althaus et al. propose a branch-and-cut algorithm for the multiple sequence alignment problem based on an integer linear programming (ILP) formulation. As solving the LP-relaxation is by far the most expensive part of the algorithm and even not possible for moderately large instances, we propose a Lagrangian approach to approximate the linear program and utilize the resulting bounds on the optimal value in a branch-and-bound framework. We assume that the reader is familiar with the Lagrangian relaxation approach to approximate linear programs.

The paper is organized as follows. In Section 2 we review the ILP formulation of the multiple sequence alignment problem, whose Lagrangian relaxation is described in section 3. Our algorithm for solving the resulting problem is introduced in section 4. Finally, computational experiments on a set of real-world instances are reported in section 5.

2   Previous Work

In [3] Althaus et al. use a formulation for the multiple sequence alignment problem as an ILP given by Reinert in [18].

For ease of notation, they define the *gapped trace graph*, a mixed graph whose node set corresponds to the characters of the strings and whose edge set is partitioned in undirected alignment edges and directed positioning arcs as follows: $G = (V, E_A \cup A_P)$ with $V = V^i \cup \cdots \cup V^k$ and $V^i = \{u^i_j \mid 1 \le j \le |s^i|\}$, $E_A = \{uv \mid u \in V^i, v \in V^j, i \ne j\}$ and $A_P = \{(u^i_l, u^i_{l+1}) \mid 1 \le i \le k \text{ and } 1 \le l < |s^i|\}$ (see figure 1). Furthermore, we denote with $\mathcal{G} = \{(u, v, j) \mid u, v \in V^i, j \ne i\}$ the set of all possible gaps.

The ILP formulation uses a variable for every possible alignment edge $e \in E_A$, denoted by x_e, and one variable for every possible gap $g \in \mathcal{G}$, denoted by y_g. Reinert [18] showed that solutions to the alignment problem are the $\{0,1\}$-assignments to the variables such that

1. we have pairwise alignments between every pair of strings,
2. there are no mixed cycles, i.e. in the subgraph of the gapped trace graph consisting of the positioning arcs A_P and the edges $\{e \in E_A \mid x_e = 1\}$ there is no cycle that respects the direction of the arcs of A_p (and uses the edges of E_A in either direction) and contains at least one arc of A_P (see figure 1),
3. transitivity is preserved, i.e. if u is aligned with v and v with w then u is aligned with w, for $u, v, w \in V$.

Fig. 1. The graph in the middle is the gapped trace graph for the alignment problem given in the left part. The thick edges specify the alignment given in the left part. The alignment edges in the right part can not be realized at the same time in an alignment. Together with appropriate arcs of A_P, they form a mixed cycle.

These three conditions are easily formulated as linear constraints. Given weights w_e associated with variables x_e, $e \in E_A$, and gap costs w_g associated with variables y_g, we denote the problem of finding the optimal alignment (whose overall weight is maximized) satisfying conditions (1)-(3) as (P) and its optimal value as $v(P)$. As the number of those inequalities is exponential Althaus et al. use a cutting plane framework to solve the LP relaxation (all inequalities have a polynomial separation algorithm). In their experiments they observed that the number of iterations in the cutting plane approach can be reduced, if we use additional variables $z_{(u,v)}$ for $u \in V^i, v \in V^j, i \ne j$, with the property that $z_{(u,v)} = 1$ iff at least one character of the string of u lying behind u is aligned to a character of the string of v lying before v, i.e. $z_{(u^i_l, v^j_m)} = 1$, iff there is $l' \ge l$ and $m' \le m$ with $x_{v^i_{l'}, v^j_{m'}} = 1$. This condition is captured by the inequalities

$$0 \le z \le 1, \quad z_{(u_l^i, v_m^j)} \ge z_{(u_{l+1}^i, v_m^j)} + x_{u_l^i, v_m^j} \quad \text{and} \quad z_{(u_l^i, v_m^j)} \ge z_{(u_l^i, v_{m-1}^j)} + x_{u_l^i, v_m^j}. \tag{4}$$

In the following, we describe the inequalities used in [3] to enforce (2). We resign to explicitly specify the inequalities enforcing (1) and (3), as they are not crucial for the understanding of our approach.

Using these additional variables, we can define facets that guarantee (2) as follows. Let $A_A = \{(u, v) \mid u \in V^i, v \in V^j, i \ne j\}$, i.e. for each undirected edge $uv \in E_A$, we have the two directed arcs (u, v) and (v, u) in A_A. Let $M \subseteq A_A \cup A_P$ be a cycle in $(V, A_A \cup A_P)$ that contains at least one arc of A_P. We call such a cycle a *mixed cycle*. The set of all mixed cycle inequalities is denoted by $\mathcal{M}$. For a mixed cycle $M \in \mathcal{M}$ the inequality

$$\sum_{e \in M \cap A_A} z_e \le |M \cap A_A| - 1 \tag{5}$$

is valid and defines a facet under appropriate technical conditions. In particular, there is exactly one arc of A_P in M. These inequalities are called *lifted mixed cycle inequalities*. The constraints can be formulated similarly without using the additional z-variables.

3   Outline

Our Lagrangian approach is based on the integer linear program outlined above. Hence we have three classes of variables, X, Y and Z. Notice that a single variable x_{uv}, $y_{(u,v,j)}$, or $z_{(u,v)}$ involves exactly two sequences. Let $X^{i,j}$, $Y^{i,j}$, and $Z^{i,j}$ be the set of variables involving sequences i and j. If we restrict our attention to the variables in $X^{i,j}$, $Y^{i,j}$ and $Z^{i,j}$, for a specific pair of sequences i, j, a solution of the ILP yields a description of a pairwise alignment between sequences i and j, along with appropriate values for the $Z^{i,j}$ variables. The constraints (2) and (3) are used to guarantee that all pairwise alignments together form a multiple sequence alignment. We call an assignment of $\{0, 1\}$-values to $(X^{i,j}, Y^{i,j}, Z^{i,j})$ such that $(X^{i,j}, Y^{i,j})$ imposes a pairwise alignment and $Z^{i,j}$ satisfies inequalities (4), an *extended pairwise alignment*. Given weights for the variables in $X^{i,j}$, $Y^{i,j}$ and $Z^{i,j}$, we call the problem of finding an extended pairwise alignment of maximum weight the *extended pairwise alignment problem*.

In our Lagrangian approach we dualize the constraints for condition (2) and relax conditions (3) (during experiments it turned out that relaxing condition (3) is more efficient in practice as dualizing them). Hence our Lagrangian subproblem is an extended pairwise alignment problem. More precisely, if λ_M is the current multiplier for the mixed cycle inequality of $M \in \mathcal{M}$, we have to solve the Lagrangian relaxation problem

$$\sum_{M \in \mathcal{M}} \lambda_M (|M \cap A_A| - 1) \quad +$$

$$\max \sum_{e \in E_A} w_e x_e + \sum_{g \in \mathcal{G}} w_g y_g - \sum_{M \in \mathcal{M}} \lambda_M \sum_{e \in M \cap A_A} z_e \tag{LR_λ}$$

$$\text{s.t.} (X^{i,j}, Y^{i,j}, Z^{i,j}) \text{ forms an extended pairwise alignment for all } i, j.$$

We denote its optimal value with $v(LR_\lambda)$. Our approach to obtain tighter bounds efficiently, e.g. to determine near-optimal Lagrangian multipliers to the minimum of the *Lagrangian function* $f(\lambda) = v(LR_\lambda)$, is based on the iterative subgradient method proposed by Held and Karp [13]. Similarly to [5], we experienced a faster convergence if we modify the adaption of scalar step size θ in the subgradient formula in the following way. Instead of simply reducing θ when there is no upper bound improvement for too long, we compare the best and worst upper bounds computed in the last p iterations. If they differ by more than 1%, we suspect that we are "overshooting" and thus we halve the current value of θ. If, in contrast, the two values are within 0.1% from each other, we overestimate $v(LR_{\lambda^*})$, where λ^* is an optimal solution to (LR), and therefore increase θ by a factor of 1.5. As the number of inequalities that we dualize is exponential, we modify the subgradient method in a relax-and-cut fashion, as proposed by [9]. Due to lack of space, we resign to give details and refer to [1] for a complete description.

4   Solving the Extended Pairwise Alignment Problem

Recall how a pairwise alignment with gap cost is computed for two strings s and t of length n_s and n_t, respectively (without loss of generality we assume $n_t \leq n_s$). By a simple dynamic programming algorithm, we compute for every $1 \leq l \leq n_s$ and every $1 \leq m \leq n_t$ the optimal alignment of prefixes $s_1 \ldots s_l$ and $t_1 \ldots t_m$ that aligns s_l and t_m and whose score is denoted by $D(l, m)$. This can be done by comparing all optimal alignments for strings $s_1 \ldots s_{l'}$ and $t_1 \ldots t_{m'}$ for $l' < l$ and $m' < m$, adding the appropriate gap cost to the score of the alignment (s_l, t_m). Then the determination of the optimal alignment value $D(n_s, n_t)$ takes time $\mathcal{O}\left(n_s^2 n_t^2\right)$[1].

In the affine gap weight model we can restrict the dependence of each cell in the dynamic programming matrix to adjacent entries in the matrix by associating more than one variable to each entry as follows. Besides computing $D(l, m)$, we compute the score of the optimal alignment of these substrings that aligns character s_l to a character t_k with $k < m$, denoted by $V(l, m)$, and the one that aligns t_m to a character s_k with $k < l$, denoted by $H(l, m)$. Hence, in a node $V(l, m)$, we have already paid the opening cost for the gap in t and we can traverse from $V(l, m)$ to $V(l, m + 1)$ by just adding w_{ext}, but not w_{open}. Each of the terms $D(l, m)$, $V(l, m)$ and $H(l, m)$ can be evaluated by a constant number of references to previously determined values and thus the running time reduces to $\mathcal{O}\left(n_s n_t\right)$.

The pairwise alignment problem can be interpreted as a longest path problem in an acyclic graph, having three nodes $D(l, m)$, $V(l, m)$ and $H(l, m)$ for every pair of characters $s_l \in s$, $t_m \in t$, i.e. in all cells (l, m). We call this graph the dynamic programming graph. Each pairwise alignment corresponds to a unique path through this graph, with every arc of the path representing a certain kind

---

[1] The running time can be reduced to $\mathcal{O}\left(n_s^2 n_t\right)$ by distinguishing three different types of alignments [21].

of alignment, determined by the type of its target node (figure 2). An *alignment arc* from an arbitrary node in cell $(l-1, m-1)$ to node $D(l, m)$ corresponds to an alignment of characters s_l and t_m. Accordingly, a *gap arc* has a target node $V(l, m)$ or $H(l, m)$ and represents a gap opening (source node is $D(l, m-1)$ or $D(l-1, m)$, respectively) or a gap extension (source node is $V(l, m-1)$ or $H(l-1, m)$, respectively).

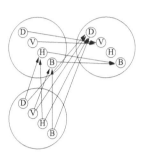

Fig. 2. Three cells of the dynamic programming matrix, with four values (nodes) associated to each of them. Note that arcs (dependencies) are between certain values D, V, H and B, the target node determines the type of the partial alignment.

Now assume some variable $z_{(u,v)}$ is multiplied by a non-zero value in the objective function, as the arc (u, v) is used in at least one mixed cycle inequality, to which a non-zero Lagrangian multiplier λ_M is associated. Recall that the multiplier of the variable $z_{(u,v)}$ in the objective function is $-\sum_{M \in \mathcal{M}|(u,v) \in M} \lambda_M$ (see (LR_λ)). Then we have to pay the multiplier as soon as our path traverses at least one alignment arc that enforces $z_{(u,v)} = 1$. Assume $s = s^i$, $t = s^j$, $u = u_l^i$ and $v = u_m^j$. Then $z_{(u,v)} = 1$, iff there is $l' \geq l$ and $m' \leq m$ such that $x_{u_{l'}^i u_{m'}^j} = 1$ (see definition of variables $z_{(u,v)}$ in (4)). In the dynamic program graph, this corresponds to alignment arcs whose target lies in the lower right rectangle from cell (l, m). Analogously, if u lies in string s_j and v in string s_i, this corresponds to alignment arcs whose target lies in an upper left rectangle. We call these rectangles *blue* and *red obstacles* and denote them by $\mathscr{O}_b(l, m)$ and $\mathscr{O}_r(l, m)$, respectively.

Let the set of all blue and red obstacles be denoted by $\mathscr{O}_b$ and $\mathscr{O}_r$, respectively, and let $\mathscr{O} = \mathscr{O}_b \cup \mathscr{O}_r$. Then the extended pairwise alignment problem is solvable by a dynamic program in $\mathcal{O}\left(n_s^2 n_t^2 |\mathscr{O}|\right)$ time, following the same approach as above: we compute the best alignment of all pairs of prefixes $s_1 \ldots s_l$ and $t_1 \ldots t_m$ that aligns s_l and t_m, based on on all best alignments of strings $s_1 \ldots s_{l'}$ and $t_1 \ldots t_{m'}$, for $l' < l$ and $m' < m$. We add the appropriate gap weight to the score of the alignment (s_l, t_m) and subtract all Lagrangian multipliers that are associated with obstacles enclosing (s_l, t_m), but not $(s_{l'}, t_{m'})$.

Definition 1 (Enclosing Obstacles). *The set of enclosing blue obstacles $\mathcal{Q}_b(p)$ of a cell $p = (x, y)$ contains all blue obstacles $\mathscr{O}_b(l, m)$ with $l \leq x, m > y$. Accordingly, $\mathcal{Q}_r(p) = \{\mathscr{O}_r(s, t) \mid s > x, t \leq y\}$. Furthermore we define $\mathcal{Q}(p) = \mathcal{Q}_b(p) \cup \mathcal{Q}_r(p)$.*

We reduce the complexity of the dynamic program by again decreasing the alignment's history, necessary to determine the benefit of any possible continuation in a partial alignment. The determination of the set of obstacles, whose associated penalty we have to pay when using an alignment arc, poses the major problem. For that we have to know the last alignment arc that has been used on our path. However, this arc can not be precomputed in a straightforward way, since the longest path in this context does not have optimal substructure. The key idea is to charge the cost of a Lagrangian multiplier λ as soon as we enter the corresponding obstacle o, i.e. if the target node of the arc is enclosed by o, no matter whether we enter it along an alignment arc or a gap arc. Hence, we have to ensure that we are able to bypass obstacles we do not have to pay, i.e. obstacles that are not enclosing any target node of an alignment arc traversed by the optimal path. We accomplish this by adding new nodes and arcs to the dynamic programming graph. Additionally we compute, for every pair of characters $s_l \in s$, $t_m \in t$, a fourth value $B(l, m)$ denoting the value of the optimal alignment that aligns either character s_l to "-" strictly left from t_m or character t_m to "-" strictly left from s_l. In other words, every cell (l, m) contains a fourth node $B(l, m)$ in the dynamic programming graph.

Before we introduce the new nodes and edges formally, we need some basic definitions. We call a pair of a blue obstacle $\mathcal{O}_b(l, m)$ and a red obstacle $\mathcal{O}_r(l', m')$ conflicting, if $l' \geq l$ and $m' \leq m$. The base $\mathit{b}(\mathcal{O}_b(l, m), \mathcal{O}_r(l', m'))$ of a pair of conflicting obstacles is defined as cell $(l - 1, m' - 1)$, the target $t(\mathcal{O}_b(l, m), \mathcal{O}_r(l', m'))$ as cell (l', m). We say a cell (l, m) dominates a cell (l', m'), denoted by $(l, m) < (l', m')$, if $l < l'$ and $m < m'$. Similarly, a blue (red) obstacle $\mathcal{O}_{b(r)}(l, m)$ dominates an obstacle $\mathcal{O}_{b(r)}(l', m')$, iff $(l, m) < (l', m')$. A blue (red) obstacle is minimal in $\hat{\mathcal{O}}_b \subseteq \mathcal{O}_b$ ($\hat{\mathcal{O}}_r \subseteq \mathcal{O}_r$), if it is not dominated by any other obstacle in $\hat{\mathcal{O}}_b$ ($\hat{\mathcal{O}}_r$). We denote the set of obstacles that are dominated by a given obstacle o, by $\mathcal{D}(o)$.

It is not difficult to see, that the insertion of arcs from the four nodes of every base b to the B-node of every target t such that $\mathit{b} < t$, would enable us to "jump over" obstacles that we do not have to pay. The weights for these arcs are determined by the cost of the gaps leading from b to t plus the penalties implied by obstacles enclosing t, but not b.

As the number of conflicting obstacles is at most $|\mathcal{O}|^2$, the number of additional arcs is at most $\mathcal{O}\left(|\mathcal{O}|^4\right)$ and hence the running time is $\mathcal{O}\left(n_s n_t + |\mathcal{O}|^4\right)$. To further reduce the number of additional arcs (dependencies) in our dynamic programming graph, we introduce the bypass graph, which is correlated to the transitive reduction of the induced subgraph on the set of newly added arcs.

Definition 2 (Bypass Graph). *We define the Bypass Graph (bpg) $G = (\mathcal{V}, \mathcal{E}, l)$ with edge set $\mathcal{E} \subset \mathcal{V} \times \mathcal{V}$ and length function $l \colon \mathcal{E} \to \mathbb{R}$ as follows. The vertex set $\mathcal{V}$ contains all pairs v of conflicting obstacles. Let v^b and v^r denote the blue and red obstacle of v, respectively. $\mathcal{E} = \mathcal{E}_b \cup \mathcal{E}_r$, where $\mathcal{E}_b = \{(v, w) \mid w^b \text{ is minimal in } \mathcal{D}(v^b)\}$ and $\mathcal{E}_r = \{(v, w) \mid w^r \text{ is minimal in } \mathcal{D}(v^r)\}$.*

The length l of edges in the bypass graph is chosen appropriately such that there exists a path from any node of base b to the B-node of every target t with $\mathit{b} < t$, that implies the correct score and such that the length of any such path is upper bounded by that score. We connect the bypass graph to the dynamic programming graph by arcs from all four nodes of every base to all its engendering vertices in $\mathcal{V}$ and by arcs from all $v \in \mathcal{V}$ to the B-node of their target $t(v)$. The length of the former kind of arcs satisfies the same requirements as l, the latter ones are defined to be of length 0.

Concerning the correctness of the dynamic program, we refer to a technical report [1] for details.

4.1   Complexity

Obviously there are at most $|\mathcal{O}|^2$ conflicting pairs of obstacles and hence the number of additional nodes $|\mathcal{V}|$ is at most $|\mathcal{O}|^2$. From definition 2 it follows immediately that the number of additional arcs $|\mathcal{A}|$ is at most $\mathcal{O}\left(|\mathcal{O}|^3\right)$, as an edge of the bypass graph is defined by three obstacles. Therefore the running time to compute an optimal solution to the extended pairwise alignment problem is $\mathcal{O}\left(nm + |\mathcal{O}|^3\right)$.

We improve the practical performance of our algorithm for solving the extended pairwise alignment problem by applying an A^*-approach: Notice that the scores $D(l, m), V(l, m), H(l, m)$ and $B(l, m)$ during an iteration of the subgradient optimization can be at most the scores of the first iteration, i.e. when all multipliers λ are set to 0. Then it is easy to see, that the length of a longest path from any node (l, m) to (n_s, n_t) determined in the first iteration provides a heuristic estimate for all other iterations, which is monotonic and thus the first path found from $(0, 0)$ to (n_s, n_t) is optimal.

5   Experiments

We have implemented our Lagrangian approach in C++ using the LEDA-library [17] and have embedded it into a branch-and-bound framework. The lower bounds in each bb node are computed by selecting, in a greedy fashion, edges from the set $\{e \in E_A \mid \bar{x}_e = 1\}$ that satisfy conditions (1)-(3). The weights for the alignment edges were obtained by the BLOSUM62 amino acid substitution matrix, whereas the gap arcs were assigned a weight that was computed as $4l+6$, where l is the number of characters in the corresponding gap.

We tested our implementation on a set of instances of the BAliBASE library. The benchmark alignments from reference 1 (R1) contain 4 to 6 sequences and are subdivided into three groups of different length (short, medium, long). They are further categorized into three subgroups by the degree of similarity between the sequences (group V1: identity $< 25\%$, group V2: identity $20 - 40\%$, group V3: identity $> 35\%$).

We compared our implementation, which we will call LASA (LAgrangian Sequence Alignment), with MSA [16] and COSA[3]. The multiple sequence alignment program MSA is based on dynamic programming and uses the so called quasi-affine gap cost model, a simplification of the (natural) affine gap cost model. The branch-and-cut algorithm COSA is based on the same ILP formulation and uses CPLEX as LP-solver. We ran the experiments on a system with a 2,39 GHz AMD Opteron Processor with 8 GB of RAM. Any run that exceeded a CPU time limit of 12 hours was considered unsuccessful.

Table 1 reports our results on short and medium sized instances from reference 1. As LASA was able to solve only three of the long instances (and no other program could solve any), we resign to show these results.

The columns in table 1 have the following meaning: *Instance:* Name of the instance, along with an indication (k, n) of the number of sequences and the overall number of characters; *Heur:* Value of the initial feasible solution found by COSA or MSA; *PUB:* Pairwise upper bound; *Root:* Value of the Lagrangian upper bound at the root node of the branch-and-bound tree; *Opt:* Optimal solution value; *#Nodes:* Number of branch-and-bound subproblems solved; *#Iter:* Total number of iterations during the subgradient optimization; *Time:* Total running time;

Although MSA reduces the complexity of the problem by incorporating quasi-affine gap costs into the multiple alignment, it could hardly solve instances with a moderate degree of similarity. In contrast, our preliminary implementation outperforms the CPLEX based approach COSA, the only method known till now to solve the MSA problem exactly. COSA was not able to solve any of the medium sized or long benchmark alignments, while LASA found the optimal solution within minutes. This is mainly because the LPs are quite complicated to solve. Moreover, one instance crashed as an LP could not be solved by CPLEX.

The running time of LASA and COSA strongly depends on tight initial lower bounds. For example, LASA takes about 13 hours for the long instance 3pmg with the bound obtained by the heuristic and only about one hour with the optimal value used as a lower bound.

Finally, we give computational evidence for the effectiveness of our novel approach to select violated inequalities to be added to our constraint pool. Considering the average of the last h solutions of the Lagrangian relaxation instead of looking only at the current solution ($h = 1$) dramatically reduces the number of iterations (see table 2). Only short sequences of high identity (short, V3) could be solved for $h = 1$. Furthermore, this table shows that the extended pairwise alignment problems are solved at least twice as fast when using the A^* approach.

The columns in table 2 have the following meaning: *Instance:* Name of the instance, along with an indication (k, n) of the number of sequences and the overall number of characters; $h = \cdot$: The number of solutions that were considered to compute an average Lagrangian solution; *LASA:* Default version of LASA, i.e. $h = 10$ and using the A^* approach; *DynProg:* LASA without using the A^* approach; *#Iter:* Number of iterations needed by a specific version of LASA; *Time:* Total running time in seconds needed by a specific version of LASA;

6   Conclusion

We have constructed a Lagrangian relaxation of the multiple sequence alignment ILP formulation that allowed us to obtain strong bounds by solving a generalization of the pairwise alignment problem. By utilizing these bounds in a branch-and-bound manner we achieved running times that outperform all other

Table 1. Results on instances from reference 1. Results on group medium V1 and two instances of medium V2 are omitted, since no program was able to solve these instances in the allowed time frame. We removed all instances that are solved by LASA within a second. *: With the COSA-code, the instance 3cyr crashed as the LP-solver was not able to solve the underlying LP.

Instance	Heur	PUB	Root	Opt	LASA #Nodes	#Iter	Time	COSA Time	MSA Time
Reference 1 Short, V3									
1dox (4/374)	749	782	751	750	3	253	3	30	<1
1fkj (5/517)	1,578	1,675	1,585	1,578	3	348	13	6:04	-
1plc (5/470)	1,736	1,824	1,736	1,736	1	218	6	4:24	20:14
2mhr (5/572)	2,364	2,406	2,364	2,364	1	65	3	2	17
Reference 1 Short, V2									
1csy (5/510)	649	769	649	649	1	393	17	3:01	-
1fjlA (6/398)	674	731	676	674	5	561	12	34	-
1hfh (5/606)	903	1,067	911	903	3	411	33	-	-
1hpi (4/293)	386	439	386	386	1	298	4	53	7
1pfc (5/560)	994	1,139	1,004	994	11	1,387	1:48	37:46	-
1tgxA (4/239)	247	317	247	247	1	566	9	53	-
1ycc (4/426)	117	309	202	200	7	1,865	2:19	-	-
3cyr (4/414)	515	615	522	515	7	983	38	-*	45
Reference 1 Short, V1									
1aboA (5/297)	-685	-476	-604	-676	3,497	417,260	11:04:02	-	-
1tvxA (4/242)	-409	-260	-358	-405	777	122,785	1:59:44	-	-
1idy (5/269)	-420	-273	-356	-414	4,193	678,592	12:00:48	-	-
1r69 (4/277)	-326	-207	-289	-326	253	54,668	58:40	-	-
1ubi (4/327)	-372	-246	-330	-372	215	43,620	1:12:57	-	-
1wit (5/484)	-198	-25	-186	-197	15	4,221	7:42	-	-
2trx (4/362)	-182	-88	-178	-182	5	2,186	3:04	-	-
Reference 1 Medium, V3									
1amk (5/1241)	5,668	5,728	5,669	5,669	1	60	8	-	-
1ar5A (4/794)	2,303	2,357	2,304	2,303	3	262	20	-	-
1ezm (5/1515)	8,378	8,466	8,378	8,378	1	105	23	-	-
1led (4/947)	2,150	2,282	2,158	2,150	33	1,435	3:54	-	-
1ppn (5/1083)	4,718	4,811	4,729	4,724	23	925	3:10	-	-
1pysA (4/1005)	2,730	2,796	2,732	2,730	3	223	28	-	-
1thm (4/1097)	3,466	3,516	3,468	3,468	3	233	30	-	-
1tis (5/1413)	5,854	5,999	5,874	5,856	83	2,993	18:31	-	-
1zin (4/852)	2,357	2,411	2,361	2,357	13	625	1:03	-	-
5ptp (5/1162)	4,190	4,329	4,233	4,205	193	8,337	35:48	-	-
Reference 1 Medium, V2									
1ad2 (4/828)	1,195	1,270	1,197	1,195	7	419	42	-	-
1aym3 (4/932)	1,544	1,664	1,551	1,544	17	1,060	2:37	-	-
1gdoA (4/988)	980	1,201	1,003	984	459	31,291	2:38:36	-	-
1ldg (4/1240)	1,526	1,640	1,539	1,526	41	2,160	8:32	-	-
1mrj (4/1025)	1,461	1,608	1,473	1,464	27	1,681	5:29	-	-
1pgtA (4/828)	683	808	691	690	9	926	2:05	-	-
1pii (4/1006)	1,099	1,256	1,103	1,100	23	1,320	4:54	-	-
1ton (5/1173)	1,550	1,898	1,609	1,554	807	44,148	5:32:47	-	-

Table 2. We give the number of iterations needed by our approach for different numbers h of solutions that were considered to compute the average Lagrangian solution. The default is $h = 10$. The last column gives the time spent in the root node if we resign to use the A^* approach.

Instance	$h = 1$ #Iter	$h = 2$ #Iter	$h = 20$ #Iter	$h = 30$ #Iter	LASA (A^*, $h = 10$) #Iter	Time	DynProg, $h = 10$ Time
1aho (5/320)	748,470	2,496	1,194	1,283	1,089	10	22
1csp (5/339)	17	14	19	19	17	<1	<1
1dox (4/374)	80,001	271	211	207	253	1	5
1fkj (5/517)	316,072	849	707	676	348	9	25
1fmb (4/400)	1,372	14	14	14	13	<1	<1
1krn (5/390)	191,281	634	148	155	104	1	8
1plc (5/470)	232,591	489	642	513	218	6	14
2fxb (5/287)	16,425	15	11	11	11	<1	<1
2mhr (5/572)	60,005	93	116	177	65	3	8
9rnt (5/499)	54	49	40	40	39	1	3

exact or almost exact methods. We plan to integrate our implementation into the software project SEQAN currently developed by the free university of Berlin.

Besides optimizing our implementation for speed an important issue in our future work will be to extend the scheme to volume and to bundle algorithms. A more sophisticated Lagrangian heuristic for computing lower bounds in the bb nodes will be necessary to be able to solve instances of larger size.

References

1. Althaus, E., Canzar, S.: A lagrangian relaxation approach for the multiple sequence alignment problem. Technical Report MPI-I-2007-1-001, Max-Planck-Institut für Informatik, 66123 Saarbrücken, Germany (May 2007)
2. Althaus, E., Caprara, A., Lenhof, H.-P., Reinert, K.: Multiple sequence alignment with arbitrary gap costs: Computing an optimal solution using polyhedral combinatorics. In: Lengauer, T., Lenhof, H.-P. (eds.) Proceedings of the European Conference on Computational Biology, vol. 18 of Bioinformatics, pp. S4–S16, Saarbrücken, Oxford University Press, Oxford (2002)
3. Althaus, E., Caprara, A., Lenhof, H.-P., Reinert, K.: Aligning multiple sequences by cutting planes. Mathematical Programming 105, 387–425 (2006)
4. Altschul, S.F., Gish, W., Miller, W., Myers, E.W., Lipman, D.J.: Basic local alignment search tool. J. Mol. Biol. 215, 403–410 (1990)
5. Caprara, A., Fischetti, M., Toth, P.: A heuristic method for the set cover problem. Operations Research 47, 730–743 (1999)
6. Carrillo, H., Lipman, D.J.: The multiple sequence alignment problem in biology. SIAM J. Appl. Math. 48(5), 1073–1082 (1988)
7. Delcher, A., Kasif, S., Fleischmann, R., Peterson, W.O.J., Salzberg, S.: Alignment of whole genomes. Nucleic Acids Research 27, 2369–2376 (1999)
8. Eppstein, D.: Sequence comparison with mixed convex and concave costs. Journal of Algorithms 11, 85–101 (1990)

9. Fisher, M.: Optimal solutions of vehcile routing problems using minimum k-trees. Operations Research 42, 626–642 (1994)
10. Garey, M., Johnson, D.: Computers and Intractability: A Guide to the Theory of NP-Completeness. W.H. Freeman (1979)
11. Gupta, S., Kececioglu, J., Schaeffer, A.: Improving the practical space and time efficiency of the shortest-paths approach to sum-of-pairs multiple sequence alignment. J. Comput. Biol. 2, 459–472 (1995)
12. Gusfield, D.: Algorithms on strings, trees and sequences: computer science and computational biology. Cambridge University Press, Cambridge (1997)
13. Held, M., Karp, R.: The traveling salesman problem and minimum spanning trees: part ii. Mathematical Programming 1, 6–25 (1971)
14. Larmore, L., Schieber, B.: Online dynamic programming with applications to the prediction of rna secondary structure. In: Proceedings of the First Symposium on Discrete Algorithms, pp. 503–512 (1990)
15. Lermen, M., Reinert, K.: The practical use of the $\mathcal{A}^*$ algorithm for exact multiple sequence alignment. Journal of Computational Biology 7(5), 655–673 (2000)
16. Lipman, D., Altschul, S., Kececioglu, J.: A tool for multiple sequence alignment. Proceedings of the National Academy of Sciences of the United States of America 86, 4412–4415 (1989)
17. Mehlhorn, K., Näher, S.: The LEDA Platform of Combinatorial and Geometric Computing. Cambridge University Press, Cambridge (1999), See also http://www.mpi-sb.mpg.de/LEDA/
18. Reinert, K.: A Polyhedral Approach to Sequence Alignment Problems. PhD thesis, Universität des Saarlandes (1999)
19. Reinert, K., Lenhof, H.-P., Mutzel, P., Mehlhorn, K., Kececioglu, J.: A branch-and-cut algorithm for multiple sequence alignment. In: Proceedings of the First Annual International Conference on Computational Molecular Biology (RECOMB-97), pp. 241–249 (1997)
20. Reinert, K., Stoye, J., Will, T.: An iterative methods for faster sum-of-pairs multiple sequence alignment. BIOINFORMATICS 16(9), 808–814 (2000)
21. Sankoff, D., Kruskal, J.B.: Time Warps, String Edits and Macromolecules: the Theory and Practice of Sequence Comparison. Addison-Wesley, Reading (1983)
22. Wang, L., Jiang, T.: On the complexity of multiple sequence alignment. J. Comput. Biol. 1, 337–348 (1994)

Single Machine Common Due Window Scheduling with Controllable Job Processing Times

Guohua Wan

College of Management, Shenzhen University, Shenzhen 518060, China
gh_wan@china.com

Abstract. We consider a nonpreemptive single machine common due window scheduling problem where the job processing times are controllable with linear costs and the due window is movable. The objective is to find a job sequence, a processing time for each job, and a position of the common due window to minimize the total cost of weighted earliness/tardiness and processing time compression. We discuss some properties of the optimal solutions and provide a polynomial time algorithm to solve the problem.

1 Introduction

In the pursuit of high production quality and short lead time, just-in-time (JIT) sequencing and scheduling models, which regard both job earliness and tardiness as penalties, received much attention from both researchers and practitioners, see Baker and Scudder [3] for an extensive review. In the past, most of JIT sequencing and scheduling research focuses on models with the following assumptions: (1) all job processing times are fixed; and (2) all jobs have a common due date. In this paper, we consider a single machine JIT scheduling model with both assumptions relaxed. The problem can be stated as follows. A set of n jobs are to be processed on a single machine with no preemption allowed. There is a common due window for all the jobs and it is movable. The processing times of all the jobs are under managerial control with costs proportional to their compression. The objective of the problem is to find a job sequence, a processing time for each job, and a position of the common due window to minimize the total cost of weighted earliness/tardiness and processing time compression.

First, we assume that job processing times are under managerial control. In most production scheduling research, job processing times are either treated as data known in advance or as random variables following some probability distribution. In practice, however, jobs may be completed in a shorter or longer duration by increasing or decreasing the resource units required. Studies of production scheduling problems involving controllable job processing times were initiated by Vickson [25], [26]. Nowicki and Zdrzalka [21] presented a survey of scheduling models with controllable job processing times ca. 1990. Alidaee

A. Dress, Y. Xu, and B. Zhu (Eds.): COCOA 2007, LNCS 4616, pp. 279–290, 2007.

and Kochenberger [2] presented a framework for scheduling models with various objective functions and controllable processing times on a single machine and on parallel machines, and solved them efficiently by transforming the problems into transportation problems. Recently, various scheduling problems involving controllable job processing times are investigated, see, for examples, Jansen *et al.* [10], Ng *et al.* [20], Kaspi and Shabtay [12], among others.

Among the studies closely related to our model, Panwalkar and Rajagopalan [22] considered a single machine sequencing problem with controllable job processing times and a common due date as a decision variable. Biskup and Jahnke [4] considered a problem of assigning a common due date to a set of jobs and scheduling them on a single machine with controllable processing times by the same proportional amount.

Second, we assume that there is a common due window, instead of a common due date, for the jobs. The importance of due window results from uncertainty and tolerance of due dates in many practical situations. In this vein of research, Cheng [5] studied a problem with a common due window that is small enough so that at most one job can be completed within the window. Dickman *et al.* [7] extended this model and showed that for any given job sequence, an optimal window can be determined as an interval depending on the number of jobs. Lee [15] studied the problem of minimizing the maximum earliness subject to no tardy jobs. He showed that for an arbitrary window size the problem is NP-hard; however, if the window size is given in advance, then the problem is polynomially solvable. Liman and Ramaswamy [16] considered the minimization of weighted sum of earliness and the number of tardy jobs. Kramer and Lee ([13] studied the problem of minimizing earliness and tardiness penalties. They also generalized the models to the cases with parallel machines (Kramer and Lee, [14]). Thongmee and Liman [24] considered the problem of minimizing the weighted sum of earliness, tardiness and window size penalties in which the beginning of the window is given but the window size is to be determined. Liman *et al.* [17] also studied a variant of the problem where the window size is given but the location of the window is not. Liman *et al.* [19] further generalized their models to the cases where both the location and size of the window are to be determined. Wan and Yen [28] studied a general single machine scheduling problem with distinct due windows and earliness/tardiness costs, and developed a tabu search procedure to solve the problem.

Relaxing both assumptions of fixed job processing times and a common due date, Liman *et al.* [18] studied a single machine scheduling problem with a common due window and controllable processing times, where the processing time of jobs, the location and the size of the due window are decision variables. They formulated the problem as an assignment problem and solved them efficiently. In this paper, we study a single machine common due window scheduling problem in which job processing times are controllable with linear costs and the location of the due window is to be determined but the window size is a given parameter. The objective is to find the processing time for each job, the position of the common due window and the job sequence in order to minimize the total cost of weighted earliness/tardiness and compression of processing times. This is motivated by and is also a generalization of the problems studied by Kramer and Lee [13], and Panwalkar and Rajagopalan [22]. Our model is different from the one studied by Liman *et al.* [18] in that the window size is given as a parameter, which makes the problem more difficult.

It is natural to study scheduling problems where earliness, tardiness and job processing costs are taken into consideration simultaneously. For instance, consider an assembly line with a CNC machine, where job processing time (i.e., cutting speed and feed rate) can be adjusted by allocating more or less resources (parts and tools). By selecting job processing times appropriately, system performance may be improved. On the other hand, the set of jobs are to be processed for a batch delivery thus has a common due date. Furthermore, manufacturers often consider a due date as an interval rather than a point in time so as to deal with uncertainty and tolerance of due dates. Any job finished after its latest due date is considered tardy, and no job can be delivered before its earliest due date. If it is finished earlier, a job must be held until its earliest due date, thus incurring a holding cost. The period between its earliest and latest due date is the due window. Due to managerial control, the size of the due window is not a decision variable but a given parameter. Hence, if the objective of the scheduling problem is to minimize the weighted sum of earliness/tardiness cost and job processing cost, then the described situation can be suitably modelled by our model.

The remainder of the paper is organized as follows. The problem is formally described in Section 2, followed by discussion of the properties of optimal solutions with both fixed and controllable job processing times in Section 3. In Section 4, a polynomial algorithm is developed together with an illustrative example. Conclusion and suggestions for future research are presented in Section 5.

2   Problem Formulation

Consider a scheduling problem with n jobs to be processed on a single machine with the following assumptions:

(1) all the jobs are available at time zero;

(2) the machine can process at most one job at a time;

(3) no preemption is allowed; and

(4) the processing time of a job can be compressed to a minimum with a linear compression cost; and

(5) there is a common due window for all the jobs. The window is movable while its size is fixed.

The objective of this scheduling problem is to find a job sequence, a processing time for each job, and a position of the common due window to minimize the total cost, including earliness/tardiness cost and processing compression cost. For convenience, we list the symbolic notations used throughout the paper as follows (without loss of generality, assume that all the parameters are non-negative integers except the unit costs).

$N = \{1, 2, ..., n\}$: the set of jobs to be processed on the machine;
Π: the set of all permutations of N;
$\pi \in \Pi$: a permutation of N defining the job sequence;
p_j: normal processing time of job j;
u_j: maximum compression of job j $(0 \leq u_j < p_j)$;
x_j: actual compression of job j $(0 \leq x_j \leq u_j)$;
$x = (x_1, ..., x_n)$: vector of actual compressions for all the n jobs;
$X = \{x : 0 \leq x_j \leq u_j, j \in N\}$: all the possible vectors of actual compressions;
$[e, d]$: a common due window of all the jobs, where e is the earliest due date and d is the latest due date;
α: unit earliness cost for all jobs;
β: unit tardiness cost for all jobs;
γ_j: unit cost of compressing the processing time of job j;
$C_j(\pi)$: completion time of job j in π;
$S_j(\pi)$: Starting time of job j in π;
$E_j(\pi) = max\{0, e - C_j(\pi)\}$: earliness of job j in schedule π;
$T_j(\pi) = max\{0, C_j(\pi) - d\}$: tardiness of job j in schedule π;
$E(\pi) = \{j : C_j(\pi) \leq e\}$: the set of early jobs in schedule π, a job in $E(\pi)$ is called an E-job;
$W(\pi) = \{j : S_j(\pi) < d$ and $e < C_j(\pi)\}$: the set of jobs in the common due window of schedule π, a job in $W(\pi)$ is called a W-job;
$T(\pi) = \{j : S_j(\pi) \geq d\}$: the set of tardy jobs in schedule π, a job in $T(\pi)$ is called a T-job;
We note that $E(\pi) \bigcup W(\pi) \bigcup T(\pi) = N$.

Based on the above notations, we have the following mathematical formulation of the problem (denoted by (**P**)):

$$Min \sum_{j=1}^{n} [\gamma_j x_j + \alpha E_j(\pi) + \beta T_j(\pi)] \qquad \textbf{(P)}$$

Subject to: $x \in X$ and $\pi \in \Pi$

3 Properties of an Optimal Solution for the Problem

The difficulty of this problem is to determine the W-job set. In Liman *et al.* [18], the size of the common due window is a decision variable, there is no need to determine the W-job set thus it makes the prolem relatively easy. To solve problem (**P**), we first state several well-known properties for common due date/window problems (Proposition 1-5 and Theorem 1), which also apply to problem (**P**) (see Kanet [11], Hall and Posner [8], Hall *et al.* [9], Kramer and Lee [13], and Weng and Ventura [29] for detailed proofs). Then we discuss some special properties of problem (**P**).

Proposition 1. There exists an optimal schedule of problem (**P**) such that there is no idle time between any two adjacent jobs.   □

Remark 1. Proposition 1 implies that a job sequence, a starting time and a position of the common due window are sufficient to determine a schedule for problem (**P**).

Proposition 2. There exists an optimal schedule of problem (**P**) such that either $C_j = e$ or $C_j = d$, for some job j.   □

Remark 2. Henceforth, it suffices to consider only two cases in which there is a job that can be completed at time e or d.

Proposition 3. V-shape property of an optimal schedule for problem (**P**):
(1) If $C_j = e$ for some j, then there exists an optimal schedule such that jobs in $E(\pi)$ are ordered in LPT, and the remaining jobs are ordered in SPT (called V-shape about e).
(2) If $C_j = d$ for some j, then there exists an optimal schedule such that jobs in $T(\pi)$ are ordered in SPT, and the remaining jobs are ordered in LPT (called V-shape about d).   □

In order to simplify the cost expressions below, we now introduce the following definition (c.f. Kramer and Lee [13]).

Definition 1. Given a schedule π for problem (**P**), the cumulative weight $cw(j, \pi)$ of job j with respect to a schedule π is defined as follows:
(1) If $j \in E(\pi)$ and $j = [k]$ (kth job in $E(\pi)$), then $cw(j, \pi) = (k - 1)\alpha$.
(2) If $j \in T(\pi)$ and $j = [k]$ (kth last job in $T(\pi)$), then $cw(j, \pi) = (k - 1)\beta$.
(3) If $j \in W(\pi)$ and $C_k = d$ for some k, then $cw(j, \pi) = |E(\pi)|\alpha$. If $j \in W(\pi)$ and $C_k = e$ for some k, then $cw(j, \pi) = (|T(\pi)| + 1)\beta$.
(4) Define $cw(W, \pi) = cw(j, \pi)$ since the cumulative weight of the W-jobs is the same for each W-job. where $| \bullet |$ denotes the cardinality of a set.

Proposition 4. Assume that the processing times of problem (**P**) are fixed and sorted in non-decreasing order, i.e., $p_1 \leq p_2 \leq ... \leq p_n$, then $W(\pi)$ can be determined as follows:
Let $nw^* = min\{l : \sum_{i=1}^{l} p_i \geq d - e\}$, then $W(\pi) = \{1, 2, ..., nw^*\}$.   □

Remark 3. From this proposition it easy to known that the processing time of a W-job is less than or equal to that of any E-job or T-job.

Remark 4. Problem (**P**) with controllable job processing times may have a minimal and maximal number of W-jobs as follows:
$MIN = min\{l : \sum_{i=1}^{l} p_i \geq d - e\}$, where p_i ($i = 1, 2, ..., n$), are the normal processing times in non-decreasing order.
$MAX = min\{l : \sum_{i=1}^{l} (p_i - u_i) \geq d - e\}$, where $(p_i - u_i)$ ($i = 1, 2, ..., n$), are the fully compressed processing times in non-decreasing order.

Proposition 5. An optimal schedule π of problem (**P**) must satisfy the following condition:
$max\{(|E(\pi)| - 1)\alpha, |T(\pi)|\beta\} \leq min\{|E(\pi)|\alpha, (|T(\pi)| + 1)\beta\}$.   □

Below, we introduce the algorithm for problem (**P**) with fixed job processing times and state the correctness of the algorithm in Theorem 1.

Algorithm A. (Kramer and Lee [13])
Step 1. Find $W(\pi)$ and sequence the jobs in $W(\pi)$ in any order.
Step 2. Assign the job with the longest processing time to $E(\pi)$ and set $u = 1, v = 0$.
Step 3. From the remaining jobs, assign the job with the longest processing time to $E(\pi)$ if $\alpha * u < \beta * (v + 1); u = u + 1$.
Step 4. From the remaining jobs, assign the job of longest processing time to $T(\pi)$ if $\alpha * u \geq \beta * (v + 1); v = v + 1$.
Then go to Step (3) if no job can be assigned, otherwise go to Step (5).
Step 5. Jobs in $E(\pi)$ are ordered in LPT, and jobs in $T(\pi)$ are ordered in SPT. The final sequence is organized as $(E(\pi), W(\pi), T(\pi))$ such that either the last job in $E(\pi)$ completes at e, if $\alpha * u \geq \beta * (v + 1)$, or the first job in $T(\pi)$ starts at d, if $\alpha * u < \beta * (v + 1)$.

Theorem 1. (Kramer and Lee [13]) **Algorithm A** finds an optimal schedule for problem (**P**) with fixed processing times. □

Remark 5. Theorem 1 implies that it is trivial to get the schedule for the problem with fixed job processing times after determining the W-job set. It also can be shown by proposition 5 that algorithm **A** generates an optimal schedule with $|E(\pi)| = \lceil \beta(n - |W(\pi)|)/(\alpha + \beta) \rceil$ and $|T(\pi)| = n - |E(\pi)| - |W(\pi)|$, where for a real number z, $\lceil z \rceil$ denotes the largest integer less than or equal to z.

Now we consider the impact of controllable job processing times on problem (**P**).

Proposition 6. There exists an optimal schedule in which a job is either fully compressed or uncompressed, except that one job in the window, at most, can be partially compressed.
Proof. Consider a job j in some position of an optimal sequence. Suppose that the processing time of this job is compressed by x_j. Then the contribution of this job to the total cost is:

$$cw(j, \pi) * (p_j - x_j) + \gamma_j x_j = cw(j, \pi) * p_j + (\gamma_j - cw(j, \pi)) * x_j.$$

It is obvious that compression of job j is only dependent on $cw(j, \pi)$, i.e., its position in the sequence. Now we consider the following three cases:

(1) If this job is either an E-job or a T-job, then it will be fully compressed if $cw(j, \pi) \geq \gamma_j$, or not compressed at all, otherwise.
(2) If this job is a W-job and there is at least one E-job or T-job after this job is fully compressed (from proposition 2, there is at most one such job), then it will be fully compressed if $cw(j, \pi) \geq \gamma_j$.
(3) If this job is a W-job, then it may be partially compressed until there is no processing of this job outside the common due window if $cw(j, \pi) \geq \gamma_j$ and

$cw'(j, \pi) \leq \gamma_j$, where $cw'(j, \pi)$ is the new cumulative weight with one less outside job (the partially compressed job). If there is more than one partially compressed job, then the benefits from compression of all these jobs should be the same. Thus at most one can be left to be partially (or possibly fully) compressed (and the other partially compressed jobs should totally uncompressed, if any).

Hence, the proposition holds.   □

Proposition 7. The cumulative weights $cw(j, \pi)$ will not be affected by the introduction of controllable job processing times and the corresponding cost provided that the number of W-jobs is given in advance.

Proof. Since the number of the W-jobs is known in advance, it can also be ascertain the number of jobs outside the due window, i.e., the total number of E-jobs and T-jobs. Therefore, the proposition follows from the results of Panwalkar *et al.* [23] and Kramer and Lee [13], which state that the earliness and tardiness cost can be converted to a positional penalty that is independent of the job processing times.   □

4   A Solution Algorithm

Combining the analysis in last two sections, below we present an algorithm to solve problem (**P**).

Algorithm B
Step 1. Determine the minimum and maximum numbers of W-jobs as follows:
$MIN = min\{l : \sum_{i=1}^{l} p_i \geq d - e\}$, where p_i $(i = 1, 2, ..., n)$ are the normal processing times in non-decreasing order.
$MAX = min\{l : \sum_{i=1}^{l} (p_i - u_i) \geq d - e\}$, where $(p_i - u_i)$ $(i = 1, 2, ..., n)$ are the fully compressed processing times in non-decreasing order.
Step 2. For k from MIN to MAX:
(1). Construct the matrix Q defined in the proof of theorem 2.
(2). Solve the corresponding assignment problem and record the best solution found so far.
Step 3. The final best solution is the optimal solution for the problem and the final sequence is organized as $(E(\pi), W(\pi), T(\pi))$ such that the last job in $E(\pi)$ is completed at e if $\alpha * u \geq \beta * (v + 1)$, and the first job in $T(\pi)$ starts at d if $\alpha * u < \beta * (v + 1)$. The jobs in $E(\pi)$ are ordered in LPT while the jobs in $T(\pi)$ are ordered in SPT. The jobs in $W(\pi)$ are in any order.
Step 4. Fully compress the job j in ($E(\pi)$ or $T(\pi)$) if $cw(j, \pi) \geq \gamma_j$; Fully compress job j in $W(\pi)$ if $cw(j, \pi) \geq \gamma_j$ and $cw'(j, \pi) \leq \gamma_j$; Partially compress job i in $W(\pi)$ if $cw(j, \pi) \geq \gamma_j$, but $cw'(j, \pi) \leq \gamma_j$; where $cw'(j, \pi)$ is the new cumulative weight with one less outside job (the partially compressed job).

Theorem 2. Algorithm B finds an optimal solution for problem (**P**).

Proof. The objective function of the problem can be written as follows:

$$\sum_{j=1}^{n}\gamma_j x_j + \sum_{j=1}^{n}(\alpha E_j(\pi) + \beta T_j(\pi)) = \sum_{j=1}^{n}\gamma_j x_j + \sum_{j=1}^{|E|}(j-1)*\alpha(p_{e_j} - x_j)$$

$$+ \sum_{j=1}^{|T|} j*\beta(p_{t_j} - x_j) + cw(W,\pi)*[\sum_{j=1}^{|W|}(p_{w_j} - x_j) - (d-e)]$$

$$= \{\sum_{j=1}^{|E|}[\gamma_{e_j} - (j-1)*\alpha]*x_{e_j} + \sum_{j=1}^{|T|}[\gamma_{t_j} - j*\beta]*x_{t_j} + \sum_{j=1}^{|W|}[\gamma_{w_j} - cw(W,\pi)]*x_{w_j}\}$$

$$+\{\sum_{j=1}^{|E|}(j-1)*\alpha*p_{e_j} + \sum_{j=1}^{|T|}j*\beta*p_{t_j} + cw(W,\pi)*[\sum_{j=1}^{|W|}p_{w_j} - (d-e)]\}$$

where e_j, t_j, w_j denote the jth job in E-job set, T-job set, and W-job set, respectively.

If the number of W-jobs, $|W(\pi)|$, is given, then by **Proposition 7**, the minimization of the objective function of the problem can be achieved by solving an assignment problem as follows, since the cost of a job can be transformed into its positional penalty. Now consider two cases.

Case 1: If $|W(\pi)| = n$, then all the jobs (either compressed or uncompressed) finish within the common due window. If the total processing time is larger than the latest due day, since all jobs have the same position penalty in this case, thus it suffices to compress from the jobs with the smallest unit compression cost to the jobs with the largest unit compression cost until the total processing time is less than or equal to the latest due date of the common due window. Note that possibly the last job is partially compressed.

Case 2: If $|W(\pi)| < n$, define an $n \times n$ matrix as follows:

$$Q = [Q_1 Q_2 Q_3]$$

where $Q_1 = [c_{ij}^1]_{n \times |E(\pi)|}, Q_2 = [c_{ij}^2]_{n \times |T(\pi)|}$, and $Q_3 = [c_{ij}^3]_{n \times |W(\pi)|}$.

By Remark 5, it is easy to obtain $|E(\pi)|$ and $|T(\pi)|$, and:

(1) For $i = 1, ..., n, j = 1, ..., |E(\pi)|$:
$c_{ij}^1 = \gamma_i * u_i + \alpha * (j-1) * (p_i - u_i)$, if $\alpha * (j-1) > \gamma_i$, and $c_{ij}^1 = \alpha * (j-1) * p_i$, otherwise.

(2) For $i = 1, ..., n, j = 1, ..., |T(\pi)|$:
$c_{ij}^2 = \gamma_i * u_i + \beta * j * (p_i - u_i)$, if $\beta * j > \gamma_i$, and $c_{ij}^2 = \beta * (j-1) * p_i$, otherwise.

(3) For $i = 1, ..., n, j = 1, ..., |W(\pi)|$:
$c_{ij}^3 = \gamma_i * u_i + min\{|E(\pi)|\alpha + (|T(\pi)|+1)\beta\} * (p_i - u_i)$, if $min\{|E(\pi)|\alpha + (|T(\pi)|+1)\beta\} > \gamma_i$,
and $c_{ij}^3 = min\{|E(\pi)|\alpha + (|T(\pi)|+1)\beta\} * p_i$, otherwise, where $|E(\pi)| = \lceil \beta(n - |W(\pi)|)/(\alpha + \beta)\rceil$ and $|T(\pi)| = n - |E(\pi)| - |W(\pi)|$.

Now choose n elements from the matrix so that:
(1) there is exactly one element from each row;
(2) there is exactly one element from each column; and
(3) the sum of the elements is minimized.

By propositions 4, 6, and 7, it is obvious that solving this assignment problem provides a feasible solution with minimal cost for the scheduling problem with a fixed $|W(\pi)|$.

Algorithm B goes through all the possible $|W(\pi)|$ and finds an optimal solution for the problem.   □

Remark 6. The computational complexity of **Algorithm B** is $O(n^3 \log n)$, since the complexity of the algorithm for the assignment problem is $O(n^2 \log n)$ (Ahuja *et al.* 1993), and this algorithm may be carried out $O(n)$ times.
    The following example is used to illustrate the algorithm.

Example 1: Consider an instance of problem (**P**) with $n = 5$, a common due window size=5, earliness penalty $\alpha = 5$, tardiness penalty $\beta = 7$, and the other job data as shown in Table 1.

Table 1. Job processing times and compression costs in Example 1

Job j	1	2	3	4	5
Normal processing time p_j	3	5	6	8	10
Minimal processing time $p_j - u_j$	2	2	4	5	3
Unit compression cost γ_j	11	9	8	4	12

Then we know that:

$$MIN = min\{l : \sum_{i=1}^{l} p_i \geq d - e\} = min\{l : \sum_{i=1}^{l} p_i \geq 5\} = 2$$

$$MAX = min\{l : \sum_{i=1}^{l} (p_i - u_i) \geq d - e\} = min\{l : \sum_{i=1}^{l} t_i \geq 5\} = 3$$

where p_i is the normal processing time in non-decreasing order and $(p_i - u_i)$ is the fully compressed processing time in non-decreasing order.
    (1). When $|W(\pi)| = MIN = 2$, then:

$$|E(\pi)| = \lceil \beta(n - |W(\pi)|)/(\alpha + \beta) \rceil = \lceil 7(5 - 2)/(5 + 7) \rceil = 2,$$
$$|T(\pi)| = n - |E(\pi)| - |W(\pi)| = 5 - 2 - 2 = 1,$$

$$Q_1 = \begin{bmatrix} 0 & 15 \\ 0 & 25 \\ 0 & 30 \\ 0 & 37 \\ 0 & 50 \end{bmatrix}, Q_2 = \begin{bmatrix} 21 \\ 35 \\ 42 \\ 47 \\ 70 \end{bmatrix} \text{ and, } Q_3 = \begin{bmatrix} 30 & 30 \\ 47 & 47 \\ 56 & 56 \\ 62 & 62 \\ 100 & 100 \end{bmatrix}$$

(2).When $|W(\pi)| = MAX = 3$, then:

$$|E(\pi)| = \lceil \beta(n - |W(\pi)|)/(\alpha + \beta)\rceil = \lceil 7(5 - 3)/(5 + 7)\rceil = 2,$$
$$|T(\pi)| = n - |E(\pi)| - |W(\pi)| = 5 - 3 - 2 = 0,$$

$$Q_1 = \begin{bmatrix} 0 \ 15 \\ 0 \ 25 \\ 0 \ 30 \\ 0 \ 37 \\ 0 \ 50 \end{bmatrix}, Q_2 \text{ is empty and, } Q_3 = \begin{bmatrix} 21 \ 21 \ 21 \\ 35 \ 35 \ 35 \\ 42 \ 42 \ 42 \\ 47 \ 47 \ 47 \\ 70 \ 70 \ 70 \end{bmatrix}$$

Solving these two assignment problems and choosing the solution with the lower cost, we can get the optimal solution for the scheduling problem as follows:
(1) The sequence is: $\pi = (5, 3, 1, 2, 4)$, where job 5 and 3 are E-jobs, job 1 and 2 are W-jobs, and job 4 is a T-job.
(2) Job 2 and 4 are fully compressed to processing times 2 and 5, respectively. Other jobs are uncompressed.
(3) Job 5 starts at time zero and the completion time of job 2 coincides with the end of the common due window, i.e., the end of the common due window is set at time 21.
(4)The total cost of this schedule is $(5*6+7*5)+9*(5-2)+4*(8-5)=104$ (see the Gantt chart in Fig. 1).

Fig. 1. Gantt chart of the schedule in Example 1

5   Conclusions

We have studied a single machine common due window scheduling problem with controllable job processing times and a movable due window to minimize the total costs of weighted earliness/tardiness and compression cost of processing times. After presenting the mathematical formulation and discussing some properties of optimal solutions, we developed a polynomial algorithm to solve this problem. The algorithm is based on the algorithm for the classical assignment problem. We also used an example to illustrate the application of the algorithm.

Further research can be undertaken to investigate the case with a given position of the common due window or distinct due windows, and the case on parallel machines.

Acknowledgement

The author would like to thank the anonymous reviewers for helpful comments. This work was supported in part by NSFC (70372058) and Guangdong NSF (031808).

References

1. Ahuja, R.K., Magnanti, T.L., Orlin, J.B.: Network Flows: Theory, Algorithms, and Applications. Prentice-Hall, Englewood Cliffs (1993)
2. Alidaee, B., Kochenberger, G.A.: A framework for machine scheduling problems with controllable processing times. Production and Operations Management 5, 391–405 (1996)
3. Baker, K., Scudder, G.: Sequencing with earliness and tardiness penalties: A review. Operations Research 38, 22–36 (1990)
4. Biskup, D., Jahnke, H.: Common due date assignment for scheduling on a single machine with jointly reducible processing times. International Journal of Production Economics 69, 317–322 (2001)
5. Cheng, T.C.E.: Optimal common due-date with limited completion time deviation. Computers and Operations Research 15, 91–96 (1988)
6. Daniels, R.L., Sarin, R.K.: Single machine scheduling with controllable processing times and number of jobs tardy. Operations Research 37, 981–984 (1989)
7. Dickman, B., Wilamowsky, Y., Epstain, S.: Optimal common due-date with limited completion time. Computers and Operations Research 39, 125–127 (1991)
8. Hall, N.G., Posner, M.E.: Earliness-tardiness scheduling problems (I). Operations Research 39, 836–846 (1991)
9. Hall, N.G., Kubiak, W., Sethi, S.P.: Earliness-tardiness scheduling problems (II). Operations Research 39, 847–856 (1991)
10. Jansen, K., Mastrolilli, M., Solis-Oba, R.: Approximation schemes for job shop scheduling problems with controllable processing times. European Journal of Operational Research 167, 297–319 (2005)
11. Kanet, J.J.: Minimize the average deviation of job completion times about a common due date. Naval Research Logistics 28, 643–651 (1981)
12. Kaspi, M., Shabtay, D.: A bicriterion approach to time/cost trade-offs in scheduling with convex resource-dependent job processing times and release dates. Computers and Operations Research 33, 3015–3033 (2006)
13. Kramer, F.J., Lee, C.Y.: Common due window scheduling. Production and Operations Management 2, 262–275 (1993)
14. Kramer, F.J., Lee, C.Y.: Due window scheduling for parallel machines. Mathematics and Computer Modelling 20, 69–89 (1994)
15. Lee, C.Y.: Earliness-tardiness scheduling problems with constant size of due window. Research Report, Dept. of Industrial and Systems Engineering, University of Florida (1991)
16. Liman, S.D., Rawaswamy, S.: Earliness-tardiness scheduling problems with a common delivery window. Operations Research Letters 15, 195–203 (1994)
17. Liman, S.D., Panwalkar, S.S., Thongmee, S.: Determination of common due window location in a single machine scheduling problem. European Journal of Operational Research 93, 68–74 (1996)
18. Liman, S.D., Panwalkar, S.S., Thongmee, S.: A single machine scheduling problem with common due window and controllable processing times. Annals of Operations Research 70, 145–154 (1997)
19. Liman, S.D., Panwalkar, S.S., Thongmee, S.: Common due window size and location determination in a single machine scheduling problem. Journal of the Operational Research Society 49, 1007–10 (1998)
20. Ng, C.T., Cheng, T.C.E., Janiak, A., Kovalyov, M.Y.: Group Scheduling with Controllable Setup and Processing Times: Minimizing Total Weighted Completion Time. Annals of Operations Research 133, 163–174 (2005)

21. Nowicki, E., Zdrzalka, S.: A survey of results for sequencing problems with controllable processing times. Discrete Applied Mathematics 26, 271–287 (1990)
22. Panwalkar, S.S., Rajagopalan, R.: Single machine sequencing with controllable processing times. European Journal of Operational Research 59, 298–302 (1992)
23. Panwalkar, S.S., Smith, M.L., Seidmann, A.: Common due date assignment to minimize total penalty for the one machine scheduling problem. Operations Research 30, 391–399 (1982)
24. Thongmee, S., Liman, S.D.: Common due window size determination in a single machine scheduling problem. Research Report, Dept. of Industrial Engineering, Texas Tech University (1995)
25. Vickson, R.G.: Two single machine sequencing problem involving controllable job processing times. AIIE Transactions 12, 258–262 (1980a)
26. Vickson, R.G.: Choosing the sequence and processing times to minimize total processing plus flow cost on a single machine. Operations Research 28, 1155–1167 (1980b)
27. Van Wassenhove, L.N., Baker, K.: A Bicriterion approach to time/cost trade-offs in sequencing. European Journal of Operational Research 11, 48–54 (1982)
28. Wan, G., Yen, B.P.C.: Tabu search for single machine scheduling with distinct due windows and weighted earliness/tardiness penalties. European Journal of Operational Research 142, 271–281 (2002)
29. Weng, M.X., Ventura, J.A.: A note on Common due window Scheduling. Production and Operations Management 5, 194–200 (1996)

A Lower Bound on Approximation Algorithms for the Closest Substring Problem*

Jianxin Wang[1], Min Huang[1], and Jianer Chen[1,2]

[1] School of Information Science and Engineering, Central South University,
Changsha 410083, China
jxwang@mail.csu.edu.cn
[2] Department of Computer Science, Texas A&M University,
College Station, TX 77843-3112, USA
chen@cs.tamu.edu

Abstract. The Closest Substring problem (CSP), where a short string is sought that minimizes the number of mismatches between it and each of a given set of strings, is a minimization problem with polynomial time approximation schemes. In this paper, a lower bound on approximation algorithms for the CSP problem is developed. We prove that unless the Exponential Time Hypothesis (ETH Hypothesis, i.e., not all search problems in SNP are solvable in subexponential time) fails, the CSP problem has no polynomial time approximation schemes of running time $f(1/\varepsilon)|x|^{O(1/\varepsilon)}$ for any function f, where $|x|$ is the size of input instance. This essentially excludes the possibility that the CSP problem has a practical polynomial time approximation scheme even for moderate values of the error bound ε. As a consequence, it is unlikely that the study of approximation schemes for the CSP problem in the literature would lead to practical approximation algorithms for the CSP problem for small error bound ε.

1 Introduction

The Closest Substring problem was introduced by Lanctot *et al.* [1] and is a key theoretical problem in molecular biology applications such as genetic drug design, creating diagonal probes, and creating universal PCR primers. Li *et al.* [2] defined the problem as follows:

Definition 1 (Closest Substring Problem (CSP)). *Given a set $S = \{s_1, s_2, ..., s_n\}$ of strings each of length m, and an integer L, find a string s of length L minimizing d such that for each $s_i \in S$ there is a length L substring t_i of s_i with $D(s, t_i) \leq d$.*

---

* This work is supported by the National Natural Science Foundation of China (60433020), the Program for New Century Excellent Talents in University (NCET-05-0683) and the Program for Changjiang Scholars and Innovative Research Team in University (IRT0661).

A. Dress, Y. Xu, and B. Zhu (Eds.): COCOA 2007, LNCS 4616, pp. 291–300, 2007.

Herein, $D(s,t)$ represents the Hamming distance between two strings s and t. Closest Substring problem is an NP-hard minimization problem [1].

One interesting direction of the research on CSP is the design and analysis of approximation algorithms for the problem. Lanctot *et al.* [1] proposed a straightforward ratio-2 approximation algorithm for the problem. Later, Ma [3] further studied the approximation algorithms for the problem. In 2002, Li *et al.* [2] developed a polynomial time approximation scheme (PTAS) for CSP. However, the PTAS presented in [2] has a very high computational complexity and is not practical even for moderate values of the error bound ϵ. Therefore, whether there is an efficient PTAS for CSP becomes an interesting problem in current research.

The theory of parameterized computation and complexity [4,5] is a recently developed subarea in theoretical computer science. The theory is aimed at practically solving a large number of computational problems that are theoretically intractable. The theory is based on the observation that many intractable computational problems in practice are associated with a parameter that varies within a small or moderate range. Therefore, by taking the advantages of the small parameters, many theoretically intractable problems can be solved effectively and practically. On the other hand, the theory of parameterized computation and complexity has also offered powerful techniques that enable us to derive strong computational lower bounds for many computational problems [6,7], thus explaining why certain theoretically tractable problems cannot be solved effectively and practically.

Fellows *et al.* analyzed the parameterized complexity of the Closest Substring problem [8], and proved that the problem is W[1]-hard. Therefore, it is deduced that unless an unlikely collapse occurs in parameterized complexity theory, CSP does not have a PTAS of running time $f(1/\varepsilon)|x|^{O(1)}$ for any function f, where $|x|$ is the size of input instance.

However, this does not completely exclude the possibility that the problem may have feasible PTAS. For instance, if the problem could be solvable by a PTAS running in time $|x|^{\log\log(1/\varepsilon)}$, then such an algorithm is still feasible for moderately values of ε (i.e. $\varepsilon = 0.01\%$).

In this paper, we prove that unless the Exponential Time Hypothesis (ETH Hypothesis, i.e., not all search problems in SNP are solvable in subexponential time, [9,10]) fails, the CSP problem has no polynomial time approximation schemes of running time $f(1/\varepsilon)|x|^{o(1/\varepsilon)}$ for any function f, where $|x|$ is the size of input instance. The class SNP introduced by Papadimitriou and Yannakakis [9] contains many well-known NP-hard problems including, for any fixed integer $q \geq 3$, CNF q-SAT, q-COLORABILITY, q-SET COVER, VERTEX COVER, CLIQUE, and INDEPENDENT SET, etc. It is commonly believed that it is unlikely that all problems in SNP are solvable in subexponential time. So our result is a stronger lower bound of PTAS for CSP. This essentially excludes the possibility that the CSP problem has a practically efficient PTAS, even for those with running time like $O(|x|^{\log\log(1/\varepsilon)})$. In particular, it indicates that it is unlikely that the PTAS of Li *et al.* [2] can be significantly improved to become practical.

2   Preliminaries

In this section, we give brief review and description on the fundamentals of parameterized computation and complexity, and of approximation algorithms for NP optimization problems.

2.1   Parameterized Complexity and W-Hardness Under Linear FPT-Reductions

A *parameterized problem* Q is a subset of $\Omega^* \times N$, where Ω is a fixed alphabet and N is the set of all non-negative integers. Therefore, each instance of Q is a pair (x, k), where the non-negative integer k is called the *parameter*. The parameterized problem Q is *fixed-parameter tractable* [4] if there is an algorithm that decides if an input (x, k) is a yes-instance of Q in time $f(k)|x|^c$, where c is a fixed constant and $f(k)$ is an arbitrary function. The complexity class FPT consists of all fixed-parameter tractable problems.

Computational practice has shown that certain parameterized problems seem not fixed-parameter tractable. To reflect this fact, a hierarchy of fixed-parameter intractability, the W-*hierarchy* $\cup_{t \geq 0} W[t]$, where $W[t] \subseteq W[t+1]$ for all $t \geq 0$, has been introduced, in which the 0-th level $W[0]$ is the class FPT. The hardness and completeness have been defined for each level $W[i]$ of the W-hierarchy for $i \geq 1$ [4]. It is commonly believed that $W[1] \neq FPT$. Thus, $W[1]$-hardness has served as the hypothesis for fixed-parameter intractability.

Chen *et al.* [6,7] introduced the concepts of linear fpt-reduction and W-hardness under linear fpt-reductions.

Definition 2. *A parameterized problem Q is linear fpt-reducible, shortly fpt_l-reducible, to a parameterized problem Q' if there exist a function f and an algorithm A of running time $f(k)|x|^{O(1)}$ that, on each instance (x, k) of Q, produces an instance (x', k') of Q', where $k' = O(k)$, $|x'| = |x|^{O(1)}$, and (x, k) is a yes-instance of Q if and only if (x', k') is a yes-instance of Q'.*

Thus, an fpt_l-reduction is a regular fpt-reduction with additional constraints on the parameter value and the instance size. From the definition, the transitivity of the fpt_l-reduction can be easily deduced [7]:

Lemma 1. *Let Q_1, Q_2, and Q_3 be three parameterized problems. If Q_1 is fpt_l-reducible to Q_2, and Q_2 is fpt_l-reducible to Q_3, then Q_1 is fpt_l-reducible to Q_3.*

The definition of $W[1]$-hardness under the fpt_l-reduction, shortly $W_l[1]$-*hardness*, is given by the fpt_l-reduction from the Clique problem. We first give the definition of the Clique problem.

Definition 3 (The Clique Problem)
Instance: A graph $G = (V, E)$ and a parameter k.
Question: Is there a set V' of k vertices in G such that for any two vertices u and v in V', we have $[u, v] \in E$?

Definition 4. *A parameterized problem Q is $W[1]$-hard under the fpt_l-reduction, shortly $W_l[1]$-hard, if the Clique problem is fpt_l-reducible to Q.*

It has been proved [6,7] that the following parameterized problems are $W_l[1]$-hard: WCNF-SAT, HITTING SET, DOMINATING SET, RED-BLUE DOMINATING SET, DOMINATING CLIQUE, PRECEDENCE CONSTRAINED PROCESSOR SCHEDULING, FEATURE SET, and WEIGHTED BINARY INTEGER PROGRAMMING, WCNF q-SAT for any integer $q \geq 2$, CLIQUE, INDEPENDENT SET, SET COVER, and SET PACKING.

2.2   NP Optimization Problems and Approximation Algorithms

We now review the basic concepts for NP optimization problems and approximation algorithms. More discussions can be found in [11]. We will also discuss the relationships between approximability and parameterized complexity of NP optimization problems.

Definition 5. *An NP optimization problem Q is a 4-tuple (I_Q, S_Q, f_Q, opt_Q), where*

1. *I_Q is the set of input instances. It is recognizable in polynomial time;*
2. *For each instance $x \in I_Q$, $S_Q(x)$ is the set of feasible solutions for x, which is defined by a polynomial p and a polynomial time computable predicate π (p and π only depend on Q) as $S_Q(x) = \{y : |y| \leq p(|x|) \text{ and } \pi(x, y)\}$;*
3. *$f_Q(x, y)$ is the objective function mapping a pair $x \in I_Q$ and $y \in S_Q(x)$ to a non-negative integer. The function f_Q is computable in polynomial time;*
4. *$opt_Q \in \{max, min\}$. Q is called a maximization problem if $opt_Q = max$, and a minimization problem if $opt_Q = min$.*

An *optimal solution* y_0 for an instance $x \in I_Q$ is a feasible solution in $S_Q(x)$ such that $f_Q(x, y_0) = opt_Q\{f_Q(x, z) \mid z \in S_Q(x)\}$. Denote the value $opt_Q\{f_Q(x, z) \mid z \in S_Q(x)\}$ by $opt_Q(x)$.

An algorithm A is an *approximation algorithm* for an NP optimization problem Q if, for each input instance x in I_Q, the algorithm A returns a feasible solution $y_A(x)$ in $S_Q(x)$. The solution $y_A(x)$ has an *approximation ratio* $r(n)$ if it satisfies the following condition:

$$opt_Q(x)/f_A(x, y_A(x)) \leq r(|x|) \quad \text{if } Q \text{ is a maximization problem}$$
$$f_A(x, y_A(x))/opt_Q(x) \leq r(|x|) \quad \text{if } Q \text{ is a minimization problem}$$

The approximation algorithm A has an approximation ratio $r(m)$ if for any instance x in I_Q, the solution $y_A(x)$ constructed by the algorithm A has an approximation ratio bounded by $r(|x|)$.

An NP optimization problem Q has a *polynomial time approximation scheme* (PTAS) if there is an algorithm A_Q that takes a pair (x, ε) as input, where x is an instance of Q and $\varepsilon > 0$ is a real number, and returns a feasible solution y for x such that the approximation ratio of the solution y is bounded by $1 + \varepsilon$, and for each fixed $\varepsilon > 0$, the running time of the algorithm A_Q is bounded by a polynomial of $|x|$.

The definition of parameterization of NP optimization problems is given as follows [7,5].

Definition 6. *Let $Q = (I_Q, S_Q, f_Q, opt_Q)$ be an NP optimization problem. The parameterized version of Q is defined as follows:*

1. *If Q is a maximization problem, then the parameterized version of Q is defined as $Q_\geq = \{(x, k) \mid x \in I_Q \text{ and } opt_Q(x) \geq k\}$;*
2. *If Q is a minimization problem, then the parameterized version of Q is defined as $Q_\leq = \{(x, k) \mid x \in I_Q \text{ and } opt_Q(x) \leq k\}$.*

The above definition offers the possibility to study the relationship between the approximability and the parameterized complexity of NP optimization problems. Chen *et al.* proved the following theorem (Theorem 6.1 in [6]):

Theorem 1. *Let Q be an NP optimization problem. If the parameterized version of Q is $W_l[1]$-hard, then Q has no PTAS of running time $f(1/\varepsilon)|x|^{o(1/\varepsilon)}$ for any function f, unless the ETH Hypothesis fails.*

3   A Lower Bound on PTAS for the CSP Problem

Since the definitions of the Closest Substring problem in [1,3,2] are not standard NP optimization problem definitions, we define the NP optimization version of the Closest Substring problem as follows:

Definition 7. *The Closest Substring Problem (CSP) is a tuple (I_C, S_C, f_C, opt_C), where*

1. *I_C is the set of all instances $x = (S, L)$, where L is an integer, $S = \{s_1, s_2, ..., s_n\}$ is a set of n strings each of length m;*
2. *For an instance $x = (S, L)$ in I_C, $S_C(x)$ is the set of all strings of length L;*
3. *For an instance $x = (S, L)$ in I_C and a string $s \in S_C(x)$, define $D(s, s_i) = \min\{D(s, t_i) \mid t_i \text{ is a length } L \text{ substring of } s_i\}$ and $dist(s, S) = \max\{D(s, s_i) \mid s_i \in S\}$. The objective function $f_C(x, s) = dist(s, S)$;*
4. *$opt_C = min$.*

Denote the value $opt_C\{f_C(x, s) \mid s \in S_C(x)\}$ by $opt_C(x)$. With the standard definition, the approximation algorithm A for CSP developed by Li *et al* in [2] can be described in the following sense: for a given instance $x = (S, L)$ for CSP and a small constant $\varepsilon > 0$, the approximation algorithm A produces a string s of length L such that $dist(s, S) \leq (1 + \varepsilon)opt_C(x)$, and the running time of A is $O(n^2 m)^{O(1/\varepsilon^4)}$. Thus A is a PTAS for CSP.

CSP is a minimization problem, so we denote the parameterized version of it by $CSP_\leq$ and define it as follows:

$$CSP_\leq = \{(x, k) \mid x \in I_C \text{ and } opt_C(x) \leq k\}$$

Our main result in this paper is the following lemma on the parameterized complexity of the problem $CSP_\leq$.

Lemma 2. *The parameterized problem $CSP_\leq$ is $W_l[1]$-hard.*

Proof. We prove the lemma by an fpt$_l$-reduction from the $W_l[1]$-hard problem Clique to the $CSP_\leq$ problem. The proof is given by the following discussion. □

The linear fpt-reduction here is similar to the fpt-reduction presented in [8]. However, from the reduction in [8], we could not get Lemma 2 directly (or it needs difficult mathematical analysis and proof that were not given in [8]). Instead, we give a different reduction here, which seems easier to verify.

Consider two instances of the CSP problem:

$$x_f = (\{AAA, BBB\}, 3) \quad \text{and} \quad x_t = (\{AAA, AAB\}, 3)$$

Obviously, for x_f, the Hamming distance between any length-3 string s and one of the two strings AAA and BBB is at least 2. In fact, if the distance between s and AAA is less than 2, then there must be at least 2 A's in s, which means the distance between s and BBB is at least 2. So we have $opt_C(x_f) = 2$. Similarly, we get $opt_C(x_t) = 1$.

Therefore, $(x_f, 1)$ is a no-instance for the $CSP_\leq$ problem, while $(x_t, 1)$ is a yes-instance for the $CSP_\leq$ problem.

The fpt$_l$-reduction from Clique to $CSP_\leq$

Input: (G, k), G is a graph with n vertices $v_1, v_2, ..., v_n$ and m edges $e_1, e_2, ..., e_m$

Output: An instance of the $CSP_\leq$ problem

1. If $k \leq 4$, decide whether (G, k) is a yes-instance of Clique by brute-force. If (G, k) is a yes-instance, then let $(x_G, k') = (x_t, 1)$, otherwise let $(x_G, k') = (x_f, 1)$; Output (x_G, k'); Stop.

2. If $k \geq 5$, construct an instance (x_G, k') of $CSP_\leq$, where $x_G = (S, L)$, as follows:

2-1. $k' = k - 2$;

2-2. $L = k + 2$;

2-3. S contains $n' = \binom{k}{2} = k(k-1)/2$ strings:

$$S = \{s_{1,2}, s_{1,3}, ..., s_{1,k}, s_{2,2}, s_{2,3}, ..., s_{2,k}, ..., s_{k-1,k}\}$$

The length of $s_{i,j}(1 \leq i < j \leq k)$ is $m' = m(k+2) + 2k(m-1)$, and

$$s_{i,j} = \langle block(i, j, e_1) \rangle (\psi_{i,j})^{2k} \langle block(i, j, e_2) \rangle (\psi_{i,j})^{2k} ... (\psi_{i,j})^{2k} \langle block(i, j, e_m) \rangle$$

where $\psi_{i,j}$ is the unique symbol for the string $s_{i,j}$, called a *separator*; substring $\langle block(i, j, e_l) \rangle$, $1 \leq l \leq m$, is called a *block* of $s_{i,j}$ and defined as follows: let symbol σ_i denote vertex v_i in G. Suppose edge e_l connects vertices v_r and v_s, $r < s$, then $\langle block(i, j, e_l) \rangle$ is a length-L string:

$$\langle block(i, j, e_l) \rangle = \#(\psi_{i,j})^{i-1} \sigma_r (\psi_{i,j})^{j-i-1} \sigma_s (\psi_{i,j})^{k-j} \#$$

Fig. 1. The fpt$_l$-reduction from Clique to $CSP_\leq$

Consider the fpt reduction from the Clique problem to the $\text{CSP}_\leq$ problem given in Figure 1. The length of each string $s_{i,j}$ is $m' = m(k+2) + 2k(m-1) = O(mn) = O(n^3)$, the number of strings in S is $n' = \binom{k}{2} = k(k-1)/2 = O(n^2)$. Therefore, the size of $x_G = (S, L)$ is bounded by a polynomial of the size of the graph G. Meanwhile, $k' = k - 2 = O(k)$. Thus, (x_G, k') makes an instance for the $\text{CSP}_\leq$ problem, satisfying the restrictions of the linear fpt reduction. The running time of the reduction in Figure 1 is no more than $O(n^5)$ (the first step in the reduction can be solved by enumerating all the subgraphs of G with at most 4 vertices).

Now we only need to verify that (G, k) is a yes-instance for the Clique problem if and only if (x_G, k') is a yes-instance for the $\text{CSP}_\leq$ problem.

In step 1 of the reduction in Figure 1, we can easily conclude that when $k \leq 4$, (G, k) is a yes-instance for Clique if and only if (x_G, k') is a yes-instance for $\text{CSP}_\leq$. So we only need to give the proof when $k \geq 5$.

Proposition 1. *If (G, k) is a yes-instance for the Clique problem, then (x_G, k') is a yes-instance for the $\text{CSP}_\leq$ problem.*

Proof. Suppose that the graph G has a clique Q of k vertices. Let $h_1, h_2, ..., h_k$ denote the indices of the vertices in the clique Q, $1 \leq h_1 < h_2 < \cdots < h_k \leq n$. Consider the string $s = \#\sigma_{h_1}\sigma_{h_2}...\sigma_{h_k}\#$ of length L. Since Q is a clique, for any two indices i and j, $1 \leq i < j \leq k$, $[v_{h_i}, v_{h_j}]$ is an edge in G. Now consider the string $s_{i,j}$. Because the string $s_{i,j}$ encodes all edges in G, we can find a length-L substring $t_{i,j}$, i.e., $t_{i,j} = \langle block(i, j, [v_{h_i}, v_{h_j}]) \rangle$ in $s_{i,j}$, where $[v_{h_i}, v_{h_j}]$ is an edge in G, such that $D(s, t_{i,j}) = k - 2$. Note that there are m length-L substrings in $s_{i,j}$ of the form $\langle block(i, j, e_l) \rangle$, in which only the substring $t_{i,j}$ and the string s satisfy $D(s, t_{i,j}) \leq k - 2$. Moreover, it is easy to verify that for any other length-L substring $t'_{i,j}$ that is not a block of $s_{i,j}$, we always have $D(s, t'_{i,j}) > k - 2$. Thus $D(s, s_{i,j}) = k - 2$. Since this is true for all $1 \leq i < j \leq k$, we conclude $f_C(x, s) = dist(s, S) = k - 2 = k'$. This verifies that $opt_C(x) \leq f_C(x, s) = k'$. In consequence, (x_G, k') is a yes-instance of the $\text{CSP}_\leq$ problem. $\square$

Proposition 2. *If (x_G, k') is a yes-instance for the $\text{CSP}_\leq$ problem, then (G, k) is a yes-instance for the Clique problem.*

Proof. If (x_G, k') is a yes-instance for the $\text{CSP}_\leq$ problem, then there is a length-L string s such that for each string $s_{i,j}$ in S, $D(s, s_{i,j}) \leq k'$. We observe that the string s has the following properties:

Observation 1. *Suppose $t_{i,j}$ is a length-L substring of $s_{i,j}$, and $D(s, t_{i,j}) \leq k' = k - 2$. If the separator symbol $\psi_{i,j}$ of $s_{i,j}$ does not appear in s, then $t_{i,j}$ must be a block of $s_{i,j}$, i.e., $t_{i,j} = \langle block(i, j, e_l) \rangle$ for some edge e_l, and the 4 non-separator symbols in the block $t_{i,j}$ must coincide with the respective positions in s.*

Proof of Observation 1. There are $k + 2$ symbols in a block of $s_{i,j}$, where 4 of them are not separator symbols: two $\#$ symbols and two σ symbols. Between two blocks there are $2k$ separator symbols $\psi_{i,j}$. Therefore, if there are 4 non-separator symbols within a length-L substring $t_{i,j}$ of $s_{i,j}$, then $t_{i,j}$ must be a block. To satisfy $D(s, t_{i,j}) \leq k' = k - 2$, there must be 4 symbols in $t_{i,j}$ coincide with the respective positions in s (note that the length of both $t_{i,j}$ and s is $k + 2$). Because the separator symbol $\psi_{i,j}$ of $s_{i,j}$ does not appear in s, these 4 symbols are non-separator symbols. In summary, $t_{i,j}$ is a block of $s_{i,j}$, and the 4 non-separator symbols in the block coincide with the respective positions in s. This completes the proof of Observation 1   □

Now we prove Proposition 2. Suppose (x_G, k') is a yes-instance of the CSP$_\leq$ problem, then there is a string s of length L such that $f_C(x_G, s) = dist(s, S) \leq k' = k - 2$. This means that for each $s_{i,j} \in S$, we should find a length-L substring $t_{i,j}$ in $s_{i,j}$, such that $D(s, t_{i,j}) \leq k - 2$.

When $k \geq 5$, the number of strings in S is $n' = k(k-1)/2 > k + 2$. The length of s is $L = k + 2$, which implies that there are at most $k + 2$ different symbols in s. So we can safely say that there is at least one string $s_{i,j}$ in S, whose separator symbol $\psi_{i,j}$ does not appear in s. For this string $s_{i,j}$, because there is a length-L substring $t_{i,j}$ in $s_{i,j}$ satisfying $D(s, t_{i,j}) \leq k - 2$, according to Observation 1, $t_{i,j}$ must be a block of $s_{i,j}$, i.e., $t_{i,j} = \langle block(i, j, e_l) \rangle$, and the 4 non-separator symbols in the block $t_{i,j}$ must coincide with the respective positions in s. Therefore, the first and last symbols in s are $\#$, and there are two σ symbols in s. There are at most $k - 2$ separator symbols in the length-L string s.

Now for any given position p, $1 \leq p \leq k$, consider a subset of strings in S: $S' = \{s_{i,p} \mid 1 \leq i < p\} \cup \{s_{p,j} \mid p < j \leq k\}$. There are totally $k - 1$ strings in S'. According to the analysis above, there are at most $k - 2$ separator symbols in s. Thus there is at least one string in S', say $s_{p,j}$ (it is similar to verify the case of $s_{i,p}$), whose separator symbol $\psi_{p,j}$ does not appear in s. For this string $s_{p,j}$, because there is a length-L substring $t_{p,j}$ in $s_{p,j}$ satisfying $D(s, t_{p,j}) \leq k - 2$, according to Observation 1, $t_{p,j}$ must be a block of $s_{p,j}$, and the 4 non-separator symbols in the block must coincide with the respective positions in s. Note that the $(p+1)$-st symbol in $t_{p,j}$ is a σ symbol. In consequence, the $(p+1)$-st symbol in s is a σ symbol, too. The position p could be any value in $[1, k]$, so we have proved that s has $\#$ symbols at its first and last positions, and σ symbols at any other positions. Therefore, there is no separator symbol in s. In consequence, Observation 1 is true for any string $s_{i,j}$ in S.

We further prove that the σ symbols in s are all different from each other. Suppose that the $(i+1)$-st and $(j+1)$-st symbols in s are σ_r and σ_s respectively, $1 \leq i < j \leq k$. Because there is a length-L substring $t_{i,j}$ in $s_{i,j}$ satisfying $D(s, t_{i,j}) \leq k - 2$, according to Observation 1, $t_{i,j}$ must be a block of $s_{i,j}$, and the 4 non-separator symbols in the block must coincide with the respective positions in s. Note that the $(i + 1)$-st and $(j + 1)$-st symbols in $t_{i,j}$ are non-separator symbols, so they must be σ_r and σ_s. In addition, because the $(i+1)$-st and $(j + 1)$-st symbols in each block of $s_{i,j}$ represent two different vertices in G that are connected by an edge, we know that σ_r and σ_s are two different symbols

$(r \neq s)$. Since the values of i and j are arbitrary, we have verified that all the σ symbols in the string s are different from each other.

Therefore, the string s has k different σ symbols, corresponding to k different vertices in G. Let $s = \#\sigma_{h_1}\sigma_{h_2}\cdots\sigma_{h_k}\#$. For any two indices i and j, $1 \leq i < j \leq k$, since $D(s, s_{i,j}) \leq k' = k - 2$, there is a length-$L$ substring $t_{i,j}$ in $s_{i,j}$ such that $D(s, t_{i,j}) \leq k' = k - 2$. By Observation 1, the substring $t_{i,j}$ is a block in $s_{i,j}$ of the form

$$t_{i,j} = \langle block(i, j, e_l)\rangle = \#(\psi_{i,j})^{i-1}\sigma_r(\psi_{i,j})^{j-i-1}\sigma_s(\psi_{i,j})^{k-j}\#$$

for an edge $e_l = [v_r, v_s]$ in the graph G, and $v_r = v_{h_i}$ and $v_s = v_{h_j}$. In consequence, $[v_{h_i}, v_{h_j}]$ is an edge in the graph G. Since this is true for all i and j, we conclude that the vertices $v_{h_1}, v_{h_2}, \ldots, v_{h_k}$ induce a clique of size k in the graph G. Therefore, (G, k) is a yes-instance for the Clique problem.   □

This completes the proof that the problem Clique is ftp$_l$-reducible to the problem CSP$_\leq$. In consequence, CSP$_\leq$ is $W_l[1]$-hard, and Lemma 2 is proved.

From Lemma 2 and Theorem 6.1 in [6] , we immediately get:

Theorem 2. *Unless the ETH Hypothesis fails, the optimization problem CSP has no PTAS of running time $f(1/\varepsilon)|x|^{o(1/\varepsilon)}$ for any function f.*

Theorem 2 implies that any PTAS for CSP cannot run in time $f(1/\varepsilon)|x|^{o(1/\varepsilon)}$ for any function f. Thus essentially, no PTAS for CSP can be practically efficient even for moderate values of the error bound ε.

4   Conclusion

The parameterized complexity is a powerful tool to derive strong computational lower bounds. In this paper, we proved that unless the ETH Hypothesis fails, the CSP problem has no PTAS of running time $f(1/\varepsilon)|x|^{o(1/\varepsilon)}$ for any function f. The result is obtained through a proof of the $W_l[1]$-hardness for the parameterized version of the CSP problem.

According to the new lower bound, the CSP problem has no practical polynomial time approximation scheme even for moderate values of the error bound $\varepsilon > 0$. As a consequence, it is unlikely that the study of approximation schemes for the CSP problem in the literature would lead to practical approximation algorithms for the CSP problem for small error bound $\varepsilon > 0$.

We remark that our result is based on unbounded alphabet. The lower bound of the CSP problem over constant size alphabet is still open. In [12], Gramm *et al.* wrote: "It was conjectured that (for constant alphabet size) Closest Substring is also fixed-parameter intractable with respect to the distance parameter, but it is an open question to prove (or disprove) this statement." We conjecture that for constant alphabet size, the CSP problem still has no PTAS of running time $f(1/\varepsilon)|x|^{o(1/\varepsilon)}$ for any function f, unless the ETH Hypothesis fails.

References

1. Lanctot, J.K., Li, M., Ma, B., Wang, S., Zhang, L.: Distinguishing string selection problems. Information and Computation 185(1), 41–55 (2003)
2. Li, M., Ma, B., Wang, L.: On the closest string and substring problems. Journal of the ACM 49(2), 157–171 (2002)
3. Ma, B.: A polynomial time approximation scheme for the closest substring problem. In: Giancarlo, R., Sankoff, D. (eds.) CPM 2000. LNCS, vol. 1848, pp. 99–107. Springer, Heidelberg (2000)
4. Downey, R., Fellows, M.: Parameterized Complexity. Springer, New York (1999)
5. Chen, J.: Parameterized computation and complexity: a new approach dealing with NP-hardness. Journal of Computer Science and Technology 20(1), 18–37 (2005)
6. Chen, J., Huang, X., Kanj, I., Xia, G.: Linear FPT reductions and computational lower bounds. In: Proceedings of the 36th Annual ACM Symposium on Theory of Computing (STOC 04), pp. 212–221. ACM Press, New York (2004)
7. Chen, J., Huang, X., Kanj, I., Xia, G.: W-hardness under linear FPT-reductions: structural properties and further applications. In: Wang, L. (ed.) COCOON 2005. LNCS, vol. 3595, pp. 16–19. Springer, Heidelberg (2005)
8. Fellows, M., Gramm, J., Niedermeier, R.: On the parameterized intractability of Closest Substring and related problems. In: Alt, H., Ferreira, A. (eds.) STACS 2002. LNCS, vol. 2285, pp. 262–273. Springer, Heidelberg (2002)
9. Papadimitriou, C., Yannakakis, M.: Optimization, approximation, and complexity classes. In: Proceedings of the 12th Annual ACM Symposium on Theory of Computing (STOC 88), pp. 229–234. ACM Press, New York (1988)
10. Impagliazzo, R., Paturi, R., Zane, F.: Which problems have strongly exponential complexity? Journal of Computer and System Sciences 63(4), 512–530 (2001)
11. Garey, M., Johnson, D.: Computers and Intractability. A Guide to the Theory of NP-Completness. W.H. Freeman and Company, New York (1979)
12. Gramm, J., Guo, J., Niedermeier, R.: On exact and approximation algorithms for distinguishing substring selection. In: Lingas, A., Nilsson, B.J. (eds.) FCT 2003. LNCS, vol. 2751, pp. 195–209. Springer, Heidelberg (2003)

A New Exact Algorithm for the Two-Sided Crossing Minimization Problem[*]

Lanbo Zheng[1,2] and Christoph Buchheim[3]

[1] School of Information Technologies, University of Sydney, Australia
[2] IMAGEN program, National ICT Australia
lzheng@it.usyd.edu.au
[3] Computer Science Department, University of Cologne, Germany
buchheim@informatik.uni-koeln.de

Abstract. The *Two-Sided Crossing Minimization* (TSCM) problem calls for minimizing the number of edge crossings of a bipartite graph where the two sets of vertices are drawn on two parallel layers and edges are drawn as straight lines. This well-known problem has important applications in VLSI design and automatic graph drawing. In this paper, we present a new branch-and-cut algorithm for the TSCM problem by modeling it directly to a binary quadratic programming problem. We show that a large number of effective cutting planes can be derived based on a reformulation of the TSCM problem. We compare our algorithm with a previous exact algorithm by testing both implementations with the same set of instances. Experimental evaluation demonstrates the effectiveness of our approach.

1 Introduction

Real world information is often modeled by abstract mathematical structures so that relationships between objects are easily visualized and detected. Directed graphs are widely used to display information with hierarchical structures which frequently appear in computer science, economics and social sciences.

Sugiyama, Tagawa, and Toda [14] presented a comprehensive approach to draw directed graphs. First, vertices are partitioned and constrained to a set of equally spaced horizontal lines, called *layers*, and edges are straight lines connecting vertices from adjacent layers. They then select a permutation of the vertices in each layer to reduce the number of crossings between the edges. The second step is very important as it is generally accepted that drawings with less crossings are easier to read and understand. This problem attracted a lot of studies in graph drawing and is usually solved by considering two neighboring layers at a time. The resulting problem is generally called the *two-layer crossing minimization* (TLCM) problem. Another motivation comes from a layout problem in VLSI design [12]. A recent study shows that solutions of the two-layer crossing

---

[*] This work was partially supported by the Marie Curie Research Training Network 504438 (ADONET) funded by the European Commission.

A. Dress, Y. Xu, and B. Zhu (Eds.): COCOA 2007, LNCS 4616, pp. 301–310, 2007.

minimization problem can be used to solve the rank aggregation problem that has applications in meta-search and spam reduction on the Web [2].

Given a bipartite graph $G = (V_1 \cup V_2, E)$, a two-layer drawing consists of placing the vertices from V_1 on distinct positions on a straight line L_1 and placing the vertices from V_2 on distinct positions on a parallel line L_2. Each edge is drawn using a straight line segment connecting the positions of the end vertices of the edge. Clearly, the number of edge crossings in a drawing only depends on the permutations of the vertices on L_1 and L_2. The two-layer crossing minimization problem asks to find a permutation π_1 of vertices on L_1 and a permutation π_2 of vertices on L_2 so that the number of edge crossings is minimized. This problem was first introduced by Harary and Schwenk [7] and has two different versions. The first one is called *two-sided* crossing minimization (TSCM), where vertices of the two vertex sets can be permuted freely. For multi-graphs, Garey and Johnson proved the NP-hardness of this problem by transforming the *Optimal Linear Arrangement* problem to it [5]. The *one-sided* crossing minimization (OSCM) problem is more restricted; here the permutation of one vertex set is given. However, this problem is also NP-hard [4], even for forests of 4-stars [10].

It is obvious from the literature that the one-sided crossing minimization problem has been intensively studied. Several heuristic algorithms deliver good solutions, theoretically or experimentally. The barycenter heuristic [14] is an $O(\sqrt{n})$-approximation algorithm, while the median heuristic [4] guarantees 3-approximative solutions. Yamaguchi and Sugimoto [15] gave a 2-approximation algorithm for instances where the maximum degree of vertices on the free side is not larger than 4. A new approximation algorithm presented by Nagamochi [11] has an approximation ratio of 1.4664.

Jünger and Mutzel [8] used integer and linear programming methods to solve the TLCM problem exactly for the first time. For the one-sided version, they reduced it to a linear arrangement problem and used the branch-and-cut algorithm published in [6] to solve it. For the two-sided version, an optimal solution was found by enumerating all permutations of one part of the vertices for a given graph. A good starting solution and a good theoretical lower bound were used to make the enumeration tree small. They did extensive experiments to compare the exact algorithm with various existing heuristic algorithms. They found that if one layer is fixed, then the branch-and-cut algorithm is very effective and there is no need to use heuristics in practice. But for the TSCM problem, in the worst case, the algorithm enumerates an exponential number of solutions. For some instances whose optimal solutions could not be computed, we found that the gaps between the optima and the results approached by iterated heuristic algorithms are not negligible. See Fig. 1 for an example.

In this paper, we directly model the TSCM problem as a binary quadratic programming (BQP) problem. Because all variables are binary, this model can be easily transformed into an integer linear programming (ILP) model so that general optimization methods can be applied. In particular, branch-and-cut is one of the most successful methods in solving ILP problems. The performance of a branch-and-cut algorithm often depends on the number and quality of cutting

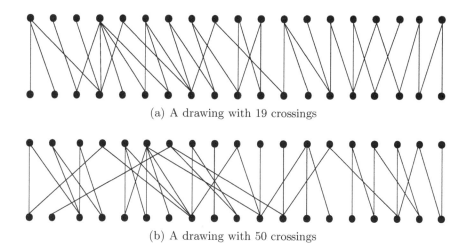

(a) A drawing with 19 crossings

(b) A drawing with 50 crossings

Fig. 1. A 20+20 graph with 40 edges. Drawing (a) has a minimum number of crossings, drawing (b) is the best drawing found by the iterated barycenter heuristic.

planes generated within the algorithm. Unfortunately, we do not know many classes of cutting planes for the TSCM problem from the literature. Our approach is based on a reformulation of the TSCM problem such that all valid inequalities for a *maximum cut* problem become valid. The maximum cut problem has been well-studied and many classes of cutting planes are known. We conjecture that these cutting planes could be helpful to solve our problem. We compared our approach with a previous exact algorithm by testing it with the same instances. Experimental evaluation positively proves our conjecture.

This paper is organized as follows. In Sect. 2, the problem under consideration is formalized and necessary notation is introduced. In Sect. 3, we describe how to reformulate the TSCM problem and present a corresponding branch-and-cut algorithm. Experimental results are analyzed in Sect. 4, and in Sect. 5, we summarize and conclude our work.

2   Preliminaries

For a bipartite graph $G = (V_1 \cup V_2, E)$, let $V = V_1 \cup V_2$, $n_1 = |V_1|$, $n_2 = |V_2|$, $m = |E|$ and let $N(i) = \{j \in V \mid \{i, j\} \in E\}$ denote the set of neighbors of $i \in V$ in G. For $k \in \{1, 2\}$, a *vertex ordering* (or *vertex permutation*) π_k of V_k is a bijection $\pi_k \colon V_k \to \{1, 2, \ldots, n_k\}$. For a pair of vertices $(i, j) \in V_k$, we write $i < j$ instead of $\pi_k(i) < \pi_k(j)$. Any solution of TLCM is obviously completely specified by a permutation π_1 of V_1 and a permutation π_2 of V_2. The formulation system given in [8] can be applied directly to our problem: let $\delta^k_{ij} = 1$ if $\pi_k(i) < \pi_k(j)$ and 0 otherwise. Then π_k is characterized by the binary vector

$$\delta^k \in \{0, 1\}^{\binom{n_k}{2}} .$$

Given π_1 and π_2, the induced number of crossings is:

$$C(\pi_1, \pi_2) = C(\delta^1, \delta^2) = \sum_{i=1}^{n_2-1} \sum_{j=i+1}^{n_2} \sum_{s \in N(i)} \sum_{t \in N(j)} (\delta_{st}^1 \cdot \delta_{ji}^2 + \delta_{ts}^1 \cdot \delta_{ij}^2) \qquad (1)$$

$$= \sum_{s=1}^{n_1-1} \sum_{t=s+1}^{n_1} \sum_{i \in N(s)} \sum_{j \in N(t)} (\delta_{st}^1 \cdot \delta_{ji}^2 + \delta_{ts}^1 \cdot \delta_{ij}^2) \qquad (2)$$

In the one-sided crossing minimization problem, the permutation π_1 of V_1 is fixed, thus δ^1 is a constant vector. Hence the objective functions (1) and (2) are linear in δ^2 in this case. On contrary, in the two-sided crossing minimization problem, both δ^1 and δ^2 are vectors of binary variables, so that (1) and (2) become quadratic functions.

In the following, for simplicity, we write $\sum_{s<t}$ instead of $\sum_{s \in N(i)} \sum_{t \in N(j), s<t}$ and $\sum_{s>t}$ instead of $\sum_{s \in N(i)} \sum_{t \in N(j), s>t}$. Using $\delta_{ji}^k = 1 - \delta_{ij}^k$ and $\delta_{ii}^k = 0$, we can reformulate our objective function as:

$$C(\pi_1, \pi_2) = C(\delta^1, \delta^2)$$
$$= \sum_{i=1}^{n_2-1} \sum_{j=i+1}^{n_2} \left(\sum_{s<t}(\delta_{st}^1 + \delta_{ij}^2 - 2\delta_{st}^1 \delta_{ij}^2) + \sum_{s>t}(2\delta_{ts}^1 \delta_{ij}^2 - \delta_{ts}^1 - \delta_{ij}^2 + 1) \right) \qquad (3)$$

Note that there are different ways to formulate the objective function. We use (3) because it has an advantage when solving dense graphs, see Sect. 4 for details.

As the next step, all quadratic terms $\delta_{st}^1 \cdot \delta_{ij}^2$ in the objective function (3) can be linearized by introducing new variables

$$\beta_{stij} = \delta_{st}^1 \cdot \delta_{ij}^2 \qquad (4)$$

that are zero-one valued and satisfy the following sets of constraints

$$\delta_{st}^1 + \delta_{ij}^2 - \beta_{stij} \leq 1 \qquad (5)$$
$$-\delta_{st}^1 + \beta_{stij} \leq 0 \qquad (6)$$
$$-\delta_{ij}^2 + \beta_{stij} \leq 0 \qquad (7)$$
$$\delta_{st}^1, \delta_{ij}^2, \beta_{stij} \in \{0, 1\} \qquad (8)$$

This standard linearization technique is well-known from the literature.

Combining the above results with the linear inequalities for the linear ordering problem in [8], we obtain the following integer linear programming model for the TSCM problem:

$$\text{Minimize} \quad \sum_{i=1}^{n_2-1} \sum_{j=i+1}^{n_2} \left(\sum_{s<t}(\delta_{st}^1 + \delta_{ij}^2 - 2\beta_{stij}) + \sum_{s>t}(2\beta_{tsij} - \delta_{ts}^1 - \delta_{ij}^2 + 1) \right)$$

Subject to
$$0 \le \delta_{ij}^2 + \delta_{jk}^2 - \delta_{ik}^2 \le 1 \qquad \text{for } 1 \le i < j < k \le n_2$$

$$0 \le \delta_{st}^1 + \delta_{tl}^1 - \delta_{sl}^1 \le 1 \qquad \text{for } 1 \le s < t < l \le n_1$$

$\delta_{st}^1, \delta_{ij}^2, \beta_{stij}$ satisfy (6), (7), (8) for coef(β_{stij}) < 0
$\qquad\qquad\qquad 1 \le i < j \le n_2$
$\qquad\qquad\qquad s \in N(i), t \in N(j), s < t$

$\delta_{st}^1, \delta_{ij}^2, \beta_{stij}$ satisfy (5), (8) for coef(β_{stij}) > 0
$\qquad\qquad\qquad 1 \le i < j \le n_2$
$\qquad\qquad\qquad s \in N(i), t \in N(j), s < t$

Here coef(β_{stij}) is the coefficient of variable β_{stij} in the objective function. If the variable β_{stij} has a negative coefficient, Constraint (5) is not necessary because it tightens the linear relaxation on a side that is not relevant for our optimization objective. Similarly, Constraints (6) and (7) are unnecessary for variables with positive coefficients. The 3-dicycle inequalities in the first two lines of the above formulation are necessary to ensure that the vectors δ^1 and δ^2 indeed correspond to permutations of V_1 and V_2, i.e., that integer solutions of our ILP correspond to solutions of the TSCM problem.

3 Our Algorithm

3.1 Branch-and-Cut

The exact algorithm used by Jünger and Mutzel [8] to solve the TSCM problem becomes unpractical as graphs are growing larger and the theoretical lower bound is no longer effective to bound the enumeration tree. However, the basic idea of branch-and-bound is the pruning of branches in this tree: at some node of the tree, the ordering of a certain set of vertices is already fixed. According to this information, one can derive a lower bound on the number of crossings subject to these fixed vertices. If the lower bound is at most as good as a feasible solution that has already been found, e.g., by some heuristics, it is clear that the considered subtree cannot contain a better solution, so it does not have to be explored.

Through the integer programming model we formulated in the previous section, we can solve the TSCM problem directly with an LP-based branch-and-cut algorithm. The basic structure of this approach is an enumeration tree rooted at an LP relaxation of the original problem, i.e., the integrality constraints are relaxed. LPs are solved very quickly in practice. If the LP-solution is integer, we can stop. Otherwise, we try to add cutting planes that are valid for all integer solutions of the ILP but not necessary for (fractional) solutions of the LP. If such cutting planes are found, they are added to the LP and the process is reiterated. We resort to the branching part only if no more cutting planes are found. High quality cutting planes that cause big changes to the objective function can be crucial to make the enumeration tree small. However, finding them is usually a sophisticated problem.

In the following, we describe our approach to resolve the above issue. We show that the TSCM problem is, in fact, a cut problem with additional constraints. We then describe how to generate a set of cutting planes which may help to improve the performance of our algorithm.

3.2   Generating Cutting Planes

If the 3-dicycle inequalities are relaxed, the TSCM problem is an *unconstrained binary quadratic programming* (UBQP) problem. We denote it as TSCM*. Every polytope corresponding to an UBQP problem is isomorphic to a cut polytope by [13], thus TSCM* can be reduced to a maximum cut problem.

The corresponding graph $H = (A \cup B \cup r, E_1 \cup E_2)$ is defined as follows. Every vertex $a_{st} \in A$ ($b_{ij} \in B$) corresponds to a variable δ^1_{st} (δ^2_{ij}) for $1 \leq s < t \leq n_1$ ($1 \leq i < j \leq n_2$) and every edge $e = (a_{st}, b_{ij}) \in E_1$ to a variable β_{stij}. Edges in E_2 join vertex r with all vertices in A and B. Now for some cut in H defined by a (possibly empty) set $S \subseteq A \cup B \cup r$, let $\gamma_{v,w}$ be 1 if precisely one of the vertices v and w belongs to S, and 0 otherwise. Then the connection between the original UBQP problem and the maximum cut problem on H is given by the equations

$$\delta^1_{st} = \gamma_{r,a_{st}} \qquad\qquad \text{for all vertices } a_{st} \in A$$

$$\delta^2_{ij} = \gamma_{r,b_{ij}} \qquad\qquad \text{for all vertices } b_{ij} \in B$$

$$\beta_{stij} = \tfrac{1}{2}(\gamma_{r,a_{st}} + \gamma_{r,b_{ij}} - \gamma_{a_{st},b_{ij}}) \quad \text{for all edges } (a_{st}, b_{ij}) \text{ in } E_1$$

Thus, the TSCM problem can be considered as a maximum cut problem with the following additional constraints:

$$0 \leq \gamma_{v,a_{st}} + \gamma_{v,a_{tl}} - \gamma_{v,a_{sl}} \leq 1 \quad \text{for } 1 \leq s < t < l \leq n_1$$

$$0 \leq \gamma_{v,b_{ij}} + \gamma_{v,b_{jk}} - \gamma_{v,b_{ik}} \leq 1 \quad \text{for } 1 \leq i < j < k \leq n_2$$

which are transformed from the 3-dicycle inequalities. Then it is not hard to see that cutting planes for maximum cut problems are all valid for the transformed TSCM problem. The cut polytope has been investigated intensively in the literature and many classes of cutting planes are known, see [3] for a survey.

In our algorithm, we concentrate on *odd-cycle* inequalities because in general a large number of such inequalities can be found and separated in polynomial time [1]. The validity of odd-cycle inequalities is based on the observation that the intersection of a cut and a cycle in a graph always contains an even number of edges. This leads to the following formulation of the odd-cycle inequalities:

$$\gamma(F) - \gamma(C \setminus F) \leq |F| - 1 \quad \text{for } F \subseteq C, |F| \text{ odd, and } C \text{ a cycle in } H$$

However, due to the 3-dicycle inequalities, it is hard to determine which odd-cycle inequalities are useful for improving the performance of our branch-and-cut algorithm. Moreover, the exact separation routine in [1] requires an $O(n^3)$

running time, where n is the number of vertices of H. For practical purposes this is rather slow.

In our algorithm, we augment H to $H' = (A \cup B \cup r, E_1 \cup E_2 \cup E_3)$ by adding new edges $e_1 = (a_{st}, a_{tl})$, $e_2 = (a_{tl}, a_{sl})$ and $e_3 = (a_{st}, a_{sl})$ for $1 \leq s < t < l \leq n_1$ and assigning them weights of zero. A similar process is also applied to vertices in B. We scan through all triangle sets (r, a_{st}, b_{ij}), (a_{st}, a_{tl}, b_{ij}) and (b_{ij}, b_{jk}, a_{st}) and check each of the four associated cycle inequalities for violation. This can be done in $O(n^2)$ time.

The main procedure of our branch-and-cut algorithm is performed as follows:

1. Solve an initial LP without the 3-dicycle inequalities,
2. If the solution is infeasible, try to find violated cutting planes. The separation is performed in the order: 3-dicycle inequalities, odd-cycle inequalities associated with triangle sets (v, a_{st}, b_{ij}) and odd-cycle inequalities associated with triangle sets (a_{st}, a_{tl}, b_{ij}) and (b_{ij}, b_{jk}, a_{st}),
3. Revise the LP and resolve it,
4. Repeat step 2 and 3. If no cutting planes are generated, branch.

Besides the triangle inequalities described above, we also tried to generate other odd-cycle inequalities by using the algorithm in [1]. However, this approach is time-consuming and does not remarkably improve the performance of our algorithm; so we do not include it in our experimental evaluation. Moreover, in our experiments we found that our cutting plane algorithm performs very well with the set of cutting planes described above. For some instances, we do not have to go to the branching step at all, see Sect. 4.

4   Experimental Results

In order to evaluate the practical performance of our approach presented in the previous section, we performed a computational evaluation. In this section, we report the results and compare them to the results obtained with the branch-and-bound algorithm in [8]. We tested the performance of

- B&C 1: Our branch-and-cut algorithm using odd-cycle inequalities,
- B&C 2: The CPLEX MIP-solver with default options,
- JM: The exact algorithm used by Jünger and Mutzel [8].

These algorithms have been implemented in C++ using the Mixed Integer Optimizer of CPLEX 9.0. For a better comparison, our experiments for all three algorithms were carried out on the same machine, a standard desktop computer with a 2.99 GHz processor and 1.00 GB of RAM running Windows XP. We reimplemented the algorithm of [8] using the ILP solver of CPLEX to solve a subproblem on the enumeration tree. In the remainder of this section, all running times are given in seconds. The test suite of our experiments is the same as used in [8]. It is generated by the program **random_bigraph** of the Stanford GraphBase by Knuth [9].

The main results from our experiments are reported in Table 1. We give the results for "10+10-graphs", i.e., bipartite graphs with 10 nodes on each layer, with increasing edge densities up to 90%. We compare our results with those of the exact algorithm presented in [8]. The notation used in Table 1 is as follows:

- n_i: number of nodes on layer i for $i = 1, 2$
- m: number of edges
- time: the running time of each algorithm
- value: the minimum number of crossings computed by each algorithm
- nodes: the number of generated nodes in the branch-and-cut tree
- cuts: the number of user cuts generated by our branch-and-cut algorithm
- Gom.: the number of Gomory cuts generated by CPLEX with default options
- cliq.: the number of clique cuts generated by CPLEX with default options

Table 1. Results for "10+10-graphs" with increasing density

n_i	m	B&C 1				B&C 2					JM [8]	
		nodes	cuts	time	value	nodes	Gom.	cliq.	time	value	time	value
10	10	2	120	0.50	1	0	0	0	**0.03**	1	0.20	1
10	20	2	670	2.12	11	2	4	3	**0.30**	11	0.53	11
10	30	0	1787	**2.86**	52	159	4	19	1.45	52	3.92	52
10	40	0	9516	**8.20**	142	3610	2	48	26.16	142	8.52	142
10	50	0	14526	**12.52**	276	8938	1	112	101.03	276	19.41	276
10	60	0	19765	**21.23**	459	14154	2	236	220.46	459	39.65	459
10	70	9	37857	245.88	717	[38044]	[1]	[427]	[1000]	[717]	103.11	717
10	80	0	25553	**24.05**	1037	19448	3	813	642.87	1037	216.26	1037
10	90	0	5468	**9.67**	1387	3354	3	1334	**158.96**	1387	234.62	1387

Notice that in our approach we did not use any initial heuristics, in order to give clearer and independent runtime figures. Nevertheless, as obvious from Table 1, our approach is much faster than the previous algorithm. This is particular true for dense graphs, e.g., graphs with more than 80 edges (bold figures indicate that optimal solutions have been found earlier than by the previous algorithm). Compared to B&C 2, there is a much smaller number of subproblems to be solved in our branch-and-cut algorithm. It is exciting to see that many instances have been solved to optimality in the cutting plane phase.

For sparse graphs, the branch-and-cut algorithm with default options of CPLEX performs better. It has Gomory cuts and clique cuts that could be helpful in reducing the effort required to complete the enumeration. However for dense graphs, i.e., graphs with 70 edges, the instances are too large for the default branch-and-cut algorithm, we set a general time limit of 1000 seconds. Whenever this limit was reached, we report the best result and the numbers of nodes and cuts generated so far; the figures are then put into brackets. Notice that both branch-and-cut algorithms perform worse than the previous algorithm when the testing graph has 70 edges.

In Sect. 2 we have mentioned that there is an advantage of the objective function used in our ILP model. Now it is clearly visible in Table 1: some very dense graphs are solved even faster than graphs with less edges. The reason is that, in our formulation, variables generated from a subgraph $K_{2,2}$ are implicitly substituted by 1, as the two corresponding terms in (3) sum up to 1 then. This is possible since every $K_{2,2}$ induces exactly one crossing in any solution. This helps to reduce the LP size and allows to solve very dense instances quickly.

5 Conclusion and Future Work

We have studied the two-sided crossing minimization problem by modeling it directly to a binary quadratic programming problem. We have described a strategy to generate effective cutting planes for the TSCM problem. This is based on reformulating it to a cut problem with additional constraints. We have shown that these cutting planes can remarkably improve the performance of our branch-and-cut algorithm. Our computational results show that our algorithm runs significantly faster than earlier exact algorithms even without using any heuristic algorithm for computing starting solutions, in particular for dense graphs. Nevertheless, our approach for solving the TSCM problem has transcended its original application. In graph drawing, many combinatorial optimization problems can be modeled as binary quadratic programming problems in a natural way. This is particularly true for various types of crossing minimization problems.

Our encouraging computational results are obtained for small-scale graphs. In the future, we plan to test the performance of our algorithm with larger graphs. Beyond the sets of inequalities used in this paper, it would be interesting to identify classes of facet-defining inequalities for the TSCM problem. We are now making investigations in this direction.

References

1. Barahona, F., Mahjoub, A.R.: On the cut polytope. Mathematical Programming 36, 157–173 (1986)
2. Biedl, T., Brandenburg, F.J., Deng, X.: Crossings and permutations. In: Healy, P., Nikolov, N.S. (eds.) Proceedings of Graph Drawing 2005, pp. 1–12. Limerick, Ireland (2005)
3. Boros, E., Hammer, P.L.: The max-cut problems and quadratic 0-1 optimization; polyhedral aspects, relaxations and bounds. Annals of Operations Research 33, 151–180 (1991)
4. Eades, P., Wormald, N.C.: Edge crossings in drawing bipartite graphs. Algorithmica 11, 379–403 (1994)
5. Garey, M.R., Johnson, D.S.: Crossing number is NP-complete. SIAM Journal on Algebraic and Discrete Methods 4, 312–316 (1983)
6. Grötschel, M., Jünger, M., Reinelt, G.: A cutting plane algorithm for the linear ordering problem. Operations Research 32, 1195–1220 (1984)
7. Harary, F., Schwenk, A.J.: Trees with hamiltonian square. Mathematika 18, 138–140 (1971)

8. Jünger, M., Mutzel, P.: 2-layer straight line crossing minimization: performance of exact and heuristic algorithms. Journal of Graph Algorithms and Applications 1–25 (1997)
9. Knuth, D.: The Stanford GraphBase: A platform for combinatorial computing (1993)
10. Muñoz, X., Unger, W., Vrt'o, I.: One sided crossing minimization is NP-hard for sparse graphs. In: Mutzel, P., Jünger, M., Leipert, S. (eds.) Proceedings of Graph Drawing 2001, Vienna, Austria, pp. 115–123 (2001)
11. Nagamochi, H.: An improved approximation to the one-sided bilayer drawing. In: Liotta, G. (ed.) Proceedings of Graph Drawing 2003, Perugia, Italy, pp. 406–418 (2003)
12. Sechen, C.: VLSI Placement and Global Routing using Simulated Annealing. Kluwer Academic Publishers, Boston (1988)
13. De Simone, C.: The cut polytope and the boolean quadric polytope. Discrete Mathematics 79, 71–75 (1990)
14. Sugiyama, K., Tagawa, S., Toda, M.: Methods for visual understanding of hierarchical systems. IEEE Transactions on Systems, Man, and Cybernetics (SMC) 11(2), 109–125 (1981)
15. Yamaguchi, A., Sugimoto, A.: An approximation algorithm for the two-layered graph drawing problem. In: 5th Annual International Conference on Computing and Combinatorics, pp. 81–91 (1999)

Improved Approximation Algorithm for Connected Facility Location Problems
(Extended Abstract)

Mohammad Khairul Hasan, Hyunwoo Jung, and Kyung-Yong Chwa

Division of Computer Science, Korea Advanced Institute of Science and Technology,
Daejeon, Republic of Korea
{shaon,solarity,kychwa}@tclab.kaist.ac.kr

Abstract. We study the *Connected Facility Location* problems. We are given a connected graph $G = (V, E)$ with non-negative edge cost c_e for each edge $e \in E$, a set of clients $D \subseteq V$ such that each client $j \in D$ has positive demand d_j and a set of facilities $F \subseteq V$ each has non-negative opening cost f_i and capacity to serve all client demands. The objective is to open a subset of facilities, say $\hat{F}$, to assign each client $j \in D$ to exactly one open facility $i(j)$ and to connect all open facilities by a Steiner tree T such that the cost $\sum_{i \in \hat{F}} f_i + \sum_{j \in D} d_j c_{i(j)j} + M \sum_{e \in T} c_e$ is minimized. We propose a LP-rounding based 8.29 approximation algorithm which improves the previous bound 8.55. We also consider the problem when opening cost of all facilities are equal. In this case we give a 7.0 approximation algorithm.

1 Introduction

Facility location is one of the prominent fields of research now a days. Ease of formulation using integer linear program and enormous application demand have made facility location problems a central area of research specially in Operational Research community. In *Connected Facility Location (ConFL)* problem, we are given a connected graph $G = (V, E)$ with non-negative edge cost c_e for each edge $e \in E$. A set of facilities $F \subseteq V$ and a set of clients $D \subseteq V$ such that each client $j \in D$ has positive demand d_j and each facility $i \in F$ has non-negative opening cost f_i with capacity to serve all client demands. Let c_{ij} be the cost of shortest path between vertices i and j in the graph. We assume that this cost is non-negative, symmetric and obeys triangular inequality, i.e., $c_{ij} \geq 0$, $c_{ij} = c_{ji}$ and $c_{ik} + c_{kj} \geq c_{ij}$ where k is a vertex of G. For each client j, the cost of serving d_j demand from facility i is $d_j c_{ij}$, where i be an open facility such that client j is assigned to it. In a solution of ConFL, all open facilities must be connected among themselves. We say that we have *bought* an edge when we have selected that edge to connect open facilities. For each bought edge, we pay M times of its actual cost, where $M \geq 1$ is an input parameter. Our objective is to open a set of facilities, $\hat{F} \subseteq F$, assign each client j to exactly one open facility $i(j) \in \hat{F}$ and buy a set of edges that forms Steiner tree T to connect all facilities of $\hat{F}$ such that total cost, which is $\sum_{i \in \hat{F}} f_i + \sum_{j \in D} d_j c_{i(j)j} + M \sum_{e \in T} c_e$ is minimized.

A. Dress, Y. Xu, and B. Zhu (Eds.): COCOA 2007, LNCS 4616, pp. 311–322, 2007.
© Springer-Verlag Berlin Heidelberg 2007

1.1 Related Works

Connected facility location problem has been studied first by Karger and Minkoff [1]. The same authors gave the first constant factor approximation algorithm. Gupta et al. [2] improved the previous result and gave a 10.66 factor approximation algorithm for ConFL. Their algorithm is based on LP-rounding. Swamy and Kumar [3] gave first primal-dual algorithm with 9 approximation factor and later same authors [4] improved the factor to 8.55.

A special case of ConFL problem is Rent-or-Buy problem where each facility has zero opening cost. Swamy and Kumar in [3] and in [4] gave primal-dual based algorithms for Rent-or-Buy problem with approximation factors 5 and 4.55 respectively. Gupta, Kumar and Roughgarden in [5] addressed two special cases of ConFL problem. They gave a 3.55 factor randomized approximation algorithm for the case where each facility opening cost is zero (Rent-or-Buy problem) and a 5.55 factor randomized approximation algorithm for the case where each facility opening cost is either zero or infinite. Although these two cases are categorized as "CONNECTED FACILITY LOCATION" problems in [5], one can easily observe that these two problems are actually two special cases of ConFL since none of them can handle arbitrary non-negative facility opening cost.

1.2 Our Results

We propose a 8.29-approximation algorithm for general ConFL problem. When all facility opening costs are equal we give a 7.0-approximation algorithm. We use LP-rounding method to solve these two problems. First, we get an optimum fractional solution to the linear programming relaxation. Then, we filter the solution using Lin & Vitter [6] technique. Finally, we round the filtered solution.

If we consider all polynomial approximation algorithm for ConFL problem, our algorithm outperforms the best known algorithm (8.55 factor) in terms of approximation ratio. For the special case when all facilities have equal opening cost, no polynomial algorithm with better than 8.55-approximation factor exists as far we know. Thus our 7.0 factor algorithm is a much improvement in this case.

2 Relaxed Linear Program Formulation

We can assume a particular facility $r_0 \in F$ has zero opening cost and belongs to the Steiner tree connecting the open facilities in optimum solution. This assumption is WLOG and does not affect the approximation ratio [3]. Let us consider $F_v = F - \{r_0\}$. So, the relaxed linear program (LP) for ConFL, which is equivalent to that used in [2] is as follows:

$$\min \sum_{i \in F_v} f_i y_i + \sum_{j \in D} d_j \sum_{i \in F} c_{ij} x_{ij} + M \sum_{e \in E} c_e z_e \qquad \text{(P1)}$$

$$s.t \sum_{i \in F} x_{ij} = 1 \qquad\qquad \forall j \in D \qquad\qquad (1)$$

$$x_{ij} \leq y_i \qquad\qquad \forall i \in F_v, j \in D \qquad\qquad (2)$$

$$x_{r_0 j} \leq 1 \qquad\qquad \forall j \in D \qquad\qquad (3)$$

$$\sum_{i \in S} x_{ij} \leq \sum_{e \in \delta(S)} z_e \qquad\qquad \forall S \subseteq V - \{r_0\}, j \in D \qquad\qquad (4)$$

$$x_{ij}, y_i, z_e \geq 0$$

3 Filtering

Let (x, y, z) be an optimum fractional solution to LP-P1. This solution can be found in polynomial time using ellipsoid algorithm [2]. First we want to get a $g - close$ solution. A feasible solution $(\bar{x}, \bar{y}, \bar{z})$ to LP-P1 is g-clsoe if it satisfies the property: given g_j value for each client $j \in D$, $\bar{x}_{ij} > 0 \Rightarrow c_{ij} \leq g_j$

Suppose α is a fixed value in the range $(0, 1)$. For a fixed client $j \in D$, let π be the permutation on facility set such that $c_{\pi(1)j} \leq c_{\pi(2)j} \leq c_{\pi(3)j} \leq \cdots \leq c_{\pi(n)j}$. For each client j, we set $c_j(\alpha) = c_{\pi(i_{min})j}$, where $i_{min} = \min\{i' : \sum_{i=1}^{i'} x_{\pi(i)j} \geq \alpha\}$. The following lemma shows that a g-close solution to LP-P1 can be found in polynomial time. This lemma is similar to Lemma 1 in [7] and its proof has been omitted.

Lemma 1. *Given a feasible solution* (x, y, z), *for any fixed value* α *within the range (0, 1) we can find a g-close solution* $(\bar{x}, \bar{y}, \bar{z})$ *in polynomial time, such that*

1. $g_j \leq c_j(\alpha)$, *for each* $j \in D$
2. $\sum_{i \in F_v} f_i \bar{y}_i \leq (1/\alpha) \sum_{i \in F_v} f_i y_i$
3. $M \sum_{e \in E} c_e \bar{z}_e \leq (1/\alpha) M \sum_{e \in E} c_e z_e$

4 Algorithm Design

Definition 1. *A collection of facilities and clients is called a* Group. *There is a special group called* "Root Group" *which contains* r_0. *Every group except "Root Group" must have a representative of that group which must be a member client of the same group (see Figure 1).*

There are two types of group - (i) *Primary Group* where there is exactly one member facility and (ii) *Secondary Group* where there are more than one member facilities. The "Root Group" is always a primary group. For all groups except "Root Group", we use $\mathcal{G}_j$, $\mathcal{F}_j$ and $\mathcal{D}_j$ to represent a group with representative j, its facility set and its client set respectively. For "Root Group", $\mathcal{G}_0, \mathcal{F}_0$ and $\mathcal{D}_0$ represent the group itself, its facility set and its client set respectively.

Given a g-close solution $(\bar{x}, \bar{y}, \bar{z})$ to LP-P1, the algorithm *CreateGroups* will create some groups such that facility sets of these groups are disjoint, each client

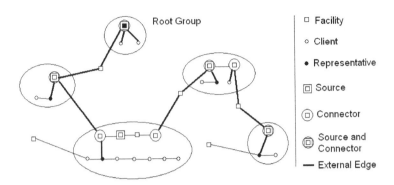

Fig. 1. (a) Example of groups, connectors, sources and external edges connecting all groups

is assigned to exactly one group and one of these groups is "Root Group", $\mathcal{G}_0$. Throughout the algorithm, $\eta \geq 3$ be a parameter whose value will be decided later.

Algorithm *CreateGroups*
Begin
    $R \leftarrow D$
    Create $\mathcal{G}_0$ with $\mathcal{F}_0 = \{r_0\}$
    $\mathcal{D}_0 \leftarrow \{k \in R : c_{r_0 k} \leq \eta c_k(\alpha)\}$
    $R \leftarrow R - \mathcal{D}_0$
    While $\exists_{(i,j) \in F_v \times R} : \bar{x}_{ij} > 0$ and $\bar{y}_i = 1$ do
        Create primary group $\mathcal{G}_j$ with $\mathcal{F}_j = \{i\}$ s.t. $\bar{x}_{ij} > 0$ and $\bar{y}_i = 1$
        $\mathcal{D}_j \leftarrow \{k \in R : c_{jk} \leq \eta c_k(\alpha)\}$
        $R \leftarrow R - \mathcal{D}_j$
    Repeat
    While $R \neq \phi$ do
        Let $l \in R$ be the client with smallest $c_l(\alpha)$
        Create secondary group $\mathcal{G}_l$ with $\mathcal{F}_l = \{i : c_{il} \leq c_l(\alpha)\}$
        $\mathcal{D}_l \leftarrow \{k \in R : c_{lk} \leq (\eta - 2)c_k(\alpha)\}$
        $R \leftarrow R - \mathcal{D}_l$
    Repeat
End

Definition 2. *An edge $(p, q) \in E$ is called an* external edge *iff for each group $\mathcal{G}_j$, $|\{p, q\} \cap F_j| < 2$. An edge $(p, q) \in E$ is called a* cut-edge *of group $\mathcal{G}_j$ iff $|\{p, q\} \cap \mathcal{F}_j| = 1$.*

Definition 3. *Given a set of external edges A, for each group $\mathcal{G}_j$, a set of vertices $\mathcal{T}_j(A)$ or $\mathcal{T}_j$ when A is clearly specified, is called the* connector set *of A in $\mathcal{G}_j$ iff each vertex $v \in \mathcal{T}_j(A)$ is also present in $\mathcal{F}_j$ and it is an endpoint of a* *cut-edge e of $\mathcal{G}_j$ such that $e \in A$. Each element of $\mathcal{T}_j(A)$ is called a* connector of A in $\mathcal{G}_j$ *(see Figure 1).*

Definition 4. *For each group $\mathcal{G}_j$, the minimum opening cost facility in $\mathcal{F}_j$ is called the* source r_j *of $\mathcal{G}_j$.*

Definition 5. *Given graph $G_{E'} = (V, E')$ with $E' \subseteq E$, a* group-identified graph $Gc_{E'} = (V_c, E'_c)$ *of $G_{E'}$ can be created by this: For each group $\mathcal{G}_j$, we identify all vertices of $\mathcal{F}_j$ into one single vertex $\hat{v}_j$. Each cut-edge (p, q) of $\mathcal{G}_j$ with $p \in \mathcal{F}_j$ is replaced by an edge $(\hat{v}_j, q)$ with same cost. Notice that there may be loops and multiple edges. In that case delete all loops and for each pair of neighboring vertices with multiple edges keep the edge of minimum cost.*

Definition 6. *A set of edges $A \subseteq E$ externally connects all groups iff all group-vertices are connected in group-identified graph Gc_A.*

Definition 7. *A set of facilities S is called* valid set *iff it contains all facilities of at least one group and it does not contain r_0. It is easy to see that a set of edges, $A \subseteq E$ externally connects all groups iff for every valid set S, $\delta(S) \cap A \neq \phi$, where $\delta(S)$ is the set of edges such that exactly one endpoint of each edge is inside S.*

Suppose $V_G = \{S \subseteq V - \{r_0\} : S$ is a valid set$\}$; that is, V_G is the set of all valid sets. Let us consider the following LP:

$$\min \sum_e c_e \hat{z}_e \tag{P2}$$

$$s.t \sum_{e \in \delta(S)} \hat{z}_e \geq 1 \quad \forall S \in V_G$$

The dual of this LP can be defined as:

$$\max \sum_{S \in V_G} \theta_S \tag{D2}$$

$$\sum_{S \in V_G : e \in \delta(S)} \theta_S \leq c_e \quad \forall e \in E$$

Primal-dual based algorithm *BuyEdges* creates two sets of edges A and B such that sources of all groups are connected in graph $G_{A \cup B} = (V, A \cup B)$. Here A is a minimal feasible solution to LP-P2 and B is the set of edges required to make path among connectors of A in $\mathcal{G}_j$ and r_j for each and every group.

Algorithm *BuyEdges*

Begin

    $\theta \leftarrow 0$ Comment: Phase - 1

    $A \leftarrow \phi$

    $l \leftarrow 0$

    Let $\zeta \leftarrow \{\mathcal{F}_j : \mathcal{G}_j$ group exists and $\mathcal{G}_j \neq \mathcal{G}_0\}$

$RootComp \leftarrow \{r_0\}$
While $\zeta \neq \phi$ do
    $l \leftarrow l + 1$
    Increase θ_S for all $S \in \zeta$ until $\exists_{e_l \in \delta(T), T \in \zeta} : \sum_{S : e_l \in \delta(S)} \theta_S = c_{e_l}$
    $A \leftarrow A \cup \{e_l\}$
    $\zeta \leftarrow \zeta - \{T\}$
    Let $e_l = (p, q)$ such that $p \in T$
    If $\exists_{T' \in \zeta} : q \in T'$
    then $\zeta \leftarrow \zeta \cup \{T \cup T'\} - \{T'\}$
    else if $q \in RootComp$
        then $RootComp \leftarrow RootComp \cup T$
        else $\zeta \leftarrow \zeta \cup \{T \cup \{q\}\}$
Repeat
For $t \leftarrow l$ downto 1   Comment: Reverse delete Step
    if $A - \{e_t\}$ is feasible then $A \leftarrow A - \{e_t\}$
$B \leftarrow \phi$ Comment: Phase - 2
For each secondary group $\mathcal{G}_j$ do
    $B_j = \phi$
    For each $i \in \mathcal{T}_j(A)$ do
        Let S be the set of edges of shortest path from i to r_j
        $B_j \leftarrow B_j \cup S$
    Repeat
    $B \leftarrow B \cup B_j$
Repeat
Output A, B and θ
End

Phase-1 of algorithm *BuyEdges* actually simulates the primal-dual algorithm for rooted Steiner tree problem on group-identified graph Gc_E of $G = (V, E)$ with r_0 as the root. Phase-1 generates the set of edges A which externally connects all groups. The set of edges B, generated in phase-2 connects the sources of each group to r_0 (see Figure 2).

Once we get the grouping and sets A and B, we can easily get an integer solution to LP-P1. For each group $\mathcal{G}_j$ we open r_j and close other facilities of $\mathcal{F}_j$. Edges of A and B actually connects r_j of each group $\mathcal{G}_j$ with r_0. We buy the edges of $A \cup B$. For each group $\mathcal{G}_j$, we assign all clients of $\mathcal{D}_j$ to r_j. Thus each client gets assigned to exactly one open facility and all open facilities are connected with the root facility using bought edges. In the next section we will prove feasibility and bound the cost of this integer solution.

5   Analysis

Lemma 2. *Each client $k \in D$ is assigned to exactly one open facility such that $c_{ik} \leq \eta c_k(\alpha)$.*

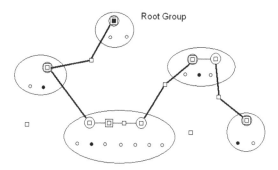

Root Group

Fig. 2. Example of algorithm *BuyEdges* implementation: thick edges form set A and thin edges form set B

Proof. In algorithm *CreateGroups*, initially $R = D$. Each client leaves R only when it is assigned to some group and once assigned, it is never considered for any other group. In each group exactly one facility is opened and all clients are assigned to that only open facility. It remains to prove that $c_{ik} \leq \eta c_k(\alpha)$. Let us consider a client k. If it is assigned to some primary group then the condition is trivial. So, let us assume that it has been assigned to some secondary group $\mathcal{G}_j$. If $j = k$ then for each facility $i \in \mathcal{F}_j$, $c_{ik} = c_{ij} \leq c_j(\alpha) = c_k(\alpha)$. So, let us consider $j \neq k$. Let $i' \in \mathcal{F}_j$ such that $c_{i'k} \leq (\eta - 2)c_k(\alpha)$. We can find such i' since $k \in \mathcal{D}_j$. Now, $c_{ii'} \leq c_{ij} + c_{i'j} \leq 2c_j(\alpha)$. So, $c_{ik} \leq c_{ii'} + c_{i'k} = 2c_j(\alpha) + (\eta - 2)c_k(\alpha)$. Now, since j has been selected as representative of $\mathcal{G}_j$ while creating $\mathcal{G}_j$ and at that time k is present in R, $c_j(\alpha) \leq c_k(\alpha)$. Thus, $c_{ik} \leq 2c_j(\alpha) + (\eta - 2)c_k(\alpha) \leq 2c_k(\alpha) + (\eta - 2)c_k(\alpha) = \eta c_k(\alpha)$. □

Lemma 3. *For each secondary group $\mathcal{G}_j$, $f_{r_j} \leq \sum_{i \in F_j} f_i \bar{y}_i$, where f_{r_j} is the source of $\mathcal{G}_j$.*

Proof. Here, $f_{r_j} = \min_{i \in F_j} f_i \leq \sum_{i \in F_j} f_i \bar{x}_{ij}$, where the last inequality holds since $\sum_{i \in F_j} \bar{x}_{ij} = 1$. Since $(\bar{x}, \bar{y}, \bar{z})$ is a feasible solution to LP-P1, from Equation 2 we get, $f_{r_j} \leq \sum_{i \in F_j} f_i \bar{y}_i$. □

Lemma 4. *Suppose A be the set of external edges returned by algorithm* BuyEdges. *Then $\sum_{e \in A} c_e \leq 2 \sum_{S \in V_G} \theta_S$.*

Proof. In Phase 1 of *BuyEdges* each edge is taken in A only when $\sum_{S \in V_G : e \in \delta(S)} \theta_S = c_e$.
So, $\sum_{e \in A} c_e = \sum_{e \in A} \sum_{S \in V_G : e \in \delta(S)} \theta_S = \sum_{S \in V_G} |A \cap \delta(S)| \theta_S$
Then, we only need to show that:

$$\sum_{S \in V_G} |A \cap \delta(S)| \theta_S \leq 2 \sum_{S \in V_G} \theta_S \qquad (5)$$

The idea of this proof is very similar to that of Primal-Dual Prize Collecting Steiner Tree Problem described in [8]. We can prove this by induction on the iteration in phase 1 of algorithm *BuyEdges*. Initially $\theta_S = 0$. So the basic step is trivial.

Let us consider an iteration where the set of minimal *violating valid set* is ζ. We can construct a graph H like this: We take vertices of V and edges of final solution A. For each *violating valid set* $S \in \zeta$ we contract the vertices of V those are in S into single vertex u_s. We contract vertices of $RootComp$ of this iteration into u_0. We remove all the vertices with degree zero. For simplicity we assume that H denotes both the graph and its vertices set. Since algorithm *BuyEdges* uses reverse delete step at the end of phase-2 and since we removed all isolated vertices while creating graph H, H is a tree. If the uniform increase in this iteration be ϵ, it is easy to see that in this iteration the left side of the Equation 5 increases by $\epsilon(\sum_{S \in \zeta} d_H(u_s))$ where $d_H(u_s)$ is the degree of vertex u_s in H. The right side increases $\epsilon 2|\zeta|$. We only need to show that $\sum_{S \in \zeta} d_H(u_s) \leq 2|\zeta|$. Let U be the set of vertices of H such that each has been created by contracting corresponding vertices of $S \in \zeta$. Let $U' = H - U - \{u_0\}$. We claim that for each $u \in U'$, $d_H(u) \geq 2$. For contradiction suppose $d_H(u') = 1$ such that $u' \in U'$. Since u' is not a result of contraction of some vertices, its degree is also 1 in graph $G = (V, A)$ as well as in *group-identified graph* Gc_A. Let e' be the only edge incident to u'. Then in $Gc_{A-\{e'\}}$ all *group-vertices* are still connected. So $A - \{e\}$ is a feasible solution to LP-P2. Then reverse delete step must have removed e' and thus degree of u' is zero. This is a contradiction. Now,

$$\sum_{u_s \in U} d_H(u_s) = \sum_{u \in H} d_H(u) - \sum_{u \in U'} d_H(u) - d_H(u_0) \leq 2(|H| - 1) - 2|U'| - d_H(u_0) = 2|H| - 2 - 2(|H| - |U| - 1) - d_H(u_0) = 2|U| - d_H(u_0) \leq 2|U| = 2|\zeta|. \quad \square$$

Lemma 5. $\sum_{e \in A} c_e \leq 2 \sum_{e \in E} c_e \bar{z}_e$

Proof. We can assume that in optimum fractional solution (x, y, z), for each facility $i \in F_v$, $\exists_{j \in D} : x_{ij} = y_i$. This is trivial when each facility in F_v has positive opening cost since otherwise we can reduce y_i to get better solution. When some facility in F_v has zero opening cost, we can ensure this property by taking optimum solution and reducing y_i as much as possible for each facility with zero opening cost and this can be done in polynomial time. So we can assume that this property holds. Next, we are going to prove that $\bar{z}$ is a feasible solution to LP-P2. Since $\sum_{S \in V_G} \theta_S$ is the lower bound of optimum solution to LP-P2, feasibility of $\bar{z}$ in LP-P2 and the result of previous lemma proves this lemma.

Let us consider any valid set S'. Then by definition, $r_0 \notin S'$ and $\exists_{g_j} : \mathcal{F}_j \subseteq S'$. Suppose $\mathcal{G}_k$ is such a group. There are two cases depending on if $\mathcal{G}_k$ is a primary or a secondary group:

Case 1. ($\mathcal{G}_k$ is a primary group)
Let i' be the only facility of $\mathcal{F}_k$. Definitely $i' \neq r_0$, because in that case S' can not be a valid set. Then, $\exists_j : x_{i'j} = y_{i'}$. Let us consider j' be the client such that $x_{i'j'} = y_{i'} \geq \alpha$, where the last inequality holds since $\mathcal{G}_k$ is a primary group and thus $\bar{y}_i = \min(y_{i'}/\alpha, 1) = 1$. Now Equation 4 of LP-P1 is true for $S = S'$ and $j = j'$. That is, $\sum_{e \in \delta(S')} z_e \geq \sum_{i \in S'} x_{ij'} \geq x_{i'j'} \geq \alpha$. Thus, $\sum_{e \in \delta(S')} \bar{z}_e \geq 1$.

Case 2. ($\mathcal{G}_k$ is a secondary group)

In this case $\sum_{i \in \mathcal{F}_k} \bar{x}_{ik} = 1$. Equation 4 of LP-P1 is true for $S = S'$ and $j = k$. That is, $\sum_{e \in \delta(S')} \bar{z}_e \geq \sum_{i \in S'} \bar{x}_{ik} \geq \sum_{i \in \mathcal{F}_k} \bar{x}_{ik} = 1$.

Thus, for any choice of valid set S', we are getting $\sum_{e \in \delta(S')} \bar{z}_e \geq 1$. This implies that $\bar{z}$ is a feasible solution to LP-P2. □

Lemma 6. *Suppose $\mathcal{G}_m$ and $\mathcal{G}_n$ be two distinct groups created by the algorithm CreateGroups. If $i_1 \in \mathcal{F}_m$ and $i_2 \in \mathcal{F}_n$ then $c_{i_1 i_2} \geq ((\eta - 3)/2)I_m$ and $c_{i_1 i_2} \geq ((\eta - 3)/2)I_n$. Where I_j is the maximum distance between any two facilities in group $\mathcal{G}_j$*

Proof. This is trivial when both are primary groups because $I_m = I_n = 0$ in that case. So let us consider two cases: (1) when both are secondary groups and (2) one is a primary and the other one is a secondary group.

Case 1. (Both are secondary groups)

In this case, it is enough to prove that $c_{i_1 i_2} \geq (\eta - 3)c_m(\alpha)$ and $c_{i_1 i_2} \geq (\eta - 3)c_n(\alpha)$. WLOG, let us assume that group $\mathcal{G}_m$ has been created first. Then $c_m(\alpha) \leq c_n(\alpha)$. For contradiction let us assume that the lemma is not true for this case. Then definitely $c_{i_1 i_2} < (\eta - 3)c_n(\alpha)$. Now, $c_{i_1 n} \leq c_{i_2 n} + c_{i_1 i_2} \leq c_n(\alpha) + c_{i_1 i_2} < c_n(\alpha) + (\eta - 3)c_n(\alpha) = (\eta - 2)c_n(\alpha)$. Then, when $\mathcal{G}_m$ has been created, client n has been assigned to $\mathcal{G}_m$. But in that case, $\mathcal{G}_n$ has not been created. It is a contradiction.

Case 2. (One is a primary and other one is a secondary group)

WLOG, let us assume that $\mathcal{G}_m$ is a primary and $\mathcal{G}_n$ is a secondary group. Then $c_{i_1 i_2} \geq ((\eta - 3)/2)I_m$ is trivial. For contradiction let us assume that $c_{i_1 i_2} < (\eta - 3)c_n(\alpha)$. Then with similar argument as shown in previous case we can show that $c_{i_1 n} \leq (\eta - 2)c_n(\alpha) \leq \eta c_n(\alpha)$ Then, when $\mathcal{G}_m$ has been created, client n has been assigned to $\mathcal{G}_m$. This is a contradiction. □

The algorithm *BuyEdges* generates two sets of edges: A and B. Lemma 7 bounds the cost of edges of set B by the cost of edges of set A. We omit the proof of this lemma.

Lemma 7. *Suppose $A, B \subseteq E$ are two sets of edges generated by algorithm BuyEdges. Let $\mathcal{G}_{j_1}, \mathcal{G}_{j_2}, ..., \mathcal{G}_{j_s}$ are the groups created by the algorithm Create-Groups. Let $Gc_A = (V_c, A_c)$ be the group-identified graph of $G_A = (V, A)$ as described in Definition 5. In each group-vertex $\hat{v}_{j_k} \in V_c$ we put weight: $W(\hat{v}_{j_k}) = d_{Gc}(\hat{v}_{j_k})I_{j_k}$, where $d_{Gc}(\hat{v}_{j_k})$ is the degree of $\hat{v}_{j_k}$ in Gc_A. The weights of non group-vertices are zero. Then*

(1) $\sum_{e \in B} c_e \leq \sum_{v \in V_c} W(v)$

(2) $\sum_{v \in V_c} W(v) \leq \frac{4}{\eta - 3} \sum_{e \in A} c_e$.

Theorem 1. *Given optimum fractional solution (x, y, z). Grouping over filtered solution yields a feasible integer solution to LP-P1 with cost at most -*
$$\frac{1}{\alpha} \sum_{i \in F_v} f_i y_i + \eta \sum_{j \in D} d_j c_j(\alpha) + \frac{2\eta + 2}{(\eta - 3)\alpha} M \sum_{e \in E} c_e z_e$$

Proof. We construct integer solution from filtered solution $(\bar{x}, \bar{y}, \bar{z})$ like this: (1) for each secondary group $\mathcal{G}_j$, we open r_j completely and close all other facilities of this group, (2) for primary group facilities, we do nothing since they are already open, (3) we close all facilities which are unassigned to any group, (4) we assign all clients of each group to the only open facility of that group and finally, (5) we buy the edges of the set $A \cup B$ to connect open facilities.

Total cost of the integer solution is composed of (1) facility opening cost, (2) client servicing cost and (3) facility connection cost.

By Lemma 3 and Lemma 1, facility opening cost is at most $\frac{1}{\alpha} \sum_{i \in F_v} f_i y_i$. From Lemma 2, the client servicing cost is at most $\eta \sum_{j \in D} d_j c_j(\alpha)$. By Lemma 7 we get, $\sum_{e \in B} c_e \leq \frac{4}{\eta-3} \sum_{e \in A} c_e$. That is, total buying cost is $M \sum_{e \in A \cup B} c_e \leq \frac{\eta+1}{\eta-3} M \sum_{e \in A} c_e \leq \frac{2\eta+2}{\eta-3} M \sum_{e \in E} \bar{z}_e$. The last inequality holds because of Lemma 5. Lemma 1 shows that this cost is at most $\frac{2\eta+2}{(\eta-3)\alpha} M \sum_{e \in E} z_e$.

Thus, cost of integer solution, $Cost \leq \frac{1}{\alpha} \sum_{i \in F_v} f_i y_i + \eta \sum_{j \in D} d_j c_j(\alpha) + \frac{2\eta+2}{(\eta-3)\alpha} M \sum_{e \in E} c_e z_e$. □

Lemma 8. *For each client $j \in D$, $\int_{\alpha=0}^{1} c_j(\alpha) = \sum_{i=1}^{n} c_{ij} x_{ij}$ where n is the total number of facilities.*

Proof. The proof of this Lemma is exactly same as that of Lemma 10 in [7] and has been omitted.

Lemma 9. *Suppose the value of α has been chosen uniformly from the interval $(\beta, 1)$, where β is a parameter in the interval (0, 1) and its value will be decided later. Then the expected cost of integer solution found as described above is at most $\frac{\frac{2\eta+2}{\eta-3} \ln(1/\beta)}{1-\beta} (\sum_{i \in F_v} f_i y_i + M \sum_{e \in E} c_e z_e) + \frac{\eta}{1-\beta} \sum_{j \in D} d_j \sum_{i \in F} c_{ij} x_{ij}$.*

Proof. This proof is similar to Theorem 11 in [7]. From the result of Theorem 1, we find, $Exp(Cost)$
$\leq Exp[\frac{2\eta+2}{(\eta-3)\alpha} (\sum_{i \in F_v} f_i y_i + M \sum_{e \in E} c_e z_e) + \eta \sum_{j \in D} d_j c_j(\alpha)]$
$= \frac{2\eta+2}{\eta-3} Exp[\frac{1}{\alpha}](\sum_{i \in F_v} f_i y_i + M \sum_{e \in E} c_e z_e) + \eta \sum_{j \in D} d_j Exp[c_j(\alpha)]$
$= \frac{2\eta+2}{\eta-3} (\int_{\alpha=\beta}^{1} \frac{1}{1-\beta} \frac{1}{\alpha} d\alpha)(\sum_{i \in F_v} f_i y_i + M \sum_{e \in E} c_e z_e) + \eta \sum_{j \in D} d_j (\int_{\alpha=\beta}^{1} \frac{1}{1-\beta} c_j(\alpha) d\alpha)$
$\leq \frac{2\eta+2}{\eta-3} \frac{\ln(1/\beta)}{1-\beta} (\sum_{i \in F_v} f_i y_i + M \sum_{e \in E} c_e z_e) + \frac{\eta}{1-\beta} \sum_{j \in D} d_j \int_{\alpha=0}^{1} c_j(\alpha) d\alpha$
$= \frac{\frac{2\eta+2}{\eta-3} \ln(1/\beta)}{1-\beta} (\sum_{i \in F_v} f_i y_i + M \sum_{e \in E} c_e z_e) + \frac{\eta}{1-\beta} \sum_{j \in D} d_j \sum_{i \in F} c_{ij} x_{ij}$
where the last line follows form Lemma 8. □

Theorem 2. *For* Connected Facility Location *problem, grouping over filtered solution yields a 8.29-approximation algorithm.*

Proof. We can choose α from the interval $(e^{-\frac{\eta^2-3\eta}{2\eta+2}}, 1)$ uniformly and execute our algorithm. Using the result of Lemma 9 where $\beta = e^{-\frac{\eta^2-3\eta}{2\eta+2}}$, it can be shown that expected cost is at most $\psi(\eta) \times OPT$ where, $\psi(\eta) = \eta/(1 - e^{-\frac{\eta^2-3\eta}{2\eta+2}})$ and $OPT = \sum_{i \in F_v} f_i y_i + \sum_{j \in D} d_j \sum_{i \in F} c_{ij} x_{ij} + M \sum_{e \in E} c_e z_e$ is the cost of optimum fractional solution. Finally we can use *Derandomization* technique described in [7] to get a deterministic solution of cost at most $\psi(\eta) \times OPT$. For $\eta = 6.1168$, we get a solution with cost at most $8.2883 \times OPT \leq 8.29 \times OPT$. Here, 6.1168 is an approximate value of η_{min} such that $\eta_{min} = \arg\min_{\eta \geq 3} \psi(\eta)$. This value can be found by differentiating $\psi(\eta)$ with respect to η and equating the the resulting function to zero. In our case, we used MATLAB to get approximate value for η_{min} in the range of $[3, \inf)$.   □

Theorem 3. *For* Connected Facility Location *problem, grouping over filtered solution yields a 7.0 approximation algorithm, when opening costs of all facilities are equal.*

Proof. When all facilities have equal opening cost, we create the groups using the same algorithm *CreateGroups*. We generate the set of edges A that externally connects all groups. For each secondary group $\mathcal{G}_j$, we choose any facility of $\mathcal{T}_j(A)$ arbitrarily and open it completely. We close other facilities of this group. The rest of the steps of getting integer solution is same as that of previous case. It can be shown that the cost of integer solution in this case is at most $\eta/(1 - e^{-\frac{\eta^2 - 3\eta}{2\eta - 2}}) \times OPT$(detail has been omitted). Finally, for $\eta = 5.195$, randomization and derandomization yield integer solution with cost at most $7.0\ OPT$.   □

6   Discussion and Future Work

In this paper we propose a 8.29 approximation algorithm for *Connected Facility Location* problem. For special case when all facility opening costs are equal we propose a 7.0 approximation algorithm. Our algorithm is very simple and easy to implement. Although we have not bound the complexity, it is definitely polynomial.

Our algorithm works when each facility has capacity to serve all demands. A prominent future work could be to extend this algorithm to handle capacitated facilities.

References

1. Karger, D.R., Minkoff, M.: Building steiner trees with incomplete global knowledge. In: FOCS '00: Proceedings of 41st Annual Symposiam on Foundations of Computer Science, Washington, DC, USA, 2000, p. 613. IEEE Computer Society Press, Los Alamitos (2000)
2. Gupta, A., Kleinberg, J.M., Kumar, A., Rastogi, R., Yener, B.: Provisioning a virtual private network: a network design problem for multicommodity flow. In: ACM Symposium on Theory of Computing, pp. 389–398. ACM Press, New York (2001)
3. Swamy, C., Kumar, A.: Primal-dual algorithms for connected facility location problems. In: Jansen, K., Leonardi, S., Vazirani, V.V. (eds.) APPROX 2002. LNCS, vol. 2462, pp. 256–269. Springer, Heidelberg (2002)
4. Swamy, C., Kumar, A.: Primal-dual algorithms for connected facility location problems. Algorithmica, 40, 245–269 (2004)

5. Gupta, A., Kumar, A., Roughgarden, T.: Simpler and better approximation algorithms for network design. In: STOC '03: Proceedings of the thirty-fifth annual ACM symposium on Theory of Computing, pp. 365–372. ACM Press, New York, NY, USA (2003)
6. Lin, J., Vitter, J.S.: ϵ-approximations with minimum packing constraint violation. In: Proceedings of 24th STOC, pp. 771–782 (1992)
7. Shmoys, D.B., Tardos, É., Aardal, K.: Approximation algorithms for facility location problems (extended abstract). In: STOC '97: Proceedings of the twenty-ninth annual ACM symposium on Theory on Computing, pp. 265–274. ACM Press, New York (1997)
8. Goemans, M.X., Williamson, D.P.: A general approximation technique for constrained forest problems. SIAM Journal of Computing, 24, 296–317 (1995)

The Computational Complexity of Game Trees by Eigen-Distribution

ChenGuang Liu[*] and Kazuyuki Tanaka

Mathematical Institute, Tohoku University, Sendai 980-8578, Japan
liu@mail.tains.tohoku.ac.jp, tanaka@math.tohoku.ac.jp

Abstract. The AND-OR tree is an extremely simple model to compute the read-once Boolean functions. For an AND-OR tree, the eigen-distribution is a special distribution on random assignments to the leaves, such that the distributional complexity of the AND-OR tree is achieved. Yao's Principle[8] showed that the randomized complexity of any function is equal to the distributional complexity of the same function. In the present work, we propose an eigen-distribution-based technique to compute the distributional complexity of read-once Boolean functions. Then, combining this technique and Yao's Principle, we provide a unifying proof way for some well-known results of the randomized complexity of Boolean functions.

1 Introduction

Computational complexity aims to understand "how much" computation is necessary and sufficient to perform certain computational tasks. Usually, we take a simpler and more limited model of computation to analyze the original, more difficult problems. In various models of computation, perhaps the simplest one is the game tree.

A game tree is a rooted tree in which each leaf (i.e., external node) has a real value, all the internal nodes are labeled with MIN or MAX gates in the alternating way from the root to the leaves. A game tree is uniform if the internal nodes have the same number of children and the root-leaf paths are of the same length. The value of MIN-gates (resp. MAX-gates) is recursively defined as the minimum (resp. maximum) of the values of its children.

By "computing a game tree" we mean evaluating the value of the root. At the beginning of computing, the leaves are assigned with real values, but are "covered" so that one cannot see how they are labeled. In computing, a basic step consists of querying the value of one of the leaves, and the operations repeat until the value of the root can be determined. The cost (complexity) associated with this computation is only the number of leaves queried; all the other computations are cost free.

In the present work, we are interested in the AND-OR trees, a large class of game trees in which the leaf value is either 0 or 1. Such a tree is an extremely simple model to compute the read-once Boolean functions. By T^k and T^k_m, we denote

---

[*] Corresponding author.

A. Dress, Y. Xu, and B. Zhu (Eds.): COCOA 2007, LNCS 4616, pp. 323–334, 2007.

an AND-OR tree and a uniform m-ary AND-OR tree with k rounds, respectively. One round is one level AND(OR) gates followed by one level OR(AND) gates. By $f_{(T^k)}$, we denote the read-once Boolean function $f : \{0,1\}^n \to \{0,1\}$ with respect to the AND-OR tree T^k.

Over the years a number of game tree algorithms have been invented. Among them, the Alpha-Beta pruning algorithm has been proven quite successful, and no other algorithm has achieved such a wide-spread use in practical applications as it. A precise formulation of the Alpha-Beta pruning algorithm can be found in [3]. For the purposes of our discussion, we restrict the deterministic algorithm used in this paper to the Alpha-Beta pruning algorithm.

Let A_D be a deterministic algorithm and $\omega = \omega_1\omega_2\cdots\omega_n$ an assignment of Boolean values to the input variables $\{v_1, v_2, ..., v_n\}$ of function f. By $C(A_D, \omega)$, we denote the number of input variables queried by A_D computing f on ω. By $\mathcal{W}$ and $\mathcal{A}_D(f)$, we denote the set of assignments for f and the family of deterministic algorithms computing f, respectively. The **deterministic complexity** $D(f)$ of function f, is defined as the minimal cost of a deterministic algorithm computing f for the worst assignment, that is,

$$D(f) = \min_{A_D \in \mathcal{A}_D(f)} \max_{\omega \in \mathcal{W}} C(A_D, \omega).$$

It is easy to see that to query all input variables is always enough to compute f by the deterministic algorithms, thus for every function f on n variables, $D(f) \leq n$. For every read-once function $f_{(T^k)}$, a simple adversary argument shows that $D(f_{(T^k)}) = n$.

A randomized algorithm, denoted by A_R, is a probability distribution over the family of deterministic algorithms. For an assignment ω and a randomized algorithm A_R that has probability p_{A_D} to proceed exactly as a deterministic algorithm A_D, the cost of A_R for ω is defined as

$$C(A_R, \omega) = \sum_{A_D \in \mathcal{A}_D(f)} p_{A_D} C(A_D, \omega).$$

We denote by $\mathcal{A}_R(f)$ the family of randomized algorithms computing f. For function f, the **randomized complexity** $R(f)$ is defined by

$$R(f) = \min_{A_R \in \mathcal{A}_R(f)} \max_{\omega \in \mathcal{W}} C(A_R, \omega).$$

By the definitions of deterministic and randomized complexity, we have the obvious trivial relationship $R(f) \leq D(f)$. The first nontrivial result $R(f) \geq \sqrt{D(f)}$ was observed independently in [1], [2] and [6]. After that, several strong results were obtained for certain classes of functions. For the read-once Boolean functions $f_{(T_2^k)}$ with respect to uniform binary AND-OR trees T_2^k, where the subscript 2 means "binary", Saks and Wighderson[5] have shown that $R(f_{(T_2^k)}) = \Theta(D(f_{(T_2^k)})^\alpha)$, $\alpha = \log_2(\frac{1+\sqrt{33}}{4})$. Moreover, Saks and Wigderson[5] conjectured that:

Conjecture 1 (Saks and Wighderson[5]). For any Boolean function f,

$$R(f) = \Omega\left(D(f)^{\log_2\left(\frac{1+\sqrt{33}}{4}\right)}\right).$$

This conjecture is still wide open at the moment.

Given $\mathcal{W}$, the set of assignments, let p_ω^d be the probability of ω over $\mathcal{W}$ with respect to the distribution d. The average complexity $C(A_D, d)$ of a deterministic algorithm A_D on the assignments with distribution d is defined by

$$C(A_D, d) = \sum_{\omega \in \mathcal{W}} p_\omega^d C(A_D, \omega).$$

Let $\mathcal{D}$ be the set of distributions and $\mathcal{A}_D(f)$ the set of deterministic algorithms computing f. The **distributional complexity** $P(f)$ of function f is defined by

$$P(f) = \max_{d \in \mathcal{D}} \min_{A_D \in \mathcal{A}_D(f)} C(A_D, d).$$

In the set $\mathcal{D}$, the distribution δ such that the distributional complexity is achieved, that is,

$$\min_{A_D \in \mathcal{A}_D(f)} C(A_D, \delta) = P(f),$$

is called an **eigen-distribution** on assignments for f in this paper.

Yao[8] initially considered two roles of randomness in algorithms, randomness inside the algorithm itself, and randomness on the inputs, and constructed a bridge between these two randomness. Yao showed the well-known Minimax Theorem[7] by von Neumann implies that:

Theorem 1 (Yao's Principle[8]). *For any function f,*

$$R(f) = P(f).$$

Yao's Principle is a very useful and valid tool in randomized complexity analysis. Since its introduction, it has been extensively investigated in the literature. In applying Yao's Principle, the most key step is to compute the distributional complexity. However, no effective computing method for distributional complexity has been reported at the moment. It is one of the main aims for doing the present work to propose a technique that can effectively compute the distributional complexity of read-once Boolean functions $f_{(T^k)}$ with respect to the AND-OR trees T^k.

In the present work, we propose an eigen-distribution-based technique to compute the distributional complexity of read-once Boolean functions. Then, combining this technique and Yao's Principle, we provide a unifying proof way for some well-known results of the randomized complexity of Boolean functions. The paper is organized as follows: We introduce the main idea in Section 2. The general computing formulae are described in Section 3. Section 4 is devoted to the applications of our proposed technique. Section 5 concludes this paper.

2   Main Idea

In order to eliminate the unnecessary assignments in computing the distributional complexity, we have the following technique to form two particular sets of assignments to leaves of T^k, namely 1-set and 0-set:

Methodology 1 (General reverse assigning technique: GRAT). *The technique to form 1-set (resp. 0-set) includes three stages:*

1) Assign a 1 (resp. 0) to the root of tree T^k.
2) From the root to the leaves, assign a 0 or a 1 to each child of any internal node as the follows:
   - *for AND gate with value 1, assign 1s to all its children;*
   - *for OR gate with value 0, assign 0s to all its children;*
   - *for AND gate with value 0, assign at random a 0 to one of its children and 1s to the other children;*
   - *for OR gate with value 1, assign at random a 1 to one of its children and 0s to the other children;*
3) Form the 1-set (resp. 0-set) by collecting all possible assignments to the leaves.

It is not hard to see that, following from this technique, we preclude the possibility of more inputs to an AND gate being 0 and to a OR gate being 1 for an AND-OR tree T^k.

Let $\beta(f_{(T^k)})$ and $\alpha(f_{(T^k)})$ denote the particular distributional complexity of function $f_{(T^k)}$ such that the assignments are restricted to the 1-set and 0-set, respectively. By the E^1-**distribution** (resp. E^0-**distribution**) we denote a unique distribution on the assignments of 1-set (resp. 0-set) such that $\beta(f_{(T^k)})$ (resp. $\alpha(f_{(T^k)})$) is achieved. It is easy to see that, for the function $f_{(T^k)}$, when $\beta(f_{(T^k)}) \geq \alpha(f_{(T^k)})$, the E^1-distribution is the eigen-distribution, and when $\beta(f_{(T^k)}) \leq \alpha(f_{(T^k)})$, the E^0-distribution is the eigen-distribution.

Let a read-once Boolean function $f_{(T_{z:m_1,m_2,\dots,m_z})} = f_{(T_{m_1})} \wedge f_{(T_{m_2})} \wedge \dots \wedge f_{(T_{m_z})}$, where $f_{(T_{m_1})} = x_1 \vee x_2 \vee \dots \vee x_{m_1}$, $f_{(T_{m_2})} = x_{m_1+1} \vee x_{m_1+2} \vee \dots \vee x_{m_1+m_2}$, $\dots$, $f_{(T_{m_z})} = x_{m_1+m_2+\dots+m_{z-1}+1} \vee x_{m_1+m_2+\dots+m_{z-1}+2} \vee \dots \vee x_{m_1+m_2+\dots+m_{z-1}+m_z}$, and variable $x_i \in \{0,1\}$ for all i. For such a function, the corresponding AND-OR tree can be represented as Figure 1.

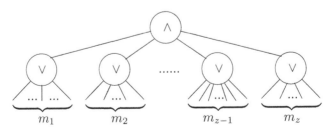

Fig. 1. The AND-OR tree for function $f_{(T_{z:m_1,m_2,\dots,m_z})}$

Fix $m_{max} = \max\{m_1, m_2, ..., m_z\}$. By $p^0_{f_{(T_{m_k})}}$, we denote the probability that $f_{(T_{m_k})}$ returns a 0.

Theorem 2. *For the read-once Boolean function $f_{(T_{z:m_1,m_2,...,m_z})}$, we have*

1) The complexity $\beta\left(f_{(T_{z:m_1,m_2,...,m_z})}\right) = \frac{m_1+m_2+...+m_z+z}{2}$, and in the E^1-distribution, the probability of each assignment of 1-set is equal to $\frac{1}{m_1 \cdot m_2 \cdot ... \cdot m_z}$.

2) The complexity $\alpha\left(f_{(T_{z:m_1,m_2,...,m_z})}\right)$ is given by

$$\max\left\{m_{max}, \frac{1}{m_1+m_2+...+m_z+z} \cdot \sum_{i=1}^{z-1}\sum_{j=2}^{z} \frac{(m_i+1)(m_j+1)}{2} + \sum_{k=1}^{z} p^0_{f_{(T_{m_k})}} \cdot m_k\right\}, \text{ and in}$$

the E^0-distribution:

- *if $m_{max} < \frac{1}{m_1+m_2+...+m_z+z} \cdot \sum_{i=1}^{z-1}\sum_{j=2}^{z} \frac{(m_i+1)(m_j+1)}{2} + \sum_{k=1}^{z} p^0_{f_{(T_{m_k})}} \cdot m_k$, the probability $p^0_{f_{(T_{m_k})}} = \frac{m_k+1}{m_1+m_2+...+m_z+z}$.*

- *if $m_{max} \geq \frac{1}{m_1+m_2+...+m_z+z} \cdot \sum_{i=1}^{z-1}\sum_{j=2}^{z} \frac{(m_i+1)(m_j+1)}{2} + \sum_{k=1}^{z} p^0_{f_{(T_{m_k})}} \cdot m_k$, the probability of each assignment of 0-set such that $f_{(T_{m_{max}})}$ returns a 0 is equal to $\frac{m_{max}}{m_1 m_2 ... m_z}$, and the probability of any assignment of 0-set such that $f_{(T_{m_{max}})}$ returns a 1 is equal to 0.*

3) The eigen-distribution is
- *the E^0-distribution when $m_{max} \geq \frac{m_1+m_2+...+m_z+z}{2}$, and in this case, the distribution complexity $P\left(f_{(T_{z:m_1,m_2,...,m_z})}\right) = m_{max}$.*
- *the E^1-distribution when $m_{max} < \frac{m_1+m_2+...+m_z+z}{2}$, and in this case, the distribution complexity $P\left(f_{(T_{z:m_1,m_2,...,m_z})}\right) = \frac{m_1+m_2+...+m_z+z}{2}$.*

Proof. Following from the above technique GRAT, the 1-set and 0-set for this read-once Boolean function can be formed as follows (see Table 1).

A little more consideration shows that the complexity $\beta\left(f_{(T_{z:m_1,m_2,...,m_z})}\right)$ is achieved if and only if all the assignments of 1-set have the same probability. Hence, in the E^1-distribution, the probability of each assignment of 1-set is equal to $\frac{1}{m_1 m_2 ... m_z}$. In this case, it is easy to see that $\beta\left(f_{(T_{z:m_1,m_2,...,m_z})}\right) = \frac{m_1+m_2+...+m_z+z}{2}$.

Clearly, there also exits a unique distribution on the assignments of 0-set such that all the deterministic algorithms have the same complexity equal. In such a distribution, for any assignment such that $f_{(T_{m_k})}$ returns a 0, where $k \in \{1, 2, ..., z\}$, the probability is equal to $p^0_{f_{(T_{m_k})}} \cdot \frac{1}{m_1 ... m_{k-1} m_{k+1} ... m_z}$ (see Table 1). When $m_{max} < \frac{1}{m_1+m_2+...+m_z+z} \cdot \sum_{i=1}^{z-1}\sum_{j=2}^{z} \frac{(m_i+1)(m_j+1)}{2} + \sum_{k=1}^{z} p^0_{f_{(T_{m_k})}} \cdot m_k$, such a distribution is just the E^0-distribution.

Table 1. Each assignment and its probability for function $f_{\left(T_{z:m_1,m_2,\ldots,m_z}\right)}$

probability	assignments of 1-set	probability	assignments of 0-set
$\frac{1}{m_1 m_2 \ldots m_z}$	00...0100...01......00...01	$p^0_{f_{(T_{m_1})}} \cdot \frac{1}{m_2 m_3 \ldots m_z}$	00...0000...01......00...01
$\frac{1}{m_1 m_2 \ldots m_z}$	00...0100...01......00...10	$p^0_{f_{(T_{m_1})}} \cdot \frac{1}{m_2 m_3 \ldots m_z}$	00...0000...01......00...10
⋮	⋮	⋮	⋮
$\frac{1}{m_1 m_2 \ldots m_z}$	00...0100...01......01...00	$p^0_{f_{(T_{m_1})}} \cdot \frac{1}{m_2 m_3 \ldots m_z}$	00...0010...00......01...00
$\frac{1}{m_1 m_2 \ldots m_z}$	00...0100...01......10...10	$p^0_{f_{(T_{m_1})}} \cdot \frac{1}{m_2 m_3 \ldots m_z}$	00...0010...00......10...00
$\frac{1}{m_1 m_2 \ldots m_z}$	00...0100...10......00...01	$p^0_{f_{(T_{m_2})}} \cdot \frac{1}{m_1 m_3 \ldots m_z}$	00...0100...00......00...01
$\frac{1}{m_1 m_2 \ldots m_z}$	00...0100...10......00...10	$p^0_{f_{(T_{m_2})}} \cdot \frac{1}{m_1 m_3 \ldots m_z}$	00...0100...00......00...10
⋮	⋮	⋮	⋮
$\frac{1}{m_1 m_2 \ldots m_z}$	00...0100...10......01...00	$p^0_{f_{(T_{m_2})}} \cdot \frac{1}{m_1 m_3 \ldots m_z}$	10...0000...00......01...00
$\frac{1}{m_1 m_2 \ldots m_z}$	00...0100...10......10...00	$p^0_{f_{(T_{m_2})}} \cdot \frac{1}{m_1 m_3 \ldots m_z}$	10...0000...00......10...00
⋮	⋮	⋮	⋮
⋮	⋮	⋮	⋮
$\frac{1}{m_1 m_2 \ldots m_z}$	10...0010...00......00...01	$p^0_{f_{(T_{m_z})}} \cdot \frac{1}{m_1 m_2 \ldots m_{z-1}}$	00...0100...01......00...00
$\frac{1}{m_1 m_2 \ldots m_z}$	10...0010...00......00...10	$p^0_{f_{(T_{m_z})}} \cdot \frac{1}{m_1 m_2 \ldots m_{z-1}}$	00...0100...10......00...00
⋮	⋮	⋮	⋮
$\frac{1}{m_1 m_2 \ldots m_z}$	10...0010...00......01...00	$p^0_{f_{(T_{m_z})}} \cdot \frac{1}{m_1 m_2 \ldots m_{z-1}}$	10...0001...00......00...00
$\frac{1}{m_1 m_2 \ldots m_z}$	10...0010...00......10...00	$p^0_{f_{(T_{m_z})}} \cdot \frac{1}{m_1 m_2 \ldots m_{z-1}}$	10...0010...00......00...00
	$\underbrace{\quad}_{m_1}\ \underbrace{\quad}_{m_2}\ \underbrace{\quad}_{m_z}$		$\underbrace{\quad}_{m_1}\ \underbrace{\quad}_{m_2}\ \underbrace{\quad}_{m_z}$

When $m_{max} \geq \frac{1}{m_1 + m_2 + \ldots + m_z + z} \cdot \sum_{i=1}^{z-1} \sum_{j=2}^{z} \frac{(m_i+1)(m_j+1)}{2} + \sum_{k=1}^{z} p^0_{f_{(T_{m_k})}} \cdot m_k$, we

have that $\alpha\left(f_{\left(T_{z:m_1,m_2,\ldots,m_z}\right)}\right)$ is achieved if and only if $f_{(T_{m_{max}})}$ returns a 0. Hence, in this case, the probability of each assignment of 0-set such that $f_{(T_{m_{max}})}$ returns a 0 is equal to $\frac{m_{max}}{m_1 m_2 \ldots m_z}$, and the probability of any assignment of 0-set such that $f_{(T_{m_{max}})}$ returns a 1 is equal to 0.

Therefore, we have that the complexity $\alpha\left(f_{\left(T_{z:m_1,m_2,\ldots,m_z}\right)}\right)$ is given by

$$\max\left\{ m_{max}, \frac{1}{m_1 + m_2 + \ldots + m_z + z} \cdot \sum_{i=1}^{z-1} \sum_{j=2}^{z} \frac{(m_i+1)(m_j+1)}{2} + \sum_{k=1}^{z} p^0_{f_{(T_{m_k})}} \cdot m_k \right\}.$$

By the definitions of distributional complexity and eigen-distribution, it is easy to see that $P(f) = \max\{\alpha(f), \beta(f)\}$. Hence, the distributional complexity

$$P\left(f_{\left(T_{z:m_1,m_2,\ldots,m_z}\right)}\right) = m_{max} \text{ for } m_{max} \geq \frac{m_1+m_2+\ldots+m_z+z}{2}, \text{ and the distribu-}$$

tional complexity $P\left(f_{\left(T_{z:m_1,m_2,\ldots,m_z}\right)}\right) = \frac{m_1+m_2+\ldots+m_z+z}{2}$ for m_{max}

$< \frac{m_1+m_2+\ldots+m_z+z}{2}$. □

3   General Computing Formulae

By $f_{(T_m^\wedge)}$ (resp. $f_{(T_m^\vee)}$), we denote a read-once Boolean function corresponding to an AND-OR tree with the root labeled by AND gate (resp. OR gate) and on m sub-trees or leaves. Without any loss of generality, let a function $f_{(T_m^\wedge)} = f_{(T_{n_1}^\vee)} \wedge f_{(T_{n_2}^\vee)} \wedge \ldots \wedge f_{(T_{n_m}^\vee)}$, and fix $\alpha\left(f_{(T_{max}^\vee)}\right) = \max\left\{\alpha\left(f_{(T_{n_1}^\vee)}\right), \alpha\left(f_{(T_{n_2}^\vee)}\right), \ldots, \alpha\left(f_{(T_{n_m}^\vee)}\right)\right\}$.

Theorem 3. *For the AND-OR tree function* $f_{(T_m^\wedge)} = f_{(T_{n_1}^\vee)} \wedge f_{(T_{n_2}^\vee)} \wedge \ldots \wedge f_{(T_{n_m}^\vee)}$, *we have*

1) The complexity $\beta\left(f_{(T_m^\wedge)}\right) = \sum_{k=1}^m \beta\left(f_{(T_{n_k}^\vee)}\right)$.

2) The complexity $\alpha\left(f_{(T_m^\wedge)}\right)$ *is given by*

$$\max\left\{\alpha\left(f_{(T_{max}^\vee)}\right), \frac{\sum_{i=1}^{m-1}\sum_{j=2}^m \beta\left(f_{(T_{n_i}^\vee)}\right)\beta\left(f_{(T_{n_j}^\vee)}\right) + \sum_{k=1}^m \alpha\left(f_{(T_{n_k}^\vee)}\right)\beta\left(f_{(T_{n_k}^\vee)}\right)}{\sum_{k=1}^m \beta\left(f_{(T_{n_k}^\vee)}\right)}\right\}.$$

3) The distributional complexity $P\left(f_{(T_m^\wedge)}\right) = \max\left\{\alpha(f_{(T_{max}^\vee)}), \sum_{k=1}^m \beta\left(f_{(T_{n_k}^\vee)}\right)\right\}$.

Proof. We sketch the proof here.

1) From the above technique GRAT, $\beta\left(f_{(T_m^\wedge)}\right) = \sum_{k=1}^m \beta\left(f_{(T_{n_k}^\vee)}\right)$ is clear.

2) For the 0-set of $f_{(T_m^\wedge)}$, there exits a special distribution on assignments

$$\beta\left(f_{(T_{n_1}^\vee)}\right)\beta\left(f_{(T_{n_2}^\vee)}\right)\beta\left(f_{(T_{n_3}^\vee)}\right)\ldots\beta\left(f_{(T_{n_{m-1}}^\vee)}\right)\alpha\left(f_{(T_{n_m}^\vee)}\right)$$
$$\beta\left(f_{(T_{n_1}^\vee)}\right)\beta\left(f_{(T_{n_2}^\vee)}\right)\beta\left(f_{(T_{n_3}^\vee)}\right)\ldots\alpha\left(f_{(T_{n_{m-1}}^\vee)}\right)\beta\left(f_{(T_{n_m}^\vee)}\right)$$
$$\vdots \qquad \vdots \qquad \vdots \qquad \vdots \qquad \vdots \qquad \vdots$$
$$\beta\left(f_{(T_{n_1}^\vee)}\right)\beta\left(f_{(T_{n_2}^\vee)}\right)\alpha\left(f_{(T_{n_3}^\vee)}\right)\ldots\beta\left(f_{(T_{n_{m-1}}^\vee)}\right)\beta\left(f_{(T_{n_m}^\vee)}\right)$$
$$\beta\left(f_{(T_{n_1}^\vee)}\right)\alpha\left(f_{(T_{n_2}^\vee)}\right)\beta\left(f_{(T_{n_3}^\vee)}\right)\ldots\beta\left(f_{(T_{n_{m-1}}^\vee)}\right)\beta\left(f_{(T_{n_m}^\vee)}\right)$$
$$\alpha\left(f_{(T_{n_1}^\vee)}\right)\beta\left(f_{(T_{n_2}^\vee)}\right)\beta\left(f_{(T_{n_3}^\vee)}\right)\ldots\beta\left(f_{(T_{n_{m-1}}^\vee)}\right)\beta\left(f_{(T_{n_m}^\vee)}\right)$$

such that all deterministic algorithms computing $f_{(T_m^\wedge)}$ have the same complexity. In such a distribution, let $p^0_{f_{(T_{n_k}^\vee)}}$ denote the probability of the assignment

$$\beta\left(f_{(T_{n_1}^\vee)}\right)...\beta\left(f_{(T_{n_{k-1}}^\vee)}\right)\alpha\left(f_{(T_{n_k}^\vee)}\right)\beta\left(f_{(T_{n_{k+1}}^\vee)}\right)...\beta\left(f_{(T_{n_m}^\vee)}\right).$$ Then, we have

$$p^0_{f_{(T_{n_k}^\vee)}} = \frac{\beta\left(f_{(T_{n_k}^\vee)}\right)}{\sum\limits_{i=1}^{m}\beta\left(f_{(T_{n_i}^\vee)}\right)}.$$

Given any deterministic algorithm, we can obtain the complexity

$$\alpha\left(f_{(T_m^\wedge)}\right) = \frac{\sum\limits_{i=1}^{m-1}\sum\limits_{j=2}^{m}\beta\left(f_{(T_{n_i}^\vee)}\right)\beta\left(f_{(T_{n_j}^\vee)}\right) + \sum\limits_{k=1}^{m}\alpha\left(f_{(T_{n_k}^\vee)}\right)\beta\left(f_{(T_{n_k}^\vee)}\right)}{\sum\limits_{k=1}^{m}\beta\left(f_{(T_{n_k}^\vee)}\right)}.$$

When $\alpha\left(f_{(T_{max}^\vee)}\right) < \dfrac{\sum\limits_{i=1}^{m-1}\sum\limits_{j=2}^{m}\beta\left(f_{(T_{n_i}^\vee)}\right)\beta\left(f_{(T_{n_j}^\vee)}\right) + \sum\limits_{k=1}^{m}\alpha\left(f_{(T_{n_k}^\vee)}\right)\beta\left(f_{(T_{n_k}^\vee)}\right)}{\sum\limits_{k=1}^{m}\beta\left(f_{(T_{n_k}^\vee)}\right)}$, such a

distribution is the worst one for any deterministic algorithm. Therefore, when

$\alpha\left(f_{(T_{max}^\vee)}\right) < \dfrac{\sum\limits_{i=1}^{m-1}\sum\limits_{j=2}^{m}\beta\left(f_{(T_{n_i}^\vee)}\right)\beta\left(f_{(T_{n_j}^\vee)}\right) + \sum\limits_{k=1}^{m}\alpha\left(f_{(T_{n_k}^\vee)}\right)\beta\left(f_{(T_{n_k}^\vee)}\right)}{\sum\limits_{k=1}^{m}\beta\left(f_{(T_{n_k}^\vee)}\right)}$, such a distribu-

tion with $p^0_{f_{(T_{n_k}^\vee)}} = \dfrac{\beta\left(f_{(T_{n_k}^\vee)}\right)}{\sum\limits_{i=1}^{m}\beta\left(f_{(T_{n_i}^\vee)}\right)}$ is the E^0-distribution for function $f_{(T_m^\wedge)}$.

When $\alpha\left(f_{(T_{max}^\vee)}\right) \geq \dfrac{\sum\limits_{i=1}^{m-1}\sum\limits_{j=2}^{m}\beta\left(f_{(T_{n_i}^\vee)}\right)\beta\left(f_{(T_{n_j}^\vee)}\right) + \sum\limits_{k=1}^{m}\alpha\left(f_{(T_{n_k}^\vee)}\right)\beta\left(f_{(T_{n_k}^\vee)}\right)}{\sum\limits_{k=1}^{m}\beta\left(f_{(T_{n_k}^\vee)}\right)}$, the

complexity $\alpha\left(f_{(T_m^\wedge)}\right)$ is achieved if and only if the probability of the assignment $\beta\left(f_{(T_{n_1}^\vee)}\right)...\beta\left(f_{(T_{n_{k-1}}^\vee)}\right)\alpha\left(f_{(T_{max}^\vee)}\right)\beta\left(f_{(T_{n_{k+1}}^\vee)}\right)...\beta\left(f_{(T_{n_m}^\vee)}\right)$ is 1. In such a distribution, $\alpha\left(f_{(T_m^\wedge)}\right) = \alpha\left(f_{(T_{max}^\vee)}\right)$. By the definition of $\alpha\left(f_{(T_m^\wedge)}\right)$, we have

$$\alpha\left(f_{(T_m^\wedge)}\right) = \max\left\{\alpha\left(f_{(T_{max}^\vee)}\right), \frac{\sum\limits_{i=1}^{m-1}\sum\limits_{j=2}^{m}\beta\left(f_{(T_{n_i}^\vee)}\right)\beta\left(f_{(T_{n_j}^\vee)}\right) + \sum\limits_{k=1}^{m}\alpha\left(f_{(T_{n_k}^\vee)}\right)\beta\left(f_{(T_{n_k}^\vee)}\right)}{\sum\limits_{k=1}^{m}\beta\left(f_{(T_{n_k}^\vee)}\right)}\right\}.$$

3) Considering $\sum\limits_{k=1}^{m}\beta\left(f_{(T_{n_k}^\vee)}\right) \geq \dfrac{\sum\limits_{i=1}^{m-1}\sum\limits_{j=2}^{m}\beta\left(f_{(T_{n_i}^\vee)}\right)\beta\left(f_{(T_{n_j}^\vee)}\right) + \sum\limits_{k=1}^{m}\alpha\left(f_{(T_{n_k}^\vee)}\right)\beta\left(f_{(T_{n_k}^\vee)}\right)}{\sum\limits_{k=1}^{m}\beta\left(f_{(T_{n_k}^\vee)}\right)}$,

we have $P\left(f_{(T_m^\wedge)}\right) = \max\left\{\alpha(f_{(T_{max}^\vee)}), \sum\limits_{k=1}^{m}\beta\left(f_{(T_{n_k}^\vee)}\right)\right\}$ by the definition of distributional complexity. $\square$

Let a read-once function $f_{(T_m^\vee)} = f_{(T_{n_1}^\wedge)} \vee f_{(T_{n_2}^\wedge)} \vee ... \vee f_{(T_{n_m}^\wedge)}$. Fix $\beta(f_{(T_{max}^\wedge)}) = \max\left\{\beta\left(f_{(T_{n_1}^\wedge)}\right), \beta\left(f_{(T_{n_2}^\wedge)}\right), ..., \beta\left(f_{(T_{n_m}^\wedge)}\right)\right\}$. By the same way, we have

Theorem 4. *For the AND-OR tree function* $f_{(T_m^\vee)} = f_{(T_{n_1}^\wedge)} \vee f_{(T_{n_2}^\wedge)} \vee ... \vee f_{(T_{n_m}^\wedge)}$,

1) *The complexity* $\alpha\left(f_{(T_m^\vee)}\right) = \sum\limits_{k=1}^{m}\alpha\left(f_{(T_{n_k}^\wedge)}\right)$.

2) The complexity $\beta\left(f_{(T_m^\vee)}\right)$ is given by

$$\max\left\{\beta\left(f_{(T_{max}^\wedge)}\right), \frac{\sum_{i=1}^{m-1}\sum_{j=2}^{m}\alpha\left(f_{(T_{n_i}^\wedge)}\right)\alpha\left(f_{(T_{n_j}^\wedge)}\right) + \sum_{k=1}^{m}\beta\left(f_{(T_{n_k}^\wedge)}\right)\alpha\left(f_{(T_{n_k}^\wedge)}\right)}{\sum_{k=1}^{m}\alpha\left(f_{(T_{n_k}^\wedge)}\right)}\right\}.$$

3) The distributional complexity $P\left(f_{(T_m^\vee)}\right) = \max\left\{\beta(f_{(T_{max}^\wedge)}), \sum_{k=1}^{m}\alpha\left(f_{(T_{n_k}^\wedge)}\right)\right\}.$

It is not hard to see that the distributional complexity of any read-once Boolean function $f_{(T^k)}$ can be computed by applying Theorem 3 and Theorem 4 recursively. Then, by using Yao's Principle, we can obtain its randomized complexity.

4 Applications

In this section, we provide a unifying proof way for the randomized complexity of some read-once Boolean functions, including two well-known results derived by Saks and wigderson. We start with the simplest case: the OR's and AND's functions. Then we turn to the more general cases. At the same time, we present a simplified method to compute the distributional complexity of read-once Boolean functions that have the iterated properties.

Theorem 5. For the read-once Boolean function $f_{(T_n^\wedge)} = x_1 \wedge x_2 \wedge ... \wedge x_n$,
1) there is only one assignment in the 1-set. Therefore, in the E^1-distribution, the probability of the unique assignment is equal to 1. The complexity $\beta\left(f_{(T_n^\wedge)}\right) = n$.
2) there are n assignments in the 0-set, and in the E^0-distribution, the probability of each assignment is equal to $\frac{1}{n}$. The complexity $\alpha\left(f_{(T_n^\wedge)}\right) = \frac{1+n}{2}$.
3) the E^1-distribution is the eigen-distribution for $f_{(T_n^\wedge)}$, and the distributional complexity $P\left(f_{(T_n^\wedge)}\right) = n$.

Proof. This is a special case of Theorem 3 where $f_{(T_{n_i}^\vee)}$ is a function with a single variable x_i. $\qquad\square$

Theorem 6. For the read-once Boolean function $f_{(T_n^\vee)} = x_1 \vee x_2 \vee ... \vee x_n$,
1) there are n assignments in the 1-set, and in the E^1-distribution, the probability of each assignment is equal to $\frac{1}{n}$. The complexity $\beta\left(f_{(T_n^\vee)}\right) = \frac{1+n}{2}$.
2) there is only one assignment in the 0-set. Therefore, in the E^0-distribution, the probability of the unique assignment is equal to 1. The complexity $\alpha\left(f_{(T_n^\vee)}\right) = n$.
3) the E^0-distribution is the eigen-distribution for $f_{(T_n^\vee)}$, and the distributional complexity $P\left(f_{(T_n^\vee)}\right) = n$.

Proof. This is a special case of Theorem 4 where $f_{(T_{n_i}^\wedge)}$ is a function with a single variable x_i. $\qquad\square$

Together with Yao's Principle, Theorem 5 (resp. Theorem 6) shows that, for the read-once AND's (resp. OR's) function on n variables, the randomized complexity is n.

Theorem 7. *For the read-once Boolean function* $f_{(T_m^1)} = (x_1 \vee x_2 \vee ... \vee x_m) \wedge (x_{m+1} \vee x_{m+2} \vee ... \vee x_{2m}) \wedge ... \wedge (x_{m^2-m+1} \vee x_{m^2-m+2} \vee ... \vee x_{m^2})$, *where* $x_k \in \{0,1\}$, *we have*

1) The complexity $\beta\left(f_{(T_m^1)}\right) = \frac{m(m+1)}{2}$, *and in the* E^1-*distribution, the probability of each assignment of 1-set is equal to* $\frac{1}{m^m}$.

2) The complexity $\alpha\left(f_{(T_m^1)}\right) = \frac{m^2+4m-1}{4}$, *and in the* E^0-*distribution, the probability of each assignment of 0-set is equal to* $\frac{1}{m^m}$.

3) the distributional complexity $P\left(f_{(T_m^1)}\right) = \frac{m(m+1)}{2}$.

Proof. This is a special case of Theorem 2 on the function $f_{(T_{z:m_1,m_2,...,m_z})}$ where $m_1 = m_2 = ... = m_z = z = m$. For the function $f_{(T_{z:m_1,m_2,...,m_z})}$, when $m_1 = m_2 = ... = m_z = z = m$, it is easy to see that $m_{max} < \frac{1}{m_1+m_2+...+m_z+z} \cdot$

$$\sum_{i=1}^{z-1}\sum_{j=2}^{z}\frac{(m_i+1)(m_j+1)}{2} + \sum_{k=1}^{z} p_{f_{(T_{m_k})}}^0 \cdot m_k \text{ for any } m. \qquad \square$$

Let $f_{(T_m^k)}$ be a fully iterated read-once Boolean function on m^{2k} variables with base function $f_{(T_m^1)} = (x_1 \vee x_2 \vee ... \vee x_m) \wedge (x_{m+1} \vee x_{m+2} \vee ... \vee x_{2m}) \wedge ... \wedge (x_{m^2-m+1} \vee x_{m^2-m+2} \vee ... \vee x_{m^2})$, where $x_i \in \{0,1\}$. Such a Boolean function is corresponding to a uniform m-ary AND-OR tree with k rounds. Given a k, we can compute the randomized complexity of $f_{(T_m^k)}$ by directly applying the results of Theorem 3 and Theorem 4. But, when k is very large, such a computing work is complex since there are more variables. In fact, we can simplify the computation by making use of the iterated properties of functions. We refer to Liu and Tanaka[4] for a thorough discussion of such a simplified computing method for the special function $f_{(T_2^k)}$.

Theorem 8. *For the fully iterated read-once Boolean function* $f_{(T_m^k)}$ *on* $n = m^{2k}$ *variables with base function* $f_{(T_m^1)} = (x_1 \vee x_2 \vee ... \vee x_m) \wedge (x_{m+1} \vee x_{m+2} \vee ... \vee x_{2m}) \wedge ... \wedge (x_{m^2-m+1} \vee x_{m^2-m+2} \vee ... \vee x_{m^2})$, *we have*

$$R\left(f_{(T_m^k)}\right) = \Theta\left(n^{\log_m\left(\frac{m-1+\sqrt{m^2+14m+1}}{4}\right)}\right).$$

Proof. Select at random a deterministic algorithm, e.g., the algorithm that reads the nodes from left to right. To get a recurrence equation for $\beta\left(f_{(T_m^k)}\right)$ and $\alpha\left(f_{(T_m^k)}\right)$, we associate a 1 (resp. 0) occurring in the assignments of 1-set and 0-set for base function $f_{(T_m^1)}$ with the $\beta(f_{(T_m^{k-1})})$ (resp. $\alpha\left(f_{(T_m^{k-1})}\right)$) for $f_{(T_m^k)}$. Considering the results obtained in Theorem 7 that the probability of each assignment in the E^1-distribution and E^0-distribution for $f_{(T_m^1)}$ is equal to $\frac{1}{m^m}$, we have the recurrence

$$\begin{cases} \alpha\left(f_{(T_m^k)}\right) = \frac{m-1}{2}\beta\left(f_{(T_m^{k-1})}\right) + \frac{(m+1)^2}{4}\alpha\left(f_{(T_m^{k-1})}\right) \\ \beta\left(f_{(T_m^k)}\right) = m\beta\left(f_{(T_m^{k-1})}\right) + \frac{m(m-1)}{2}\alpha\left(f_{(T_m^{k-1})}\right) \end{cases}$$

with the initial conditions

$$\begin{cases} \alpha\left(f_{(T_m^0)}\right) = 1 \\ \beta\left(f_{(T_m^0)}\right) = 1 \end{cases}$$

Solving this recurrence, we have

$$\alpha\left(f_{(T_m^k)}\right), \beta\left(f_{(T_m^k)}\right) = \left(\frac{m^2 + 6m + 1 + (m-1)\sqrt{m^2 + 14m + 1}}{8}\right)^k$$

Since $k = \log_{(m^2)} n = \frac{1}{2}\log_m n$ for function $f_{(T_m^k)}$, we have

$$P\left(f_{(T_m^k)}\right) = \Theta\left(n^{\log_m\left(\frac{m-1+\sqrt{m^2+14m+1}}{4}\right)}\right).$$

By Yao's principle, we obtain $R\left(f_{(T_m^k)}\right) = \Theta\left(n^{\log_m\left(\frac{m-1+\sqrt{m^2+14m+1}}{4}\right)}\right).$ □

Note that when $m = 2$ this result specializes to

$$R\left(f_{(T_2^k)}\right) = \Theta\left(n^{\log_2\left(\frac{1+\sqrt{33}}{4}\right)}\right) = \Theta\left(n^{0.7537\ldots}\right).$$

These two well-known results of $R\left(f_{(T_m^k)}\right) = \Theta\left(n^{\log_m\left(\frac{m-1+\sqrt{m^2+14m+1}}{4}\right)}\right)$
and $R\left(f_{(T_2^k)}\right) = \Theta\left(n^{\log_2\left(\frac{1+\sqrt{33}}{4}\right)}\right)$ were first proved by Saks and Wigderson[5]
in another way, but our proof based on eigen-distribution seems to be much
simpler than their original.

5   Conclusions

In this paper, we proposed an eigen-distribution-based technique to compute
the distributional complexity of read-once Boolean function $f_{(T^k)}$ with respect
to any AND-OR tree T^k. We claim that such a technique is useful to prove
the well-known conjecture of Saks and Wigderson. Further investigation on this
subjective will be appeared in the future literature.

Moreover, the randomized complexity and the distributional complexity of a
function f can be defined not only in the Las Vegas case, but also in the Monte
Carlo case with one-sided error and the Monte Carlo case with two-sided error.
However, our work in the present paper is strictly restricted to the Las Vegas
case. We suggest to investigate the eigen-distribution in the Monte Carlo case
as a future research topic.

Acknowledgments

The authors would like to thank Professor Takeshi Yamazaki for fruitful discussions, and the anonymous referees for many useful and constructive comments.

References

1. Blum, M., Impagliazzo, R.: Generic Oracles and Oracle Classed. In: Proceedings of the 28th Annual Symposium on Foundations of Computer science(FOCS), pp. 118–126 (1987)
2. Hartmanis, J., Hemachandra, L.A.: Complexity classes without machines: on complete languages for UP. Theoretical Computer Science 58(1-3), 129–142 (1988)
3. Knuth, D.E., Moore, R.W.: An analysis of alpha-beta pruning. Artificial Intelligence. 6, 293–326 (1975)
4. Liu, C.G., Tanaka, K.: The Complexity of Algorithms Computing Game Trees on Random Assignments. In: Proceedings of the 3rd International Conference on Algorithmic Aspects in Information and Management, Portland,OR, USA, June, 2007. LNCS, vol. 4508, pp. 241–250 (2007)
5. Saks, M., Wigderson, A.: Probabilistic Boolean Decision Trees and the Complexity of Evaluating Game Trees. In: Proceedings of the 27th Annual IEEE Symposium on Foundations of Computer science(FOCS), pp. 29–38. IEEE Computer Society Press, Los Alamitos (1986)
6. Tardos, G.: Query complexity, or why is it difficult to separate $NP^A \cap coNP^A$ from P^A by random oracles A. Combinatorica 9, 385–392 (1990)
7. von Neumann, J.: Zur Theorie der Gesellschaftsspiele. Mathematische Annalen 100, 295–320 (1928)
8. Yao, A.C.-C.: Probabilistic Computations: Towards a Unified Measure of Complexity. In: Proceedings of the 18th Annual IEEE Symposium on Foundations of Computer science(FOCS), pp. 222–227. IEEE Computer Society Press, Los Alamitos (1977)

The Minimum All-Ones Problem for Graphs with Small Treewidth

Yunting Lu[1,2] and Yueping Li[2]

[1] Shenzhen Institute of Information Technology, Shenzhen 518029, P.R. China
[2] SunYat-sen University, Guangzhou 510275, P.R. China

Abstract. The minimum all-ones problem is applied in linear cellular automata. It is NP-complete for general graphs. In this paper, we consider the problem for graphs with small treewidth≤ 4. We give an $O(|V|)$ algorithm.

1 Introduction and Terminology

Suppose each square is pressed, the light of that square will change from off to on, and vice versa; the same happens to the lights of all the edge-adjacent squares. Initially all the lights are off. The minimum all-ones problem is to find a solution that press as few buttons as possible such that all the lights are off at the end. It was introduced by Sutner[5] and is applied in linear cellular automata. A button lights not only its neighbors but also its own light is σ^+ rule of the all-ones problem. If a button lights only its neighbors but not its own, this is called σ rule. In this paper, we consider the σ^+ rule. For the terminology and notation not defined in this paper, reader can refer to [1].

The all-ones problem has been extensively studied recently; see Sutner[7,8], Barua and Ramakrishnan[2], and Dodis and Winkler[3]. Sutner[6] proved that it is always possible to light every lamp in any graphs by σ^+ rule, using linear algebra. Galvin[4] gave a graph-theoretic algorithm of linear time to find solutions for trees. Chen and Li[10] gave a linear algorithm for the minimum all-ones problem for trees and also got a linear algorithm to find solutions to the all-ones problem in a unicyclic graph.

The notions of "tree-decomposition" and "treewidth" have received much attention recently. The treewidth is thus a parameter that measures how close to a tree a graph is. Several classes of graphs, which are important in practice, have constant bounded treewidth. For example, trees and forests have treewidth ≤ 1, series-parallel graphs and outerplanar graphs ≤ 2, Halin graphs ≤ 3, members of k-terminal recursive graph families have treewidth $\leq k$.

Let $G = (V, E)$ be a graph. A tree decomposition TD of G is a pair (T, X), where $T = (I, F)$ is a tree, and $X = \{X_i | i \in I\}$ is a family of subsets of V, one for each node (vertex) of T, such that:

1. $\bigcup_{i \in I} X_i = V$

2. for every edge $v, w \in E$, there is an $i \in I$ with $v \in X_i$ and $v \in X_i$, and

A. Dress, Y. Xu, and B. Zhu (Eds.): COCOA 2007, LNCS 4616, pp. 335–342, 2007.

3. for all $i, j, k \in I$, if j is on the path from i to k in T, then $X_i \bigcap X_k \subseteq X_j$.

The treewidth of a tree decomposition $((I, F), \{X_i | i \in I\})$ is max $|X_i| - 1$ ($i \in I$). The treewidth of a graph G, denoted by $tw(G)$, is the minimum width over all possible tree decomposition of G. The vertices of a tree in a tree decomposition are usually called nodes to avoid confusion with the vertices of a graph. If a vertex v or the end points of an edge e are contained in X_i for some node i of a tree decomposition, we also say node i contains v or e. An example of a graph G of treewidth two and a tree decomposition of width two of the graph is given in Fig.1.

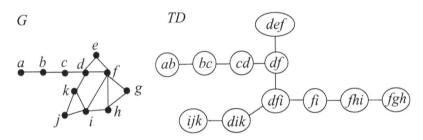

Fig. 1. A graph G of treewidth two and a tree decomposition of width two of G

2   Preliminary Results

Lemma 1. *(Fluiter[11]) Let G be a graph and $TD = (T, X)$ a tree decomposition of G.*

1. Let $u, v \in V(G)$ and let $i, j \in I$ be such that $u \in X_i$ and $v \in X_j$. Then each node on the path from i to j in T contains a vertex of every path from u to v in G.

2. For each connected subgraph G' of G, the nodes in T which contain a vertex of G' induce a subtree of T.

Lemma 2. *(Fluiter[11]) Let G be a graph. There is a rooted binary tree decomposition of minimum width of G with $O(n)$ nodes.*

3   Analysis

When $k \leq 4$, Sanders[9] gave a linear-time algorithm to determine whether a given graph $G = (V, E)$ has treewidth at most k and, if so, outputing a tree decomposition of G. Let $TD = (T, X)$ be a tree decomposition of width k for G, with $T = (I, F)$ and $X = \{X_i | i \in I\}$. The proof of Lemma 1 indicates that we can turn TD into a rooted binary tree decomposition of width k for G[11].

The contraction operation removes two adjacent vertices v and w and replaces them with one new vertex that is made adjacent to all vertices that were adjacent to v and w.

A tree decomposition (X, T) of treewidth k is smooth if for all $i \in I$, $|X_i| = k + 1$ and for all $(i, j) \in F$, $|X_i \cap X_j| = k$. Any tree decomposition of a graph G can be transformed to a smooth tree decomposition of G with the same treewidth[11]. Apply the following operations until none is possible:

1. If for $(i, j) \in F$, $X_i \subseteq X_j$, then contract the edge (i, j) in T and take as the new node $X_{j'} = X_j$.

2. If for $(i, j) \in F$, $X_i \subseteq X_j$ and $|X_j| < k+1$, then choose a vertex $v \in X_i - X_j$ and add v to X_j.

3. If for $(i, j) \in F$, $|X_i| = |X_j| = k + 1$ and $|X_i - X_j| > 1$, then subdivide the edge (i, j) in T; let i' be the new node; choose a vertex $v \in X_i - X_j$ and a vertex $w \in X_j - X_i$, and let $X_{i'} = X_i - \{v\} \bigcup \{w\}$.

After performing the above two operations, we can turn TD into a smooth rooted binary tree decomposition of width k for G. See Fig.2.

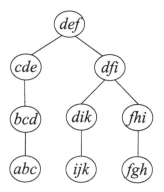

Fig. 2. A smooth binary tree decomposition of the above graph G

Lemma 3. *(Fluiter[11]) If (X, T) is a smooth tree-decomposition of $G = (V, E)$ with treewidth k, then $|I| = |V| - k$.*

Theorem 1. *(X, T) is a smooth tree-decomposition of $G = (V, E)$ with treewidth k, if for $\forall\, i, j \in I$, j is a i's child, $v = X_j - X_i$, then v must not be included in X_m, where $m \in I$ such that $m \neq j$ and m is not a j's descendant.*

Proof. Obviously.

We have a rooted binary tree decomposition $TD = (T, X)$ of width k for G with $T = (I, F)$ and $X = \{X_i | i \in I\}$. Let r denote the root of T. For each i, let
$\quad Y_i = \{v \in X_j | j = i \text{ or } j \text{ is a descendant of } i \text{ in } T\}$,
$\quad Y_i' = \{v \in X_j | j = i \text{ or } j \text{ is } i\text{'s left child in } T \text{ or } j \text{ is a descendant of } i\text{'s left child in } T\}$
and let $G_i = G[Y_i]$, $G_i' = G[Y_i']$. Note that $G_r = G$. For each $i \in I$, a table S_i is computed which contains information about the graph G_i. Obviously S_i contains the following properties:

1. For each node $i \in I$, the minimum all-ones problem can be solved for G_i solely from the information in table S_i when S_i is calculated out.

2. For each leaf node $i \in I$, S_i can be computed from $G[X_i]$.

3. For each internal node $i \in I$, S_i can be computed from $G[X_i]$ and the tables of i's children in the tree.

Suppose $i_1, i_2, ..., i_{k+1}$ are the vertices in node i. S_i is a set of all the tuples $(x(i_1), x(i_2), ..., x(i_{k+1}), s(i_1), s(i_2), ..., s(i_{k+1}), num)$. We use $x(v)$ to denote whether v is pressed or not, $v \in \{i_1, i_2, ..., i_{k+1}\}$. The denotation $x(v) = Y$ means v is pressed and $x(v) = N$ means v is not pressed. And $s(v) = 1$ denotes v is on and 0 denotes off in G_i under the choice $(x(i_1), x(i_2), ..., x(i_{k+1}))$. The value of num records the number of the vertices in G_i that are pressed. So one tuple in S_i records one choice of $i_1, i_2, ..., i_{k+1}$ and the corresponding state $s(i_1), s(i_2), ..., s(i_{k+1})$ in G_i. It is straightforward that for every $i \in I$ S_i has the above three properties.

Now we describe how to calculate S_i. Dynamic programming on the tree decomposition T is applied. If the nodes in T have the largest depth, they are called nodes with level zero. We divide the nodes in T into the following three types.

Type I. Nodes with no children(leaves). Suppose i is such a node. Enumerate all the combinations of every vertices in node i that are pressed or not and record the states correspondingly.

$S_i = \{t | t.x(v) \in \{N, Y\}$, if the number of the vertices in G_i which are pressed adjacent or equal to v is odd $t.s(v) = 1$ otherwise $t.s(v) = 0$, $t.num$ is the number of v that $t.x(v) = y$. $v \in \{i_1, i_2, ..., i_{k+1}\}\}$

There are at most $2^{k+1} = 2^5 = 32$ tuples in S_i ($k \leq 4$).

Type II. Nodes with only one child. Suppose i is such a node, j is a i's child. Let $X_j = \{i_1, i_2, ..., i_{k+1}\}$, $X_i = \{i_2, i_3, ..., i_{k+2}\}$. $M(t) = \{v | t \in S_j$ such that $t.x(v) = Y$ and $t.s(i_1) = 1\}$. Since $X_j - X_i = \{i_1\}$, then there must exit t such that $M(t)$ belong to the optimal solution to the all-ones problem to G_j since $s(i_1) = 1$. So we remain all the values of $x(v)$ in X_j where the corresponding $s(i_1) = 1$, $v \in X_i \cap X_j = \{i_2, ..., i_{k+1}\}$.

$S_i = \{t | t.x(i_j) = t'.x(i_j)$ for all $2 \leq j \leq k + 1$, $(t.x(i_{k+2}) = N$ and $t.num = t'.num)$ or $(t.x(i_{k+2}) = Y$ and $t.num = t'.num + 1)$ where there exits a tuple $t' \in S_j$ and $t'.s(i_1) = 1$. For all $2 \leq j \leq k + 2$, $t.s(i_j) = 1$ if the number of the vertices in G_i which are pressed adjacent or equal to i_j is odd otherwise $t.s(i_j) = 0$. $\}$.

Type III. Nodes with two children. Suppose i is such a node, il is its left child and ir is the right child. Of course, S_{il}, S_{ir} have been calculated already. The procedure of computing S'_i which contains information about the graph G'_i is

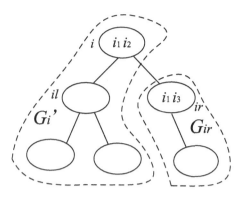

Fig. 3. G'_i and G_{ir}

the same as Type II. Then we describe how to combine S'_i and S_{ir} to get S_i, denoted by $S_i = S'_i + S_{ir}$.

(1). $k = 1$. $|X_i| = 2(i \in I)$. Suppose $X_i = \{i_1, i_2\}$, $X_{ir} = \{i_1, i_3\}$, so $X_i \cap X_{ir} = \{i_1\}$.

$S_i = S'_i + S_{ir} = \{t | \exists a \in S'_i$ and $b \in S_{ir}$ such that $a.x(i_1) = b.x(i_1)$ and $b.s(i_3) = 1$, then $t.x(i_1) = a.x(i_1)$, $t.x(i_2) = a.x(i_2)$, $t.s(i_1) = (a.x(i_1), a.s(i_1)) \oplus (b.x(i_1), b.s(i_1))$, $t.s(i_2) = a.s(i_2)$ and $t.num = a.num \}$

We define $(a.x(i_1), a.s(i_1)) \oplus (b.x(i_1), b.s(i_1))$ as the following:
$(N, 0) \oplus (N, 0) = 0$, $(N, 0) \oplus (N, 1) = 1$,
$(N, 1) \oplus (N, 0) = 1$, $(N, 1) \oplus (N, 1) = 0$,
$(Y, 0) \oplus (Y, 0) = 1$, $(Y, 0) \oplus (Y, 1) = 0$,
$(Y, 1) \oplus (Y, 0) = 0$, $(Y, 1) \oplus (Y, 1) = 1$.

Note that $i_1 \in G'_i$ and $i_1 \in G_{ir}$. The state of i_1 after combination of S_i and S'_i is according to different cases in G_i and G'_i that whether i_1 is pressed or not and its states. For example $(Y, 0) \oplus (Y, 1) = 0$ means that i_1 in G'_i is pressed and the state is off, i_1 in G_{ir} is pressed and the state is on and the state of i_1 after combination is off. Obviously only when i_1 are both pressed or both not pressed in G'_i and G_{ir}, they can combine. In G'_i, i_1 is pressed and off, so in G'_i the number of vertices which are pressed adjacent to i_1 is odd. In G_{ir}, i_1 is pressed and on, so in G_{ir} the number of vertices which are pressed adjacent to i_1 is even. Then in G_i, the number of vertices which are pressed adjacent to i_1 is odd. Considering that i_1 itself is pressed, i_1 is off. That is $(Y, 0) \oplus (Y, 1) = 0$.

(2). $k = 2$. $|X_i| = 3(i \in I)$. Suppose $X_i = \{i_1, i_2, i_3\}$, $X_{ir} = \{i_1, i_2, i_4\}$, so $X_i \cap X_{ir} = \{i_1, i_2\}$.

(2.1)i_1, i_2 are adjacent
$S_i = \{t | \exists a \in S'_i$ and $b \in S_{ir}$ such that $a.x(i_1) = b.x(i_1)$, $a.x(i_2) = b.x(i_2)$ and $b.s(i_4) = 1$, then $t.x(i_1) = a.x(i_1)$, $t.x(i_2) = a.x(i_2)$, $t.x(i_3) = a.x(i_3)$, $t.s(i_1)$

$= (a.x(i_1), a.x(i_2), a.s(i_1)) \oplus (b.x(i_1), b.x(i_2), b.s(i_1)), t.s(i_2) = (a.x(i_1), a.x(i_2),$
$a.s(i_2)) \oplus (b.x(i_1), b.x(i_2), b.s(i_2)), t.s(i_3) = a.s(i_3)$ and $t.num = a.num.\}$

We define $(a.x(i_1), a.x(i_2), a.s(i_1)) \oplus (b.x(i_1), b.x(i_2), b.s(i_1))$ and $(a.x(i_1),$
$a.x(i_2), a.s(i_2)) \oplus (b.x(i_1), b.x(i_2), b.s(i_2))$ as the following:

$(N, N, 0) \oplus (N, N, 0) = 0, (N, N, 0) \oplus (N, N, 1) = 1,$
$(N, N, 1) \oplus (N, N, 0) = 1, (N, N, 1) \oplus (N, N, 1) = 0,$
$(Y, N, 0) \oplus (Y, N, 0) = 1, (Y, N, 0) \oplus (Y, N, 1) = 0,$
$(Y, N, 1) \oplus (Y, N, 0) = 0, (Y, N, 1) \oplus (Y, N, 1) = 1,$
$(N, Y, 0) \oplus (N, Y, 0) = 1, (N, Y, 0) \oplus (N, Y, 1) = 0,$
$(N, Y, 1) \oplus (N, Y, 0) = 0, (N, Y, 1) \oplus (N, Y, 1) = 1,$
$(Y, Y, 0) \oplus (Y, Y, 0) = 0, (Y, Y, 0) \oplus (Y, Y, 1) = 1,$
$(Y, Y, 1) \oplus (Y, Y, 0) = 1, (Y, Y, 1) \oplus (Y, Y, 1) = 0$

$(2.2)i_1, i_2$ are not adjacent
$S_i = \{t| \exists a \in S'_i$ and $b \in S_{ir}$ such that $a.x(i_1) = b.x(i_1), a.x(i_2) = b.x(i_2)$ and
$b.s(i_4) = 1$, then $t.x(i_1) = a.x(i_1), t.x(i_2) = a.x(i_2), t.x(i_3) = a.x(i_3), t.s(i_1) =$
$(a.x(i_1), a.x(i_2), a.s(i_1)) \otimes (b.x(i_1), b.x(i_2), b.s(i_1)), t.s(i_2) = (a.x(i_1), a.x(i_2),$
$a.s(i_2)) \otimes (b.x(i_1), b.x(i_2), b.s(i_2)), t.s(i_3) = a.s(i_3)$ and $t.num = a.num. \}$

We define $(a.x(i_1), a.x(i_2), a.s(i_1)) \otimes (b.x(i_1), b.x(i_2), b.s(i_1))$ and $(a.x(i_1),$
$a.x(i_2), a.s(i_2)) \otimes (b.x(i_1), b.x(i_2), b.s(i_2))$ as the following:

$(N, N, 0) \otimes (N, N, 0) = 0, (N, N, 0) \otimes (N, N, 1) = 1,$
$(N, N, 1) \otimes (N, N, 0) = 1, (N, N, 1) \otimes (N, N, 1) = 0,$
$(Y, N, 0) \otimes (Y, N, 0) = 1, (Y, N, 0) \otimes (Y, N, 1) = 0,$
$(Y, N, 1) \otimes (Y, N, 0) = 0, (Y, N, 1) \otimes (Y, N, 1) = 1,$
$(N, Y, 0) \otimes (N, Y, 0) = 0, (N, Y, 0) \otimes (N, Y, 1) = 1,$
$(N, Y, 1) \otimes (N, Y, 0) = 1, (N, Y, 1) \otimes (N, Y, 1) = 0,$
$(Y, Y, 0) \otimes (Y, Y, 0) = 1, (Y, Y, 0) \otimes (Y, Y, 1) = 0,$
$(Y, Y, 1) \otimes (Y, Y, 0) = 0, (Y, Y, 1) \otimes (Y, Y, 1) = 1$

$(3)k = 3$. Similarly to the above two conditions.

$(4)k = 4$. Similarly to the above two conditions.

To sum up, if treewidth equals k, then $|X_i| = k + 1 (i \in I)$. Suppose $X_i =$
$\{i_1, i_2, ..., i_k, i_{k+1}\}$, $X_{ir} = \{i_1, i_2, ..., i_k, i_{k+2}\}$, so $X_i \bigcap X_{ir} = \{i_1, i_2, ..., i_k\}$.

The tuple t in $S_i = S'_i + S_{ir}$ satisfies that there exit one tuple a in S'_i and one
tuple b in S_{ir} such that $a.x(i_j) = b.x(i_j)$ where $1 \leq j \leq k$ and $b.x(i_{k+2}) = 1$, then
$t.x(i_j) = a.x(i_j)$ where $1 \leq j \leq k + 1$, $t.s(i_j) = (a.x(i_1), ..., a.x(i_k), a.s(i_j)) \oplus$
$(b.x(i_1), ..., b.x(i_k), b.s(i_j)) = 1$ if the number of the vertices in G_i which are
pressed adjacent or equal to i_j is odd otherwise $t.s(i_j) = 0$ where $1 \leq j \leq k$,
$t.s(i_{k+1}) = a.s(i_{k+1})$ and $t.num = a.num$.

Our algorithm can be depicted as follows: First compute the tables S_i for all
nodes i on level zero in T. Next use these tables to compute the tables of all

nodes on level one and so on until, finally S_r is computed. According to S_r, perform a up-bottom way of T to find the optimal solution.

4   Procedure of the Algorithm

Algorithm to Solve the Minimum all-ones problem with small treewidth

Input: A smooth rooted binary tree decomposition of graph G with tree width $k \leq 4$.

Output: An optimal solution to the all-ones problem

1. The root of T, denoted by v_{root}.
2. Compute S_i for all the nodes i on the level zero in T.
3. Perform a bottom-up way of T, for each i do
4. If $i \neq v_{root}$

If i is type II then
calculate S_i as type II;
If i is type III then
calculate S_i as type III;
5. If $i = v_{root}$

Choose the tuples t in S_i that $t.s(v) = 1$ and with the minimum $t.num$, for every $v \in X_i$.

If there are more than one situation obtain the minimum value, choose one arbitrarily.

Perform a up-bottom way of T to find the optimal solution.

Find the optimal solution

5   The Correctness and the Time Complexity

Theorem 2. *The above algorithm outputs an optimal solution to the all-ones problem and the time complexity is linear.*

Proof. If to every nodes we have enumerated all the conditions of the vertices in them, then the above algorithm outputs an optimal solution to the all-ones problem.

All the leaf nodes have enumerated all the situations.

All the type II nodes are those that have only one child. Suppose i is a type II node, j is i's child. Suppose $X_j = \{i_1, i_2, ..., i_{k+1}\}$, $X_i = \{i_2, i_3, ..., i_{k+2}\}$. $M(t) = \{v | t \in S_j$ such that $t.x(v) = Y$ and $t.s(i_1) = 1\}$. By Theorem 1, i_1 must not appear in any other nodes in the tree T. So M may belong to the solution to the all-ones problem to G_j since $s(i_1) = 1$. We keep down the values of $x(v)$ in X_j where the corresponding $s(i_1) = 1$, $v \in X_i \cap X_j = \{i_2, ..., i_{k+1}\}$. $x(i_{k+2})$ can be either 'N' or 'Y'. Then compute the corresponding $s(v)$ where $v \in \{i_2, i_3, ..., i_{k+2}\}$. So to all the type II nodes, we have enumerated all the situations.

Type III nodes are those that have two children. Suppose i is such a node, il is its left child and ir is the right child. Note that S_{il}, S_{ir} have been calculated already. According to S_{il}, compute S'_i as type II. Then combine S'_i and S_{ir} to get S_i. If S_i has included all the situations, then the algorithm is correct. According to the definition of S'_i and S_{ir}, S'_i records all the tuples of X_i that make the vertices only appear in G_{il} on. Let vertices $u_1, u_2...u_n$ be all the vertices in all the children nodes of X_{ir}. Denote the induced subgraph $H[\{u_1, u_2...u_n\}]$ by G'_r. S_{ir} records all the tuples of X_{ir} that make the vertices only appear in G'_r on. S_i records all the tuples of X_i that make all the vertices in G_{il} and G_{ir} on. Then choose the tuples t in S_i that $t.s(v) = 1$ and with the minimum $t.num$, $v \in X_i$. Perform a up-bottom way of T. Then algorithm can find the optimal solution.

Since $|X_i| \leq k + 1$, each table S_i has size $O(2^{k+1})$. Each table of a node can be computed in $O(2^{k+1})$ as the adjacency of two vertices can be checked in a constant time. After a bottom-up way of T, it needs $O(2^{k+1}n)$ time totally. Performing a up-bottom way of T also needs $O(2^{k+1}n)$ time. So the time complexity is $O(n)$.

References

1. Bondy, J.A., Murty, U.S.R.: Graph theory with application. Macmillan, London (1976).
2. Barua,R., Ramakrishnan,S.: σ-game, σ^+-game and two dimensional additive cellular automata. Theoret. Comput. Sci. **154** (1996) 349–366.
3. Dodis,Y., Winkler, P.: Universal configurations in light-flipping games. Proceedings of 12-th Annuaql ACM/SIAM Symposium on Discrete Algorithms(SODA), January 2001. 926–927.
4. Galvin, F.: Solution to problem 88-8. Math.Intelligencer. **11** (1989) 31–32.
5. Sutner,K.: Problem 88-8 , Math.Intelligencer. **10**(1988).
6. Sutner,K.: Linear Cellular automata and the Garden-of-Eden. Math.Intelligencer. **11** (1989) 49–53.
7. Sutner,K.: The σ game and cellular automata. Amer.Math.Monthly. **97** (1990) 24–34.
8. Sutner,K.: σ-automata and Chebyshev-polynominals. Theoret. Comput. Sci. **230** (2000) 49–73.
9. Daniel, P., Sanders: On linear recognition of tree-width at most four. SIAM J.Discrete Math. **9** (1995) 101–117.
10. Chen,Y.C., Li,X., Wang,C., Zhang X.: The minimum all-ones problem for trees. SIAM J.COMPUT. **33** (2003) 379–392.
11. Babette de Fluiter: Algorithms for Graph of Small Treewidth. France (1997).

An Exact Algorithm Based on Chain Implication for the Min-CVCB Problem[*]

Jianxin Wang, Xiaoshuang Xu, and Yunlong Liu

College of Information Science and Engineering, Central South University
Changsha 410083, China
jxwang@mail.csu.edu.cn, xxshuang@126.com, ylonglew@yahoo.com.cn

Abstract. The constrained minimum vertex cover problem on bipartite graphs (the Min-CVCB problem), with important applications in the study of reconfigurable arrays in VLSI design, is an NP-hard problem and has attracted considerable attention in the literature. Based on a deeper and more careful analysis on the structures of bipartite graphs, we develop an exact algorithm of running time $O((k_u+k_l)|G|+1.1892^{k_u+k_l})$, which improves the best previous algorithm of running time $O((k_u + k_l)|G| + 1.26^{k_u+k_l})$ for the problem.

1 Introduction

With the development of VLSI technology, the scale of electric circuit chip becomes larger and larger, and the possibility of introducing defects also increases along with the manufacture craft. With the increasing in the chip integration, it is not allowed that the wrong memory element appears in the manufacture process. A better solution is to use reconfigurable arrays. A typical reconfigurable memory array consists of a rectangular array plus a set of k_u spare rows and k_l spare columns. A defective element is repaired by replacing the row or the column containing the element with a spare row or a spare column. Therefore, to repair a reconfigurable array with defective elements, we need to decide how the rows and columns in the array are selected and replaced by spare rows and columns. The constraint here is that we only have k_u spare rows and k_l spare columns. It has now become well-known that this problem can be formulated as a constrained minimum vertex cover problem on bipartite graphs [1], as follows.

Definition 1 (Constrained minimum vertex cover in bipartite graphs (Min-CVCB)). *Given a bipartite graph $G = (V, E)$ with the vertex bipartition $V = U \cup L$ and two integers k_u and k_l, determine whether there is a minimum vertex cover of G with at most k_u vertices in U and at most k_l vertices in L.*

[*] This research is supported by the National Natural Science Foundation of China (60433020), the Program for New Century Excellent Talents in University (NCET-05-0683) and the Program for Changjiang Scholars and Innovative Research Team in University (IRT0661).

A. Dress, Y. Xu, and B. Zhu (Eds.): COCOA 2007, LNCS 4616, pp. 343–353, 2007.
© Springer-Verlag Berlin Heidelberg 2007

The problem is NP-complete [9], therefore has no efficient algorithms in general. On the other hand, in practice the number of spare rows and spare columns is much smaller than the size of the reconfigurable array: typically, a reconfigurable array is a 1000×1000 matrix plus 20 spare rows and 20 spare columns [1]. Therefore, it is practically important, and theoretically interesting, to develop efficient algorithms for the Min-CVCB problem, assuming k_u and k_l much smaller than the size of the graph G.

Hasan and Liu [1] introduced the concept of critical set to develop a branch-and-bound algorithm for solving the Min-CVCB problem, based on the A^* algorithm [2]. No explicit analysis was given in [1] for the running time of the algorithm, but it is not hard to see that in the worst-case the running time of the algorithm is at least of order of $2^{k_u+k_l} + mn^{1/2}$. Following the work in [1], the Min-CVCB problem has been extensively studied in last two decades. Most of these studies were focused on heuristic algorithms for the problem [3-6].

More recently, people have become interested in developing parameterized algorithms for the Min-CVCB problem [8-9]. Fernau and Niedermeier [8] used a branching search technology and developed an algorithm with running time $O((k_u + k_l)n + 1.3999^{k_u+k_l})$ for the problem. Chen and Kanj [9] proved that the Min-CVCB problem is NP-complete, and developed an improved algorithm of running time $O((k_u + k_l)|G| + 1.26^{k_u+k_l})$ for the problem. The algorithm given in [9] made use of a number of classical results in matching theory and recently developed techniques in parameterized algorithms, which is currently the best algorithm for the problem.

In this paper, we perform a deeper and more careful analysis on related structures of bipartite graphs. Based on the analysis, we effectively integrate the techniques of chain implication, branching search, and dynamic programming, and develop an improved parameterized algorithm EACI of running time $O((k_u + k_l)|G| + 1.1892^{k_u+k_l})$ for the Min-CVCB problem.

2   Related Lemmas

For further discussion of our algorithm EACI, we first give some definitions and describe certain known results that are related to the Min-CVCB problem and to our algorithm.

Definition 2 (Bipartite graph). *A graph G is bipartite if its vertex set can be partitioned into two sets U (the "upper part") and L (the "lower part") such that every edge in G has one endpoint in U and the other endpoint in L. A bipartite graph is written as $G = (U \cup L, E)$ to indicate the vertex bipartition. The vertex sets U and L are called the U-part and the L-part of the graph. A vertex is a U-vertex (resp. an L-vertex) if it is in the U-part (resp. the L-part) of the graph.*

Let $G = (U \cup L, E)$ be a bipartite graph with a perfect matching. The graph G is *elementary* if every edge in G is contained in a perfect matching in G. It is known that an elementary bipartite graph has exactly two minimum vertex covers, namely U and L, without any other possibility [10].

Lemma 1. [9] *The time complexity for solving an instance* $< G; k_u, k_l >$ *of Min-CVCB problem, where G is a bipartite graph of n vertices and m edges, is bounded by $O(mn^{1/2} + t(k_u + k_l))$, where $t(k_u + k_l)$ is the time complexity for solving an instance $< G'; k'_u, k'_l >$ of Min-CVCB, with $k'_u < k_u$, $k'_l < k_l$ and G' having perfect matchings and containing at most $2(k'_u + k'_l)$ vertices.*

Lemma 2. (The Dulmage-Mendelsohn Decomposition theorem [10]). *A bipartite graph $G = (U \cup L, E)$ with perfect matchings can be decomposed and indexed into elementary subgraphs $B_i = (U_i \cup L_i, E_i)$, $i = 1, 2, \ldots r$, such that every edge in G from a subgraph B_i to a subgraph B_j with $i < j$ must have one endpoint in the U-part of B_i and the other endpoint in the L-part of B_j. Such a decomposition can be constructed in time $O(|E|^2)$.*

The elementary subgraphs B_i will be called (elementary) *blocks*. The block B_i is a *d-block* if $|U_i| = |L_i| = d$. Edges connecting vertices in two different blocks will be called *inter-block edges*. Let B_i be a block. The number λ_{in} of blocks B_j such that $i \neq j$ and there is an inter-block edge from the U-part of B_i to the L-part of B_j is called the *in-degree* of B_i. Similarly, the number λ_{out} of blocks B_j such that $i \neq j$ and there is an inter-block edge from the U-part of B_j to the L-part of B_i is called the *out-degree* of B_i.

Lemma 3. [10] *Let G be a bipartite graph with perfect matchings, and let B_1, ..., B_r be the blocks of G given by the Dulmage-Mendelsohn Decomposition. Then any minimum vertex cover for G is the union of minimum vertex covers of the blocks $B_1, B_2, \ldots, B_r$.*

By Lemma 1, in order to solve a general instance $\langle G; k_u, k_l \rangle$ of the Min-CVCB problem, we only need to concentrate on a "normalized" instance $\langle G'; k'_u, k'_l \rangle$ of the problem, in which G' has a perfect matching and contains at most $2(k'_u + k'_l)$ vertices. By Lemma 2, the graph G' with perfect matchings can be decomposed and represented as a directed acyclic graph (DAG) D in which each node corresponds to a block in G' and each edge corresponds to a group of inter-block edges from the U-part of a block to the L-part of another block. By Lemma 3, a minimum vertex cover of the graph G' is the union of minimum vertex covers of the blocks B_1, ..., B_r. All these are very helpful and useful when we construct a desired minimum vertex cover in the originally given bipartite graph G.

3 The Strategy for Reducing the Search Space in Algorithm EACI

Algorithm EACI is based on the DAG D constructed above and its execution is depicted by a search tree whose leaves correspond to the potential constrained minimum vertex covers K (shortly K) of the graph G with at most k_u U-vertices and at most k_l L-vertices. For a given instance of Min-CVCB problem, let $f(k_u + k_l)$ be the number of leaves in the search tree, if in a step we can break the original problem into two sub-problems, and in each sub-problem the parameter scale can reduce a and b respectively, then we would establish a recurrence relation $f(k_u + k_l) \leq f(k_u + k_l - a) + f(k_u + k_l - b)$. When constructing search tree, we could include some blocks' U-part or L-part into K, until in a certain step breaks DAG D's NP-Hard structure, then uses dynamic programming technology to solve the surplus partial in the polynomial time.

In order to speed up the searching process, we will apply the technology of chain implication[9] , which makes full use of the block's adjacency relations to speed up the searching process significantly. Let $[B'_1, B'_2, \ldots, B'_h]$ be a path in the DAG D. If we include the L-part of the block B'_1 in K, then the U-part of the B'_1 must be excluded from K. Since there is an edge in G from the U-part of B'_1 to L-part of the block B'_2, we must also include the L-part of the block B'_2 in K, which, in consequence, will imply that the L-part of the block B'_3 must be in K, and so on. In particular, the L-part of the block B'_1 in K implies that the L-parts of all blocks $B'_2, \ldots, B'_h$ on the path must be in K. Similarly, the U-part of the block B'_h in K implies that U-parts of all blocks $B'_1, .., B'_{h-1}$ must be in K. This technology enables us to handle many cases very efficiently.

The particular operation of the algorithm is to list all the possible adjacency of the blocks in which we branch in the search process. First we analysis the corresponding branching of the blocks whose weight is no less than 4, then analysis all the possible joint of block whose weight is 3 to establish the searching tree. For the block whose weight is 3, first listing the possible joint of the block in a case-by-case exhaustive manner, and then makes the best of bounded search-trees technology to construct new recurrence relations. Let $\lambda_{in}(B_i)$ be the in-degree of the block B_i, $\lambda_{out}(B_i)$ be the out-degree of B_i, $w(B_i)$ be the weight of B_i, and $w(P_{B_i})$ be the weight of all the blocks that have a directed path to the block B_i. We would divide it into two situations as follows according to the block B_0's weight.

1. $w(B_0) \geq 4$. Since the constrained minimum vertex cover K of the DAG D either contains the entire U_i-part and is disjoint from the L_i-part, or contains the entire L_i-part and is disjoint from the U_i-part of the block B_i, we branch in this case by either including the entire U_i-part in K (and remove the L_i-part from the graph) or including the entire L_i-part in K (and removing the U_i-part from the graph). In each case, we add at least 4 vertices in K and remove block B_0 from DAG D. Thus, this branch satisfies the recurrence relation

$$f(k_u + k_l) \leq 2f(k_u + k_l - 4) \tag{1}$$

2. $w(B_0) = 3$. According to the value of in-degree and out-degree of block B_0, we would divide it into four situations as follows.

2.1 $\lambda_{in}(B_0) \geq 1$ and $\lambda_{out}(B_0) \geq 1$. If we include the U-part of B_0 in K, it forces at least $3 + \lambda_{in}(B_0)$ vertices in K by the chain implication. If we include the L-part of B_0 in K, it also forces at least $3 + \lambda_{out}(B_0)$ vertices in K. Thus in this case, the branching satisfies recurrence relation (1).

2.2 $\lambda_{out}(B_0) \geq 1$ and $\lambda_{in}(B_0) = 0$. According to the out-degree of B_0 and $w(P_{B_0})$, we would divide it into three situations as follows.

2.2.1 $w(P_{B_0}) \geq 3$. If we included the U-part of B_0 in K, it forces at least 3 vertices in K by the "chain implication". If we include the L-part of B_0 in K, it forces at least 6 vertices in K. Thus, this branching satisfies the recurrence relation

$$f(k_u + k_l) \leq f(k_u + k_l - 3) + f(k_u + k_l - 6) \qquad (2)$$

2.2.2 $w(P_{B_0}) = 2$. In this case, all the connections of block B_0 have eight cases shown in Fig.1, after excluding B_0 's isolated connections. From Fig.1(a) to Fig.1(g), let the two connected blocks of B_0 be B_1 and B_2. When $\lambda_{in}(B_1) \geq 2$, the connected blocks is B_3, when $\lambda_{in}(B_2) \geq 2$, the connected blocks is B_4. In Fig.1(h), let the blocks connected with B_1 is B_3. We'll give an analysis of how to establish a bounded search tree in the following.

2.2.2.1 In Fig.1(a), B_0 is connected with two connected blocks B_1 and B_2 whose weight are 1, and $\lambda_{in}(B_1) \geq 2$, $\lambda_{in}(B_2) \geq 3$. When $w(B_3) > 1$, the time complexity of the branching is lower than the one when $w(B_3) = 1$, so we only need to consider the situation when $w(B_3) = 1$. It is also the same in the following context. In general, we only have to analyze the equal situation. When the situation of $\lambda_{in}(B_1) \geq 1$ is analyzed, the time complexity of branching is also lower than the situation when $\lambda_{in}(B_1) = 1$. Also, in the following, if it is required to analyze the in-degree or out-degree of a block whether the value is larger or equal to a constant, we only have to analyze the equal situation is enough.

Let the block B_1 be the core of branching: if the U-part of block B_1 is in K, it can be concluded by the chain implication that: the U-part of the block B_0 and B_3 are also in K, thus it equals that 5 vertices are included in the K. If the L-part vertices of block B_1 are in K, the block B_0 and B_2 become "isolated block". Thus, it equals that 5 vertices are included in the K. So, the branching is at least (5, 5), and the corresponding recurrence is just as formula

$$f(k_u + k_l) \leq 2f(k_u + k_l - 5) \qquad (3)$$

2.2.2.2 In Fig.1(b), B_0 is connected with two connected blocks B_1 and B_2 whose weight are 1, and $\lambda_{in}(B_1) \geq 2, \lambda_{in}(B_2) = 2$.

Let B_2 be the core of branching, the problem under this situation is exactly the same as the (2.2.2.1), so the analysis is identical, and it can be branched at least (6, 5), and the corresponding recurrence is just as formula

$$f(k_u + k_l) \leq f(k_u + k_l - 6) + f(k_u + k_l - 5) \qquad (4)$$

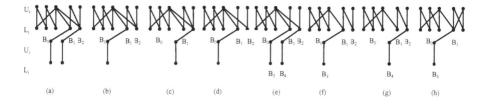

Fig. 1. All possible connections in DAG D when $w(P_{B_0}) = 2$

2.2.2.3 In Fig.1(c), B_0 is connected with two connected blocks B_1 and B_2 whose weight are 1, and $\lambda_{in}(B_1) = 1, \lambda_{in}(B_2) \geq 3$.

Let B_2 be the core of branching, the problem under this situation is exactly the same as the (2.2.2.1), so the analysis is identical, so it can be branched at least (6, 5), and the corresponding recurrence is just as formula (4).

2.2.2.4 In Fig.1(d), B_0 is connected with two blocks B_1 and B_2 with no connections whose weight are 1, and $\lambda_{in}(B_1) \geq 2$, $\lambda_{in}(B_2) = 1$.

Let B_1 be the core of branching, if U-part of block B_1 is in K, it can be concluded from the chain implication that: the U-part of the block B_0 and B_4 are also in K. thus, the K contains at least 5 vertices and the block B_2 becomes the "isolated block". Thus it equals that 6 vertices are included in the K. If the L-part vertices of block B_1 are in K, the block B_0 and B_2 become "isolated block". Thus, it means that 5 vertices are included in the K. So the branching is at least (6, 5), and the corresponding recurrence is just as formula (4).

2.2.2.5 In Fig.1(e), B_0 is connected with two blocks B_1 and B_2 whose weight are 1 with no connections, and $\lambda_{in}(B_1) \geq 2$, $\lambda_{in}(B_2) \geq 2$.

From the "vertex folding" in Ref.[11], to contain the edges among the blocks B_0, B_1, B_2, B_3, B_4, one is to make the U-part vertices be included in K, which is equal to putting at least 4 vertices into K(corresponding to the situation that $B_3$and B_4 are in the same block, and Fig.1(de) gives an exact connection); the other is to make the L-part vertices be included in K, and it will make the block B_0 become the "isolated block", thus, it equals to includes at least 5 vertices into K. so branch is (4, 5), and the corresponding recurrence relation is formula

$$f(k_u + k_l) \leq f(k_u + k_l - 4) + f(k_u + k_l - 5) \tag{5}$$

2.2.2.6 In Fig.1(f), B_0 is connected with two blocks B_1 and B_2 which has no connections, and $\lambda_{in}(B_1) \geq 2$, $\lambda_{in}(B_2) = 2$.

Let B_1 be the core of branching, if the U-part of block B_1 are in K, it can be concluded by the chain implication that: the U-part of the block B_0 and $B_2$are also in K. Thus, the K contains at least 5 vertices and the block B_2 becomes the "isolated block". Thus it equals that 6 vertices are included in the K. If the L-part vertices of block B_1 are in K, it can be contained at least 2 vertices in K and B_0 becomes "isolated block". Thus it equals that 5 vertices are included in K. So it can be branched at least (6, 5), and the corresponding recurrence is just as formula (4).

2.2.2.7 In Fig.1(g), B_0 is connected with two blocks B_1 and B_2 whose weight is 1 and has no connections, and $\lambda_{in}(B_1) = 1$, $\lambda_{in}(B_2) \geq 2$.

To make the block B_2 as the core of branching, the problem under this situation is exactly the same as the (2.2.2.6), so the analysis is identical, so the branching is at least (6, 5), and the corresponding recurrence is just as formula (4).

2.2.2.8 In Fig.1(h), B_0 is connected with a blocks B_1 whose weight is 2, and $\lambda_{in}(B_1) \geq 2$.

Let B_1 be the core of branching, the problem under this situation is exactly the same as the (2.2.2.6), so the analysis is identical, and the branching is at least (6, 5), the corresponding recurrence is just as formula (4).

2.2.3 $w(P_{B_0}) = 1$

From $w(P_{B_0}) = 1$, we know that B_0 is connected with a blocks B_1 whose weight is 1, let another block that connects with B_1 is B_2 (when $\lambda_{in}(B_1)$, block B_0, B_1 become "isolated block"). Let B_1 be the core of branching, the problem under this situation is exactly the same as (2.2.2.6), so the analysis is identical, and the branching is at least (4, 5), the corresponding recurrence is just as formula (5).

2.3 $\lambda_{in}(B_0) \geq 1$ and $\lambda_{out}(B_0) = 0$

Under this situation, all kinds of connection in the DAG D is entirely symmetry like (2.2), so the handling method is just the same and we can get the same recurrence relation.

2.4 $\lambda_{in}(B_0) = 0$ and $\lambda_{out}(B_0) = 0$

The block B_0 becomes the "isolated block" and we can make full use of the dynamic programming technology to solve it in polynomial time in the fourth part.

Considering all the recurrence relations above, it is obvious that formula (1) is the strictest one. So a theorem can be presented as follows:

Theorem 1. *When a block B_0 in the DAG D satisfies the inequality $w(B_0) \geq 3$, the branching recurrence relation brought out by the branching process at least satisfies the formula $f(k_u + k_l) \leq 2f(k_u + k_l - 4)$.*

4   Algorithm EACI-dyn

After processing the blocks of weight larger than 3, the remain DAG D contains only isolated blocks of weight 3 and connected subgraphs that are composed by blocks of weight 1 or 2. We can solve the Min-CVCB problem on this structure by dynamic programming. The corresponding algorithm is EACI-dyn. Let the connected subgraphs in the remaining DAG D be G'_i, $1 \leq i \leq r$ (r be the number of the connected subgraphs). Let $G_0 = G'_1 + G'_2 + \ldots + G'_r$, and let the number of vertices in the connected subgraph G_i be $2n_i$. Therefore, the total number of vertices in the graph G_0 is $2n_0 = 2n_1 + \cdots + 2n_r$. We show that all the possible minimum vertex covers in each connected subgraph can be enumerated in polynomial time. Then the dynamic programming algorithm is used to find the minimum vertex cover in G_0 satisfying the constraints.

After enumerating all possible minimum vertex covers in each connected subgraph G'_i, the next step is to find a minimum vertex cover of size (k_u, k_l) in the graph G_0. Obviously, G_0 has the minimum vertex cover of size (k_u, k_l) if and only if each connected subgraph G'_i has a minimum vertex cover of size $(k_u^{(i)}, k_l^{(i)})$, such that $k_u^{(1)} + \ldots + k_u^{(r)} \leq k_u$, and $k_l^{(1)} + \ldots + k_l^{(r)} \leq k_l$.

The procedure that finds a minimum vertex cover of size (k_u, k_l) in the graph G_0 is as follows: let $\bar{c} = c_1 + \ldots + c_i$, $1 \leq i \leq r$, and $A[1 \ldots r, 0 \ldots k_u]$ be a matrix of size $r^*(k_u + 1)$. Each element $A[i, j]$ in the matrix is to record a minimum vertex cover of size $(j, \bar{c} - j)$ in the graph $G'_1 + \cdots + G'_i$. The matrix A can be constructed by the dynamic programming algorithm in Fig. 2.

Input: the connected graphs of $G'_1, G'_2 \ldots G'_r$ after section 3's branching
Output: a minimum vertex cover K of G with at most k_u U-vertices and at
       most k_l L-vertices if such a minimum vertex cover exists
1. list all the possible minimum vertex cover of $G'_1, G'_2 \ldots G'_r$;
2. **foreach** $1 \leq i \leq r, 0 \leq j \leq k_u$ **do**
     $A[i, j] = \phi$;
3. **foreach** $(k_u^{(1)}, k_l^{(1)})$-*minimum vertex cover of C'_1 of G'_1* **do**
     $A[1, k_u^{(1)}] = C'_1$;
4. **for** $i = 1 \ldots r - 1$ **do**
     for $j = 0 \ldots k_u$ **do**
       if $A[i, j] \neq \phi$ **then**
         let $[i, j] = V_u \cup V_l, V_u \subseteq U, V_l \subseteq L$);
         foreach $(k_u^{(i+1)}, k_l^{(i+1)})$-*minimum vertex cover*, $C'_{i+1} = V_u^{(i+1)} \cup V_l^{(i+1)}$ *of*
         G'_{i+1} *in the list* L_{i+1} **do**
           $A[i+1, j + k_u^{(i+1)}] = (V_u \cup V_u^{(i+1)}) \cup (V_l \cup V_l^{(i+1)})$;
5. **for** $j = 0 \ldots k_u$ **do**
     if $(j \leq k_u) \& (n_0 - j \leq k_l) \& [r, j] \neq \phi)$ **then**
       then return $A[r, j]$;
6. **return** ϕ;

Fig. 2. Algorithm. EACI-dyn.

Theorem 2. *The time complexity of the algorithm EACI-dyn is $O((k_u + k_l)k_u^2)$.*

Proof. After the branching process in section 3, the remaining DAG D is composed of isolated blocks of weight 3 and blocks of weight 1 or 2. First, all possible minimum vertex covers of each connected subgraph G'_i, $1 \leq i \leq r$, can be listed in linear time, then the matrix A can be constructed by the dynamic programming algorithm to find the constrained minimum vertex cover. In the dynamic programming algorithm, the number of the minimum vertex covers in every row L_i of the matrix A is at most k_u, and the value of the next row depends on the value of the above one, so the time complexity of constructing the matrix A is $O(rk_u^2)$, Since r be the number of the connected subgraphs, and $r \leq (k_u + k_l)$, So, the running time of the algorithm EACI-dyn is bounded by $O((k_u + k_l)k_u^2)$.

5   Putting All Together

With all the previous discussions combined, an algorithm EACI is given in Fig.3, which solves the Min-CVCB problem. We explain the steps of the algorithm as follows.

Step 1 is the initialization of the vertex cover K. Steps 2 and 3 make immediate decisions on high-degree vertices. If a U-vertices u of degree larger than k_l is not in the minimum vertex cover K, then all neighbors of u should be in K, which would exceed the bound k_l. Thus, every U-vertex of degree larger than k_l should

---

Input: a bipartite graph $G = (U, L, E)$ and two integers k_u and k_l

Output: a minimum vertex cover K of G with at most k_u U-vertices and at most k_l L-vertices, or report no such a vertex cover exists

1. $K = \phi$;
2. **foreach** U-*vertex* u *of degree larger than* k_l **do**
   include u in K and remove u from G; $k_u = k_u - 1$;
3. **foreach** L-*vertex* v *of degree larger than* k_u **do**
   include v in K and remove v from G; $k_l = k_l - 1$;
4. apply lemma 1 to reduce the instance so that G is a bipartite graph with perfect matching and with at most $2(k_u + k_l)$ vertices (with the integers k_u and k_l and the minimum vertex cover K also properly updated);
5. apply lemma 2 to decompose the graph G into elementary blocks $B_1, B_2, \ldots, B_r$, sorted topologically;
6. for connections that contain the block B_i in DAG D has weight at least 3, branching it according in section 3;
7. All other cases not in section3, we can use algorithm EACI-dyn to solve it in polynomial time in section 4;

Fig. 3. Algorithm. EACI.

be automatically included in K. Similar justification applies to L-vertices of degree larger than k_u. Of course, if k_u or k_l becomes negative in step 2 or step 3, then we should stop and claim the nonexistence of the desired minimum vertex cover. After these steps, the degree of the vertices in the graph is bounded by $k' = \max\{k_u, k_l\}$. Since now each vertex can cover at most k' edges, the number of edges in the resulting graph must be bounded by $k'(k_u + k_l) \leq (k_u + k_l)^2$, otherwise the graph cannot have a minimum vertex cover of no more than $k_u + k_l$ vertices. In step 4, Lemma 1 allows us to further reduce the bipartite graph G so that G has a perfect matching (the integers k_u and k_l are also properly reduced). The number of vertices in the graph G now is bounded by $2(k_u + k_l)$. Step 5 applies Lemma 2 to decompose the graph G into blocks. Step 6 is to analyze all the possible minimum vertex covers on the condition that the weight of the blocks in the connected sub-graphs is no less than 3, then use "chain implication" and bounded search technology to reduce the searching space in order to construct the bounded-search tree. Step 7 further analyzes the possible minimum vertex cover of the connected sub-graphs after step 6, and then applies algorithm EACI-dyn to search for the constraint minimum vertex cover.

Theorem 3. *The algorithm EACI runs in time* $O((k_u + k_l)|G| + 1.1892^{k_u+k_l})$, *i.e, the Min-CVCB problem is solvable in time* $O((k_u + k_l)|G| + 1.1892^{k_u+k_l})$.

Proof. As explained above, the algorithm EACI solves the Min-CVCB problem correctly. Thus, we only need to verify the running time of the algorithm.

It is easy to verify that the total running time of steps 1-3 of the algorithm is bounded by $O((k_u + k_l)|G|)$. Step 4 applies Lemma 1 to further reduce the bipartite graph G, and the running time of this step is bounded by $(k_u + k_l)^3$

(note that in this step, the number m of edges in the graph G is bounded by $(k_u + k_l)^2$ and the number n of vertices in the graph G is bounded by $2(k_u + k_l)$). Step 5 applies Lemma 2 to decompose the graph G into elementary bipartite subgraphs and it takes time $O(|E|^2)$. Since $|E|$ is the number of edges in G, and $|E| \leq (k_u + k_l)^2$, step 5 takes time $O((k_u + k_l)^4)$. In step 7, by Theorem 2, the running time of the algorithm EACI-dyn is bounded by $O((k_u + k_l)k_u^2)$.

The only place the algorithm EACI branches is in step 6. Let $f(k_u + k_l) = x^{k_u + k_l}$ be the function in Theorem 1. By Theorem 1, we have

$$f(k_u + k_l) \leq 2f(k_u + k_l - 4)$$

Solving this recurrence relation gives us $f(k_u + k_l) \leq 1.1892^{k_u + k_l}$. Combining all steps together, we derive that the running time of the algorithm EACI is bounded by $O((k_u + k_l)|G|) + (k_u + k_l)^3 + (k_u + k_l)^4 | + 1.1892^{k_u + k_l} + (k_u + k_l)k_u^2) = O((k_u + k_l)|G| + 1.1892^{k_u + k_l})$, i.e., the Min-CVCB problem could be solved in $O((k_u + k_l)|G| + 1.1892^{k_u + k_l})$.

6   Conclusions

In this paper, we study the Min-CVCB problem that has important applications in the area of VLSI manufacturing. We develop an improved parameterized algorithm for the problem based on a deeper and more careful analysis on the structures of bipartite graphs. We propose new techniques to handle blocks of weight bounded by 3, and use new branch search technology to reduce searching space. Our improved algorithm is achieved by integrating these new techniques with the known techniques developed by other researchers. The running time of our algorithm is $O((k_u + k_l)|G| + 1.1892^{k_u + k_l})$, compared to the previous best algorithm for the problem of running time $O((k_u + k_l)|G| + 1.26^{k_u + k_l})$.

References

1. Hasan, N., Liu, C.L.: Minimum Fault Coverage in Reconfigurable Arrays. In: Pro of the 18th International Symposium on Fault-Tolerant Computing (FTCS'88), pp. 348–353. IEEE Computer Society Press, Los Alamitos,CA (1988)
2. Niosson, N.J.: Principles of Artificial Intelligence. Tioga Publishing Co., Palo Alto, CA (1980)
3. Kuo, S.Y., Fuchs, W.: Efficient spare allocation for reconfigurable arrays. IEEE Des. Test 4, 24–31 (1987)
4. Blough, D.M., Pelc, A.: Complexity of fault diagnosis in comparison models. IEEE Trans.Comput. 41(3), 318–323 (1992)
5. Low, C.P., Leong, H.W.: A new class of efficient algorithms for reconfiguration of memery arrays. IEEE Trans. Comput. 45(1), 614–618 (1996)
6. Smith, M.D., Mazumder, P.: Generation of minimal vertex cover for row/column allocation in self-repairable arrays. IEEE Trans.Comput. 45, 109–115 (1996)
7. Downey, R., Fellows, M.: Parameterized Complexity. Springer, New York (1999)

8. Fernau, H., Niedermeier, R.: An efficient exact algorithm for constraint bipartite vertex cover. J. Algorithms 38, 374–410 (2001)
9. Chen, J., Kanj, I.A.: Constrained minimum vertex cover in bipartite graphs: complexity and parameterized algorithms. Journal of Computer and System Science 67, 833–847 (2003)
10. Lovasz, L., Plummer, M.D.: Matching Theory. In: Annals of Discrete Mathematics, vol. 29, North-Holland, Amsterdam (1986)
11. Chen, J., Kanj, I.A., Jia, W.: Vertex cover: further observations and further improvements. Journal of Algorithms, 280–301 (2001)

Arc Searching Digraphs Without Jumping

Brian Alspach[1], Danny Dyer[2], Denis Hanson[1], and Boting Yang[3]

[1] Department of Mathematics and Statistics, University of Regina
{alspach,dhanson}@math.uregina.ca
[2] Department of Mathematics and Statistics, Memorial University of Newfoundland
dyer@math.mun.ca
[3] Department of Computer Science, University of Regina
boting@cs.uregina.ca

Abstract. The arc-searching problem is a natural extension of the edge searching problem, which is to determine the minimum number of searchers (search number) needed to capture an invisible fast intruder in a graph. We examine several models for the internal arc-searching problem, in which the searchers may not "jump" from a vertex to a non-adjacent vertex. We will explore some elementary results and characterize directed graphs with small search number for the various models.

1 Introduction

Searching a graph was introduced by Parsons [12]. In Parson's *general searching model*, graphs are considered to be embedded in 3-space, and the searchers' and intruders' movements in the graph are described by continuous functions. A successful strategy is a collection of continuous functions such that for each intruder there is a time t for which one of the searchers and the intruder have the same function value. Searching graphs serve as models for important applied problems (see [3], [4] and [8]. A survey of graph searching results can be found in [1].

The concept of searching may be extended to directed graphs. With regard to definitions, we will follow [5], except as noted. We consider searching directed graphs D and the minimum number of searchers required. In most models for searching digraphs, the searchers are allowed to jump from a vertex to a non-adjacent vertex. Such games are generally characterized in terms of one of the "width" parameters, such as directed treewidth [7], Kelly-width [6], or DAG-width [11]. In this paper, we will only consider *internal* searching models in which jumping is forbidden. Other work relating to searching digraphs could be found in [10,13,14,15]. In [13,14,15], the searchers are allowed to jump.

In the following models, the specifics of searching a digraph D are as follows. Initially, all arcs of D are *contaminated* (may contain an intruder). A *search strategy* is a sequence of actions designed so that the final action leaves all arcs of D *uncontaminated* or *cleared* (do not contain an intruder). Initially, the searchers are placed on some subset of the vertex set of D. The only action available to a searcher is to move from a vertex u to a vertex v along an arc (u, v) (or possibly

A. Dress, Y. Xu, and B. Zhu (Eds.): COCOA 2007, LNCS 4616, pp. 354–365, 2007.
© Springer-Verlag Berlin Heidelberg 2007

(v, u)). If the searchers and the intruder must move in the direction of the arcs, we call this a *directed search* and the minimum number of searchers needed to clear D is the *directed search number* $s_d(D)$. If the searchers and the intruder can move with or against the direction of the arcs, we call this an *undirected search* and obtain the *undirected search number* $s_u(D)$. If the intruder must move in the direction of the arcs, but the searchers need not, we call this a *strong search* and obtain the *strong search number* $s_s(D)$. Finally, if the searchers must move in the direction of the arcs, but the intruder need not, we call this a *weak search* and obtain the *weak search number* $s_w(D)$.

We define a digraph D to be k-directed-searchable if and only if $s_d(D) \leq k$. Similarly, we define k-undirected-searchable, k-strongly-searchable, and k-weakly-searchable. Of the various models, the least interesting is the undirected digraph search, as this is identical to a search in the underlying undirected graph. It is included only for completeness.

The methods by which arcs are cleared vary by model. In a strong search, an arc (u, v) can be cleared in one of three ways: at least two searchers are placed on vertex u of arc (u, v), and one of them traverses the arc from u to v while the others remain at u; a searcher is placed on vertex u, where all incoming arcs incident with u already are cleared, with the searcher then moving from u to v; or, a searcher is placed on vertex v, and traverses the arc (u, v) in reverse, from v to u.

In a directed search, an arc (u, v) can be cleared in one of two ways: at least two searchers are placed on vertex u of arc (u, v), and one of them traverses the arc from u to v while the others remain at u; or a searcher is placed on vertex u, where all incoming arcs incident with u already are cleared, and the searcher moves from u to v.

Finally, in a weak search, an arc (u, v) can be cleared in one of two ways: at least two searchers are placed on vertex u of arc (u, v), and one of them traverses the arc from u to v while the others remain at u; or a searcher is placed on vertex u, where all arcs incident with u other than (u, v) already are cleared, and the searcher moves from u to v.

It also should be mentioned that in a strong search or a directed search, a cleared arc (u, v) is *recontaminated* if u is the head of a contaminated arc and contains no searcher. In an undirected search or a weak search, a cleared arc (u, v) is recontaminated if either of u and v is the head or tail of a contaminated arc and contains no searcher.

In a strong search or a directed search, a vertex v is *clear* if all of the incoming arcs with v as head are clear. In an undirected or weak search, a vertex v is clear if all of the arcs incident with v are clear. The following result appeared in [10].

Theorem 1. *If D is an acyclic directed graph, then $s_w(D)$ equals the minimum number of directed paths in the arc digraphs $A(D)$ of D that covers the vertices of $A(D)$.*

If vertices of the strong components D_1, D_2, ..., D_m of D partition V into sets, then this partition is called the strong decomposition of D. The *strong component digraph* or *condensation* $S(D)$ is obtained by contracting each of the strong components of D to a single vertex and deleting any parallel arcs formed. In particular, the strong component digraph is an acyclic digraph.

In Section 2 of this paper, we examine some elementary bounds on the search models introduced here. Section 3 examines the relationships between the search numbers of a digraph and the search numbers of its subdigraphs and digraph minors. In Sections 4 and 5 we characterize the 1- and 2-searchable digraphs for the various searching models, with the exception of those digraphs D with $s_s(D) = 2$, which remains open. Section 6 contains several results bounding the search number of strong digraphs. Finally, we give some further directions in Section 7.

2   Elementary Bounds

Theorem 2 follows directly from the definitions of the searching models.

Theorem 2. *If D is a digraph, then $s_s(D) \leqq s_d(D) \leqq s_w(D)$ and $s_s(D) \leqq s_u(D) \leqq s_w(D)$.*

All of these equalities can be achieved by considering a directed path. It is easy to see that $s_s(\overrightarrow{P_n}) = s_w(\overrightarrow{P_n}) = 1$. The inequalities can also be strict. In fact, there exist digraphs X and Y (see Figure 1) such that $s_s(X) < s_u(X) < s_d(X) < s_w(X)$ and $s_s(Y) < s_d(Y) < s_u(Y) < s_w(Y)$. For X, $s_s(X) = 1$, $s_u(X) = 3$, $s_d(X) = 4$, and $s_w(X) = 6$. For Y, $s_s(Y) = 2$, $s_d(Y) = 4$, $s_u(Y) = 5$, and $s_w(X) = 10$.

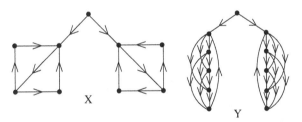

Fig. 1. The digraphs X and Y

We introduce some elementary lower bounds involving minimum indegree, paralleling similar results from searching graphs [16,17].

Theorem 3. *If D is a digraph, then $s_w(D) \geqq s_d(D) \geqq \delta^-(D) + 1$ and $s_s(D) \geqq \delta^-(D)$.*

It is natural to consider the outdegree sequence. However, this is not a useful parameter to consider from the perspective of the directed search number. An *almost transitive tournament $TT_n^\star$* is a tournament that differs from the transitive tournament TT_n only in that arc (v_n, v_1) is included instead of (v_1, v_n). For example, the almost transitive tournament $TT_7^\star$ on 7 vertices has score sequence

$1, 1, 2, 3, 4, 5, 5,$ and $s_d(TT_7^\star) = 2$. Let S be a tournament on the vertices v_i, $1 \leq i \leq 7$, with arcs (v_i, v_j) for all $i < j$, except the arcs (v_7, v_5), (v_5, v_3), and (v_3, v_1) replace the arcs (v_5, v_7), (v_3, v_5), and (v_1, v_3), respectively. Then S also has score sequence $1, 1, 2, 3, 4, 5, 5$, but $s_d(S) = 3$.

3   Subdigraphs and Digraph Minors

We begin by recalling a classical result from graph searching.

Theorem 4. *If G' is a minor of a multigraph G, then the search number of G' is less than or equal to the search number of G.*

This gives an automatic corollary for digraphs and motivates an examination of similar results for other digraph searching models.

Corollary 1. *If D' is a connected subdigraph of a digraph D, then $s_u(D') \leq s_u(D)$.*

Theorem 5. *If D' is a connected subdigraph of a digraph D, then $s_s(D') \leq s_s(D)$.*

Unlike strong searching, directed and weak searching allow for subdigraphs to have a larger search number than their superdigraphs. Consider the almost transitive tournament $TT_4^\star$. The directed search number of this tournament is 2. If we remove the arc (v_4, v_1), the resulting subdigraph has directed search number 3.

We construct a digraph G to see that the same is true for weak searching. Let u and v be two vertices. Add arcs such that there are 4 directed paths from u to v, each of which shares only the start and end vertices u and v. Then add the arc (v, u) to obtain G. It is not hard to see that $s_w(G) = 3$. However, if we remove the arc (v, u), the resulting digraph has weak search number equal to 4.

However, restricting ourselves to strong subdigraphs, we obtain the following.

Theorem 6. *If D' is a strong subdigraph of a digraph D, then $s_d(D') \leq s_d(D)$ and $s_w(D') \leq s_w(D)$.*

From Theorem 4, we also obtain the following corollary.

Corollary 2. *If D' is a digraph minor of a digraph D, then $s_u(D') \leq s_u(D)$ and $s_w(D') \leq s_w(D)$.*

The analogue of Corollary 2 does not hold for strong or directed searching. The digraph W, as pictured in Figure 2, has strong search number 1. However, if the arc (a, b) is contracted, the resulting digraph minor has strong search number 2. Similarly, the digraph Z has directed search number 3, but if the arc (c, d) is contracted, the resulting digraph minor has directed search number 4.

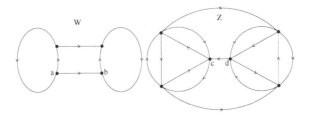

Fig. 2. The digraphs W and Z

4   Characterizing 1-Searchable Digraphs

In the case of graphs, characterizations are known for graphs that are k-searchable, for $k \leq 3$ [9]. It is natural to consider analogous characterizations for digraphs with respect to the various search models. We begin by characterizing those digraphs that are 1-searchable.

In contrast to the characterization of the weak search number for acyclic digraphs in Theorem 1, the next result shows the strong search number is much smaller in general.

Theorem 7. *If D is an acyclic digraph, then* $s_s(D) = 1$.

Interestingly, acyclic digraphs are not the only digraphs for which a strong search strategy exists using exactly one searcher. For example, consider the directed cycle $\overrightarrow{C_n}$. Place a single searcher anywhere on $\overrightarrow{C_n}$, and begin moving against the direction of arcs. This clears arcs, and since the intruder cannot move against arc direction, the arcs must remain clear. Thus, $s_s(\overrightarrow{C_n}) = 1$.

Theorem 8. *If D is a strong digraph of order n and $\sum_{v \in V(D)} d^+(v) \geq n + 2$, then* $s_s(D) \geq 2$.

Proof. Assume that $s_s(D) = 1$. We claim that every arc of D that is cleared by the single searcher γ must be cleared by traversing the arc in the direction opposite the direction of the arc. Since D is strong, every vertex has non-zero indegree and outdegree. Thus, the claim certainly is true for the first arc cleared by γ.

Suppose there exists an arc (u, v) of D that is cleared by γ traversing the arc from u to v. Consider the subdigraph D' of all currently cleared arcs. There is a directed path P from v to u because D is strongly connected. The directed path must use arcs of D' because all incoming arcs at u belong to D' or else γ cannot clear (u, v) by traversing the arc from u to v. On the other hand, the arc from u to v currently is contaminated and γ is located at u. Thus, all the arcs along P also are contaminated. This is a contradiction and our claim follows.

Also note that since $\sum_{v \in V(D)} d^+(v) \geq n + 2$, either two or more vertices have indegree at least two, or exactly one vertex has indegree at least three.

Case 1. The digraph D has at least two vertices with indegree at least two. Let u and v be two vertices such that $d^-(u) \geq 2$ and $d^-(v) \geq 2$. Consider a strong search of D that uses exactly one searcher. If the search begins at vertex $w \notin \{u, v\}$, then the searcher begins clearing arcs in reverse until finally it reaches (without loss of generality) u. At this point, when the searcher clears any in-arcs incident with u, the entire digraph cleared to this point becomes recontaminated. Thus, the search must begin at u.

If the search begins at u, then the searcher clears arcs until it reaches v. Then if the searcher clears any in-arc incident with v, all of the digraph cleared to this point becomes recontaminated. Thus, there cannot be two vertices with indegree at least two.

Case 2. The digraph D has exactly one vertex u with indegree at least three. Consider a strong search of D that uses exactly one searcher. If the search does not begin at u, then it must eventually reach u, at which point no further arcs can be cleared without recontaminating all arcs previously cleared. Thus, the search must begin at u. A search beginning at u must eventually reach u again, at which point there are at least two in-arcs that are contaminated. Neither of these arcs can be cleared without recontaminating all the arcs previously cleared. Thus, there cannot be a vertex with indegree at least three.

So, for a strong digraph D with $s_s(D) = 1$ that contains arcs, either $\sum_{v \in V(D)} d^+(v) = n$ or $\sum_{v \in V(D)} d^+(v) = n + 1$. If the former holds, then D must be a directed n-cycle since every vertex has non-zero indegree and outdegree. If the latter holds, then there must be exactly one vertex with indegree two, and exactly one vertex with outdegree two. Again there are two cases. If the vertex with indegree two is the same as the vertex with outdegree two, then D is made up of two directed cycles with a single common vertex. If the vertex of indegree two is distinct from the vertex of outdegree two, then D is a directed cycle $\overrightarrow{C_m}$, $m \leq n$, with an internally disjoint directed path from one vertex of $\overrightarrow{C_m}$ to a distinct vertex of $\overrightarrow{C_m}$.

Theorem 9. *A digraph D is 1-strongly-searchable if and only if every strong component of D is one of the three digraphs described above or a single vertex.*

We now consider the other search models. From [9], we know that those graphs with search number 1 are paths. This immediately gives the following corollary.

Corollary 3. *For a digraph D, $s_u(D) = 1$ if and only if D is an orientation of a path.*

In a directed search, there are only two ways to clear an arc as described earlier. One of the ways requires two searchers, so any digraph which is 1-searchable must only use the other. That is, a digraph that is directed 1-searchable allows an arc (u, v) to be cleared only when all incoming arcs at u are already cleared. (The characterization for weak searching follows.)

Theorem 10. *For a digraph D, $s_d(D) = 1$ if and only if $D = \overrightarrow{P_n}$.*

Corollary 4. *For a digraph D, $s_w(D) = 1$ if and only if $D = \overrightarrow{P_n}$.*

5   Characterizing 2-Searchable Digraphs

To characterize 2-searchable digraphs for the various searching models, we introduce the concept of a *homeomorphic reduction* of a reflexive multidigraph X. Let V' be those vertices of X that do not have indegree and outdegree 1. The *homeomorphic reduction* of X is the reflexive multidigraph X' obtained from X with vertex set V' and the following arcs. Any loop of X incident with a vertex of V' is a loop of X' incident with the same vertex. Any arc of X joining two vertices of V' is an arc of X' joining the same two vertices. Any internally disjoint directed path of X joining two vertices of V' is replaced by a single arc in X' joining the same two vertices, with the arc following the direction of the path. Any internally disjoint directed cycle of X containing a vertex u of V' is replaced by a loop in X' incident with u. In the special case that X has connected components that are directed cycles, these cycles are replaced by loops on a single vertex. An analogous definition occurs for the homeomorphic reduction of a graph.

The next theorem comes from [9], where it originally appeared as a result on graphs. We have rewritten it slightly as a digraph result.

Theorem 11. *A digraph D is 2-undirected-searchable if and only if D is the orientation of a graph whose homeomorphic reduction consists of an undirected path $u_1, u_2, \ldots, u_n$ such that there are an arbitrary number of loops incident with each vertex, and the multiplicity of any non-loop edge is at most 2.*

Corollary 5. *For a digraph D, $s_w(D) = 2$ if and only if the homeomorphic reduction of D consists of a directed path $u_1 u_2 \ldots u_n$ such that there are an arbitrary number of loops incident with each vertex, any arc of the form (u_i, u_{i+1}) for $1 \leq i \leq n-1$ must have multiplicity at most 2, and u_1 (resp. u_n) may have two in-arcs (resp. out-arcs).*

Moving to directed searching, we introduce the following lemmas before characterizing strong digraphs that are 2-directed searchable.

Lemma 1. *If D is a digraph that contains a directed cycle as a subdigraph, then whenever D goes from having no cleared directed cycles to one cleared directed cycle in a directed search strategy, that directed cycle contains at least 2 searchers.*

Proof. Consider the moment before the first directed cycle $(c_1, c_2, \ldots, c_k, c_1)$ is cleared. Let (c_k, c_1) be the final arc of this cycle to be cleared. Then one searcher must be on c_k, about to move to c_1. Since (c_1, c_2) is a cleared arc, but (c_k, c_1) is not, another searcher must be on c_1.

Lemma 2. *If D is a strong digraph that contains two vertex-disjoint directed cycles, then $s_d(D) \geq 3$.*

Proof. Assume that this digraph has $s_d(D) = 2$. (We know it does not have directed search number 1 by Lemma 1.) Let the two vertex-disjoint directed cycles be C_1 and C_2. Since D is strong, there is a directed path P_1 from a vertex

a of C_1 to a vertex b of C_2, and a directed path P_2 from a vertex c of C_2 to a vertex d of C_1. (It is possible that $a = d$ or $b = c$.) Assume that C_1 is the first cycle cleared. By Lemma 1, we know that C_1 must contain both searchers.

Since at least one arc of C_2 is contaminated, all arcs of C_2 and P_2 are contaminated. Thus, if C_1 is to remain cleared, one of the searchers must remain at d. The path from c to d cannot be cleared by the other searcher without first clearing C_2, and a single searcher cannot clear C_2. Thus, each arc in the path from c to d cannot be cleared, a contradiction.

Theorem 12. *If D is a strong digraph, then* $s_d(D) = 2$ *if and only if there exists $v \in V(D)$ such that every directed cycle of D contains v.*

Proof. Since D is strong, it is not a directed path, and hence $s_d(D) \geq 2$. If there exists a vertex $v \in V(D)$ such that $D - \{(x,v)|(x,v) \in A(D)\}$ is acyclic, then place one searcher on v. The other searcher may clear all arcs having v as their tail, at which time the contaminated arcs form an acyclic digraph. One searcher remains on v while the other searcher clears the arcs of the acyclic digraph. The searcher doing the arc clearing uses the strong connectivity of all of D. Thus, $s_d(D) = 2$.

On the other hand, assume that $s_d(D) = 2$, but there is no v as described. Then, for any $v \in V(D)$, if the arcs having v as their head are removed, a directed cycle remains. If every directed cycle in D has two or more vertices with incoming arcs not on the directed cycle, then we claim two searchers cannot clear D.

To see this, consider the first directed cycle C that is cleared. Let u be one of the vertices of C with indegree at least 2. If there is a directed cycle disjoint from C, then by Lemma 2 we are done. Since every directed cycle does not go through u, consider a cycle C' that does not contain u. Certainly, C and C' must share at least one vertex, w. Without loss of generality, assume that the search strategy begin with vertex u. Then, until C is cleared, the searcher at u must remain there, because movement would recontaminate all of C. This leaves only one searcher to clear the arc of C with w as tail, which is impossible until C' is cleared.

Hence, some directed cycle C exists with all arcs leading into C going to a single vertex v. If we remove all the arcs with v as head, then D contains a directed cycle. But this cycle cannot then contain any vertex of C. Thus, D contains two vertex-disjoint directed cycles, and $s_d(D) \geq 3$ by Lemma 2, a contradiction.

Theorem 13. *For an acyclic digraph D, $s_d(D) = 2$ if and only if the homeomorphic reduction of D consists of a directed path $u_1 u_2 \ldots u_n$ such that any arc in the path has multiplicity at most 2, and u_1 and u_n may have two in-arcs and out-arcs, respectively.*

Proof. Such digraphs are certainly 2-directed-searchable. On the other hand, any acyclic digraph D that is 2-directed-searchable must satisfy $\delta^-(D) \leq 2$, $\delta^+(D) \leq 2$, and have at most two sources and two sinks. Consider an acyclic ordering of D, where the vertices are labelled $v_1, v_2, \ldots, v_n$ such that an arc goes from v_i to v_j only if $i < j$.

If no vertex has outdegree 2 or more, then there can be at most one vertex of indegree 2, else there are too many sources. Then certainly this digraph is of the required form.

Suppose that v_i is the first vertex under this ordering that has outdegree 2. If there is no vertex after this with indegree 2, then there can be no further vertices with outdegree 2, as this would imply at least 3 sinks. Thus these digraphs are also of the required form.

If v_i is the first vertex under this ordering that has outdegree 2 and v_j is the first vertex occurring after v_i under this ordering that has indegree 2, then we claim that $v_j = v_{i+1}$. Assume there exists v_k with $i < k < j$. If v_k is a source, then only two of the three or more arcs having v_i or v_k as their tail can be cleared. If v_k is a sink, then only two of the three or more arcs having v_k or v_j as their head can be cleared. Thus v_k has non-zero indegree and outdegree. If $d^+(v_k) > 1$, then at most three of the four arcs with v_i or v_k as tail can be cleared. Since v_j is the first vertex after v_i that has indegree 2, then $d^+(v_k) = d^-(v_k) = 1$, but since X is a homeomorphic reduction, D contains no such vertices. Thus $v_j = v_{i+1}$.

Finally, D contains two sources if and only if the initial arcs from either source do not have multiplicity 2, and D contains two sinks if and only if the final arcs to either of the sinks cannot have multiplicity 2.

Consider an acyclic digraph D as described in Theorem 13. Certainly, if the arcs are subdivided into directed paths, the resulting digraph D' remains 2-directed-searchable. Then another digraph D'' may be formed by amalgamating certain vertices in D' with vertices in strong digraphs that are 1- or 2-directed-searchable. (These are either vertices or the strong digraphs described in Theorem 12.) The vertices of D' that may be used in the amalgamation are those that correspond to vertices in D or those that are in a directed path that has replaced an arc of multiplicity 1 in D, other than those arcs of multiplicity 1 that are incident with a source or sink. If x is the head of an arc (w, x) (or tail of an arc (x, w)) and is replaced by a strong 2-searchable digraph E, add a new arc from w to any vertex of E (or from any vertex of E to w). The resulting digraphs D'' are 2-directed-searchable. We claim that these are exactly the digraphs that are 2-directed-searchable.

Assume that D is a 2-directed-searchable digraph. We consider the condensation of D, but this time do not remove multiple arcs between strong components. We denote this digraph by $S^\star$. Certainly, $S^\star$ is an acyclic digraph, and if it is not 2-directed-searchable, then D will not be 2-directed-searchable. So the homeomorphic reduction of $S^\star$ must be as in Theorem 13. If $S^\star$ has two sources, and either of the sources or any of the other vertices that occur along the directed path between the sources and the first vertex with indegree 2 corresponds to a non-trivial strong component, then D is not 2-directed-searchable. A similar argument holds for the two sinks.

Consider an arc in homeomorphic reduction of $S^\star$ that has multiplicity 2. If this arc represents an internally disjoint direct path in $S^\star$, then consider one of the internal vertices of this path. If this vertex corresponds to a non-trivial strong component, then again D is not 2-directed-searchable, as then no searcher would be available to move along the directed path in $S^\star$ that corresponds to the other multiples of that arc. Finally, if any of the remaining vertices are replaced by non-trivial strong components other than those described in Theorem 12, then these components have directed search number 3 or more, and by Theorem 6, the search number of D must be 2. So D must be exactly as posited, and we obtain the following theorem.

Theorem 14. *For a digraph D, $s_d(D) = 2$ if and only if each of the following three conditions is satisfied: (1) every non-trivial strong component C of D has the property that there exists $v \in V(C)$ such that every directed cycle in C contains v; (2) the homeomorphic reduction of $S^\star$ consists of a directed path $u_1 u_2 \ldots u_n$ such that any arc in the path has multiplicity at most 2, and u_1 (resp. u_n) may have two in-edges (resp. out-edges); and (3) the non-trivial strong components occur only at vertices u_i ($1 \leq i \leq n$), or at vertices in $S^\star$ that lie on paths between u_i and u_{i+1}, where (u_i, u_{i+1}) has multiplicity 1 in the homeomorphic reduction of $S^\star$.*

6 Strong Digraphs

Suppose that for some D we have that the condensation $S(D)$ is a directed path $\overrightarrow{P_m}$ and that $D_1, D_2, \ldots, D_m$ (allowing $\overrightarrow{P_m}$ to be a trivial path of length 0, with $m = 1$), are the strong components of D. Let the *weight* of the path $\overrightarrow{P_m}$ be $w(\overrightarrow{P_m}) = \max_{i,j} \{s_d(D_i), d^+(D_j)\}$.

Theorem 15. *If $S(D)$ is a directed path $\overrightarrow{P_m}$ and $D_1, D_2, \ldots, D_m$ are the strong components of D, then $s_d(D) = w(\overrightarrow{P_m})$.*

Combining this result with Theorem 5, we obtain the following corollary.

Corollary 6. *If $S(D)$ is a directed path $\overrightarrow{P_m}$ and $D_1, D_2, \ldots, D_m$, for some $m > 1$, are the strong components of D, then $s_s(D) = \max_i \{s_s(D_i)\}$.*

The strong decomposition of a non-strong tournament is a partition of the vertex set into strong subtournaments $T_1, T_2, \ldots, T_m$ in which all vertices in T_i have arcs to all vertices in T_j whenever $i < j$. A special case of this occurs when each T_i is of order $|T_i| = 1$, that is, we have the transitive tournament TT_m on m vertices.
    A variation of the above is the following:

Corollary 7. *If the vertices of tournament T can be partitioned into sets A and B in such a way that every vertex in A dominates every vertex in B, that is, T is not strong, then $s_d(T) \geq |A||B|$.*

In the special case of a transitive tournament, we have the following result.

Corollary 8. *If $T = TT_n$ is a transitive tournament, then $s_d(T) = \lfloor \frac{n^2}{4} \rfloor$.*

Proof. If $n = 2t$, then $s_d(T) \geqq t^2$ by choosing A in Corollary 7 to be the t vertices of highest score. By searching the vertices in decreasing order of their scores, we see that v_n requires $2t - 1$ searchers to clear all of its outgoing arcs. Vertex v_{n-1} requires $2t - 2$ searchers to clear all of its outgoing arcs but a searcher has been obtained from v_n so that we need only $2t - 3$ new searchers. Vertex v_{n-2} requires $2t - 3$ searchers to clear all of its outgoing arcs but 2 searchers have been obtained from v_n and v_{n-1}, and thus we need only $2t - 5$ new searchers. Continuing in this way, we see that the number of searchers required is t^2. If n is odd the result follows in a similar manner.

Corollary 8 is especially interesting when considered in the light of the number of paths in a *path decomposition* of TT_n. A path decomposition of a digraph D is a partition of the arcs of D into a minimum number of directed paths. We denote this minimum number of paths $m(D)$. It was shown in [2] that for any tournament T, $m(T) \leqq \lfloor \frac{n^2}{4} \rfloor$, with equality holding if T is transitive. Thus, combined with Corollary 8, we see that a searcher may search each path in the decomposition, so that every arc is traversed exactly once. Since this would be a strategy that clears the transitive tournament on n vertices with fewest arc traversals and the minimum number of searchers, this can be considered an optimal strategy. In general, for acyclic digraphs, every minimum path decomposition corresponds to a set of paths that contain all vertices in the arc digraph of D, so the result of Theorem 1 is less than $m(D)$.

In marked contrast to Corollary 8, we note that for the $TT_n^\star$ that differs from TT_n only in that vertex v_n dominates vertex v_1, we have $s_d(TT_n^\star) = 2$ — we may leave a searcher on v_n while we clear all of its outgoing arcs with just one other searcher and then move on to other vertices in order. Corollary 8 may also be considered as a special case of the following.

Corollary 9. *If T is a non-strong tournament with strong decomposition $T_1, T_2,$ $\ldots, T_m$, then $s_d(T) = \max_j \left\{ \left(\sum_{i=1}^{j} |T_i| \right) \cdot \left(\sum_{i=j+1}^{m} |T_i| \right) \right\}$.*

7   Further Directions

There are several directions that further research could take. The dramatic change in the directed search number between TT_n and $TT_n^\star$ provide a strong motivation to look further at the search numbers of tournaments. Computing the search number of the complete graph is quite dull, but it seems that the search numbers of its orientations may vary a great deal.

Similarly, we see that strong subdigraphs play an important roll in characterizations, motivating further research into the relationship between being strongly connected and the various search models.

Finally, finding a characterization of those digraphs that are 2-strongly-searchable remains open. As in the method of Theorem 9, it seems plausible that we should first consider the condensations. Since such digraphs are acyclic, and hence 1-strongly-searchable, we need only consider which strong digraphs are 2-strongly-searchable. This appears to be a difficult problem.

References

1. Alspach, B.: Searching and sweeping graphs: A brief survey. Combinatorics 04 (Catania, 2004). Matematiche (Catania), Fasc. I–II, vol. 59, pp. 5–37 (2004)
2. Alspach, B., Pullman, N.: Path decompositions of digraphs. Bull. Austral. Math. Soc. 10, 421–427 (1974)
3. Fellows, M., Langston, M.: On search, decision and the efficiency of polynomial time algorithm. In: 21st ACM Symp. on Theory of Computing (STOC 89), pp. 501–512. ACM Press, New York (1989)
4. Frankling, M., Galil, Z., Yung, M.: Eavesdropping games: A graph-theoretic approach to privacy in distributed systems. Journal of ACM 47, 225–243 (2000)
5. Gross, J.L., Yellen, J. (eds.): Handbook of Graph Theory, Discrete Mathematics and its Applications (Boca Raton). CRC Press, Boca Raton, FL (2004)
6. Hunter, P., Kreutzer, S.: Digraph measures: Kelly decompositions, games, and orderings. In: Proceedings of the Seventeenth Annual ACM-SIAM Symposium on Discrete Algorithms, SODA 2007, pp. 637–644. ACM Press, New York (2007)
7. Johnson, T., Robertson, N., Seymour, P.D., Thomas, R.: Directed tree-width. Journal of Combinatorial Theory, Series B 82, 138–154 (2001)
8. Kirousis, L.M., Papadimitriou, C.H.: Searching and pebbling. Theoret. Comput. Sci. 47, 205–218 (1986)
9. Megiddo, N., Hakimi, S.L., Garey, M., Johnson, D., Papadimitriou, C.H.: The complexity of searching a graph. Journal of ACM 35, 18–44 (1988)
10. Nowakowski, R.J.: Search and sweep numbers of finite directed acyclic graphs. Discrete Appl. Math. 41, 1–11 (1993)
11. Obdržálek, J.: DAG-width - Connectivity measure for directed graphs. In: Proceedings of the Seventeenth Annual ACM-SIAM Symposium on Discrete Algorithms, SODA 2006, pp. 814–821. ACM Press, New York (2006)
12. Parsons, T.: Pursuit-evasion in a graph. In: Theory and Applications of Graphs. LNCS, pp. 426–441. Springer, Heidelberg (1976)
13. Yang, B., Cao, Y.: Directed searching digraphs: monotonicity and complexity. In: Proceedings of the 4th Annual Conference on Theory and Applications of Models of Computation (TAMC07). LNCS, vol. 4484, Springer, Heidelberg (2007)
14. Yang, B., Cao, Y.: Digraph strong searching: monotonicity and complexity. In: Proceedings of the 3rd International Conference on Algorithmic Aspects in Information and Management (AAIM07). LNCS, vol. 4508, pp. 37–46. Springer, Heidelberg (2007)
15. Yang, B., Cao, Y.: On the Monotonicity of Weak Searching on Digraphs (submitted)
16. Yang, B., Dyer, D., Alspach, B.: Searching graphs with large clique number (submitted)
17. Yang, B., Dyer, D., Alspach, B.: Sweeping graphs with large clique number (extended abstract). In: Fleischer, R., Trippen, G. (eds.) ISAAC 2004. LNCS, vol. 3341, pp. 880–892. Springer, Heidelberg (2004)

On the Complexity of Some Colorful Problems Parameterized by Treewidth[*]

Michael Fellows[1], Fedor V. Fomin[2], Daniel Lokshtanov[2], Frances Rosamond[1],
Saket Saurabh[2,3], Stefan Szeider[4], and Carsten Thomassen[5]

[1] University of Newcastle, Newcastle, Australia
{michael.fellows,frances.rosamond}@newcastle.edu.au
[2] Department of Informatics, University of Bergen, Bergen, Norway
{fedor.fomin,Daniel.Lokshtanov,saket}@ii.uib.no
[3] Institute of Mathematical Sciences, Chennai, India
saket@imsc.res.in
[4] Department of Computer Science, Durham University, Durham, U.K.
stefan.szeider@durham.ac.uk
[5] Mathematics Institute, Danish Technical University, Lyngby, Denmark
C.Thomassen@mat.dtu.dk

Abstract. We study the complexity of several coloring problems on graphs, parameterized by the treewidth t of the graph:

(1) The *list chromatic number* $\chi_l(G)$ of a graph G is defined to be the smallest positive integer r, such that for every assignment to the vertices v of G, of a list L_v of colors, where each list has length at least r, there is a choice of one color from each vertex list L_v yielding a proper coloring of G. We show that the problem of determining whether $\chi_l(G) \leq r$, the LIST CHROMATIC NUMBER problem, is solvable in linear time for every fixed treewidth bound t. The method by which this is shown is new and of general applicability.

(2) The LIST COLORING problem takes as input a graph G, together with an assignment to each vertex v of a set of colors C_v. The problem is to determine whether it is possible to choose a color for vertex v from the set of permitted colors C_v, for each vertex, so that the obtained coloring of G is proper. We show that this problem is $W[1]$-hard, parameterized by the treewidth of G. The closely related PRECOLORING EXTENSION problem is also shown to be $W[1]$-hard, parameterized by treewidth.

(3) An *equitable coloring* of a graph G is a proper coloring of the vertices where the numbers of vertices having any two distinct colors differs by at most one. We show that the problem is hard for $W[1]$, parameterized by (t, r). We also show that a list-based variation, LIST EQUITABLE COLORING is $W[1]$-hard for trees, parameterized by the number of colors on the lists.

Topics: Parameterized Complexity, Bounded Treewidth, Graph Coloring.

---

[*] This research has been supported by the Australian Research Council through the Australian Centre in Bioinformatics. The first author also acknowledges the support provided by a Fellowship to the Institute of Advanced Studies, Durham University, and the support of the Informatics Institute at the University of Bergen during an extended visit.

A. Dress, Y. Xu, and B. Zhu (Eds.): COCOA 2007, LNCS 4616, pp. 366–377, 2007.
© Springer-Verlag Berlin Heidelberg 2007

1   Introduction

Coloring problems that involve local and global restrictions on the coloring have many important applications in such areas as operations research, scheduling and computational biology, and also have a long mathematical history. For recent surveys of the area one can turn to [Tu97, KTV98, Al00, Wo01] and also the book [JT95]. In this paper we study the computational complexity of such problems, for graphs of bounded treewidth, in the framework of parameterized complexity [DF99, Nie06], where we take the parameter to be the treewidth bound t.

Our main results are summarized:

- We show that the *list chromatic number* (also known as the *choice number* [KTV98]) of a graph can be computed in linear time for any fixed treewidth bound t. (We prove this using a new "trick" for extending the applicability of Monadic Second Order logic that is of general interest.)
- We show that LIST COLORING and PRECOLORING EXTENSION are $W[1]$-hard for parameter t.
- We show that EQUITABLE COLORING is $W[1]$-hard parameterized by t.

The problems are defined as follows.

LIST CHROMATIC NUMBER
Input: A graph $G = (V, E)$ of treewidth at most t, and a positive integer r.
Parameter: t
Question: Is $\chi_l(G) \leq r$?

LIST COLORING
Input: A graph $G = (V, E)$ of treewidth at most t, and for each vertex $v \in V$, a list $L(v)$ of permitted colors.
Parameter: t
Question: Is there a proper vertex coloring c with $c(v) \in L(v)$ for each v?

PRECOLORING EXTENSION
Input: A graph $G = (V, E)$ of treewidth at most t, a subset $W \subseteq V$ of *precolored* vertices, a *precoloring* c_W of the vertices of W, and a positive integer r.
Parameter: t
Question: Is there a proper vertex coloring c of V which extends c_W (that is, $c(v) = c_W(v)$ for all $v \in W$), using at most r colors?

EQUITABLE COLORING (ECP)
Input: A graph $G = (V, E)$ of treewidth at most t and a positive integer r.
Parameter: t
Question: Is there a proper vertex coloring c using at most r colors, with the property that the sizes of any two color classes differ by at most one?

Previous Results. LIST COLORING is NP-complete, even for very restricted classes of graphs, such as complete bipartite graphs [JS97]. Jansen and Scheffler described a dynamic programming algorithm for the problem that runs in time $O(n^{t+2})$ for graphs of treewidth at most t [JS97]. PRECOLORING EXTENSION is NP-complete, and can also be solved in time $O(n^{t+2})$ for graphs of treewidth at most t [JS97]. The LIST CHROMATIC NUMBER problem is Π_2^p-complete for any fixed $r \geq 3$, a result attributed to Gutner and Tarsi [Tu97]. There does not appear to have been any previous result on the complexity of the LIST CHROMATIC NUMBER problem for graphs of bounded treewidth.

Some Background on Parameterized Complexity

Parameterized complexity is basically a two-dimensional generalization of "P vs. NP" where in addition to the overall input size n, one studies the effects on computational complexity of a secondary measurement that captures additional relevant information. This additional information can be, for example, a structural restriction on the input distribution considered, such as a bound on the treewidth of an input graph. Parameterization can be deployed in many different ways; for general background on the theory see [DF99, FG06, Nie06].

The two-dimensional analogue (or generalization) of P, is solvability within a time bound of $O(f(k)n^c)$, where n is the total input size, k is the parameter, f is some (usually computable) function, and c is a constant that does not depend on k or n. Parameterized decision problems are defined by specifying the input, the parameter, and the question to be answered. A parameterized problem that can be solved in such time is termed *fixed-parameter tractable* (FPT). There is a hierarchy of intractable parameterized problem classes above FPT, the main ones are:

$$FPT \subseteq M[1] \subseteq W[1] \subseteq M[2] \subseteq W[2] \subseteq \cdots \subseteq W[P] \subseteq XP$$

The principal analogue of the classical intractability class NP is $W[1]$, which is a strong analogue, because a fundamental problem complete for $W[1]$ is the k-STEP HALTING PROBLEM FOR NONDETERMINISTIC TURING MACHINES (with unlimited nondeterminism and alphabet size) — this completeness result provides an analogue of Cook's Theorem in classical complexity. A convenient source of $W[1]$-hardness reductions is provided by the result that k-CLIQUE is complete for $W[1]$. Other highlights of the theory include that k-DOMINATING SET, by contrast, is complete for $W[2]$. $FPT = M[1]$ if and only if the *Exponential Time Hypothesis* fails. XP is the class of all problems that are solvable in time $O(n^{g(k)})$.

The principal "working algorithmics" way of showing that a parameterized problem is unlikely to be fixed-parameter tractable is to prove $W[1]$-hardness. The key property of a parameterized reduction between parameterized problems Π and Π' is that the input (x, k) to Π should be transformed to input (x', k') for Π', so that the receiving parameter k' is a function only of the parameter k for the source problem.

1.1 LIST CHROMATIC NUMBER Parameterized by Treewidth Is FPT

The notion of the *list chromatic number* (also known as the *choice number*) of a graph was introduced by Vizing in 1976 [Viz76], and independently by Erdös, Rubin and Taylor in 1980 [ERT80]. A celebrated result that gave impetus to the area was proved by Thomassen: every planar graph has list chromatic number at most five [Th94].

We describe an algorithm for the LIST CHROMATIC NUMBER problem that runs in linear time for any fixed treewidth bound t. Our algorithm employs the machinery of Monadic Second Order (MSO) logic, due to Courcelle [Cou90] (also [ALS91, BPT92]). At a glance, this may seem surprising, since there is no obvious way to describe the problem in MSO logic — one would seemingly have to quantify over all possible list assignments to the vertices of G, and the vocabulary of MSO seems not to provide any way to do this. We employ a "trick" that was first described (to our knowledge) in [BFLRRW06], with further applications described in [CFRRRS07, FGKPRWY07].

The essence of the trick is to construct an auxiliary graph that consists of the original input, augmented with additional *semantic vertices*, so that the whole ensemble has — or can safely be assumed to have — bounded treewidth, and relative to which the problem of interest *can* be expressed in MSO logic.

A list assignment L with $|L(v)| \geq r$ for all $v \in V$ is termed an *r-list assignment*. A list assignment L from which G cannot be properly colored is called *bad*. Thus, a graph G does not have list chromatic number $\chi_l(G) \leq r$, if and only if there is a bad r-list assignment for G.

The following lemma is crucial to the approach.

Lemma 1. *If a graph of treewidth at most t admits any bad r-list assignment, then it admits a bad list assignment where the colors are drawn from a set of $(2t+1)r$ colors.*

Proof. First of all, we may note that if G has treewidth bounded by t, then $\chi_l(G) \leq t+1$ (and similarly, the chromatic number of G is at most $t + 1$). This follows easily from the inductive definition of t-trees. We can therefore assume that $r \leq t$.

Fix attention on a width t tree decomposition $\mathcal{D}$ for G, where the bags of the decomposition are indexed by the tree T. For a node t of T, let $\mathcal{D}(t)$ denote the bag associated to the node t. Suppose that L is a bad r-list assignment for G, and let $\mathcal{C}$ denote the union of the lists of L. For a color $\alpha \in \mathcal{C}$, let T_α denote the subforest of T induced by the set of vertices t of T for which $\mathcal{D}(t)$ contains a vertex v of G, where the color α occurs in the list $L(v)$. Let $\mathcal{T}(\alpha)$ denote the set of trees of the forest T_α. Let $\mathcal{T}$ denote the union of the sets $\mathcal{T}(\alpha)$, taken over all of the colors α that occur in the list assignment L:

$$\mathcal{T} = \bigcup_{\alpha \in \mathcal{C}} \mathcal{T}(\alpha)$$

We consider that two trees T' and T'' in $\mathcal{T}$ are *adjacent* if the distance between T' and T'' in T is at most one. Note that T' and T'' might not be disjoint, so the distance between them can be zero. Let $\mathcal{G}$ denote the graph thus defined: the vertices of $\mathcal{G}$ are the subtrees in $\mathcal{T}$ and the edges are given by the above adjacency relationship.

Suppose that $\mathcal{G}$ can be properly colored by the coloring function $c' : \mathcal{T} \to \mathcal{C}'$. We can use such a coloring to describe a modified list assignment $L'[c']$ to the vertices of G in the following way: if $T' \in \mathcal{T}(\alpha)$ and $c'(T') = \alpha' \in \mathcal{C}'$, then replace each occurrence of the color α on the lists $L(v)$, for all vertices v that belong to bags $\mathcal{D}(t)$, where $t \in T'$, with the color α'.

This specification of $L'[c']$ is consistent, because for any vertex v such that $\alpha \in L(v)$, there is exactly one tree $T' \in \mathcal{T}(\alpha)$ such that v belongs to a bag indexed by vertices of T'.

Claim 1. If c' is a proper coloring of $\mathcal{G}$, and L is a bad list assignment for G, then $L'[c']$ is also a bad list assignment for G.

This follows because the trees in $\mathcal{G}$ preserve the constraints expressed in having a given color on the lists of adjacent vertices of G, while the new colors α' can only be used on two different trees T' and T'' when the vertices of G in the bags associated with these trees are at a distance of at least two in $\mathcal{G}$.

Claim 2. The graph $\mathcal{G}$ has treewidth at most $2(t+1)r - 1$.

A tree decomposition $\mathcal{D}'$ for $\mathcal{G}$ of width at most $2(t+1)r$ can be described as follows. Subdivide each edge tt' of T with a node of degree two denoted $s(t, t')$. Assign to each node t the bag $\mathcal{D}'(t)$ consisting of those trees T' of $\mathcal{G}$ that include t. There are at most $(t+1)r$ such trees. Assign to each node $s(t, t')$ the bag $\mathcal{D}'(s(t, t')) = \mathcal{D}'(t) \cup \mathcal{D}'(t')$. It is straightforward to verify that this satisfies the requirements of a tree decomposition for $\mathcal{G}$.

The lemma now follows from the fact that $\mathcal{G}$ can be properly colored with $2(t+1)r$ colors.   □

Theorem 1. *The* LIST CHROMATIC NUMBER *problem, parameterized by the treewidth bound t, is fixed-parameter tractable, solvable in linear time for every fixed t.*

Proof. The algorithm consists of the following steps.

Step 1. Compute in linear time, using Bodlaender's algorithm, a tree-decomposition for G of width at most t. Consider the vertices of G to be of *type 1*.

Step 2. Introduce $2(t+1)r$ new vertices of *type 2*, and connect each of these to all vertices of G. The treewidth of this augmented graph is at most $t + 2(t+1)r = O(t^2)$.

Step 3. The problem can now be expressed in MSO logic. That this is so, is not entirely trivial, and is argued as follows (sketch). We employ a routine extension of MSO logic that provides predicates for the two types of vertices.

If G admits a bad r-list assignment, then this is witnessed by a set of edges F between vertices of G (that is, type 1 vertices) and vertices of type 2 (that represent the colors), such that every vertex v of G has degree r relative to F. Thus, the r incident F-edges represent the colors of L_v. It is routine to assert the existence of such a set of edges in MSO logic.

The property that such a set of edges F represents a bad list assignment can be expressed as: "For every subset $F' \subset F$ such that every vertex of G has degree 1 relative to F' (and thus, F' represents a choice of a color for each vertex, chosen from its list), there is an adjacent pair of vertices u and v of G, such that the represented color choice is the same, i.e., u and v are adjacent by edges of F' to the same type 2 (color-representing) vertex." The translation of this statement into formal MSO is routine.   □

2   Some Coloring Problems That Are Hard for Treewidth

We tend to think that "all" (or almost all) combinatorial problems are easy for bounded treewidth, but in the case of structured coloring problems, the game is more varied in outcome.

2.1   LIST COLORING and PRECOLORING EXTENSION are $W[1]$-Hard, Parameterized by Treewidth

There is a relatively simple reduction to the LIST COLORING and PRECOLORING EX-TENSION problems from the MULTICOLORED CLIQUE problem. The MULTICOLORED CLIQUE problem is known to be $W[1]$-complete [FHR07] (by a simple reduction from the ordinary CLIQUE). The MULTICOLORED CLIQUE problem takes as input a graph G together with a proper k-coloring of the vertices of G, and is parameterized by k. The question is whether there is a k-clique in G consisting of exactly one vertex of each color.

As example of the reduction is shown in Figure 1. The figure shows, for the parameter value $k = 4$, the construction of an instance G' of LIST COLORING that admits a proper choice of color from each list if and only if the source instance G has a multicolor k-clique.

The general construction can be easily inferred from the example in Figure 1. The colors on the lists are in 1:1 correspondence with the vertices of G. There are k vertices $v[i]$, $i = 1, ..., k$, one for each color class of G, and the list assigned to $v[i]$ consists of the colors corresponding to the vertices in G of color i. For $i \neq j$, there are various vertices of degree two in G', each having a list of size 2. There is one such degree two vertex in G' for each pair x, y of *nonadjacent* vertices, where x has color i and y has color j.

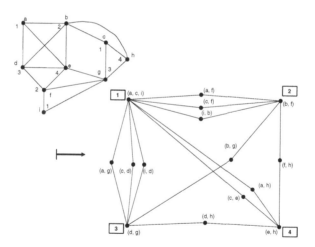

Fig. 1. Example of the reduction from MULTICOLOR CLIQUE to LIST COLORING

Verification that the reduction works correctly is easy, and is left to the reader. The treewidth of G' is bounded by $k + 1$.

Theorem 2. LIST COLORING *parameterized by treewidth is* $W[1]$-*hard.*

To see that PRECOLORING EXTENSION is also $W[1]$-hard when parameterized by treewidth, we can reduce from the LIST COLORING problem, simply using many precolored vertices of degree 1 to enforce the lists. This construction does not increase the treewidth. We have:

Theorem 3. PRECOLORING EXTENSION *parameterized by treewidth is* $W[1]$-*hard.*

2.2   EQUITABLE COLORING Is $W[1]$-Hard Parameterized by Treewidth

The notion of equitable coloring seems to have been first introduced by Meyer in 1973, where an application to scheduling garbage trucks is described [Mey73]. Recently, Bodlaender and Fomin have shown that determining whether a graph of treewidth at most t admits an equitable coloring, can be solved in time $O(n^{O(t)})$ [BF05].

We consider the parameterized complexity of EQUITABLE COLORING (ECP) in graphs with bounded treewidth. We actually prove a stronger result than the one we have so far stated. We show that when ECP is parameterized by (t, r), where t is the treewidth bound, and r is the number of color classes, then the problem is W[1]-hard.

To show the desired reduction, we introduce two more general problems. List analogues of equitable coloring have been previously studied by Kostochka, et al. [KPW03].

> The LIST EQUITABLE COLORING PROBLEM (LECP): Given an input graph $G = (V, E)$, lists L_v of colors for every vertex $v \in V$ and a positive integer r; does there exist a proper coloring f of G with r colors that for every vertex $v \in V$ uses a color from its list L_v such that for any two color class, V_i and V_j of the coloring f, $||V_i| - |V_j|| \leq 1$?
>
> The NUMBER LIST COLORING PROBLEM (NLCP): Given an input graph $G = (V, E)$, lists L_v of colors for every vertex $v \in V$, a function $h : \cup_{v \in V} L_v \to \mathbb{N}$, associating a number to each color, and a positive integer r; does there exist a proper coloring f of G with r colors that for every vertex $v \in V$ uses a color from its list L_v, such that any color class V_c of the coloring f is of size $h(c)$?

Our main effort is in the reduction of the MULTICOLOR CLIQUE problem to NLCP. Consider that the instance $G = (V, E)$ of MULTICOLOR CLIQUE has its vertices colored by the integers $1, ..., k$. Let $V[i]$ denote the set of vertices of color i, and let $E[i, j]$, for $1 \leq i < j \leq k$, denote the set of edges $e = uv$, where $u \in V[i]$ and $v \in V[j]$. We can assume that $|V[i]| = N$ for all i, and that $|E[i, j]| = M$ for all $i < j$, that is, we can assume that the vertex color classes of G, and also the edge sets between them, have uniform sizes. (For a simple justification of this assumption, we can reduce MULTICOLOR CLIQUE to itself, taking a union of $k!$ disjoint copies of G, one for each permutation of the color set.)

We will use following sets of colors in our construction of an instance of NLCP:

(1) $\mathcal{S} = \{\sigma[i, j] : 1 \leq i \neq j \leq k\}$
(2) $\mathcal{S}' = \{\sigma'[i, j] : 1 \leq i \neq j \leq k\}$
(3) $\mathcal{T} = \{\tau_i[r, s] : 1 \leq i \leq k, \ 1 \leq r < s \leq k, r \neq i, s \neq i\}$
(4) $\mathcal{T}' = \{\tau_i'[r, s] : 1 \leq i \leq k, \ 1 \leq r < s \leq k, r \neq i, s \neq i\}$
(5) $\mathcal{E} = \{\epsilon[i, j] : 1 \leq i < j \leq k\}$
(6) $\mathcal{E}' = \{\epsilon'[i, j] : 1 \leq i < j \leq k\}$

Note that $|\mathcal{S}| = |\mathcal{S}'| = 2\binom{k}{2}$, that is, there are distinct colors $\sigma[2, 3]$ and $\sigma[3, 2]$, etc. In contrast, the colors $\tau_i[r, s]$ are only defined for $r < s$.

We associate with each vertex and edge of G a pair of (unique) *identification numbers*. The *up-identification number* $v[up]$ for a vertex v should be in the range $[n^2 + 1, n^2 + n]$, if G has n vertices. Similarly, the *up-identification number* $e[up]$ of an edge e of G can be assigned (arbitrarily, but uniquely) in the range $[2n^2 + 1, 2n^2 + m]$, assuming G has m edges.

Choose a suitably large positive integer Z_0, for example $Z_0 = n^3$, and define the *down-identification number* $v[down]$ for a vertex v to be $Z_0 - v[up]$, and similarly for the edges e of G, define the *down-identification number* $e[down]$ to be $Z_0 - e[up]$.

Choose a second large positive integer, $Z_1 >> Z_0$, for example, we may take $Z_1 = n^6$.

Next we describe various gadgets and the way they are combined in the reduction. First we describe the gadget which encodes the *selection* of the edge going between two particular color classes in G. In other words, we will think of the representation of a k-clique in G as involving the selection of edges (with each edge selected twice, once in each direction) between the color classes of vertices in G, with gadgets for *selection*, and to check two things: (1) that the selections in opposite color directions match, and (2) that the edges chosen from color class $V[i]$ going to $V[j]$ (for the various $j \neq i$) all emanate from the same vertex in $V[i]$. (This is sometimes termed an *edge representation strategy* for the parameterized reduction from MULTICOLOR CLIQUE.)

There are $2\binom{k}{2}$ groups of gadgets, one for each pair of color indices $i \neq j$. If $1 \leq i < j \leq k$, then we will refer to the gadgets in the group $\mathcal{G}[i, j]$ as *forward gadgets*, and we will refer to the gadgets in the group $\mathcal{G}[j, i]$ as *backward gadgets*.

If $e \in E[i, j]$, then there is one forward gadget corresponding to e in the group $\mathcal{G}[i, j]$, and one backward gadget corresponding to e in the group $\mathcal{G}[j, i]$. The construction of these gadgets is described as follows.

The forward gadget corresponding to $e = uv \in E[i, j]$

The gadget has a root vertex $r[i, j, e]$, and consists of a tree of height 2. The list assigned to this root vertex contains two colors: $\sigma[i, j]$ and $\sigma'[i, j]$. The root vertex has $Z_1 + 1$ children, and each of these is also assigned the two-element list containing the colors $\sigma[i, j]$ and $\sigma'[i, j]$. One of the children vertices is distinguished, and has $2(k-1)$ groups of further children:

- $e[up]$ children assigned the list $\{\sigma'[i, j], \epsilon[i, j]\}$.
- $e[down]$ children assigned the list $\{\sigma'[i, j], \epsilon'[i, j]\}$.
- For each r in the range $j < r \leq k$, $u[up]$ children assigned the list $\{\sigma'[i, j], \tau_i[j, r]\}$.
- For each r in the range $j < r \leq k$, $u[down]$ children assigned $\{\sigma'[i, j], \tau_i'[j, r]\}$.
- For each r in the range $1 \leq r < j$, $u[down]$ children assigned $\{\sigma'[i, j], \tau_i[r, j]\}$.
- For each r in the range $1 \leq r < j$, $u[up]$ children assigned the list $\{\sigma'[i, j], \tau_i'[r, j]\}$.

The backward gadget corresponding to $e = uv \in E[i, j]$

The gadget has a root vertex $r[j, i, e]$, and consists of a tree of height 2. The list assigned to this root vertex contains two colors: $\sigma[j, i]$ and $\sigma'[j, i]$. The root vertex has $Z_1 + 1$ children, and each of these is also assigned the two-element list containing the colors $\sigma[j, i]$ and $\sigma'[j, i]$. One of the children vertices is distinguished, and has $2k$ groups of further children:

- $e[up]$ children assigned the list $\{\sigma'[j, i], \epsilon'[i, j]\}$.
- $e[down]$ children assigned the list $\{\sigma'[j, i], \epsilon[i, j]\}$.

- For each r in the range $i < r \leq k$, $v[up]$ children assigned the list $\{\sigma'[j,i], \tau_j[i,r]\}$.
- For each r in the range $i < r \leq k$, $v[down]$ children assigned $\{\sigma'[j,i], \tau'_j[i,r]\}$.
- For each r in the range $1 \leq r < i$, $v[down]$ children assigned $\{\sigma'[j,i], \tau_j[r,i]\}$.
- For each r in the range $1 \leq r < i$, $v[up]$ children assigned the list $\{\sigma'[j,i], \tau'_j[r,i]\}$.

The numerical targets

(1) Each color in $\mathcal{T} \cup \mathcal{T}'$ has the target: Z_0.
(2) Each color in $\mathcal{E} \cup \mathcal{E}'$ has the target: Z_0.
(3) Each color in $\mathcal{S}$ has the target: $(M-1)(Z_1 + 1) + 1$.
(4) Each color in $\mathcal{S}'$ has the target: $(M-1) + (Z_1 + 1) + (k-1)(M-1)Z_0$.

That completes the formal description of the reduction from MULTICOLOR CLIQUE to NLCP. We turn now to some motivating remarks about the design of the reduction.

Remarks on the colors, their numerical targets, and their role in the reduction

(1) There are $2\binom{k}{2}$ groups of gadgets. Each edge of G gives rise to two gadgets. Between any two color classes of G there are precisely M edges, and therefore $M \cdot \binom{k}{2}$ edges in G in total. Each group of gadgets therefore contains M gadgets. The gadgets in each group have two "helper" colors. For example, the group of gadgets $\mathcal{G}[4,2]$ has the helper colors $\sigma[4,2]$ and $\sigma'[4,2]$. The role of the gadgets in this group is to indicate a choice of an edge going *from* a vertex in the color class $V[4]$ of G *to* a vertex in the color class $V[2]$ of G. The role of the $2\binom{k}{2}$ groups of gadgets is to represent the selection of $\binom{k}{2}$ edges of G that form a k-clique, with each edge chosen twice, once in each direction. If $i < j$ then the choice is represented by the coloring of the gadgets in the group $\mathcal{G}[i,j]$, and these are the *forward* gadgets of the edge choice. If $j < i$, then the gadgets in $\mathcal{G}[i,j]$ are *backward* gadgets (representing the edge selection in the opposite direction, relative to the ordering of the color classes of G). The numerical targets for the colors in $\mathcal{S} \cup \mathcal{S}'$ are chosen to force exactly one edge to be selected (forward or backward) by each group of gadgets, and to force the gadgets that are colored in a way that indicates the edge was not selected into being colored in a particular way (else the numerical targets cannot be attained). The numerical targets for these colors are complicated, because of this role (which is asymmetric between the pair of colors $\sigma[i,j]$ and $\sigma'[i,j]$).

(2) The colors in $\mathcal{T} \cup \mathcal{T}'$ and $\mathcal{E} \cup \mathcal{E}'$ are organized in symmetric pairs, and each pair is used to transmit (and check) information. Due to the enforcements alluded to above, each "selection" coloring of a gadget (there will be only one possible in each group of gadgets), will force some numbers of vertices to be colored with these pairs of colors, which can be thought of as an information transmission. For example, when a gadget in $\mathcal{G}[4,2]$ is colored with a "selection" coloring, this indicates that the edge from which the gadget arises is selected as the edge *from* the color class $V[4]$ of G, *to* the color class $V[2]$. There is a pair of colors that handles the information transmission concerning *which edge is selected* between the groups $\mathcal{G}[2,4]$ and $\mathcal{G}[4,2]$. (Of course, something has to check that the edge selected in one direction, is the same as the edge selected in the other direction.) There is something neat about the dual-color transmission channel for this information. Each vertex and edge has two unique identification numbers, "up" and "down", that sum to Z_0. To continue the concrete example, $\mathcal{G}[4,2]$ uses the (number of vertices colored by the) pair of colors $\epsilon[2,4]$ and $\epsilon'[2,4]$ to communicate to $\mathcal{G}[2,4]$ about the edge selected. The signal from one side consists of $e[up]$ vertices colored

$\epsilon[2,4]$ and $e[down]$ vertices colored $\epsilon'[2,4]$. The signal from the other side consists of $e[down]$ vertices colored $\epsilon[2,4]$ and $e[up]$ vertices colored $\epsilon'[2,4]$. Thus the numerical targets for these colors allow us to check whether the same edge has been selected in each direction (if each color target of Z_0 is met). There is the additional advantage that the *amount* of signal in each direction is the same: in each direction a total of Z_0 colored vertices, with the two paired colors, constitutes the signal. This means that, modulo the discussion in (1) above, when an edge is *not* selected, the corresponding non-selection coloring involves uniformly the same number (i.e., Z_0) of vertices colored "otherwise" for each of the $(M-1)$ gadgets colored in the non-selection way: this explains (part of) the $(k-1)(M-1)Z_0$ term in (4) of the numerical targets.

(3) In a similar manner to the communication task discussed above, each of the $k-1$ groups of gadgets $\mathcal{G}[i, \_]$ need to check that each has selected an edge *from* $V[i]$ that originates at the same vertex in $V[i]$. Hence there are pairs of colors that provide a communication channel similar to that in (2) for this information. This role is played by the colors in $\mathcal{T} \cup \mathcal{T}'$. (Because of the bookkeeping issues, this becomes somewhat intricate in the formal definition of the reduction.)

The above remarks are intended to aid an intuitive understanding of the reduction. We now return to a more formal argument.

Claim 1. If G has a k-multicolor clique, then G' is a yes-instance to NLCP.

The proof of this claim is relatively straightforward. The gadgets corresponding to the edges of a k-clique in G are colored in a manner that indicates "selected" (for both the forward and the backward gadgets) and all other gadgets are colored in manner that indicates "non-selected". The coloring that corresponds to "selected" colors the root vertex with the color $\sigma[i, j]$, and this forces the rest of the coloring of the gadget. The coloring that corresponds to "non-selected" colors the root vertex with the color $\sigma'[i, j]$. In this case the coloring of the rest of the gadget is not entirely forced, but if the grandchildren vertices of the gadget are also colored with $\sigma'[i, j]$, then all the numerical targets will be met.

Claim 2. Suppose that Γ is a list coloring of G' that meets all the numerical targets. Then in each group of gadgets, exactly one gadget is colored in a way that indicates "selection".

We argue this as follows. There cannot be two gadgets in any group colored in the "selection" manner, since this would make it impossible to meet the numerical target for a color in $\mathcal{S}$. If no gadget is colored in the "selection" manner, then again the targets cannot be met for the colors in $\mathcal{S} \cup \mathcal{S}'$ used in the lists for this group of gadgets.

Claim 3. Suppose that Γ is a list coloring of G' that meets all the numerical targets. Then in each group of gadgets, every gadget that is not colored in a way that indicates "selection" must have all of its grandchildren vertices colored with the appropriate color in $\mathcal{S}'$.

Claim 3 follows from Claim 2, noting that the numerical targets for the $\mathcal{S}'$ colors cannot be met unless this is so.

It follows from Claims 2 and 3, that if Γ is a list coloring of G' that meets all the numerical targets, then in each group of gadgets, exactly one gadget is colored in the "selection" manner, and all other gadgets are colored in a completely determined "nonselection" manner. Each "selection" coloring of a gadget produces a numerical signal (based on vertex and edge identification numbers) carried by the colors in $T \cup T'$ and $\mathcal{E} \cup \mathcal{E}'$, with two signals per color. The target of Z_0 for these colors can only be achieved if the selection colorings indicate a clique in G.

Theorem 4. *NLCP is W[1]-hard for trees, parameterized by the number of colors that appear on the lists.*

The reduction from NLCP to LECP is almost trivial, achieved by padding with isolated vertices having single-color lists.

The reduction from LECP to ECP is described as follows. Create a clique of size r, the number of colors occuring on the lists, and connect the vertices of this clique to the vertices of G' in a manner that enforces the lists. Since G' is a tree, the treewidth of the resulting graph is at most r. We have:

Theorem 5. EQUITABLE COLORING *is* W[1]-*hard, parameterized by treewidth.*

3 Discussion and Open Problems

Structured optimization problems, such as the coloring problems considered here, have strong claims with respect to applications. A source of discussion of these applications is the recent dissertation of Marx [Ma04]. It seems interesting and fruitful to consider such problems from the parameterized point of view, and to investigate how such extra problem structure (which tends to increase both computational complexity, and real-world applicability) interacts with parameterizations (such as bounded treewidth), that frequently lead to tractability.

The outcome of the investigation here of some well-known locally and globally constrained coloring problems has turned up a few surprises: first of all, that the LIST CHROMATIC NUMBER problem is actually FPT, when we parameterize by treewidth. It is also somewhat surprising that this good news does not extend to LIST COLORING, PRECOLORING EXTENSION or EQUITABLE COLORING, all of which turn out to be hard for $W[1]$.

There are many interesting open problems concerning the parameterized complexity of "more structured" combinatorial optimization problems on graphs, parametered by treewidth. We mention the following two:

(1) Is the LIST EDGE CHROMATIC NUMBER problem fixed-parameter tractable, parameterized by treewidth?

(2) One can formulate a "list analogue" of the HAMILTONIAN PATH problem as follows: each vertex is assigned a list that is a subset of $\{1, 2, ..., n\}$ indicating the positions in the ordering of the n vertices implicit in a Hamiltonian path that are permitted to the vertex. Is the LIST HAMILTONIAN PATH problem FPT, parameterized by treewidth?

References

[Al00]   Alon, N.: Restricted colorings of graphs. In: Walker, K. (ed.) Surveys in Combinatorics 1993. London Math. Soc. Lecture Notes Series, vol. 187, pp. 1–33. Cambridge Univ. Press, Cambridge (1993)

[ALS91]   Arnborg, S., Lagergren, J., Seese, D.: Easy problems for tree-decomposable graphs. J. Algorithms 12, 308–340 (1991)

[BF05]	Bodlaender, H.L., Fomin, F.V.: Equitable colorings of bounded treewidth graphs. Theoretical Computer Science 349, 22–30 (2005)
[BFLRRW06]	Bodlaender, H.L., Fellows, M., Langston, M., Ragan, M.A., Rosamond, F., Weyer, M.: Quadratic kernelization for convex recoloring of trees. In: Proceedings COCOON 2007. LNCS, Springer, Heidelberg (to appear)
[BPT92]	Borie, R.B., Parker, R.G., Tovey, C.A.: Automatic generation of linear-time algorithms from predicate calculus descriptions of problems on recursively generated graph families. Algorithmica 7, 555–581 (1992)
[CFRRRS07]	Chor, B., Fellows, M., Ragan, M.A., Razgon, I., Rosamond, F., Snir, S.: Connected coloring completion for general graphs: algorithms and complexity. In: Proceedings COCOON 2007. LNCS, Springer, Heidelberg (to appear)
[Cou90]	Courcelle, B.: The monadic second-order logic of graphs I: Recognizable sets of finite graphs. Information and Computation 85, 12–75 (1990)
[DF99]	Downey, R.G., Fellows, M.R.: Parameterized Complexity. Springer, Heidelberg (1999)
[ERT80]	Erdös, P., Rubin, A.L., Taylor, H.: Choosability in graphs. Congressus Numerantium 26, 122–157 (1980)
[FG06]	Flum, J., Grohe, M.: Parameterized Complexity Theory. Springer, Heidelberg (2006)
[FGKPRWY07]	Fellows, M., Giannopoulos, P., Knauer, C., Paul, C., Rosamond, F., Whitesides, S., Yu, N.: The lawnmower and other problems: applications of MSO logic in geometry. Manuscript (2007)
[FHR07]	Fellows, M., Hermelin, D., Rosamond, F.: On the fixed-parameter intractability and tractability of multiple-interval graph properties. Manuscript (2007)
[JS97]	Jansen, K., Scheffler, P.: Generalized colorings for tree-like graphs. Discrete Applied Mathematics 75, 135–155 (1997)
[JT95]	Jensen, T.R., Toft, B.: Graph Coloring Problems. Wiley Interscience, Chichester (1995)
[KPW03]	Kostochka, A.V., Pelsmajer, M.J., West, D.B.: A list analogue of equitable coloring. Journal of Graph Theory 44, 166–177 (2003)
[KTV98]	Kratochvil, J., Tuza, Z., Voigt, M.: New trends in the theory of graph colorings: choosability and list coloring. In: Graham, R., et al. (eds.) Contemporary Trends in Discrete Mathematics (from DIMACS and DIMATIA to the future), AMS, Providence. DIMACS Series in Discrete Mathematics and Theoretical Computer Science, vol. 49, pp. 183–197 (1999)
[Ma04]	Marx, D.: Graph coloring with local and global constraints. Ph.D. dissertation, Department of Computer Science and Information Theory, Budapest University of Technology and Economics (2004)
[Mey73]	Meyer, W.: Equitable coloring. American Mathematical Monthly 80, 920–922 (1973)
[Nie06]	Niedermeier, R.: Invitation to Fixed Parameter Algorithms. Oxford University Press, Oxford (2006)
[Th94]	Thomassen, C.: Every planar graph is 5-choosable. J. Combinatorial Theory Ser. B 62, 180–181 (1994)
[Tu97]	Tuza, Z.: Graph colorings with local constraints — A survey. Discussiones Mathematicae – Graph Theory 17, 161–228 (1997)
[Viz76]	Vizing, V.G.: Coloring the vertices of a graph in prescribed colors (in Russian). Metody Diskret. Anal. v Teorii Kodov i Schem 29, 3–10 (1976)
[Wo01]	Woodall, D.R.: List colourings of graphs. In: Hirschfeld, J.W.P. (ed.) Surveys in Combinatorics 2001. London Math. Soc. Lecture Notes Series, vol. 288, pp. 269–301. Cambridge Univ. Press, Cambridge (2001)

A PTAS for the Weighted 2-Interval Pattern Problem over the Preceding-and-Crossing Model

Minghui Jiang

Department of Computer Science, Utah State University
Logan, Utah 84322-4205, USA
mjiang@cc.usu.edu

Abstract. The 2-Interval Pattern problem over its various models and restrictions was proposed by Vialette for RNA secondary structure prediction, and has attracted a lot of attention from the theoretical computer science community in recent years. In the framework of 2-intervals, the preceding-and-crossing model is an especially interesting model for RNA secondary structures with pseudoknots. In this paper, we present a polynomial time approximation scheme for the Weighted 2-Interval Pattern problem over the preceding-and-crossing model. Our algorithm improves the previous best 2-approximation algorithm, and closes this problem in terms of the approximation ratio.

1 Introduction

Vialette [14] proposed a geometric representation of the RNA secondary structure as a set of 2-intervals. Given a single-stranded RNA molecule, a subsequence of consecutive bases of the molecule can be represented as an interval on a single line, and a possible (stacked) pairing of two disjoint subsequences can be represented as a 2-interval, which is the union of two disjoint intervals. Given a candidate set of 2-intervals, a pairwise-disjoint subset restricted to certain pre-specified geometrical constraints gives a macroscopic approximation of the RNA secondary structure.

We review some definitions [14]. A 2-interval $D = (I, J)$ consists of two disjoint (closed) intervals I and J such that $I < J$, that is, I is completely to the left of J. Consider two 2-intervals $D_1 = (I_1, J_1)$ and $D_2 = (I_2, J_2)$. D_1 and D_2 are *disjoint* if the four intervals I_1, J_1, I_2, and J_2 are pairwise disjoint. Define three binary relations for disjoint pairs of 2-intervals:

Preceding: $D_1 < D_2 \iff I_1 < J_1 < I_2 < J_2$.
Nesting: $D_1 \sqsubset D_2 \iff I_2 < I_1 < J_1 < J_2$.
Crossing: $D_1 \between D_2 \iff I_1 < I_2 < J_1 < J_2$.

The two 2-intervals D_1 and D_2 are *R-comparable* for some $R \in \{<, \sqsubset, \between\}$ if either $(D_1, D_2) \in R$ or $(D_2, D_1) \in R$. (For example, D_1 and D_2 are $\sqsubset$-comparable if either $D_1 \sqsubset D_2$ or $D_2 \sqsubset D_1$.) Note that the set of binary relations $\{<, \sqsubset, \between\}$ is complete in the sense that any two disjoint 2-intervals are R-comparable for

A. Dress, Y. Xu, and B. Zhu (Eds.): COCOA 2007, LNCS 4616, pp. 378–387, 2007.

some $R \in \{<, \sqsubset, \between\}$. Given a *model* $\mathcal{R}$, which is a non-empty subset of $\{<, \sqsubset, \between\}$ (there are 7 such subsets), a set $\mathcal{D}$ of 2-intervals is $\mathcal{R}$-*structured* if any two distinct 2-intervals in $\mathcal{D}$ are R-comparable for some $R \in \mathcal{R}$. Given a set $\mathcal{D}$ of 2-intervals and a model $\mathcal{R}$, the 2-INTERVAL PATTERN problem is to find a maximum-size subset of $\mathcal{R}$-structured 2-intervals in $\mathcal{D}$. If each interval $D \in \mathcal{D}$ is associated with a non-negative weight $w(D)$, then we also have the WEIGHTED 2-INTERVAL PATTERN problem [7], which is to find a subset of $\mathcal{R}$-structured 2-intervals in $\mathcal{D}$ with the maximum total weight.

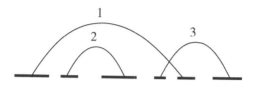

Fig. 1. Three 2-intervals

We refer to Figure 1 an example. Each interval is depicted by a thick horizontal segment; each 2-interval is depicted by an arc connecting two disjoint intervals. For the three 2-intervals D_1, D_2, and D_3 in Figure 1, we have $D_2 \sqsubset D_1$, $D_1 \between D_3$, and $D_2 < D_3$; the set $\{D_1, D_2, D_3\}$ is $\{<, \sqsubset, \between\}$-structured, the subset $\{D_1, D_2\}$ is $\{\sqsubset\}$-structured.

Beside the various models $\mathcal{R}$, various restrictions can also be imposed on the input 2-interval set $\mathcal{D}$ for the 2-INTERVAL PATTERN problem. Define the *support* of a set $\mathcal{D}$ of 2-intervals, Support($\mathcal{D}$), as the set of intervals $\{I, J \mid (I, J) \in \mathcal{D}\}$. There are four common types of restrictions:

Unlimited: No restrictions.
Balanced: Every 2-interval in $\mathcal{D}$ consists of two intervals of equal length.
Unitary: Every interval in the support of $\mathcal{D}$ has a unit length.
Point: The intervals in the support of $\mathcal{D}$ are pairwise disjoint (therefore they can be considered as intervals of unit length, or, points).

The three types of restrictions, unlimited, unitary, and point, were originally introduced by Vialette [14]. The balanced restriction was later proposed by Crochemore et al. [6] because it is natural in the biological setting: a helix of stacking base pairs in the RNA secondary structure can be represented compactly by a balanced 2-interval.

Since Vialette's pioneering work [14], the 2-INTERVAL PATTERN problem has been extensively studied. We summarize the complexities of the problem over its various models and restrictions in Table 1. Because the $\between$ relation directly models the pseudoknots in RNA secondary structures, it is not surprising that the 2-INTERVAL PATTERN problem is NP-hard or even APX-hard over the three models $\{<, \sqsubset, \between\}$, $\{\sqsubset, \between\}$, and $\{<, \between\}$; these results [2,4] are compatible with the hardness results for the other models [1,12,8,11] and are consistent with our knowledge that RNA secondary structures with pseudoknots are difficult to predict in practice. Naturally, researchers have directed their attention to the design

Table 1. The complexities of the 2-INTERVAL PATTERN problem. $^\dagger \mathcal{L} = O(n^2)$ and $d = O(n)$ [15,5].

	Unlimited Balanced Unitary	Point
$\{<,\sqsubset,\between\}$	APX-hard [2]	$O(n\sqrt{n})$ [13]
$\{\sqsubset,\between\}$	APX-hard [2]	$O(n\log n + \mathcal{L}^\dagger)$ [15,5]
$\{<,\between\}$	NP-complete [4]	complexity unknown
$\{<,\sqsubset\}$	$O(n\log n + d^\dagger n)$ [15,5]	
$\{\between\}$	$O(n\log n + \mathcal{L}^\dagger)$ [15,5]	
$\{\sqsubset\}$	$O(n\log n)$ [4]	
$\{<\}$	$O(n\log n)$ [14]	

Table 2. The best approximation ratios for the WEIGHTED 2-INTERVAL PATTERN problem. *The contributions from this paper are marked by "old $\rightarrow$ new". †For unit weight case, the best approximation ratio is $2.0 + \epsilon$ [10]; for arbitrary weight case, a $(2.5 + \epsilon)$-approximation is implicit using the same 5-claw-free technique [10].

	Unlimited	Balanced	Unitary	Point
$\{<,\sqsubset,\between\}$	4 [2]	4 [6,7]	$2.5 + \epsilon$ [10] †	N/A
$\{\sqsubset,\between\}$	4 [6,7]	4 [6,7]	$2.5 + \epsilon$ [10] †	N/A
$\{<,\between\}$	2 [9] $\rightarrow 1 + \epsilon$ *	2 [9] $\rightarrow 1 + \epsilon$ *	2 [9] $\rightarrow 1 + \epsilon$ *	2 [6,7] $\rightarrow 1 + \epsilon$ *

of efficient approximation algorithms. We refer to Table 2 for the best approximation ratios of polynomial time approximation algorithms for the WEIGHTED 2-INTERVAL PATTERN problem. In this paper, we present a polynomial time approximation scheme for the WEIGHTED 2-INTERVAL PATTERN problem over the $\{<,\between\}$ model, which improves the previous best 2-approximation [9].

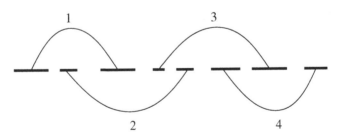

Fig. 2. A chain of pseudoknots

We note that the WEIGHTED 2-INTERVAL PATTERN problem over the $\{<,\between\}$ model is one of the most interesting variants of the 2-INTERVAL PATTERN problem. In theory, this variant is interesting because there are two open questions about it. As we can see from Table 1, the 2-INTERVAL PATTERN problem over the $\{<,\between\}$ model on point-type input is the last case whose complexity remains unknown. It was also unknown whether the 2-INTERVAL PATTERN problem over the

$\{<, \lozenge\}$ model is APX-hard [7], as the other two models $\{<, \sqsubset, \lozenge\}$ and $\{\sqsubset, \lozenge\}$. (In this paper, we settle the second open question [7] and show that the WEIGHTED 2-INTERVAL PATTERN problem over the $\{<, \lozenge\}$ model is not APX-hard.) In practice, the $\{<, \lozenge\}$ model is also interesting because it captures some of the most interesting RNA secondary structures with pseudoknots. We refer to Figure 2 for a special $\{<, \lozenge\}$-structured pattern. Bereg et al. [3] recently studied optimization problems on RNA structures with such "loop chains." Lyngsø and Pedersen [12] also noted that this "chain of pseudoknots" structure is particularly useful for comparing different types of pseudoknots that can be handled by the existing algorithms.

The rest of the paper is organized as follows. In Section 2, we introduce the idea of our algorithm. In Section 3, we present the details of our algorithm. We conclude in Section 4.

2   The Idea

For two 2-intervals $D_1 = (I_1, J_1)$ and $D_2 = (I_2, J_2)$, define a composite binary relation $\between$ such that

$$D_1 \between D_2 \iff D_1 < D_2 \text{ or } D_1 \lozenge D_2.$$

From the definitions of the two relations $<$ and $\lozenge$,

$$D_1 < D_2 \iff I_1 < J_1 < I_2 < J_2$$

and

$$D_1 \lozenge D_2 \iff I_1 < I_2 < J_1 < J_2,$$

we have

$$D_1 \between D_2 \implies I_1 < I_2 \text{ and } J_1 < J_2.$$

Just as the $<$ relation specifies a total order for disjoint intervals, the $\between$ relation specifies a total order for $\{<, \lozenge\}$-structured 2-intervals.

Let S be a set of $\{<, \lozenge\}$-structured 2-intervals. Consider S as a sequence of 2-intervals ordered by the $\between$ relation. Denote by $S[i]$ the element (2-interval) with rank i in S. Denote by $S[i, j]$ the subsequence $S[i]S[i+1]\cdots S[j]$. Define the *backbone elements* of S as follows:

1. $S[1]$ is a backbone element;
2. If $S[i]$ is a backbone element, and if $S[i] < S[j]$ and $S[i] \lozenge S[k]$ for all $i < k < j$, then $S[j]$ is also a backbone element.

For two consecutive backbone elements $S[i]$ and $S[j]$, define a *stripe* $\mathcal{T}(i, j)$ as the subsequence $S[i+1, j-1]$. By definition, each stripe is a set of $\{\lozenge\}$-structured *non-backbone* elements.

We refer to Figure 3 for an example of eight $\{<, \lozenge\}$-structured 2-intervals ordered in a sequence. The 2-intervals at indices 1, 4, 7, and 8 are the four backbone elements. The stripe between the two consecutive backbone elements

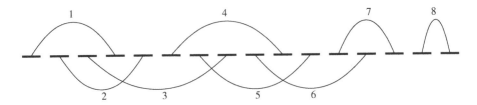

Fig. 3. Backbone elements and stripes

1 and 4 consists of two non-backbone elements 2 and 3; the stripe between 4 and 7 consists of 5 and 6; the stripe between 7 and 8 is empty.

A sequence is *c-striped* if at most c consecutive stripes of the sequence are non-empty. (The sequence in Figure 3 is 2-striped.) With the backbone elements at indices $i_1, i_2, \ldots, i_k$, the sequence $\mathcal{S}$ can be represented by backbone elements and stripes in an alternating pattern:

$$\mathcal{S}[i_1]\mathcal{T}(i_1, i_2)\mathcal{S}[i_2]\mathcal{T}(i_2, i_3)\mathcal{S}[i_3]\mathcal{T}(i_3, i_4)\mathcal{S}[i_4] \ldots$$

Although $\mathcal{S}$ itself may not be c-striped, it contains c-striped subsequences. For example, the following two subsequences of $\mathcal{S}$ are both 1-striped:

$$\mathcal{S}[i_1]\mathcal{S}[i_2]\mathcal{T}(i_2, i_3)\mathcal{S}[i_3]\mathcal{S}[i_4]\mathcal{T}(i_4, i_5)\mathcal{S}[i_5] \ldots$$

$$\mathcal{S}[i_1]\mathcal{T}(i_1, i_2)\mathcal{S}[i_2]\mathcal{S}[i_3]\mathcal{T}(i_3, i_4)\mathcal{S}[i_4]\mathcal{S}[i_5] \ldots$$

The two subsequences together cover the sequence $\mathcal{S}$: each backbone element is covered twice; each non-backbone element is covered once. This observation immediately suggests that a 2-approximation of the WEIGHTED 2-INTERVAL PATTERN problem over the $\{<, \between\}$ model can be obtained by finding a 1-striped sequence of the maximum weight. Indeed, this is exactly the idea behind the previous 2-approximation [9] for the 2-INTERVAL PATTERN problem over the $\{<, \between\}$ model. In this paper, we extend this idea further to obtain a polynomial time approximation scheme.

For each k, $0 \leq k \leq c$, we can obtain a c-striped subsequence of $\mathcal{S}$ by deleting the stripes $\mathcal{T}(i_j, i_{j+1})$ such that $j \bmod (c + 1) = k$. The $c + 1$ subsequences together cover the sequence c times: each backbone element is covered $c + 1$ times; each non-backbone element is covered c times. Therefore, the total weight of the $c + 1$ subsequences is at least c times the weight of the sequence $\mathcal{S}$. By the pigeon-hole principle, at least one of the $c + 1$ subsequences has a weight at least $\frac{c}{c+1}$ times the weight of $\mathcal{S}$. This implies that a c-striped sequence of the maximum weight is a $(1 + 1/c)$-approximation for the WEIGHTED 2-INTERVAL PATTERN problem over the $\{<, \between\}$ model.

3   The Algorithm

In this section, we design a dynamic programming algorithm to find a c-striped sequence with the maximum weight, thereby achieving a $(1+1/c)$-approximation for the WEIGHTED 2-INTERVAL PATTERN problem over the $\{<, \lozenge\}$ model. For simplicity, we only demonstrate how to compute the maximum weight of a c-striped sequence. The actual sequence with the maximum weight can be reconstructed in the same running time using standard techniques in dynamic programming.

3.1   Canonical Sequences and Chains

Given an input set $\mathcal{D}$ of 2-intervals, we first construct two zero-weight dummy elements A and Z such that $A < D < Z$ for every element $D \in \mathcal{D}$, then extend $\mathcal{D}$ with A and Z. Since the two dummy elements have zero weight, the maximum weight of a c-striped sequence remains the same after the extension, and we can conveniently assume that A and Z, respectively, are the first and the last elements of the optimal c-striped sequence.

By definition, the first element of a sequence of $\{<, \lozenge\}$-structured 2-intervals is always a backbone element. If the last element of the sequence is also a backbone element, then the sequence is *canonical*. (For example, the sequence in Figure 3 is canonical, and the sequence in Figure 2 is not.) For each element $D \in \mathcal{D}$, denote by $W[D]$ the maximum weight of a canonical c-striped sequence *anchored* between A and D (that is, with A and D, respectively, as the first and the last elements in the sequence), and denote by $W_0[D]$ the maximum weight of such a canonical sequence with the additional constraint that its last stripe (the stripe between the second-to-last and the last backbone elements) is empty. The entry $W[Z]$ denotes the maximum weight of a canonical c-striped sequence anchored between A and Z, and is the optimal solution.

We next show how to compute the two tables $W[D]$ and $W_0[D]$. Define a *chain* as a canonical sequence with at most $c+1$ backbone elements (and hence at most c stripes). For every pair of elements $C, D \in \mathcal{D}$, $C < D$, denote by $w[C, D]$ the maximum weight of a chain with C and D, respectively, as the first and the last backbone elements. Since a canonical c-striped sequence is a concatenation of (independent) chains separated by empty stripes, we can compute the tables $W[D]$ and $W_0[D]$ with the recurrence

$$
\begin{cases}
W_0[D] = \max_{C < D} \left\{ W[C] + w(D) \right\} \\
W[D] = \max \begin{cases} W_0[D] \\ \max_{C < D} \left\{ W_0[C] - w(C) + w[C, D] \right\} \end{cases}
\end{cases}
\tag{1}
$$

and the base condition $W[A] = W_0[A] = 0$. Intuitively, the element C in the recurrence is the last backbone element before D.

3.2   Computing the Chain Table

We next show how to compute the chain table $w[C, D]$. Define an *i-chain* as a chain with exactly i stripes (and exactly $i+1$ backbone elements). For each pair of elements $C, D \in \mathcal{D}$, $C < D$, and for each i, $1 \leq i \leq c$, denote by $w_i[C, D]$ the maximum weight of an i-chain with C and D as the first and the last backbone elements, respectively. We have

$$w[C, D] = \max_{1 \leq i \leq c} w_i[C, D]. \tag{2}$$

The problem reduces to computing the i-chain table $w_i[C, D]$ for each i.

Denote by $w_i[B_1, B_2, \ldots, B_{i+1}]$ the maximum weight of an i-chain with

$$B_1 < B_2 < \ldots < B_{i+1}$$

as its $i + 1$ backbone elements. We have

$$w_i[C, D] = \max_{C = B_1 < B_2 < \ldots < B_{i+1} = D} w_i[B_1, B_2, \ldots, B_{i+1}]. \tag{3}$$

The problem further reduces to computing $w_i[B_1, B_2, \ldots, B_{i+1}]$, which we will explain next.

By the definition of the backbone elements, each element D in the stripe between two consecutive backbone elements B_k and B_{k+1} of an i-chain must satisfy the constraint $B_k \between D \between B_{k+1}$. For each k, $1 \leq k \leq i$, define

$$\mathcal{D}_k = \{D \mid D \in \mathcal{D} \text{ and } B_k \between D \between B_{k+1}\}.$$

The problem then reduces to selecting a subset $\mathcal{D}'_k \subseteq \mathcal{D}_k$ for each k, $1 \leq k \leq i$, such that the set of elements

$$\mathcal{C} = \{B_1\} \cup \mathcal{D}'_1 \cup \{B_2\} \cup \ldots \cup \mathcal{D}'_i \cup \{B_{i+1}\}$$

is an i-chain with the maximum weight, which is achieved when the weight of the non-backbone elements in $\mathcal{D}'_1 \cup \ldots \cup \mathcal{D}'_k$ is maximized.

3.3   Maximizing the Weight of Non-backbone Elements with Specified Backbone Elements

We introduce some more notations. For a 2-interval D, denote by $\mathsf{L}(D)$ and $\mathsf{R}(D)$, respectively, the left and the right intervals of D. For an interval I, denote by $l(I)$ and $r(I)$, respectively, the coordinates of the left and the right endpoints of I.

To ensure that $\mathcal{C}$ is indeed an i-chain, the following two conditions are both necessary and sufficient:

1. The elements in $\mathcal{D}'_k$ are $\{\between\}$-structured, $1 \leq k \leq i$.
2. The elements in $\mathcal{D}'_k$ are disjoint from the elements in $\mathcal{D}'_{k+1}$, $1 \leq k \leq i - 1$.

To compute the maximum weight of the non-backbone elements, we again use a dynamic programming approach. Use $i + 1$ coordinates $x_1, x_2, \ldots, x_{i+1}$, where each coordinate x_k has a valid range $[x'_k, x''_k]$:

- For $k = 1$, $x'_1 = r(\mathsf{L}(B_1))$ and $x''_1 = l(\mathsf{R}(B_1))$;
- For $2 \leq k \leq i + 1$, $x'_k = r(\mathsf{R}(B_{k-1}))$ and $x''_k = l(\mathsf{R}(B_k))$.

Denote by $w[x_1, x_2, \ldots, x_{i+1}]$ the maximum weight of the subsets $\mathcal{D}'_1, \ldots, \mathcal{D}'_k$ that satisfy both the two conditions stated earlier and an additional condition 3 in the following, which limits the choice of candidate elements from $\mathcal{D}_k$:

3. Each element D in $\mathcal{D}'_k \subseteq \mathcal{D}_k$ satisfies $r(\mathsf{L}(D)) \leq x_k$ and $r(\mathsf{R}(D)) \leq x_{k+1}$, $1 \leq k \leq i$.

Finally, the table $w[x_1, x_2, \ldots, x_{i+1}]$ can be computed with the recurrence

$$w[x_1, x_2, \ldots, x_{i+1}] =$$
$$\max \begin{cases} \max_{1 \leq k \leq i+1} w[x_1, x_2, \ldots, x_k - 1, \ldots, x_{i+1}] \\ \max_{([a,x_k],[b,x_{k+1}]) \in \mathcal{D}_k} w[x_1, x_2, \ldots, a - 1, b - 1, \ldots, x_{i+1}], \end{cases} \quad (4)$$

and the base condition $w[x'_1, \ldots, x'_{i+1}] = 0$. The entry $w[x''_1, \ldots, x''_{i+1}]$ gives the maximum weight of the non-backbone elements.

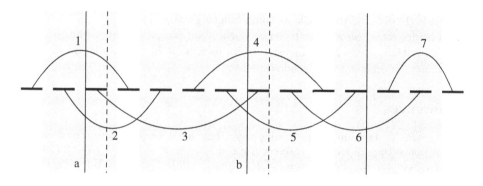

Fig. 4. Dynamic programming

We refer to Figure 4 for an example. The three solid vertical lines specify the three coordinates x_1, x_2, and x_3; the three shaded areas specify the ranges of the coordinates. The two dashed vertical lines (together with the two solid vertical lines) for x_1 and x_2 illustrate the recurrence step involving the element 3. This completes the description of our algorithm.

We now give an analysis of the running time of our algorithm. For an input of n 2-intervals, we can assume that the coordinates of the interval endpoints are between 1 and $4n$. Recurrence (4) takes $O(n^{i+2})$ time for each set of backbone elements $B_1, \ldots, B_{i+1}$. Recurrence (3) enumerates $O(n^{i+1})$ sets of backbone elements for each i, $1 \leq i \leq c$. The total time for computing the chain table $w[C, D]$ using Equation (2) is therefore $O(n^{2c+3})$. With $w[C, D]$ computed, Recurrence (1) clearly needs at most $O(n^2)$ time. We have the following theorem:

Theorem 1. *Our algorithm approximates the* WEIGHTED 2-INTERVAL PAT-TERN *problem over the* $\{<, \lozenge\}$ *model with a ratio of* $1 + 1/c$ *and runs in* $O(n^{2c+3})$ *time for any fixed integer* $c \geq 2$.

4   Concluding Remarks

The WEIGHTED 2-INTERVAL PATTERN problem over the $\{<, \emptyset\}$ model can be approximated arbitrarily well by our polynomial time approximation scheme; as a result, this variant is not APX-hard. This settles an open question by Crochemore et al. [7].

Our algorithm is mainly of theoretical interest at this time; its running time is prohibitive as c becomes large. In practice, however, the optimal $\{<, \emptyset\}$-structured pattern may be a c-striped sequence for a small c. It is quite possible that, with additional heuristics and careful implementation, our algorithm can be made practical.

Our $(1+1/c)$-approximation is achieved by solving a special c-striped sequence optimally. From this result, we gain the insight that the most difficult case of the problem happens when the optimal structure contains very long chains of pseudoknots. This is intuitively consistent with the result by Lyngsø and Pedersen [12], who showed that the "chain of pseudoknots" structure becomes increasingly difficult to handle as chain length grows.

In future work, we will try to improve the time complexity of our approximation algorithm, and we will investigate its usefulness in the prediction of RNA secondary structures with pseudoknots.

References

1. Akutsu, T.: Dynamic programming algorithms for RNA secondary structure prediction with pseudoknots. Discrete Applied Mathematics 104(1-3), 45–62 (2000)
2. Bar-Yehuda, R., Halldórsson, M.M., Naor, J.(S.), Shachnai, H., Shapira, I.: Scheduling split intervals. SIAM Journal on Computing 36(1), 1–15 (2006) (Preliminary version In: Proceedings of the 13th annual ACM-SIAM Symposium on Discrete Algorithms (SODA'02), pp. 732–741 (2002))
3. Bereg, S., Kubica, M., Waleń, T., Zhu, B.: RNA multiple structural alignment with longest common subsequences. Journal of Combinatorial Optimization 13(2), 179–188 (2007)
4. Blin, G., Fertin, G., Vialette, S.: New results for the 2-interval pattern problem. In: Sahinalp, S.C., Muthukrishnan, S.M., Dogrusoz, U. (eds.) CPM 2004. LNCS, vol. 3109, pp. 311–322. Springer, Heidelberg (2004)
5. Chen, E., Yang, L., Yuan, H.: Improved algorithms for largest cardinality 2-interval pattern problem. Journal of Combinatorial Optimization, Special Issue on Bioinformatics 13(3), 263–275 (2007)
6. Crochemore, M., Hermelin, D., Landau, G.M., Vialette, S.: Approximating the 2-interval pattern problem. In: Brodal, G.S., Leonardi, S. (eds.) ESA 2005. LNCS, vol. 3669, pp. 426–437. Springer, Heidelberg (2005)
7. Crochemore, M., Hermelin, D., Landau, G.M., Rawitz, D., Vialette, S.: Approximating the 2-interval pattern problem. Theoretical Computer Science (to appear)
8. Ieong, S., Kao, M.-Y., Lam, T.-W., Sung, W.-K., Yiu, S.-M.: Predicting RNA secondary structure with arbitrary pseudoknots by maximizing the number of stacking pairs. Journal of Computational Biology 10(6), 981–995 (2003) (Preliminary version In: Proceedings of the 2nd Annual IEEE International Symposium on Bioinformatics and Bioengineering (BIBE'01), pp. 183–190 (2001))

9. Jiang, M.: A 2-approximation for the preceding-and-crossing structured 2-interval pattern problem. Journal of Combinatorial Optimization, Special Issue on Bioinformatics 13(3), 217–221 (2007)

10. Jiang, M.: Improved approximation algorithms for predicting RNA secondary structures with arbitrary pseudoknots. In: Proceedings of the 3rd International Conference on Algorithmic Aspects in Information and Management (AAIM'07). LNCS, vol. 4508, pp. 399–410. Springer, Heidelberg (2007)

11. Lyngsø, R.B.: Complexity of pseudoknot prediction in simple models. In: Díaz, J., Karhumäki, J., Lepistö, A., Sannella, D. (eds.) ICALP 2004. LNCS, vol. 3142, pp. 919–931. Springer, Heidelberg (2004)

12. Lyngsø, R.B., Pedersen, C.N.S.: RNA pseudoknot prediction in energy-based models. Journal of Computational Biology 7(3/4), 409–427 (2000)

13. Micali, S., Vazirani, V.V.: An $O(\sqrt{|V|}|E|)$ algorithm for finding maximum matching in general graphs. In: Proceedings of the 21st Annual Symposium on Foundations of Computer Science (FOCS'80), pp. 17–27 (1980)

14. Vialette, S.: On the computational complexity of 2-interval pattern matching problems. Theoretical Computer Science 312, 223–249 (2004) (Preliminary version In: Apostolico, A., Takeda, M. (eds.) CPM 2002. LNCS, vol. 2373, pp. 53–63. Springer, Heidelberg (2002))

15. Yuan, H., Yang, L., Chen, E.: Improved algorithms for largest cardinality 2-interval pattern problem. In: Deng, X., Du, D.-Z. (eds.) ISAAC 2005. LNCS, vol. 3827, pp. 412–421. Springer, Heidelberg (2005)

Author Index

Alspach, Brian   354
Althaus, Ernst   267

Bai, Danyu   191
Behle, Markus   124
Buchheim, Christoph   301

Canzar, Stefan   267
Cao, Jiannong   20
Cao, Yi   32
Cao, Zhigang   44
Changrui, Yu   102
Chen, Jianer   291
Chen, Wenbin   163
Chen, Xujin   81
Chwa, Kyung-Yong   311

Dai, Xiaoping   220
Derrien, Vincent   52
Dress, Andreas W.M.   4
Dyer, Danny   354

Eidenbenz, Raphael   208

Fellows, Michael   366
Fomin, Fedor V.   366
Friedman, Eric J.   146, 200

Gao, Ting   72
Gu, Jian   171

Hanson, Denis   354
Hao, Bailin   3
Hao, Jin-Kao   52
Hasan, Mohammad Khairul   311
Hu, Jie   81
Hu, Xiaodong   81, 182
Huang, Min   291
Huang, Xiaobin   171
Huber, Katharina T.   4

Jia, Yan   171
Jiang, Minghui   378
Jung, Hyunwoo   311

Kara, Bahar Y.   62
Kara, İmdat   62
Koolen, Jacobus   4
Kugimoto, Yoshinori   255

Lam, Tak-Wah   242
Landsberg, Adam Scott   200
Lefmann, Hanno   230
Li, Deying   20
Li, Fengwei   91
Li, Yueping   335
Liu, ChenGuang   323
Liu, Ming   20
Liu, Yunlong   343
Lokshtanov, Daniel   366
Lu, Yunting   335

Ma, Weimin   72
Matsuhisa, Takashi   136
Mehlhorn, Kurt   1
Meng, Jiangtao   163
Mizuki, Takaaki   255
Moulton, Vincent   4

Oswald, Yvonne Anne   208

Paschos, Vangelis Th.   112
Pop, Petrica C.   154
Pop Sitar, Corina   154

Richer, Jean-Michel   52
Rosamond, Frances   366

Saurabh, Saket   366
Schmid, Stefan   208
Shang, Weiping   182
Sone, Hideaki   255
Su, Bing   11
Sung, Wing-Kin   242
Szeider, Stefan   366

Tam, Siu-Lung   242
Tanaka, Kazuyuki   323
Tang, Lixin   191
Taşcu, Ioana   154
Telelis, Orestis A.   112
Thomassen, Carsten   366

Wan, Guohua   279
Wan, Pengjun   182
Wang, Jianxin   291, 343
Wang, Ke   72
Wang, Xiaolin   220
Wattenhofer, Roger   208
Wong, Chi-Kwong   242

Xiao, Peng   11
Xu, Qingchuan   11
Xu, Xiaoshuang   343

Yan, Luo   102
Yang, Boting   32, 354

Yao, Frances   182
Ye, Qingfang   91
Yetis, M. Kadri   62
Yiu, Siu-Ming   242

Zelina, Ioana   154
Zhang, Runtao   32
Zhang, Yan   171
Zhang, Yuzhong   44
Zheng, Lanbo   301
Zheng, Yuan   20
Zhou, Bin   171
Zhou, Jianqin   220
Zissimopoulos, Vassilis   112

Lecture Notes in Computer Science

For information about Vols. 1–4527

please contact your bookseller or Springer

Vol. 4617: V. Torra, Y. Narukawa, Y. Yoshida (Eds.), Modeling Decisions for Artificial Intelligence. XII, 502 pages. 2007. (Sublibrary LNAI).

Vol. 4616: A. Dress, Y. Xu, B. Zhu (Eds.), Combinatorial Optimization and Applications. XI, 390 pages. 2007.

Vol. 4613: F.P. Preparata, Q. Fang (Eds.), Frontiers in Algorithmics. XI, 348 pages. 2007.

Vol. 4612: I. Miguel, W. Ruml (Eds.), Abstraction, Reformulation, and Approximation. XI, 418 pages. 2007. (Sublibrary LNAI).

Vol. 4611: J. Indulska, J. Ma, L.T. Yang, T. Ungerer, J. Cao (Eds.), Ubiquitous Intelligence and Computing. XXIII, 1257 pages. 2007.

Vol. 4610: B. Xiao, L.T. Yang, J. Ma, C. Muller-Schloer, Y. Hua (Eds.), Autonomic and Trusted Computing. XVIII, 571 pages. 2007.

Vol. 4609: E. Ernst (Ed.), ECOOP 2007 — Object-Oriented Programming. XIII, 625 pages. 2007.

Vol. 4608: H.W. Schmidt, I. Crnkovic, G.T. Heineman, J.A. Stafford (Eds.), Component-Based Software Engineering. XII, 283 pages. 2007.

Vol. 4607: L. Baresi, P. Fraternali, G.-J. Houben (Eds.), Web Engineering. XVI, 576 pages. 2007.

Vol. 4606: A. Pras, M. van Sinderen (Eds.), Dependable and Adaptable Networks and Services. XIV, 149 pages. 2007.

Vol. 4605: D. Papadias, D. Zhang, G. Kollios (Eds.), Advances in Spatial and Temporal Databases. X, 479 pages. 2007.

Vol. 4604: U. Priss, S. Polovina, R. Hill (Eds.), Conceptual Structures: Knowledge Architectures for Smart Applications. XII, 514 pages. 2007. (Sublibrary LNAI).

Vol. 4603: F. Pfenning (Ed.), Automated Deduction – CADE-21. XII, 522 pages. 2007. (Sublibrary LNAI).

Vol. 4602: S. Barker, G.-J. Ahn (Eds.), Data and Applications Security XXI. X, 291 pages. 2007.

Vol. 4600: H. Comon-Lundh, C. Kirchner, H. Kirchner, Rewriting, Computation and Proof. XVI, 273 pages. 2007.

Vol. 4599: S. Vassiliadis, M. Berekovic, T.D. Hämäläinen (Eds.), Embedded Computer Systems: Architectures, Modeling, and Simulation. XVIII, 466 pages. 2007.

Vol. 4598: G. Lin (Ed.), Computing and Combinatorics. XII, 570 pages. 2007.

Vol. 4597: P. Perner (Ed.), Advances in Data Mining. XI, 353 pages. 2007. (Sublibrary LNAI).

Vol. 4596: L. Arge, C. Cachin, T. Jurdziński, A. Tarlecki (Eds.), Automata, Languages and Programming. XVII, 953 pages. 2007.

Vol. 4595: D. Bošnački, S. Edelkamp (Eds.), Model Checking Software. X, 285 pages. 2007.

Vol. 4594: R. Bellazzi, A. Abu-Hanna, J. Hunter (Eds.), Artificial Intelligence in Medicine. XVI, 509 pages. 2007. (Sublibrary LNAI).

Vol. 4592: Z. Kedad, N. Lammari, E. Métais, F. Meziane, Y. Rezgui (Eds.), Natural Language Processing and Information Systems. XIV, 442 pages. 2007.

Vol. 4591: J. Davies, J. Gibbons (Eds.), Integrated Formal Methods. IX, 660 pages. 2007.

Vol. 4590: W. Damm, H. Hermanns (Eds.), Computer Aided Verification. XV, 562 pages. 2007.

Vol. 4589: J. Münch, P. Abrahamsson (Eds.), Product-Focused Software Process Improvement. XII, 414 pages. 2007.

Vol. 4588: T. Harju, J. Karhumäki, A. Lepistö (Eds.), Developments in Language Theory. XI, 423 pages. 2007.

Vol. 4587: R. Cooper, J. Kennedy (Eds.), Data Management. XIII, 259 pages. 2007.

Vol. 4586: J. Pieprzyk, H. Ghodosi, E. Dawson (Eds.), Information Security and Privacy. XIV, 476 pages. 2007.

Vol. 4585: M. Kryszkiewicz, J.F. Peters, H. Rybinski, A. Skowron (Eds.), Rough Sets and Intelligent Systems Paradigms. XIX, 836 pages. 2007. (Sublibrary LNAI).

Vol. 4584: N. Karssemeijer, B. Lieveldt (Eds.), Information Processing in Medical Imaging. XX, 777 pages. 2007.

Vol. 4583: S.R. Della Rocca (Ed.), Typed Lambda Calculi and Applications. X, 397 pages. 2007.

Vol. 4582: J. Lopez, P. Samarati, J.L. Ferrer (Eds.), Public Key Infrastructure. XI, 375 pages. 2007.

Vol. 4581: A. Petrenko, M. Veanes, J. Tretmans, W. Grieskamp (Eds.), Testing of Software and Communicating Systems. XII, 379 pages. 2007.

Vol. 4580: B. Ma, K. Zhang (Eds.), Combinatorial Pattern Matching. XII, 366 pages. 2007.

Vol. 4579: B. M. Hämmerli, R. Sommer (Eds.), Detection of Intrusions and Malware, and Vulnerability Assessment. X, 251 pages. 2007.

Vol. 4578: F. Masulli, S. Mitra, G. Pasi (Eds.), Applications of Fuzzy Sets Theory. XVIII, 693 pages. 2007. (Sublibrary LNAI).

Vol. 4577: N. Sebe, Y. Liu, Y.-t. Zhuang (Eds.), Multimedia Content Analysis and Mining. XIII, 513 pages. 2007.

Vol. 4576: D. Leivant, R. de Queiroz (Eds.), Logic, Language, Information and Computation. X, 363 pages. 2007.

Vol. 4575: T. Takagi, T. Okamoto, E. Okamoto, T. Okamoto (Eds.), Pairing-Based Cryptography – Pairing 2007. XI, 408 pages. 2007.

Vol. 4574: J. Derrick, J. Vain (Eds.), Formal Techniques for Networked and Distributed Systems – FORTE 2007. XI, 375 pages. 2007.

Vol. 4573: M. Kauers, M. Kerber, R. Miner, W. Windsteiger (Eds.), Towards Mechanized Mathematical Assistants. XIII, 407 pages. 2007. (Sublibrary LNAI).

Vol. 4572: F. Stajano, C. Meadows, S. Capkun, T. Moore (Eds.), Security and Privacy in Ad-hoc and Sensor Networks. X, 247 pages. 2007.

Vol. 4571: P. Perner (Ed.), Machine Learning and Data Mining in Pattern Recognition. XIV, 913 pages. 2007. (Sublibrary LNAI).

Vol. 4570: H.G. Okuno, M. Ali (Eds.), New Trends in Applied Artificial Intelligence. XXI, 1194 pages. 2007. (Sublibrary LNAI).

Vol. 4569: A. Butz, B. Fisher, A. Krüger, P. Olivier, S. Owada (Eds.), Smart Graphics. IX, 237 pages. 2007.

Vol. 4566: M.J. Dainoff (Ed.), Ergonomics and Health Aspects of Work with Computers. XVIII, 390 pages. 2007.

Vol. 4565: D.D. Schmorrow, L.M. Reeves (Eds.), Foundations of Augmented Cognition. XIX, 450 pages. 2007. (Sublibrary LNAI).

Vol. 4564: D. Schuler (Ed.), Online Communities and Social Computing. XVII, 520 pages. 2007.

Vol. 4563: R. Shumaker (Ed.), Virtual Reality. XXII, 762 pages. 2007.

Vol. 4562: D. Harris (Ed.), Engineering Psychology and Cognitive Ergonomics. XXIII, 879 pages. 2007. (Sublibrary LNAI).

Vol. 4561: V.G. Duffy (Ed.), Digital Human Modeling. XXIII, 1068 pages. 2007.

Vol. 4560: N. Aykin (Ed.), Usability and Internationalization, Part II. XVIII, 576 pages. 2007.

Vol. 4559: N. Aykin (Ed.), Usability and Internationalization, Part I. XVIII, 661 pages. 2007.

Vol. 4558: M.J. Smith, G. Salvendy (Eds.), Human Interface and the Management of Information, Part II. XXIII, 1162 pages. 2007.

Vol. 4557: M.J. Smith, G. Salvendy (Eds.), Human Interface and the Management of Information, Part I. XXII, 1030 pages. 2007.

Vol. 4556: C. Stephanidis (Ed.), Universal Access in Human-Computer Interaction, Part III. XXII, 1020 pages. 2007.

Vol. 4555: C. Stephanidis (Ed.), Universal Access in Human-Computer Interaction, Part II. XXII, 1066 pages. 2007.

Vol. 4554: C. Stephanidis (Ed.), Universal Acess in Human Computer Interaction, Part I. XXII, 1054 pages. 2007.

Vol. 4553: J.A. Jacko (Ed.), Human-Computer Interaction, Part IV. XXIV, 1225 pages. 2007.

Vol. 4552: J.A. Jacko (Ed.), Human-Computer Interaction, Part III. XXI, 1038 pages. 2007.

Vol. 4551: J.A. Jacko (Ed.), Human-Computer Interaction, Part II. XXIII, 1253 pages. 2007.

Vol. 4550: J.A. Jacko (Ed.), Human-Computer Interaction, Part I. XXIII, 1240 pages. 2007.

Vol. 4549: J. Aspnes, C. Scheideler, A. Arora, S. Madden (Eds.), Distributed Computing in Sensor Systems. XIII, 417 pages. 2007.

Vol. 4548: N. Olivetti (Ed.), Automated Reasoning with Analytic Tableaux and Related Methods. X, 245 pages. 2007. (Sublibrary LNAI).

Vol. 4547: C. Carlet, B. Sunar (Eds.), Arithmetic of Finite Fields. XI, 355 pages. 2007.

Vol. 4546: J. Kleijn, A. Yakovlev (Eds.), Petri Nets and Other Models of Concurrency – ICATPN 2007. XI, 515 pages. 2007.

Vol. 4545: H. Anai, K. Horimoto, T. Kutsia (Eds.), Algebraic Biology. XIII, 379 pages. 2007.

Vol. 4544: S. Cohen-Boulakia, V. Tannen (Eds.), Data Integration in the Life Sciences. XI, 282 pages. 2007. (Sublibrary LNBI).

Vol. 4543: A.K. Bandara, M. Burgess (Eds.), Inter-Domain Management. XII, 237 pages. 2007.

Vol. 4542: P. Sawyer, B. Paech, P. Heymans (Eds.), Requirements Engineering: Foundation for Software Quality. IX, 384 pages. 2007.

Vol. 4541: T. Okadome, T. Yamazaki, M. Makhtari (Eds.), Pervasive Computing for Quality of Life Enhancement. IX, 248 pages. 2007.

Vol. 4539: N.H. Bshouty, C. Gentile (Eds.), Learning Theory. XII, 634 pages. 2007. (Sublibrary LNAI).

Vol. 4538: F. Escolano, M. Vento (Eds.), Graph-Based Representations in Pattern Recognition. XII, 416 pages. 2007.

Vol. 4537: K.C.-C. Chang, W. Wang, L. Chen, C.A. Ellis, C.-H. Hsu, A.C. Tsoi, H. Wang (Eds.), Advances in Web and Network Technologies, and Information Management. XXIII, 707 pages. 2007.

Vol. 4536: G. Concas, E. Damiani, M. Scotto, G. Succi (Eds.), Agile Processes in Software Engineering and Extreme Programming. XV, 276 pages. 2007.

Vol. 4534: I. Tomkos, F. Neri, J. Solé Pareta, X. Masip Bruin, S. Sánchez Lopez (Eds.), Optical Network Design and Modeling. XI, 460 pages. 2007.

Vol. 4533: F. Baader (Ed.), Term Rewriting and Applications. XII, 419 pages. 2007.

Vol. 4531: J. Indulska, K. Raymond (Eds.), Distributed Applications and Interoperable Systems. XI, 337 pages. 2007.

Vol. 4530: D.H. Akehurst, R. Vogel, R.F. Paige (Eds.), Model Driven Architecture- Foundations and Applications. X, 219 pages. 2007.

Vol. 4529: P. Melin, O. Castillo, L.T. Aguilar, J. Kacprzyk, W. Pedrycz (Eds.), Foundations of Fuzzy Logic and Soft Computing. XIX, 830 pages. 2007. (Sublibrary LNAI).

Vol. 4528: J. Mira, J.R. Álvarez (Eds.), Nature Inspired Problem-Solving Methods in Knowledge Engineering, Part II. XXII, 650 pages. 2007.